Highways

This book is dedicated to

Teena O'Flaherty
Eva (O'Flaherty) Wallis
Nuala Rose O'Flaherty

In memoriam

Highways

The location, design, construction and maintenance of road pavements

Fourth edition

Edited by

Professor Emeritus C.A. O'Flaherty, A.M.

Contributing authors

A. Boyle, MA (*Cantab*), CEng, MICE
M.J. Brennan, BE (*NUI*), MSc (*Leeds*), DPhil (*Ulster*),
Dip Comp Eng (*Dublin*), CEng, FIEI, MIAT
J. Knapton, BSc, PhD (*Newcastle*), CEng, FICE, FIStructE, FIHT
H.A. Khalid, BSc (*Baghdad*), MSc, PhD (*Birmingham*),
CEng, MICE, MIAT, MIHT
J. McElvaney, BE (*NUI*), PhD (*Nott*), CEng, MICE
D. McMullen, BSc, MSc (*Queens*), CEng, MICE, MIHT
C.A. O'Flaherty, BE (*NUI*), MS, PhD (*Iowa State*), Hon LLD (*Tas*),
CEng, FICE, FIEI, FIEAust, FIHT, FCIT
J.E. Oliver, BSc (*Aston*), CEng, MICE, MIHT
M.S. Snaith, MA, BAI, MSc (*Dub*), PhD (*Nott*), ScD (*Dub*),
FREng, FICE, FIHT
A.R. Woodside, MPhil (*Ulster*),
CEng, FICE, FIEI, FIAT, FIQ, MIHT, MCIWEM
W.D.H. Woodward, BSc, MPhil, DPhil (*Ulster*),
MIHT, MIAT, MIQ, MIEI, PGCUT
S.E. Zoorob, MEng, PhD (*Leeds*)

OXFORD AUCKLAND BOSTON JOHANNESBURG MELBOURNE NEW DELHI SINGAPORE

Butterworth-Heinemann
Linacre House, Jordan Hill, Oxford OX2 8DP
225 Wildwood Avenue, Woburn, MA 01801–2041
A division of Reed Educational and Professional Publishing Ltd

 A member of the Reed Elsevier plc group

First published in Great Britain by Edward Arnold (Publishers) Ltd 1967
Second edition 1974
Third edition 1986
Fourth edition published by Butterworth-Heinemann 2002

British Library Cataloguing in Publication Data
A catalogue record for this book is available from the British Library

Library of Congress Cataloguing in Publication Data
A catalogue record for this book is available from the Library of Congress

ISBN 0 7506 5090 7

Typeset by Integra Software Services Pvt. Ltd., Pondicherry, India
www.integra-india.com
Printed and bound in Malta

Contents

13 Concrete pavement construction   362
C.A. O'Flaherty

14 Current British thickness design practice in relation to new bituminous and concrete pavements   377
C.A. O'Flaherty

15 Analytical design of flexible pavements   395
J. McElvaney and M.S. Snaith

About the contributors

Andrew Boyle

Andrew Boyle graduated from the University of Cambridge in 1967, and worked on motorway and trunk road design and construction, before joining the Department of Transport in 1974. There he worked on the management and maintenance of trunk roads in the West Midlands, before managing the planning, design and construction of major schemes in the Midlands and London. After managing computer software development he became Head of Engineering Policy Division at the Highways Agency with responsibility for the development of policy on Value for Money, Quality, Specifications, innovative traffic control systems and enforcement, road safety as well as European and International affairs. Since 1997 he has been an independent consultant and has worked in Poland and Palestine advising on contract matters and the establishment of new management structures for highways. He is currently advising the UK government on improvements to the regime under which utilities place apparatus in public highways.

Michael J. Brennan

Michael Brennan graduated with a 1st class honours degree in Civil Engineering from the National University of Ireland in 1970, and subsequently obtained a Diploma in Computing for Engineers from Dublin University and an MSc degree in Transportation Engineering from the University of Leeds. He worked as a Research Assistant at Leeds Institute for Transport Studies before being appointed to the staff of the National University of Ireland, Galway, where he is currently a Statutory Lecturer. He has spent sabbatical years at Purdue University and the Laboratoire Central des Ponts et Chaussées, and has consulted extensively on bituminous materials in Ireland. He was conferred with the degree of DPhil by the University of Ulster in 2000, for published research.

John Knapton

John Knapton graduated from Newcastle University with a 1st class honours degree in Civil Engineering in 1970 and with a PhD in 1973. He worked for two years as a research engineer at the Cement & Concrete Association before returning to lecture at Newcastle. Throughout the 1980s, he ran his own consulting practice specialising in pavement design in the UK, the US, Australia and The Middle East. He wrote the British Ports Association's heavy duty pavement design manual and developed design procedures for the structural design of highways, industrial pavements and aircraft pavements surfaced with pavers. He took the structures chair at Newcastle in 1991 and has undertaken research into ground bearing concrete floors and external concrete hardstandings. He is a Freeman of the Worshipful Company of Paviors and Chief of the Ghanaian village of Ekumfi-Atakwa.

Hussain A. Khalid

Hussain Khalid graduated from Baghdad University with a degree in civil engineering and from the University of Birmingham with MSc and PhD degrees. Before starting postgraduate study, he worked as a Resident Engineer on comprehensive school building sites and as a Materials Engineer involved in materials testing and site investigation. After the PhD, he spent two years as a post-doctoral Research Associate in Civil Engineering at the University of Newcastle upon Tyne. He joined the University of Liverpool in 1987 where he is currently Senior Lecturer in Civil Engineering. He is a Chartered Civil Engineer and his major interests are in Highway and Transportation Engineering.

James McElvaney

James McElvaney graduated in Civil Engineering from the National University of Ireland (UCD) in 1966, obtained his PhD degree from the University of Nottingham in 1973, and was admitted to the Institution of Civil Engineers in 1976. He has worked extensively overseas, mainly in the Middle and Far East, and from 1982 to 2000, collaborated with the Institute of Technology, Bandung, Indonesia, in implementing an MSc programme in Highway Engineering and Development. More recently, he has been involved at the Institute of Road Engineers, Bandung in a research project relating to the development of Geoguides for the design and construction of roads on soft soils and peat.

Derek McMullen

Derek McMullen graduated in Civil Engineering from the Queen's University of Belfast in 1976, completing an MSc by research (on applications of pavement analysis) in 1978. Since then he has worked for UK consulting engineers. Early experience was gained in southern England, working on the design and construction of major roads, including two years site experience on the M25 motorway. Following admission to the Institution of Civil Engineers in 1984, he was engaged on the design of pavement rehabilitation schemes and the implementation of pavement management systems in the Middle East, Africa and the UK. Since 2000 he has been based in the English Midlands, involved with maintenance assessment schemes and pavement design for D&B and DBFO road schemes in the UK and eastern Europe.

Coleman A. O'Flaherty

Coleman O'Flaherty worked in Ireland, Canada and the United States before joining the Department of Civil Engineering, University of Leeds in 1962. He was Foundation Professor of Transport Engineering and Foundation Director of the Institute for Transport Studies at Leeds University before being invited to Australia, initially as Commonwealth Visiting Professor at the University of Melbourne, and then as Chief Engineer at the National Capital Development Commission, Canberra. Since retiring as Deputy Vice-Chancellor at the University of Tasmania he has been made a Professor Emeritus of the University and awarded the honorary degree of Doctor of Laws. In 1999 he was made a member of the Order of Australia for services to education and the community.

John E. Oliver

John Oliver began his career with contractors and consultants before joining the Department of Transport in 1972. For the next ten years he worked on the design and construction of major highway projects in southern England. He then began to specialize in the management and maintenance of major roads at both regional and national levels. In 1986 he transferred to policy and programme management, and became Head of Maintenance for the Department. Following the creation of the Highways Agency in 1994 he became its Head of Operations Support before taking early retirement in 1997. Since 1997 he has been practising as an independent consultant in the fields of highway management and maintenance.

Martin S. Snaith

Martin Snaith has worked in many countries throughout the word, specializing in road maintenance and devising new methods for both the analysis of the structural performance of road pavements and the management of resources for the network as a whole. He joined the University of Birmingham in 1978 as the Overseas Development Administration Senior Lecturer to develop the highly successful Master's Programme in Highway Engineering, and later became the Professor of Highway Engineering and led the Highways Management Research Group which developed a variety of tools that are in use worldwide, e.g. BSM, HDM-4 and SPM. Currently Pro-Vice-Chancellor of Birmingham University, he was made a Fellow of the Royal Academy of Engineering in 1998.

Alan R. Woodside

Alan Woodside was employed as a civil engineer with Larne Borough Council and Ballymena Borough and Rural District Councils from 1965–1973 specializing in highway design and road reconstruction works. He then joined the lecturing staff of the Ulster College (later the Ulster Polytechnic) where he became a principal lecturer. In 1982 he was appointed Senior Lecturer in Civil Engineering at the University of Ulster and then Reader in Highway Engineering in 1990. He was awarded a personal chair in Highway Engineering within the School of the Built Environment at the University of Ulster in 1994 and promoted to Director of TRAC in 1999. Over the past 30 years he has published more than 250 technical papers, supervised many PhD students and lectured to highway engineers throughout the world on many research topics including aggregates, skidding resistance, stone mastic asphalt, recycling and durability of surfacings.

W. David H. Woodward

David Woodward completed his undergraduate studies in Geology in 1980 and then undertook his Masters and Doctoral studies by part-time research at the University of Ulster in Highway Engineering while engaged as a research officer in the Highway Engineering Research Centre. He was awarded the degree of DPhil in 1995 for research into predicting the performance of highway surfacing aggregate. In 1999 he was appointed Lecturer in Infrastructure Engineering and Head of the Highway Engineering Group within the Transport & Road Assessment Centre at the University of Ulster. He has published over 150 publications, contributed to six textbooks, supervised and examined numerous PhD students. Over the last 20 years he has lectured at many national and international conferences throughout the world. His research interests include aggregates, skidding resistance, waste materials, adhesion of bitumen and unbound roadways.

Salah E. Zoorob

Salah Zoorob completed his undergraduate and postgraduate studies at the University of Leeds, where he was awarded a PhD degree in 1995 for research into the use of fly ash in bituminous composites. In 1996 he was appointed Lecturer in Highway Materials at the University's Civil Engineering Materials Unit. In 2000 he joined the Nottingham Centre for Pavement Engineering, an initiative of the University of Nottingham, Shell Bitumen, and Shell Global Solutions(UK), as Research Coordinator. He is co-editor of the European Symposium on the Performance of Bituminous and Cement-treated Materials in Pavements, which is held in the UK every 3 years. His research interests are in the technical and environmental implications of using waste materials in roads.

Acknowledgements

My colleagues and I are indebted to the many organizations and journals which allowed us to reproduce diagrams and tables from their publications. The number in the title of each table and figure indicates the reference at the end of the relevant chapter where the source is to be found. The material quoted from government publications is Crown copyright and is reproduced by permission of the Controller of Her Majesty's Stationery Office.

In relation to the above the reader is urged to seek out the original material and, in particular, to consult the most up-to-date 'official' versions of recommended practices, standards, etc., when engaged in the actual location, design and construction of roads.

On a personal basis, I would like to express my very considerable thanks to my co-writers and to my publisher, Eliane Wigzell, for their understanding and patience during the preparation of this text. I am most grateful for this support. I would also like to thank Judy Jensen and Murray Frith for their unassuming, very professional, help in obtaining reference material and preparing drawings for this book, respectively.

Last, but not least, I pay tribute to my wife, Nuala, who helped me immeasurably in the preparation of previous editions of this book. Unfortunately, Nuala was not able to see the completion of this text but her love, support and forbearance will always be remembered by me.

Coleman O'Flaherty
Launceston
4 April 2001

INTRODUCTION

A historical overview of the development of the road

C.A. O'Flaherty

Everybody travels, whether it be to work, play, shop, do business, or simply visit people. All foodstuffs and raw materials must be carried from their place of origin to that of their consumption or adaptation, and manufactured goods must be transported to the marketplace and the consumer. Historically, people have travelled and goods have been moved: (1) by road, i.e. by walking and riding, using humans and various beasts to carry goods or to pull sleds, carts, carriages and wagons, and (since the late 19th century) using cycles and motor vehicles such as cars, buses and lorries; (2) by water, i.e. using (since early times) ships and boats on seas, rivers and canals; (3) by rail, i.e. initially using animals (in the early 19th century) and then steam-, oil- or electric-powered locomotives to pull passenger carriages and goods wagons; and (4) by air, i.e. using airships and aeroplanes (in the 20th century).

Whilst the birth of the road is lost in the mists of antiquity, there is no doubt but that the trails deliberately chosen by early man and his pack animals were the forerunners of today's road. As civilization developed and people's desire for communication increased, the early trails became pathways and the pathways evolved into recognized travelways. Many of these early travelways – termed *ridgeways* – were located high on hillsides where the underbrush was less dense and walking was easier; they were also above soft ground in wet valleys and avoided unsafe wooded areas.

The invention of the wheel in Mesopotamia in ca 5000 BC and the subsequent development of an axle that joined two wheels and enabled heavy loads to be carried more easily, gave rise to wider travelways with firmer surfacings capable of carrying concentrated loads, but with less steep connecting routes down to/up from valleys and fordable streams. Thus *trackways* evolved/were created along the contours of lower slopes, i.e. they were sufficiently above the bottoms of valleys to ensure good drainage but low enough to obviate unnecessary climbing. The trackways eventually became well-established trade routes along which settlements developed, and these gave rise to hamlets and villages – some of which, eventually, became towns and cities.

Early manufactured[1] roads were the stone-paved streets of Ur in the Middle East (ca 4000 BC), the corduroy-log paths near Glastonbury, England (ca 3300 BC), and brick pavings in India (ca 3000 BC). The oldest extant wooden pathway in Europe, the 2 km long Sweet Track, was built across (and parts subsequently preserved in) marshy ground near Glastonbury. The oldest extant stone road in Europe was built in Crete ca 2000 BC.

Notwithstanding the many examples of early man-made roads that are found in various parts of the world, it is the Romans who must be given credit for being the first 'professional' road-makers. At its peak the Roman road system, which was based on 29 major roads radiating from Rome to the outermost fringes of the Empire, totalled 52 964 Roman miles (ca 78 000 km) in length[2]. Started in 312 BC, the roads were built with conscripted labour; their purpose was to hold together the 113 provinces of the Empire by aiding the imperial administration and quelling rebellions after a region was conquered. The roads were commonly constructed at least 4.25 m wide to enable two chariots to pass with ease and legions to march six abreast. It was common practice to reduce gradients by cutting tunnels, and one such tunnel on the Via Appia was 0.75 km long. Most of the great Roman roads were built on embankments 1 m to 2 m high so as to give troops a commanding view of the countryside and make them less vulnerable to surprise attack; this had the engineering by-product of helping to keep the carriageway dry. The roads mainly comprised straight sections as they provided the most direct routes to the administrative areas; however, deviations from the straight line were tolerated in hilly regions or if suitable established trackways were available.

In the 150 years following their occupation of Britain in 55 BC the Romans built around 5000 km of major road radiating out from their capital, London, and extending into Wales and Scotland. However, the withdrawal of the legions from Britain in AD 407 foreshadowed the breakdown of the only organized road system in Europe until the advent of the 17th century.

Whilst the Roman roads in Britain continued to be the main highways of internal communication for a very long time, they inevitably began to decay and disintegrate under the actions of weather, traffic and human resourcefulness. Eventually, their condition became so deplorable that, when sections became impassable, they were simply abandoned and new trackways created about them. Most 'new roads' consisted of trackways made according to need, with care being taken to avoid obstructions, private property and cultivated land. These practices largely account for the winding character of many present-day secondary roads and lanes.

Throughout the Middle Ages through-roads in Britain were nothing more than miry tracks so that, where practicable, the rivers and seas were relied upon as the major trade arteries. During these same times, however, lengths of stone-paved streets were also built in some of the more prosperous towns, usually to facilitate their provisioning from rural hinterlands, i.e. good access roads were needed to withstand the high wheel stresses imposed by the wheels of the carts and wagons of the day.

It might be noted that the terms *road* and *street* began to come into wide usage in England in the 16th and 17th centuries[1], with the word 'road' possibly coming from the verb 'to ride' and implying a route along which one could progress by riding, whilst 'street' likely came from a latin word meaning constructed (as applied to some Roman roads, e.g. Watling Street).

Even though most roads were in a dreadful state the opportunities for overland passenger travel continued to increase, e.g. the first non-ceremonial coach was seen in London in 1555 and the first British stagecoach service to change horses at regularly-spaced posthouses was initiated between Edinburgh and Leith in 1610.

The development (in Austria in the 1660s) of the Berliner coach's iron-spring suspension system resulted in such an expansion of travel by coach that, by 1750, four-wheeled coaches and two-wheeled chaises (introduced from France) had superseded horseback-riding as the main mode of intertown travel by Britain's wealthy and its growing middle class.

The onset of the 18th century also saw foreign trade become more important to Great Britain's steadily-developing manufacturing industries, and soon long trains of carts and wagons were common sights as they laboriously dragged coal from mines to ironworks, glassworks and potteries, and manufactured goods to harbours and ports, along very inadequate ways.

Confronted by the above pressures and the terrible state of the roads Parliament passed, in 1706, the first of many statutes that eventually created over 1100 Turnpike Trusts. These Trusts, which administered some 36 800 km of road, were each empowered to construct and maintain a specified road length and to levy tolls upon certain types of traffic. The development of the toll road system, especially in the century following 1750, was important for many reasons, not least of which were: (i) it promoted the development of road-making techniques in Britain, and allowed the emergence of skilled road-makers, e.g. Thomas Telford and John Loudon McAdam; (ii) it established that road-users should pay some road costs; and (iii) it determined the framework of the 20th century's pre-motorway trunk road network.

The opening of the first steam-powered railway service (between Stockton and Darlington) in 1825 marked the beginning of the end for the Turnpike Trusts as the transfer of long-distance passengers from road to rail was almost instantaneous once towns were accessed by a railway. So many Trusts became insolvent that, in 1864, the government decided to gradually abolish them; as a consequence the final Trust (on the Anglesey portion of the London–Holyhead road) collected its last toll on 1 November 1895.

The abolition of the Trusts resulted in their roads reverting to the old system of parish maintenance. Thus, at the turn of the latter half of the 19th century, there were some 15 000 independent road boards in England and Wales alone, most of which resented having to pay for the upkeep of 'main' through roads from local funds. The situation became so chaotic that, in 1882, Parliament agreed *for the first time* to accept financial responsibility for aiding road construction and maintenance. This and subsequent governmental financial-cum-administrative reforms were very timely for, by the turn of the 20th century, the bicycle and motor vehicle had well and truly arrived. Then, the end of Word War 1 resulted in a major impetus being given to commercial road transport, when a myriad of motor trucks became available for non-military uses and thousands of trained lorry drivers were returned from the army to the civilian workforce.

Overall, the first 40 years of the 20th century were years of evolutionary development rather than revolutionary change for roads. Initially, the emphasis was on 'laying the dust' using, mainly, tar and bitumen surfacings, and then on reconstructing existing roads. Organized road research that was directly applicable to United Kingdom conditions was initiated in 1930 with the establishment of a small experimental station at Harmondsworth, Middlesex at which research was carried out into 'highway engineering, soil mechanics, and bituminous and concrete technology'; this was the start of the Transport Research Laboratory (TRL).

After World War 2, road technology took a giant step forward with the passing in the USA of the Federal Aid Highway Act of 1944 which authorized the development of the *Interstate and Defence Highway System* to connect 90 per cent of American cities with populations above 50 000, by means of some 70 000 km of motorway. Momentous research programmes, which included the development of special test tracks to study pavement materials, design and construction, were initiated in the USA as a consequence of this decision. The outcomes of these research programmes, and the development of associated road-making and traffic-management techniques, were major influences for road development on the international scene, especially in the 1950s.

In 1958, the first motorway (the 13 km Preston bypass) to open in Great Britain presaged the development of a strategic inter-urban trunk road network of over 15 000 km (including the construction of some 3100 km of new motorway and over 3500 km of dual carriageway). This also was the catalyst for the initiation of major pavement technology and traffic management research programmes by, in particular, the Transport Research Laboratory. Unlike in the USA, the TRL tended to rely upon test sections incorporated into existing main roads, rather than upon controlled test tracks, when evaluating pavement materials and design and construction criteria.

Notwithstanding the great numbers of motor cars on the road the research work clearly demonstrated that most damage to road pavements is caused by heavy goods vehicles and not by cars and, therefore, that pavement design should mainly be concerned with resisting the stresses and strains caused by commercial vehicles. Figure I.1 indicates that the demands placed upon existing and new pavements by heavy goods vehicles in Great Britain will increase substantially in future years.

The motorway and trunk road network is now the backbone of Great Britain's transport infrastructure, and the country's economic health and quality of life depend upon the system being well built, well managed, and well maintained. Increasingly, it can be expected that the governmental focus in the well-developed crowded island of Great Britain will be to maximize the use of the existing road network,

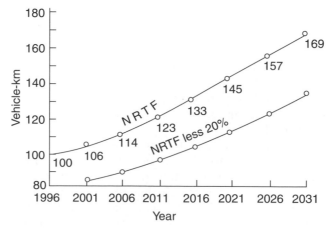

Figure I.1   National Road Transport Forecasts (NRTF) of travel by heavy (>3.5t) goods vehicles for the years 2001 to 2031. (*Note*: Index for 1996 = 100)

and most improvements will be in the form of the upgrading of existing roads, the building of bypasses about towns and villages suffering the noise, dangers and severance of inappropriate through traffic, and the widening of the most heavily trafficked and congested motorways.

Overseas, in countries that are only now experiencing the pleasures and problems associated with the motor age, the opportunities and challenges for road designers and builders are very great.

I.1 References

1. Lay, M.G., *Ways of the World*. Sydney: Primavera Press, 1993.
2. Pannell, J.P.M., *An Illustrated History of Civil Engineering*. London: Thames and Hudson, 1964.

CHAPTER 1

Road location

C.A. O'Flaherty

1.1 Complexity of the location process

As is suggested by Table 1.1, the location of a new major road can require consideration of many complex and interrelated factors, which normally utilize the skills of economists, geologists, planners and surveyors as well as those of road engineers. Before attempting to define the physical location of a new road data must be

Table 1.1 Compendium of factors commonly considered when locating a new major road (based on reference 1)

Subject area	Features	Factors
Engineering and economics	Construction costs	Topography, geology, geomorphology, materials, soils, vegetation, drainage, design criteria, safety
	Maintenance costs	Climate, traffic, soils, materials, topography, drainage, geomorphology
	User costs	Traffic, topography, travel time, safety
	Right-of-way costs*	Land values, land use, replacement costs, traffic, design criteria, utilities, tax base
	Development potential*	Agriculture, forestry, mineral extraction, trade and industry, tourism, personal mobility, political strategy
Social aspects	Neighbourhood locale	Population, culture, land use, tax rate, land value, institutions, transportation, historical sites, utilities and services, community boundaries, tax base, traffic, employment, dynamic change
Social/ecology	Recreation and conservation	Land use, vegetation, fish life, scenic areas, wildlife, drainage, topography
Ecology	Pollution**	Noise, air, water, spillage, thermal, chemical, waste
Aesthetics	Scenic value	Scenic areas, view from road, view of road, 'eye sores', topography, vegetation, drainage

*Right-of-way costs and the development potential of an area also affect the social aspects of that area
**Pollution also affects the aesthetic value of an area.

available to the road engineer regarding traffic volumes and desires, the planning intentions within the area to be traversed, and preliminary estimates of the anticipated design of the proposed road (see reference 2 for details of the development of these).

Whatever the locale being considered, location surveys provide fundamental information for the economic, environmental and social analyses which have major influences on the final location of the new road, as well as for its geometric and structural design.

Road locations are most easily determined through low-cost, relatively undeveloped, lands; in such locales basic engineering and construction cost considerations normally dominate analyses once the traffic planning need has been established and accepted – and provided that environmental issues are not of major concern. The problems become more complex and 'non-engineering' issues become more prominent as a route is sought through well-developed rural lands, and when interactions with existing roads and built-up areas have to be taken into account. The problems are normally at their most complex in and about major urban areas where community aspirations, interactions with existing roads and streets, and economic, environmental and planning issues become critical. Thus, whilst ideally a new major road needs to be located where it can best serve the traffic desire lines, be as direct as possible, and maximize its function of allowing convenient free-flowing traffic operation at minimum construction, environmental, land, traffic operations and maintenance costs, the location process in and about an urban area can, in practice, be reduced to finding an alignment that meets traffic desire lines, is acceptable to the public, and enables road construction to occur at an economical cost.

1.2 Overview of the location process

In general, once the need for a major road has been justified by the transport planning process, the approach to selecting an appropriate route location can be described[3] as a 'hierarchically structured decision process'. This classical approach is most easily explained by reference to Fig. 1.1.

The first step in the location process requires fixing the end termini, and then defining a region, A, which will include all feasible routes between these two points; in a non-urban setting this region will often be, say, one-third as wide as it is long. The region is then searched using reconnaissance techniques to obtain a limited number of broad bands, B and C, within which further (refining) searches can be concentrated; for a rural motorway, for example, such bands might be as much as 8–16 km wide. Within these bands, further reconnaissance-type searching may result in the selection of corridors D, E and F, each perhaps 3–8 km wide. A comparison of these corridors may then suggest that E will provide the best route, and route G is then generated within it; typically, this route could be 1–1.5 km wide in a rural locale. The next, preliminary location, step is to search this route and locate within it (not shown in Fig. 1.1) one or more feasible alignments, each perhaps 30 m wide and containing relatively minor design differences. These alignments are then compared during the final location phase of the analysis, and the most suitable one is selected for structural design and construction purposes.

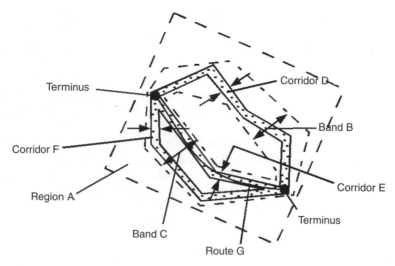

Fig. 1.1 A hypothetical route location problem showing the spatial relationships of various actions generated and evaluated[3]

Note that the above process involves continuous searching and selecting, using increasingly more detailed information and data at each decision-making stage. 'Tangible' considerations that might influence the selection process at any given instance could typically include topographic, soil and geological survey data, land usages and population distributions, travel demands and road user costs, construction and maintenance costs, and safety factors. 'Intangible' considerations of a political, social and environmental nature which require extensive public consultation may also have to be taken into account before final location decisions can be taken.

1.3 Location surveys in non-built-up locales

A conceptual survey approach has classically been recognized in relation to the gathering and analysis of 'hard' engineering data to aid in the refining of the road location decision-making process. Generally, this approach relies upon the following three types of survey: reconnaissance, preliminary location, and final location/ design. Of particular interest, are the basic approach to locating water crossings in rural areas, and the use of air photographs in location surveys, so these are therefore discussed separately.

1.3.1 Reconnaissance surveys

The first step in the reconnaissance survey is to carry out a major *Desk Study* or *Preliminary Sources Study* (*PSS*) of the bands/corridors being evaluated within the region. Much information is already publicly available at low cost (as compared with the cost of carrying out, say, new subsurface investigations) and, hence, ample budget provision should be made for this phase of the road location investigation. A good PSS will result in great savings in both time and resources, as it helps ensure the early recognition of difficult routes and allows for better planning and interpretation of the subsequent detailed site investigations. Reference 4 is an invaluable source of information regarding 'where to look' for a desk study being carried out in the UK.

The following is a useful general check-list[5] of the types of information that might be gathered for a desk study; most of these apply to a greater or lesser extent to a major road location.

1. *General land survey*: (a) location of site on published maps and charts; (b) dated air photographs; (c) site boundaries, outlines of structures and building lines; (d) ground contours and natural drainage lines; (e) above-ground obstructions to view and flying, e.g. transmission lines; (f) indications of obstructions below ground; (g) records of differences and omissions in relation to published maps; (h) positions of survey stations and benchmarks (the latter with reduced levels); and (i) appropriate meteorological information.
2. *Permitted use and restrictions*: (a) planning and statutory restrictions applying to the particular areas under Town and Country Planning Acts administered by the appropriate local authorities; (b) local authority regulations on planning restrictions, listed buildings and building by-laws; (c) Board of Trade (governmental) regulations relating to the issue of industrial development certificates; (d) rights of light, support and way including any easements; (e) tunnels, mine workings (abandoned, active and proposed), and mineral rights; (f) ancient monuments, burial grounds, etc.; (g) prior potentially contaminative uses of the site and of adjacent areas; and (h) any restrictions imposed by environmental and ecological considerations, e.g. sites of special scientific interest.
3. *Approaches and access (including temporary access for construction purposes)*: (a) road (check ownership); (b) railway (check for closures); (c) by water; and (d) by air.
4. *Ground conditions*: (a) geological maps; (b) geological memoirs; (c) flooding, erosion, landslide and subsidence history; (d) data held by central and local governmental authorities; (e) construction and investigation records of adjacent sites; and (f) seismicity.
5. *Sources of material for construction*: (a) natural materials; (b) tips and waste materials; and (c) imported materials.
6. *Drainage and sewerage*: (a) names of the sewerage, land drainage and other authorities concerned, and their by-laws; (b) locations and levels of existing systems (including fields, drains and ditches), showing sizes of pipe and whether they are foul, stormwater or combined; (c) existing flow quantities and capacity for additional flow; (d) liability to surcharging; (e) charges for drainage facilities; (f) neighbouring streams capable of taking sewage or trade effluent provided that they are purified to the required standard; (g) disposal of solid waste; and (h) flood risk to, and/or caused by, the proposed works.
7. *Water supply*: (a) names of the authorities concerned and their by-laws; (b) locations, sizes and depths of mains; (c) pressure characteristics of mains; (d) water analyses; (e) availability of water for additional requirements; (f) storage requirements; (g) water sources for fire fighting; (h) charges for connections and water; (i) possible additional sources of water; and (j) water rights and responsibilities under the Water Resources Act 1991.

 8. *Electricity supply*: (a) names of supply authorities concerned, and regulations; (b) locations, sizes and depths of mains; (c) the voltage, phases and frequency; (d) capacity to supply additional requirements; (e) transformer requirements; and (f) charges for installation and current.
 9. *Gas supply*: (a) names of the supply authorities concerned, and regulations; (b) locations, sizes and depths of mains; (c) type of gas, and thermal quality and pressure; (d) capacity to supply additional requirements; and (e) charges for installation and gas.
10. *Telecommunications*: (a) local office addresses; (b) locations of existing lines; (c) BT and other agency requirements; and (d) charges for installation.
11. *Heating*: (a) availability of fuel supplies; (b) planning restrictions (e.g. smokeless zones; Clean Air Acts administered by local authorities); and (c) district heating.
12. *Information relating to potential contamination*: (a) history of the site including details of owners, occupiers and users, and of any incidents or accidents relating to the dispersal of contaminants; (b) processes used, including their locations; (c) nature and volumes of raw materials, products, and waste residues; (d) waste disposal activities and methods of handling waste; (e) layout of the site above and below ground at each stage of development, including roadways, storage areas, hard-cover areas, and the presence of any extant structures and services; (f) the presence of any waste disposal tips, abandoned pits and quarries, and (g) the presence of nearby sources of contamination from which contaminants could migrate via air and/or groundwater onto the site.

In relation to the 'General land survey' component of the above check-list, it should be noted that the Ordnance Survey (OS) has been engaged in developing very detailed topographical information for Britain since 1805, and its maps are continually being brought up-to-date and extended. These maps, which are very useful in defining site conditions and in interpreting special maps (e.g. land use and geology maps) are available in two main groups, large-scale maps with scales of 1:1250, 1:2500, and 1:10 000, and small-scale maps with scales of 1:25 000, 1:50 000, 1:250 000, and 1:625 000. The master copies of the large-scale maps have been converted to digital data and are computer readable.

The 1:1250 scale represents 1 cm to 12.5 m (50 inches to 1 mile), and each sheet covers an area equivalent to 500 m square. These maps cover most built-up areas with populations above about 20 000.

The 1:2500 maps (1 cm to 25 m, or about 25 inches per mile) provide good cover of Great Britain's rural areas, apart from some areas of mountain and moorland which are covered by the 1:10 000 maps.

The 1:2500 map scale shows ground features such as quarries, pits, spoil tips and open cast workings, as well as industrial buildings. These maps are particularly useful for detecting land uses in built-up areas where changing industrial activity may have resulted in the release of different contaminants into the underlying soil or rock, i.e. the revisions made since the first survey in 1854 mean that the maps can be used to trace the historical usage of a site. Site types that are likely to have been contaminated by past industrial usage are: (i) asbestos works; (ii) docks and railway land, especially sidings and depots; (iii) gasworks, coal carbonization plants and by-product works; (iv) landfills and waste disposal sites; (v) metal mines, smelters, foundries, iron and steel works, and metal finishing works; (vi) motor depots and garages; (vii) munitions production works and testing sites; (viii) oil refineries, and petroleum storage and distribution works; (ix) paper and printing works; (x) heavy engineering yards, e.g. shipbuilding and shipbreaking; (xi) radioactive installations (these are subject to special regulations); (xii) scrap yards; (xiii) sewage works and farms; (xiv) tanneries; and (xv) industries making or using wood preservatives.

The 1:10 000 (1 cm to 100 m, replacing 6 inches to 1 mile) maps cover the entire country and have contour intervals of 5 m (10 m in mountainous areas). They are the smallest OS maps on which features are generally shown to the correct scale, and the largest to show contours.

The 1:25 000 Pathfinder, Outdoor Leisure and Explorer maps (1 cm to 250 m, about 2½ inches per mile) are the largest-scale OS map series in colour, are on National Grid lines, and each sheet covers an area equivalent to 20 km east-to-west by 10 km north-to-south. These maps are very useful for a first appraisal of a potential route, e.g. the colour blue is used to mark watercourses, areas of water, springs, marshes and their associated names; if a regular pattern of artificial drainage channels is shown, it suggests a high water table. Field boundaries are shown on the 1:25 000 map, and there are good indications of the types of vegetation in woodlands and uncultivated ground. The contours are at 5 m intervals, which enable steep ground that is liable to instability to be picked out. Public footpaths, bridle paths, and other rights-of-way, which may be useful in gaining access to a site, are also clearly marked.

The 1:50 000 Landranger maps (which replace the 1 inch to 1 mile maps) are useful for general planning and location purposes. The OS Gazetter of Great Britain provides National Grid references for all features named on these maps.

The 1:250 000 Travelmaster maps show motorways and classified roads (including those with dual carriageways), and other 'drivable' roads. These are revised and reprinted at regular intervals.

Specialist Travelmaster 1:625 000 (about 10 inches to the mile) maps are available for subjects such as route planning, administrative areas, archaeology and geology. The 1:625 000 route planning map shows motorways, primary routes, and many other roads; it also identifies which roads have dual carriageways.

The sites of many ancient monuments (which are generally protected by Acts of Parliament) are also marked on Ordnance Survey sheets.

Whilst the most recent OS maps are normally obtained for use in the desk study, comparing them with older maps can also help locate/suggest features that are now hidden, e.g. concealed mineshafts, adits and wells, demolished buildings with concealed cellars and foundations, filled ponds, pits and quarries, the topography of the ground surface beneath tips, made ground, embankments and artificially flooded areas, and changes in stream and river courses, coastal topography and slip areas.

In practice, of course, the decision as to which OS map to use at any stage of the reconnaissance survey depends upon the detail of the information required at the time.

The British Geological Survey (BGS), which is the national repository of geological records, is the principal source of geoscience data; its information is readily available in the form of maps and associated memoirs, geological handbooks, and papers. The BGS is often able to provide information regarding the groundwater conditions in a given area, as well as details regarding previous explorations at particular points, from its huge file records (see subsection 2.2.1) of wells, shafts and boreholes.

Published geological maps vary in scale. The 1:10 000 (and the now-being replaced 1:10 560 or 6 inches to 1 mile) maps are the base maps that are used for recording field mapping, and many smaller-scale maps are essentially reductions and simplifications of these; hence, the 1:10 000 maps are likely to be of great value to the road engineer. The positions of many boreholes, as well as brief borehole data, are marked on these maps. Whatever maps are used, they should never be interpreted without consulting their associated sheet Memoirs; these are the explanatory reports that detail the materials recorded in the field, all of which may not be shown on the map sheet. For example, valley deposits are often shown under one colour or one symbol on a map but the Memoir may show that it contains varied materials such as gravels, sands, clays, and organic matter.

The geological maps are usually available in 'Drift' and 'Solid' editions. The Drift maps show all superficial deposits, e.g. glacial drift, sand dunes, alluvium deposits, etc., as well as hard rock outcrops, that occur at ground level. The Solid maps show no superficial deposits, i.e. they show only the underlying solid rock. The amount of drift material is so small in some locations that a single (combined) Solid and Drift edition is produced for those areas; these maps and Memoirs are further discussed in subsection 2.2.1. A range of applied geology maps covering new town and development areas, as well as engineering geology maps of special local or regional significance, are also available.

A wealth of information can be obtained from various governmental authorities and universities regarding land classification and use, including the locations and extent of overhead and underground wires and piping systems, abandoned mine workings, land previously used for industrial purposes (and now contaminated) or waste disposal, etc. For example, the *Second Land Utilisation Survey of Britain*, which was initiated in 1960, has 1:25 000 maps which show 64 categories of land use; an index map, the published maps – each sheet covers an area 10 km by 20 km – and the land use survey handbook can be obtained from Kings College, University of London. A series of maps of the *Agricultural Land Classification of England and Wales* show, via five classifications, the relative value of land for agricultural usage; these may also serve as preliminary indicators of land values and engineering characteristics. Planning maps produced by the Department of Environment (DOE) often include information regarding overhead and underground power lines and cables, gas pipelines, high agricultural land and peat, conservation areas and green belts.

A useful map to a scale of 1:625 000 is published by the Ordnance Survey which shows the general areas of coal and iron, limestone (including chalk) and gravel and sand workings in Great Britain. British Geological Society maps and Memoirs show in more detail where coal and other minerals occur. The Coal Authority will usually provide advice regarding proposed developments on coal mining areas as well as information about past, current and proposed future workings, their depths as measured at the shafts, and estimates of subsidence and its effect on structures;

however, their records are not complete and many old workings dating from before 1872 are uncharted and could pose potential hazards. (Appendix E of reference 5 provides a very useful background to the problems associated with locales with underground mining of various types.)

For roadworks, in particular, aerial photography and site imagery can be most effective in supplementing the available topographical, geological and geomorphological maps, and in interpreting features or earlier uses of the site, e.g. by helping to identify man-made features such as roads and pathways, made ground (i.e. fill), buildings (existing or demolished), extensions to quarries and gravel pits, variations in field boundaries, pipelines, and movement of mining spoil tips. They are especially useful in studying potential road locations in locales where ground visibility is limited by obstructions and/or where access is difficult. Many poor drainage and marshy areas, faults and strata boundaries, unstable ground, mining subsidence, local drainage patterns, abandoned streams, swallow holes and rock outcrops that are not on maps can also be identified with the aid of these photographs. Stereoscopic pairs of aerial photographs can be very helpful in locating places where special engineering attention is necessary.

There is no central archive or complete index of aerial photographs available for the UK. Collections are held by the Ordnance Survey, by the bodies that commissioned them (e.g. most county planning authorities take photographs of their county at regular intervals), or by the commercial companies that flew them. Tracing coverage that is relevant to particular corridor routes used to be very difficult; however, in more recent years this has been much facilitated by the publication, in 1993, of the National Association of Aerial Photographic Libraries' first register of aerial photographs, the *Directory of Aerial Photographic Collections in the United Kingdom (NAPLIB)*. (Details of many other sources of air photographs are readily available in the technical literature[4].)

Information regarding air temperature, rainfall, and sunshine conditions can be obtained from the Meteorological Office, which collects and maintains historical weather data from about 6000 weather stations throughout the UK. The Office also publishes (since 1958) climatic tables of temperature, relative humidity and precipitation from some 1800 stations throughout the world. Where data are not available for particular locations in the UK, local statutory bodies such as water authorities may well hold valuable information. Useful summary data that may be obtained for selected meteorological stations include (i) monthly and annual rainfalls and their seasonal distribution, (ii) severity and incidence of storms, (iii) direction and strength of prevailing and strongest winds, and their seasonal distributions, (iv) fog liability, (v) range of seasonal and daily temperatures, (vi) the mean frost index, and (vii) humidity conditions. These weather data can be used to anticipate problems with road location, including the handling of moisture-sensitive materials, slope instability, and frost susceptibility. Evapotranspiration and soil moisture information for particular locations in Great Britain are also published by the Meteorological Office.

Surface water run-off data are collected by water authorities, private water undertakings, and sometimes by local authorities. References 6 (for surface water data) and 7 (for run-off and groundwater level data) can be useful sources of information for road location and design purposes.

When analysing the data gathered during the desk study phase of the reconnaissance survey, particular attention should be paid to features such as hills, waterways, and land uses as these impose constraints on a road's location. Low points and passes in ridges in hilly terrain, and locations with topographical features which favour the usage of flyover structures and interchanges over intersecting roads and railway lines, are potential fixed points in route location, as are river crossing points that afford desirable sites for, and approaches to, bridges. Locations that have excessively steep grades, and/or are subject to strong winds, and/or require deep cuttings, high embankments or tunnel construction, and/or are subject to subsidence or landslip, should be avoided in hilly country. Peat bogs, marshy areas and low-lying lands subject to flooding are constraints that can also be detected at this stage of the investigation; they may dictate that a route should be on higher ground rather than in a valley, in order to minimize drainage and foundation problems. Corridors that have routes located at right angles to a series of natural drainage channels also need to be looked at very carefully as they may require many (expensive) bridges. In level terrain, routes that require the expensive relocation of above- or below-ground public utilities should be avoided, if possible. Overall, however, the major physical constraints affecting the choice of route are likely to be related to land use considerations and the cost of land, the need for bridging across waterways, problems associated with existing road, canal, and railway line crossings, and the availability of convenient borrow materials for construction purposes.

Next, armed with questions generated by the desk study, the reconnaissance engineer may take to the field to seek discrete evaluations of unusual topographic, geological, drainage and/or man-made features, and answers to queries of an environmental or hydrogeological nature. Much useful information can be gathered from these direct observations. The overall objective of this *'walk-over' phase* of the reconnaissance study is to fill in omissions in information gathered during the desk study, so as to delimit corridor areas that are obviously unsuitable, to more closely define terrain areas that appear promising and, as appropriate, to provide further data that might be useful for design purposes. Recognizing that the extent of the data gathered at a given potential site will, by then, be influenced by the likelihood of it being selected as the ultimate route, typical information gathered might include the following[7] (see also subsection 2.2.1 and, for engineering geology needs, reference[8]):

General: (a) Differences and omissions from plans and maps, e.g. boundaries, buildings, roads, and transmission lines; (b) relevant details of existing structures and obstructions, e.g. mine or quarry workings (old and current), transmission lines, ancient monuments and old structures, trees subject to Preservation Orders, gas and water pipes, electricity cables, and sewers; (c) access, including the probable effects of construction traffic and heavy construction loads on existing roads, bridges and services; (d) water levels, and the rate and direction of flow in rivers, streams and canals; (e) flood levels and tidal and other fluctuations, where relevant; (f) the likelihood of adjacent property being affected by the proposed roadworks, and any adjacent activities that may have led to contamination of the proposed site; (g) obvious hazards to public health and safety (including to trespassers) or the environment; (h) areas of discoloured soil, polluted water, distressed vegetation, significant odours, and evidence of gas production or underground combustion.

Ground information: (a) Surface features, on site and nearby, such as (i) the type and variability of surface conditions that differ from those recorded, (ii) the presence of fill, erosion, or cuttings, (iii) steps in the ground surface that indicate geological faults or shatter zones, (iv) evidence of mining subsidence, e.g. steps in the ground, compression and tensile damage to brickwork, buildings and roads, structures out of plumb, and interference with drainage patterns, (v) mounds and hummocks in more or less flat country, which may indicate former glacial conditions, e.g. till and glacial gravel, (vi) broken and terraced ground on the slopes of hills, possibly due to landslips, or small steps and inclined tree trunks, which may suggest soil creep, (vii) crater-like hollows in chalk or limestone country which usually indicate swallow holes filled with soft material, and (viii) low-lying flat areas in hill country, which may be the sites of former lakes and contain soft silty soils and peat; (b) ground conditions at on-site and nearby quarries, cuttings, and escarpments; (c) groundwater levels, positions of springs and wells, and occurrences of artesian flow; (d) the nature of the vegetation, e.g. unusual green patches, reeds, rushes, willow trees and poplars usually suggest wet ground; and (e) embankments and structures with a settlement history.

Other: Ensure that a proper ground investigation can be carried out, e.g. obtain appropriate information regarding site ownership(s), site accessibility options (including noting relevant obstructions), the availability of adequate water supplies, the potential for employing temporary local labour, etc.

The importance of carrying out the reconnaissance survey in a discrete manner without causing local disquiet and annoyance in, particularly, well-developed locales cannot be too strongly emphasized. For understandable reasons, local people are often likely to draw unfavourable inferences about road impacts upon their homes and properties from obvious inspections, and this can result in inadvertent, and often wrong, long-term price blight being applied to areas that are perceived as being 'under threat' as potential road sites.

Upon completion of the low-key reconnaissance survey, the engineer should have sufficient information available to him/her which, when taken in combination with the economic, environmental, planning, social and traffic inputs, enable the selection of one or more feasible corridor routes, The results of this analysis is presented in a *reconnaissance report*. In its barest essentials, this report describes the preferred corridor route(s), states the criteria satisfied by the project, presents tentative estimates of cost, provides provisional geotechnical maps and longitudinal geological sections for the location(s) under consideration, and shows the characteristics of the more important engineering features. Special situations that may lead to design or construction problems, or which may require special attention in a subsequent subsurface investigation, are also pointed out in this report.

Major road projects deserve a comprehensive report, especially as public hearings can be anticipated. In such cases the reconnaissance report writer should have in mind that the report is likely to be used as a public-information device to inform interested persons about the advantages (and disadvantages) of particular corridor routes. In this context the report should stress the exhaustive nature of the reconnaissance survey, show concern for the public interest and emphasize the expected benefits of the proposed road to both the road user and the communities served by it, and cover matters relating to environmental, land use and social impacts.

1.3.2 Preliminary location surveys

The preliminary survey is a large-scale study of one or more feasible routes within a corridor, each typically 40–240 m wide, that is made for the purpose of collecting all physical information that may affect the location of the proposed roadway. It results in a paper location that defines the line for the subsequent final location survey. At a minimum this paper location should show enough ties to the existing topography to allow a location party to peg the centreline; in many cases, field detail that is required by the final design may also be economically obtained during the preliminary survey.

In the course of the preliminary survey the detailed relief of the ground, the locations of 'soft' ground and potential ground subsidence areas, the limits of the water catchment areas, the positions and invert levels of streams and ditches, and the positions of trees, banks and hedges, bridges, culverts, existing tracks and roads, powerlines and pipelines, houses and monuments, and other natural and man-made cultural places need to be clearly determined and noted for each route within the established corridor. These details are then translated into base topographic maps so that likely road alignments can be plotted.

A *planimetric map* is one that shows only the horizontal locations of particular features, whereas a *topographic map* shows both horizontal and vertical data, usually with the aid of contour lines, that enable the road alignment to be defined in both the horizontal and vertical planes. *Contours* are simply lines of equal elevation on the terrain, i.e. all points on the one contour have the same elevation. The main characteristics of contours are:

1. Every contour closes on itself, either within or beyond the area covered by the map. If it does not close within the map limits the contour runs to the edge of the map.
2. A contour that closes within the limits of a map indicates either a summit or a depression. A pond or a lake is usually found within a depression; where there is no water the contours are usually marked to indicate a depression.
3. Contours cannot cross each other – except in the case of an overhanging cliff where there must be two intersections, i.e. where the lower-ground contour line enters and emerges from below the overhanging contour line.
4. On a uniform slope the contours are equally spaced. However, the steeper the slope the smaller is the spacing between contour lines, whilst the flatter the slope the larger the spacing.
5. On a plane surface the contours are straight and parallel to each other.
6. When crossing a valley the contours run up the valley on one side, turn at the stream, and run back at the other side. As contours are always at right angles to the lines of steepest slope, they are always at right angles to the thread of the stream at the points of crossing. In general, the curve of the contour in a valley is convex towards the stream.

7. Contours also cross ridge lines (watersheds) at right angles.
8. All contour lines are marked in multiples of their intervals.

Two approaches are available for preliminary survey mapping, conventional ground surveys and aerial surveys. In practice a combination of these two methods may also be used.

The ground survey approach is most appropriately used when the corridor is closely-defined, narrow rights-of-way are contemplated, and problems of man-made culture and of publicity are avoidable. In such circumstances ground surveys, beginning with a traverse baseline, will usually furnish the required data quite economically. Additional operations that can be included easily are the profile levels and cross-sections, and the ties to land lines and cultural objects. The extent to which data are gathered to the left and right of each tentative centreline will vary according to the topography and the type of road; however, the survey width should never be less than the anticipated width of the proposed road's right-of-way.

The cost of an aerial survey does not increase in direct relationship to the size of the area being photographed, so this type of survey is more likely to be suitable, and economical, in the following situations: (1) where the reconnaissance survey was unable to approximate closely the final alignment, e.g. at entirely new locations in rugged terrain or where land uses and land values vary considerably; (2) where a wide right-of-way is necessitated, e.g. for a motorway; and (3) where it is desired to prevent the premature or erroneous disclosure of the road's likely location, i.e. the ground-control work required by an aerial survey reveals little of the road engineer's intentions, thereby preventing the premature awakening of local public concerns and associated land speculation and/or price blight.

What follows is concerned with the carrying out of a *conventional ground survey*, and the aerial technique is discussed separately.

The first step in the ground survey is to carry out a baseline traverse for the tentatively-accepted route(s) recommended in the reconnaissance report. This may be an open traverse consisting of a series of straight lines and deflection angles that approximately follow the recommended line(s); however, if sweeping curves with large external distances are involved, they may be included so as to avoid unnecessary ground coverage. The baseline traverse is stationed continuously from the beginning to the end of the survey. Every angle point should be referenced to at least two points outside the area likely to be occupied by the road construction.

To furnish data for a profile of the baseline, levels are at least taken at all marked stations as well as at important breaks in the ground. The elevations at cross-roads, streams, and other critical points on the line are also determined. Levels should always be referred to the standard datum plane of the country in which the roadworks are being carried out; in Britain the *Ordnance Datum*, which refers to the mean sea level at Newlyn in Cornwall, is now used for this purpose (vis-a-vis, previously, the mean sea level at Liverpool).

After the baseline has been pegged and levels run over it, critical topography elevations may be taken. The simplest way is by cross-sections which, on flat topography, are often taken by the survey party at the same time as the profile levels, i.e. observations are taken at right angles from each station so as to cover the expected construction area. Where wider deviations of the location of the centre-line from the baseline are expected, e.g. in hilly country, the topography may be more rapidly taken by tacheometry. At the same time the positions of all trees, fences, buildings, etc., are obtained so that they can be shown on the preliminary map.

Based on the survey data a preliminary survey strip map of the proposed route is developed. The minimum information shown on this map includes all tangents together with their bearings and lengths, all deflection angles, and all natural and man-made surface and subsurface features that might affect the selection of the paper alignment, e.g. the locations of all fences, properties and buildings, streams, hills, springs, lakes, swallow holes, areas of likely mining subsidence, pipelines, electricity lines, and borrow material sources. Contours are also shown, especially in hilly country and where complex structures or intersections are likely to be located.

The next step in the process is the determination of the paper location of the centreline of the proposed road so that it best fits the topography whilst meeting the intended traffic service requirements (which dictate the alignment design stand-ards). This is essentially a trial-and-error process whereby, typically, trial centre-lines are drawn on the strip map and these are then adjusted on the basis of the skills and judgement of the engineer.

Many considerations influence the location of the centreline that is finally selected. Some of these have already been discussed; others, not mentioned pre-viously, include: (1) in rural locales, locate the road along property edges rather than through them and maximize the use of existing right-of-ways (to minimize the loss of farmland and the need for subways for crossing animals and farm machinery); (2) avoid alignments that result in the motorist driving into the rising (morning) or setting (evening) sun for long periods and, to relieve the monotony of driving on long straight sections, site the road so as to view a promi-nent scenic feature; (3) minimize the destruction of man-made culture or wooded areas, and avoid cemeteries, places of worship, hospitals, old people's homes, schools and playgrounds; (4) avoid highly-developed, expensive, land areas and seek alignments that cause the least amount of environmental (visual and audio) blight; (5) in hilly terrain, maximize low-cost opportunities to pro-vide long overtaking sections (for single carriageways) and of using varying cen-tral reservation widths and separate horizontal and vertical alignments (for dual carriageways); (6) in hilly country also, avoid alignments that are shielded from the sun (so that rainwater, snow and ice on the carriageway can dissipate); (7) if a vertical curve is imposed on a horizontal curve, ensure that the horizontal curve is longer; (8) avoid introducing a sharp horizontal curve at or close to the top of a pronounced crest vertical curve or the low point of a pronounced sag vertical curve, and make horizontal and vertical curves as flat as possible at junctions with other roads (for safety reasons); (9) seek favourable sites for river crossings (preferably at right angles to stream centrelines), and avoid locating bridges or tunnels on or near curves; (10) minimize the use of alignments that require (expensive) rock excavation; and (11) try to ensure that excavation quantities are in balance with embankment quantities (so that earthworks haul-age is minimized).

1.3.3 Final location surveys

The final location survey involves fixing the final, permanent, centreline of the road, whilst at the same time gathering the additional physical data needed to prepare the construction plans. The following are the general features of the final location survey; in relation to these requirements, however, it must be emphasized that many will already have been satisfied in earlier surveys, especially during the preliminary location stage.

The centreline (including all curves) that is pegged during the final location survey should closely follow the paper location on the preliminary-survey map, conforming as much as possible to important control points (e.g. road junctions and bridge sites) and the alignments prescribed. In practice, however, this is not always possible and additional topographic data gathered at this stage may indicate that local deviations are necessary to better fit the centreline to the topography and/or to allow for previously-gathered incomplete or inaccurate data. The positions of property corners, lines, fences, buildings and other man-made improvements may also affect the final location, as will the exact extent to which property owners are affected by the proposed road right-of-way.

The levelling data obtained in relation to the final location survey are fundamental to the vertical alignment, earthworks, and drainage designs. Levels should therefore be taken at regular intervals along the centreline, at all stations, and at intermediate points where there are significant changes in ground elevation; this should be extended for, say, 175 m beyond the beginning and end of the proposed scheme so as to allow for transition connections to existing roads. Cross-section levels should be taken at right angles, on both sides, at the centreline locations previously levelled. (Where large-scale aerial photographs are available this detail work may be considerably curtailed as both profile and cross-section data can be developed in the stereo-plotter to an accuracy that is comparable with that obtained from the ground survey.)

Profile and cross-section elevations of intersecting roads should also be taken for appropriate distances on both sides of new centrelines; it is always easier to design a new junction or a 'tie-in' between a proposed road and an existing one when there are many cross-sections, rather than too few. If particular borrow pits are specified for use in construction, levels may be taken for use in later computations of the quantities of material used.

All ditches and streams within the area of construction should be carefully located with respect to the pegged centreline, and their profile elevations taken upstream and downstream. Usually about 60 m will be enough, albeit greater profile lengths and cross-sections may have to be taken of the larger streams to provide information for hydraulic design. Detailed information should also be obtained on all existing bridges or culverts, including the type, size, number of openings or spans, elevations of culvert flow lines and stream-beds under bridges and, where appropriate, high water elevations.

Benchmarks are key reference points and new permanent ones may have to be established outside the final construction area so that they can be preserved for the full period of the road contract.

The main ground investigation (see Chapter 2), which provides the detailed subsurface information required for the design and construction of the road pavement

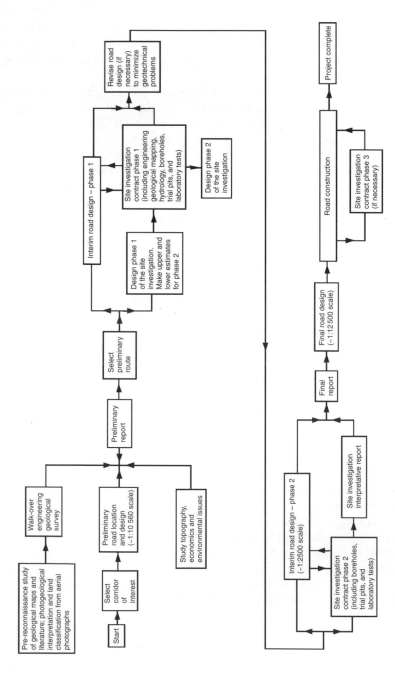

Fig. 1.2 Flow diagram showing the major site investigation activities set in their relative position to the main road design activities[8]

and of major road structures, e.g. interchanges, flyovers and bridges, is most commonly carried out during the final location survey. The subsurface investigations carried out at this stage should also provide detailed borrow pit and quarry information relating to the availability and accessibility of suitable pavement materials and concrete aggregates, earthworks for embankments and reclamation, water for compaction, and topsoil for final sodding purposes. A realistic description of the interrelationships between the surface and subsurface surveys is summarized in Fig. 1.2; this shows that the tempo and depth of detailed subsurface site investigation increases as the road location and design become more settled.

1.4 Road location in built-up areas

Determining a suitable and acceptable location for a new major road in an urban area can be most complex and challenging – and most frustrating. Because of the checks and balances that must be addressed it typically takes between 10 and 15 years to implement a major road proposal in a built-up area in the UK, i.e. from when the project is conceived to when traffic actually travels on the new road.

In concept, the search for the line of a new road involves a combination reconnaissance–preliminary survey (which is dominated by transport planning activities) and a final location survey.

The *reconnaissance–preliminary survey* involves a transport planning investigation (see reference 2) which is carried out in conjunction with a desk-based physical site survey. Simplistically, the steps involved in this survey can be summarized as follows: (1) determine the approximate traffic volume along a general corridor suggested by traffic desire lines; (2) select the road type, the number of lanes needed to carry the traffic load, and the level of service to be provided to road users; (3) establish one or more preliminary routes that meet desire-line needs, and sketch preliminary designs, including interchange and flyover locations; (4) assign traffic to one or more of the selected routes to determine design traffic volumes; (5) adjust routes and complete preliminary plans for major alternative road locations; (6) compare alternative locations using cost, environmental, road user benefit, and social analyses and select the preferred one(s) for public consultation.

The number of alternative road locations available for consideration in a built-up area are normally fairly limited. The town is an established entity and it is only in run-down areas (such as alongside old railway lines) and on the very outskirts that relatively low-cost land may be available. The existing streets are fixed in location and size by the natural topography and by the buildings that they service. Thus, the location of a new major road or the substantial upgrading of an existing one must inevitably result in changes to portions of the established city culture which, in turn, has many direct and indirect impacts upon owners and users of the affected urban infrastructure.

Whilst traffic needs and desires lines are obviously important factors affecting the location of a new major road they must always be qualified by an evaluation of the road's effect upon, and relationship with, land use and related town planning considerations. Town planning is concerned with the present and future needs of the business, industrial, residential, and recreational elements of a town so that a pleasant and functional whole is assured now and into the future. Thus, the road engineer is actively participating in town planning when locating and designing a new road and should be willing to bow to policy, social and environmental needs, where necessary, during the location process.

For example, it is obviously not desirable for new major roads to be located through residential neighbourhoods, historical areas or heavily-used public parks. On the other hand, a major road can be a desirable buffer between an industrial area and, say, a business area. The parking problem is acute in most cities; thus it is generally desirable to locate new roads as close as possible to existing and planned parking facilities. Traffic congestion in existing streets can be alleviated (or accentuated) by judicious (or poor) locations of motorway on- and off-ramps. Existing transport systems are also important controls in relation to road location, e.g. railway stations, docks and (on the outskirts of town) airports need good servicing by roads if the are to properly fulfil their functions.

Underground public utilities, e.g. stormwater pipes and sewers, and water, gas and electric lines (and their related appurtenances), can pose severe planning and cost problems in a built-up area that affect road location and design. Irrespective of whether a proposed new road is at-grade or depressed some adjustments in utilities are always necessary; in the case of a depressed road, for example, the problems may be so great as to make another site more attractive – or cause a change in design.

Whilst land and social costs are normally critical factors affecting road location in a built-up area, topography and underlying geotechnical conditions are still important controls. For example, a high water table or rock close to the surface normally precludes a depressed road, even though it may be desirable for other reasons. Similarly, poor subsoil conditions may prevent the construction of a long length of elevated roadway.

In an urban area there are usually ample data available to the road engineer from geotechnical surveys carried out previously for street improvements, building constructions, public utility installations, etc., so that there should be little need to obtain new physical information for the alternative routes being considered during the reconnaissance–preliminary survey. The use of aerial survey photographs is strongly recommended at this stage of the analysis; not only are these invaluable in detecting aspects of the urban culture that may be invisible to the completely ground-based investigator, but they enable alternative routes to be examined without alarming owners and users of properties that might/might not be eventually affected by the new road.

The *final location survey* in an urban area is similar to that described previously for a rural area, except that it is usually more difficult to carry out. For example, very rarely in an urban area can a continuous centreline be pegged directly because of the many obstructions that are encountered. Until the ground is cleared for roadworks to begin, it is always necessary to set out the centreline by means of complicated off-setting and referencing. Taking profile and cross-section levels prior to gaining control of the land upon which the road is to be built is always a complicated task; the survey team is not only faced with difficult sighting problems but its members may also be subjected to the concerns of property owners who are not terribly enthusiastic about granting them access to their properties.

1.5 Locating water crossings

For discussion purposes, water crossings can be divided into two main types; conventional high-level bridge crossings and low-water crossings. Conventional high-level bridges are of three main types, arch, beam (most common) and suspension; they are integral features of all major roads and are normally designed to cater for, at the very least, 50-year floods.

Low-water crossings are stream crossings on low-volume roads in rural-type locales that are located, and deliberately designed, to be covered with flood waters approximately once per year, usually after the maximum 1-year storm[9]. They are discussed here for completeness, albeit the type of road that they would be serving would not justify the depth of investigation described previously.

1.5.1 Bridges

Location surveys for significant bridges can be carried out also in the three classical phases of reconnaissance, preliminary, and final location surveys. Before commenting on these, however, it should be noted that Admiralty charts play the same role for coastal areas, harbours, and river estuaries that Ordnance Survey maps and plans play for land areas.

The *reconnaissance survey* is carried out to get information on the locations of all possible bridge sites, to examine the comparative suitability of each site (including obtaining preliminary estimates of construction costs), and, finally, to establish the need for the stream crossing within a limited zone. Data gathered during the desk-based component of this study (in addition to that obtained during the centreline investigation) might include[10], if appropriate: (1) requirements of statutory bodies controlling waterways, e.g. port, water and planning authorities, and fisheries; (2) topographical and marine survey data to supplement Ordnance Survey maps and Admiralty charts and publications; (3) available detailed information about rivers, the size and nature of catchment areas, tidal limits, flood levels, and their relationships to Ordnance Datum; (4) observations on tidal levels and fluctuations, velocities and directions of currents, variations in depth, and wave data; (5) information on scour and siltation, and on the movement of foreshore material by drift; (6) location and effects of existing river or marine structures, wrecks, or other obstructions above or below the waterline; and (7) observations on the conditions of existing structures, e.g. attack by marine growth and borers, corrosion of metalwork, disintegration of concrete, and attrition by floating debris or bed movements.

Most reconnaissance information can be obtained from existing records, maps and aerial photographs. In certain instances, however, it may be desirable to inspect the sites and, as necessary, carry out precursory subsurface investigations.

The *preliminary location survey* is carried out to determine the alignment within the limited zone selected by the reconnaissance survey, and to obtain the physical data needed for the acquisition of land and the design of the bridge. Ideally, this should result in the shortest crossing being selected, i.e. at right angles to the stream, as this normally results in the cheapest construction cost. In practice, however, the approach that is often taken is first to secure the most favourable location and alignment for the new road and then, recognizing that the bridge is an integral and important part of the new road but not – unless the stream is very wide – the controlling one, seeking from the bridge engineer a bridge design that fits the road's needs (rather than bending the road to fit the bridge's needs). Whilst this approach usually results in a more expensive bridge structure, i.e. skew bridges are more costly than right-angled ones and longer bridges are normally more expensive than shorter ones, the end effect is generally a better and more economically-operated road.

The preliminary location survey usually includes a preliminary triangulation to establish the starting coordinates for traverses at the bridge-heads and along the approaches, topographic surveys along these traverses, and the design of a more elaborate triangulation system that can be used later to lay out the detailed construction. In addition, a hydrographic survey is normally carried out to determine the hydraulic or flow characteristics of the water at the proposed crossing, as well as longitudinal and cross-section details of the channel bottom. Other hydrographic data collected will, of course, vary with the type of crossing and the location and exposure of the proposed structure, but it always includes engineering and geological surveys of the subsurface foundation conditions.

The *final location survey* normally has two components, the pre-construction survey and the construction survey. Pre-construction survey operations normally involve the establishment of the horizontal control stations and benchmarks that constitute the fundamental framework that defines lines and grades; much of this work may be carried out during the preliminary survey but, generally, some subsequent refinement and strengthening is needed to improve its accuracy. The construction survey operations provide the intermediate and final positioning, both horizontally and vertically, of the various components of the bridge structure. The importance of attaining high accuracy in this survey cannot be over-emphasized, if wasteful delays and needless expenditures are to be avoided.

1.5.2 Low-water crossings

It is quite reasonable nowadays to drive off-road farm and forestry vehicles, and even modern motor cars, through water that is 10–15 cm deep Thus, whilst low-level water crossings are never used in high-quality main roads, they can be very effectively and economically utilized in access roads of low priority in rural locales. For example, roads for logging operations or for recreation purposes, or those serving scattered farm residences, rarely require completely continuous access and can operate adequately with occasional road closures. Often, also, an elsewhere-located longer access road may remain open when high water temporarily closes a low-water crossing.

The construction of high-level approach roads on embankment, and a large bridge structure, may not be an environmentally-desirable or economical solution to a crossing problem in a broad flat stream valley in a rural locale, whilst the alternative solution of a smaller bridge and embankments may so constrict drainage that flooding occurs. Thus, the construction of a low-water crossing may permit the continuation of the natural waterway and minimize the potential for flooding of adjacent lands and homes.

In tropical areas, drainage-ways often carry little or no water except during sudden severe storms; in these locations also the construction of low-water crossings may be a low-cost way of providing for infrequent storms. In cold weather areas, floods are often associated with spring run-offs; the flows may be low and steady during the rest of the year and easily provided for by low-water crossings. In mountainous terrain, streams through alluvial fans have rapidly-changing flows and unstable channels; low-water crossings offer low investment solutions at these high-risk locations, and minimize the chance of being the cause of a channel change.

Selecting a location for a low-water crossing is most easily discussed according to whether the crossing is a simple ford, a vented ford, or a low-level bridge.

Fords, also known as *dips*, are formed by lowering the road surface grade to the stream-bed level, from bank to bank. Simple fords are most appropriately used across streams where the normal day-to-day flow is low or dry. Thus, ford location in this instance means finding suitable access points to/from a firm unsurfaced level crossing on the natural bed of the stream or drainage path. The ford may also be surfaced with concrete, bitumen, or gabions to protect the crossing from erosion and provide the crossing vehicle with a more stable tractive surface.

Vented fords, also known as *dips with culverts*, are formed by partially lowering the grade of the road surface so that it can be covered by water during floods, whilst culverts beneath the crossing surface handle the day-to-day stream flow. A splash apron or cut-off wall is normally placed on the downstream side of the ford, to prevent undermining of the culvert and the roadway on top. These fords are best used at locations where the normal day-to-day water flow exceeds the fordable depth by vehicles. Crossing locations for vented fords should be selected so that sloped culvert entrances, when formed, are self-cleansing.

A *low-level bridge*, usually of concrete, may be used instead of a vented ford at locations where culverts cannot handle the day-to-day stream flow. In this case a low-cost bridge may be designed so that its surface is covered by water during the design flood.

1.6 Aerial photography

Aerial photography is the science of taking a photograph from a point in the air, usually from a fixed wing aircraft, for the purpose of making some kind of study of the ground surface. The use of aerial photography (which is part of the remote sensing family of investigative techniques) in road location and geotechnical surveying is a well-accepted road engineering practice.

Aerial photographs can be taken vertically or near vertical, as high or low obliques, or horizontally or near horizontal. Vertical photographs, whereby the optical axis of the camera coincides with the direction of gravity, are normally the most useful as they are at true or near-true scale and can be quantified and qualified. High oblique photographs are taken with the axis of the camera up to about 20° from the horizontal so that the horizon is included; these are mainly used for pictorial and illustrative purposes as they provide panoramic views in familiar perspective (like a ground photograph taken from a commanding height). Low oblique photographs are taken with the camera axis inclined at such an angle that the horizon is not shown; if the angle as measured from the vertical axis is large a perspective view is obtained similar to that for a high oblique photograph, whereas if the angle is small the result very closely resembles a vertical photograph.

Aerial photographs can be studied using (a) simple optical magnifiers and stereoscopes, (b) sophisticated optical/mechanical stereoplotters that can accurately measure distances or objects in the photographs, (c) optical devices that allow the simultaneous viewing of an aerial photograph and a printed map, (d) electro-optical density-slicing systems that allow a photographic image to be separated into a number of grey levels (which can be presented in different colours on a television monitor), or (e) optical viewers that enable simultaneous viewing of up to four photographic multispectral images.

Individual vertical photographs can be used as field maps for ground familiarization, rather than for plotting or measurement. Mosaics, which are assemblies of individual aerial photographs fitted together systematically so that composite views of large areas are provided, may be used in planning and location studies, in site assessments, for general mapping, or as public relations instruments to explain proposed road schemes to the general public. Stereoscopic perception is the facility that makes it possible to see the image of a scene in three dimensions; thus, if photographs are taken along parallel flight lines so that they overlap longitudinally by about 60 per cent and laterally by about 25 per cent, and adjacent pairs are then examined with the aid of a binocular stereoscopic system (which is, in effect, a mechanical reconstruction of the air camera system) the observer is provided with a three-dimensional view of the terrain. Stereoscopic analyses are particularly valuable in mapping studies (including the preparation of contour maps and soil maps), site investigations, and the identification of landform units.

The films used in aerial photography are either black-and-white (most commonly panchromatic) or colour. Photography obtained prior to the late-1970s was predominantly black-and-white and was very often taken with large cameras at scales between 1:20 000 and 1:50 000. Since about the 1980s, the trend has been away from using black-and-white film for detailed investigations because of the additional interpretative information that can be extracted from colour photography. In this respect it should be noted that every ground feature has a distinctive 'tonal signature' which varies according to the recording medium. Thus, a cinders road will appear dark grey on both panchromatic and infra-red black-and-white photographs (and medium-light grey and light grey on thermal infra-red images (see Section 1.7) obtained at night-time and daytime, respectively); by contrast, a gravel road will appear light grey in both types of photograph (and medium-light grey and medium grey in the night and day thermal images, respectively). Colour photography contains hue and chroma data as well as tonal and textural information and is, therefore, especially valuable for construction material surveys and the study of soils, geology and erosion, as the human eye is capable of separating at least 100 times more colour combinations than grey scale values.

The scales of vertical aerial photographs currently taken from fixed-wing aircraft typically vary from 1:500 to 1:120 000. The choice of scale depends upon: (1) the intended usage of the photography; (2) the normal cloud base and the extent to which it affects flying height; and (3) problems associated with scale distortion and acceptable limits for the intended usage. Measurements taken from a vertical photograph of flat land can be as accurate as those from a printed map, whereas a similar photograph of rugged terrain can have a variety of scales depending upon the elevations of the ground area at the locations considered, and the shapes of objects can be altered and displaced about the centre of the photograph.

When planning aerial photography an average scale is used for rough calculation purposes. This average scale, S_{ave}, is given by the equation

$$S_{ave} = f(H - h_{ave}) \tag{1.1}$$

where f = focal length of the camera, H = flying height, and h_{ave} = average elevation of the terrain being photographed.

Practical scale recommendations[11] for route surveys are (a) 1:16 000 to 1:30 000 for reconnaissance photography, extended to 1:120 000 for very large projects or to 1:8000 for small projects, (b) 1:8000 to 1:16 000 for preliminary investigations, extended to 1:25 000 for large projects and 1:5000 for small projects, and (c) 1:10 000 to 1:500 for detailed investigation or monitoring, with the exact scale depending upon the type of information being sought.

If horizontal and vertical measurements are to be taken from air photographs it is first necessary to establish the positions and heights of some air-visible ground-control points on each photograph. The horizontal points are used to determine the scale and true orientation of each photograph, to provide a basis for correcting cumulative errors that arise when strips of photographs are assembled, and to link the photographs to the standard geographic (i.e. National Grid) coordinates. The vertical control points are needed before point elevations or contour lines can be derived.

The geodetic survey that establishes ground control can be very expensive and time-consuming and, hence, the accuracy to which it is carried out is generally commensurate with the desired map scale, map accuracy, and contour control, e.g. less accurate surveys may be acceptable in remote areas where the topography is rugged and land values are low. Ground-control establishment is normally divided into two parts, (a) the *basic control* which establishes a fundamental framework of pillars and benchmarks over the whole survey area, and (b) the *photo control* which establishes the horizontal locations and elevations of prominent points that can be identified on each photograph and related to the basic control points. Control points need to be easily and clearly identifiable on both the ground and the air photograph, as well as being accessible to the survey team for referencing in the field. A vertical photo-control point should always be in an area of no elevation change, e.g. at the intersection of two existing road centrelines or the base of a telephone pole in a flat field.

The number of photo-control points depends upon the purpose of the aerial survey. For example, if a controlled mosaic is being constructed no vertical control points are required and only a limited number of horizontal points may be necessary. On the other hand, if a true-to-scale topographic map is being constructed, four vertical and two horizontal control points may be required on each overlap area of the initial and final pairs of overlapping photographs, i.e. with the vertical points located at the corners of the overlap (to establish the datum above which the elevation can be measured) and the two horizontal ones located toward the centre of the overlap (so as to fix the map scale). In between, ground control will need to be maintained at the rate of one horizontal point near the centre of every fifth photograph, and one vertical point near each edge of every third or fourth photograph. Additional control points may then be 'bridged-in' to intervening photographs using a stereoplotting machine; if no bridging can be done two points will be needed on each photograph.

1.6.1 Uses of aerial photography

In general, aerial photographs greatly facilitate the location and design of a roadway. In this context their most important advantages are as follows:

1. On a large scheme the time and costs required to locate a major new road and to prepare plans for tender purposes are normally much reduced as compared with the conventional ground survey approach. The less well-mapped the terrain in which the road is being located the more the savings.
2. The ability to survey large areas of land ensures that the most suitable location is less likely to be overlooked.
3. A complete inventory of all ground features at a point in time is available for reference purposes, and subsequent analysis.
4. Topographic maps produced from aerial surveys can be more accurate than those prepared from ground surveys.
5. Road profiles and cross-sections can very often be developed without physically ·encroaching on private lands. Thus, landowners are not upset and land values are not affected during the road location investigation.
6. Skilled technical personnel can be released from repetitive work and used more profitably on more demanding problems.
7. Aerial photographs are easily intelligible to the general public, and are particularly useful as visual aids at public hearings (as well as at technical meetings).

In the hands of expert airphoto interpreters the following information[11] can be provided to the engineer engaged in selecting a suitable road route: (1) the location of sand and gravel deposits, borrow pits, and rock areas suitable for quarrying; (2) the rating of soil-bearing capacities as good, fair or poor; (3) the outlining of soils as to their texture, and the estimation of the depths of organic deposits; (4) the classification of bedrocks as to physical types and their relative depths below the surface; (5) the estimation of the depths of glacial drift and wind-blown materials; (6) the estimation of hydrological factors; (7) the delineation of active and potential areas involving problems such as sink holes, landslides, rock falls, and frost heave; and (8) the location of underground utilities placed by cut-and-cover methods.

Air photography can be most effectively used to identify ground phenomena that suggest ground instability and which might be best avoided when locating a new road, e.g. spring line and seepage areas, gully developments, eroding and accreting sections of streams, constancy of river channels, and probable flood levels. Areas susceptible to landslides are readily detected by characteristic steep scarps, hummocky surfaces, ponded depressions, and disturbed drainage conditions; also, the character of the vegetation on a moved slope will generally differ from that on an undisturbed adjacent slope. Joints, faults, shear zones, and brecciated zones

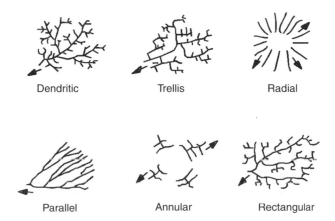

Fig. 1.3 Some basic drainage patterns

have features that reveal their presence to the skilled photogeologist, e.g. rocks on both sides of a fault will often fail to match in type or attitude, or both. Infra-red photography can be used to detect hidden features such as muck pockets, underground cavities, volcanic and hydrothermal activity, and subsurface drainage systems.

Soils can be interpreted from aerial photographs by studying the patterns created by the nature of the parent rock, the mode of deposition, and the climatic, biotic and physiographic environment. For example, erosion features such as the shapes of gullies provide clues to soil texture and structure, profile development, and other conditions. Thus, U-shaped gullies are found in wind-deposited loess soils and in stratified sandy or silty soils, i.e. the near vertical sides of the gullies reflect the fact that there is little clay present in these soils so that they crumble fairly easily. A broad saucer-shaped gully is indicative of a cohesive clay or silty clay soil, whilst V-shaped gullies are found in semi-granular soils with some cohesiveness.

Drainage patterns (see Fig. 1.3) suggest the origin and composition of the underlying geological formation. For example, a dendritic drainage pattern is commonly found in glacial topography. Trellis drainage is associated with areas of folded sedimentary rock, i.e. the characteristic pattern is due to the ridges being cut by stream gaps. A radial pattern can reflect water draining into an enclosed basin; more often, however, it indicates a high dome or hill, or a volcanic cone. A parallel drainage pattern usually indicates a regional sloping terrain condition or a system of faults or rock joints. The classical annular drainage system is usually indicative of tilted sedimentary rock about an igneous intrusion. Whenever a drainage pattern appears rectangular it is almost certain that rock is at or close to the ground surface, and that the drainage is controlled by rock joints, fissures or faults. (Reference 12 provides an excellent overview of the use of drainage patterns to identify soils and bedrocks from airphotos.)

New aerial photographs provide up-to-date information regarding current land usage, and this is especially important at the *reconnaissance* phase of a location study, particularly if it is being carried out in a built-up area. These photographs can be used to identify intermediate route-location control points so that potential band routes can be selected. At this stage of the reconnaissance study a small-scale mosaic will normally provide sufficient detail to enable elimination of all but a few potential bands; to enable the choice to be further narrowed an enlarged mosaic can be viewed in conjunction with a topographic map.

Large-scale photography is necessary at both the *preliminary* and *final location* phases of a location survey.

Airphoto interpretive methods can be used to cut down on the amount of soil investigation that needs to be carried out in the field during the final location survey. For example, as part of the routine field studies in a conventional ground survey, auger investigations are usually carried out at fixed intervals along the proposed line; these are then followed by further detailed investigations as suggested by the auger results. Much of this routine augering, and the subsequent laboratory analyses, can be avoided if the aerial photography is interpreted so that ground areas with similar physical characteristics are identified and delineated. Then it is only necessary to go into the field to test a limited number of points within each delineated area.

When large-scale airphotos are used to prepare the necessary maps, it may be possible to finish the plan preparation for a new road and let the construction contract without actually placing a centreline peg in the ground. When this occurs the final location line need not be pegged until the contractor is about to start construction – in which case the final location survey becomes a combined location and construction survey.

1.6.2 Limitations on usage of aerial photography

Whilst aerial photography generally facilitates the location and design of roads there are certain practical limitations on its usage.

First and possibly most important, the taking of quality airphotos cannot be guaranteed in inclement conditions. Good photography needs a clear atmosphere, and if it is taken on wet or misty days, in haze, or under smoky conditions, the photography may be of limited value.

Also, a topographic map cannot be accurately developed if the ground is covered by snow or obscured by the leaves of trees. Thus, for example, in temperate areas where the ground has a deciduous cover, the taking of aerial photography must await the time of leaf-fall, whilst its usage to survey heavily-wooded tropical areas may not be appropriate at all.

If the area to be studied is relatively small and well-mapped it may be more appropriate to use conventional ground survey methods rather than aerial procedures.

1.7 Other remote sensing techniques

In well-populated highly-developed countries the amount of in-field survey investigation can often be much reduced by maximizing the use of information already available from a multitude of historical sources. In newly-developing countries, as well as in poorly-explored areas in the more remote parts of large well-developed countries, e.g. Australia, Canada and the USA, historical data are often not available, with the consequence that large survey costs have to be incurred at the planning stage if subsequent major reconstruction and/or maintenance costs are to be avoided. To meet the need for data at reasonable cost in such locales, a number of remote sensing investigative techniques (in addition to conventional air photography) that can provide a wide range of different forms of imagery, are now available to the road engineer engaged in location survey work.

Multi-spectral or *multi-emulsion systems* permit simultaneous data collection over the same field of view, using four or more cameras. They allow data to be obtained throughout a broader, often more specific, band of the spectrum than conventional photography; the four most effective visual bands are the far infrared, reflective infra-red, red and green wavelength bands (in the 0.4–1.1 micron range). Data interpretation of particular features can be enhanced by additive viewing, density slicing, optical filtering and the use of electronic image-analysis devices. Tones registered on films in different spectral regions enable different ground materials to be identified; these can be quantified by optical density measurements taken by microdensitometer and converted into digital form for analysis. A major engineering application of these techniques is their ability to distinguish unstable ground and areas where movements may occur.

Infra-red linescan systems, which sample the thermal infra-red part of the spectrum, provide imagery that shows the surface pattern of temperature. Whilst such imagery can be a very useful complement to photographic imagery taken for route surveys, obtaining it can be very costly if the work has to be specially commissioned; in this context, however, it is useful to know that thermal imagery is now obtained almost as routine in surveys of geothermal features. Concealed geological features and subsurface conditions that influence road location and design can be detected by what is in effect a sensing of subtle surface temperature differences, e.g. the location of water-saturated slopes, soft organic soils, fault structures, underground cavities and subsurface voids, volcanic and hydrothermal activity, buried utilities and conduits, subsurface drainage systems, and local variations in the depth to the water table[13].

Side-looking airborne radar (SLAR) is a self-illuminating imaging sensor that operates within a wide range of frequencies in non-visible spectral bands. Thus, unlike the techniques described above, this longer-wave sensor employs antennae on planes as receivers rather than optical detectors. SLAR produces a small-scale image that is similar in appearance to a shaded relief presentation of the terrain, and the contrasts in radar return (reflectivity) values can be electronically processed to produce tones on film that can be related to various land surface conditions. Objects (including most cultural features) that are good reflectors of electromagnetic energy, and which might normally be lost on conventional small-scale mosaics of aerial photography, appear as bright spots on this imagery. SLAR has been successfully used to map structural features, to differentiate some rock types and surface materials, and to delineate old stream channels and areas of abnormal soil moisture. Its all-weather, day and night, operational capability means that it can be used to survey large areas in tropical regions that are rarely cloud-free. In forested areas the radar beam will partially penetrate the tree cover and detect topographic details beneath.

Since the launching of the first Earth Resource Technology Satellite (ERTS-1), subsequently renamed the Landsat Satellite, *multispectral scanner (MSS)* systems have been used to provide continuous images of the earth's surface to central receiving stations in many countries. Typical MSS imagery is very much like mutispectral photography in that it can record radiation in the visible and near infrared (0.1–1.1 micron) portions of the spectrum. On command, such systems are able to obtain images of ground areas of 185 km^2, with a maximum ground resolution of 79 m, and transmit them to a receiving station. Conventional images are made up of pixels, i.e. small picture elements, each of which represents an area on the earth's surface measuring 79 m × 59 m. Some Landsat data are recorded in the thermal radiation waveband with a ground resolution of 40 m and sufficient overlap to allow stereoscopic viewing. The potential for use of MSS data in poorly-explored areas is very considerable, especially if the interpretation and analysis of the data is combined with established land classification and field survey procedures. Likely uses include preparing regional inventories for road planning, route location reconnaissance studies over large areas, and the identification of significant sources of construction materials.

1.7.1 Comment on use of remote sensing techniques

When considering which remote sensing technique to use for a particular purpose professional guidance should be sought as to which type of imagery is best for that purpose.

For example, colour infra-red photography, which is generally superior to conventional colour and panchromatic photography due to its ability to penetrate haze, is normally recommended for identifying water surfaces, variations in soil moisture, texture and composition, potentially unstable water-saturated zones, and subtle differences in vegetation that may indicate variations in material types or ground conditions. Multispectral interpretation relies on the selection of the wavelength that records the greatest reflectance difference, thus providing high tonal contrast.

1.8 References

1. Beaumont, T.E., *Techniques for the Interpretation of Remote Sensing Imagery for Highway Engineering Purposes*, TRRL Report LR 753. Crowthorne, Berks: The Transport and Road Research Laboratory, 1977.
2. O'Flaherty, C.A. (ed.), *Transport Planning and Traffic Engineering*. London: Arnold, 1997.
3. Turner, A.K., Route location and selection, *Proceedings of the Conference on Computer Systems in Highway Design* held at Copenhagen on 2–9 September 1972, and organized by the Royal Technical University (Copenhagen), PTRC Ltd (London), and the Laboratory for Road Data Processing (Copenhagen). London: Planning and Transport Research and Computation (International) Company Ltd, 1972.
4. Perry, J. and West, G., *Sources of Information for Site Investigations in Britain*, TRL Report 192, Crowthorne, Berks: The Transport and Road Research Laboratory, 1996.
5. BS 5930: 1999, *Code of Practice for Site Investigations*. London: The British Standards Institution, 1999.

6. DETR Water Data Unit, *The Surface Water Year Book of Great Britain*. London: Department of the Environment, 1999.
7. *Hydrological Data UK – IH Yearbook*. London: The Institute of Hydrogeology and The British Geological Survey, 1999.
8. Fookes, P.G., Site reconnaissance and the engineering geologist, in Dumbleton, M.J. (ed.), *Available Information for Route Planning and Site Investigation*, TRRL Report LR591. Crowthorne, Berks: The Transport and Road Research Laboratory, 1973.
9. Coghlan, G. and Davis, N., Low water crossings, *Transportation Research Record 702*, 1979, pp. 98–103.
10. Committee on Engineering Surveying, *Report on Highway and Bridge Surveys*. New York: The American Society of Civil Engineers, 1962.
11. A Geological Society Working Party, Land surface evaluation for engineering practice, *Journal of Engineering Geology London*: 1982, **15**, pp. 265–316.
12. Parvis, M., Drainage pattern significance in airphoto identification of soils and bedrocks, in *Drainage and Geological Considerations in Highway Location*, Transportation Technology Support for Developing Countries Compendium 2, Text 12. Washington DC: The Transportation Research Board, 1978.
13. Beaumont, T.E., Remote sensing survey techniques, *The Highway Engineer*. 1979, **26**, No. 4, pp. 2–14.

CHAPTER 2

Subsurface investigations

C.A. O'Flaherty

2.1 Investigation aims

Overall, a subsurface or *ground* investigation is an operation of discovery[1] that is carried out prior to the design of a new road for the purpose of obtaining reliable information regarding the types, location and extent, in plan and profile, of the subsurface materials and geological structures which influence the routeing, safe design, and economic construction of the pavement. A comprehensive investigation ensures (a) the overall suitability of the road site and its environs for the proposed works, (b) that safe and economic designs can be prepared for both temporary and permanent works, (c) the determination of the changes that may arise in the ground and environmental conditions as a result of the works, (d) the planning of the best method of construction, and (e) that tender and construction needs are met.

The extent of the subsurface or *ground* investigation is determined by the character and variability of the ground and groundwater, the type and scale of the project, and the amount, quality and availability of the existing information. Planning for an investigation – which should be completed before the roadworks are finally designed – should be flexible so that the work can be changed as necessary when fresh information is obtained. The ground investigation is a process of continuous exploration and interpretation, and its scope will need regular amendment as data are interpreted. (Detailed guidance regarding the planning of a site investigation, and a general work specification, are given in references 2 and 3, respectively.)

A measure of the importance of carrying out adequate ground investigations can be gathered from a report[4] by the National Audit Office over a decade ago, in which it is pointed out that 210 premature failures during highway works cost some 260 million pounds sterling, and that geotechnical failures were a major cause for concern.

The majority of failings in relation to road construction projects come not from inadequacies with regard to the state of the art of site investigation, but rather from inadequacies in respect of the state of application of the art. Generally, good site investigation reduces ground uncertainties with consequent lower construction costs. Very often, low expenditure on site investigation results in large inaccuracies when estimating project time (scheduling) and cost (budgeting). Also, there are wide variations in the accuracy of estimation of earthworks' quantities on projects with low ground investigation expenditures; most usually, there is a substantial underestimation of the quantities or costs involved.

Classical subsurface information gathered for a major road project includes the following[5]:

1. *The depth, arrangement and nature of the underlying strata.*   The depth to the top of each underlying stratum, and the thickness of each stratum, need to be determined. Also, the materials in each stratum need to be described and classified; in this context the variability of each layer needs to be established, including the structure of the material (e.g. any fissures, laminations, interbedding, etc.). Particular attention should be paid to material changes that are near the ground surface as these may be associated with frost action, weathering, or surface movement.
2. *Strata irregularities and features of special engineering significance.*   This refers to natural or man-made discontinuities and distortions that might underlie the proposed road. Classical natural examples are fractures (including faults, fissures, and joints), tilting and distortion of beds (including dipping, folds, contortions, cambering, and valley bulging), cavities (including open fissures, solution cavities and swallow-holes), and surface movements (including solifluction and instability). Typical man-made cavities include mines, adits, shafts and cellars,
3. *Hydrological conditions.*   Ground hydrological information is required in order to resolve many stability and settlement problems, and for the design of measures to lower the groundwater level either permanently or as a construction expedient. Classically, appropriate information is gathered regarding water table levels and their seasonal and tidal variations, artesian pressures, strata permeability, wet and soft ground and peat, and springs, seepages and periodic streams.
4. *Quantitative engineering characteristics of the strata.*   The detailed engineering design may require the carrying out of special in-situ tests or the taking of disturbed and/or undisturbed samples for subsequent laboratory examination.

In addition, the information gathered should enable the determination of any environmental or ecological considerations that might impose constraints on the scope of the new roadworks.

2.2 Sequencing of the investigation

The location, design, and construction phases of a major road project occur over a long period of time, and each phase often requires information of a different type, quantity and quality. The sequence of operations used in a subsurface investigation can be divided into the following classical phases: (1) a preliminary investigation (including preliminary boreholes, if required); (2) the main subsurface or ground investigation for design (as well as any topographic and hydrographic surveying) and any special investigation(s) that may be required; and (3) a construction review, including any follow-up investigations.

It should be appreciated that whilst these phases should ideally follow each other, in practice they may overlap. The first two of these investigation phases are generally as described below.

2.2.1 Preliminary ground investigation

The preliminary investigation is mainly a *desk-based study* which searches out and gathers together as much known technical information as is possible regarding the conditions likely to be met along the alternative routes being considered for the proposed roadway. Whilst the desk study is normally supplemented by site reconnaissance trips (and sometimes by on-site subsurface explorations), the cost of the preliminary investigation generally amounts to a relatively small proportion of the overall cost of the ground investigation.

As well as helping in the selection of the preferred route for the proposed road, the preliminary investigation should reveal any parts of the selected centreline, and of adjacent borrow areas and properties, that will require special attention in the main ground investigation. It should also enable the preparation of a more accurate Bill of Quantities and costing for the main ground investigation, as there will be a greater knowledge regarding the frequency and types of exploratory holes, sampling and testing required.

Excellent sources of information for the desk-based study are well documented[1,6,7] (see also subsection 1.3.1). Annex A of reference 1 is especially useful in listing the kinds of information that may routinely be required, whilst reference 7 provides comprehensive advice regarding the most important sources of information.

As noted elsewhere, the UK is superbly endowed with respect to its coverage of geological maps, which have been the continuing work of the British Geological Survey (BGS) since the 1830s. The BGS is the national repository for geoscience data and is the custodian of an extensive collection of maps, records, materials and data relating to the geology of the UK, its continental shelf, and many countries overseas.

The main publication scale of the geological maps is 1:50 000, which is replacing the earlier 1:63 000 (1 inch to 1 mile) scale. A section showing the subsurface structural arrangement of the underlying strata is given on the margin of many of the 1:50 000 and 1:63 360 maps. These maps also show major areas of landslip (or landslide); these should be avoided when locating a road.

'Solid' geological maps are available which show the underlying solid rocks as they would appear if the cover of superficial deposits was removed. 'Drift' maps show the distribution of the various types of superficial deposits of glacial origin (see Chapter 4) throughout the UK, as well as the solid rocks exposed at the surface; details of their composition are given in accompanying explanatory sheet Memoirs. Coalfield and economic memoirs are also available for selected areas of the country.

Glacial deposits, which are distributed throughout Great Britain north of a line through the Severn and Thames estuaries, are indicated on the BGS Drift maps. Drift maps are very useful in locating sites for sand and gravel which may be potential sources of borrow material for roadbases and subbases. River terraces – these are the remnants of former river flood plains – are also shown on these maps; level, well-drained, terraces are often suitable locations for roads, as well as being potential sources for granular pavement materials.

The BGS' series of twenty handbooks, which describe the geology of individual regions of the UK, are published as *British Regional Geology Guides*. These booklets provide much information on local geology and are excellent starting points for a desk study.

The 1:10 000 and 1:10 560 BGS maps show the sites of selected boreholes, with a statement of the depths and thicknesses of the strata met in each. In addition to its published information the BGS maintains extensive unpublished records of strata encountered in boreholes, shafts, etc., and these records can be inspected, by arrangement, in the BGS libraries. The collection includes over 600 000 borehole records as well as geophysical logs, site investigation reports, road reports, mine and quarry plans and sections, field notebooks, and unpublished field reports, e.g. the BGS has statutory rights to copies of records from all sinkings and borings to depths of more than 30.5 m for minerals and of more than 15.25 m for water. From these, it may be that enough data can be obtained from the desk study to plot the approximate geological structure on a longitudinal section of the proposed road (for example, see Fig. 2.1).

Strata normally dip in the general direction in which successively younger beds outcrop at the surface, and the dip is commonly shown by arrows on each BGS map, with the direction and angle (deg) of dip at a given point. If strata dip toward the face of a cutting, there is a danger that the rock may slip into the cutting, e.g. if jointed rock is sitting on a bed of clay.

Faults, i.e. fractures of strata along which there has been differential movement, are shown on geological maps as lines with ticks on the downthrow side. Movements occur from time to time along some faults. Also, broken rock may follow the lines of faults, and these may act as routes for groundwater as well as affecting the ease of excavation and the stability of slopes and tunnels.

Dykes and sills, which are minor igneous intrusions of sheet-like form that are (more or less) vertical and horizontal, respectively, are also shown on BGS maps. Dykes, which tend to occur in swarms, are often missed by boreholes during a site investigation unless they are deliberately sought. The detection of hard dykes, and sills, is important as their presence in softer surrounding rocks normally increases excavation costs.

The BGS hydrogeological maps show the positions of springs and seepages. A spring results when water passes through a permeable formation until it meets an impermeable stratum, and then flows along the surface of that stratum to emerge at a point on a slope; if the water emerges as a line rather than a point it is termed a seepage. The identification of springs and seepages may suggest locations of potential instability in soil slopes after heavy rain.

Ground investigations previously carried out for other engineering projects in the vicinity of each route being considered also provide useful information. Local knowledge gained in the execution of existing road, railway or building works in the area should also be sought, as should the considerable amounts of data already available in local authority offices (e.g. from sewer studies) and service undertakers (e.g. gas, electricity, and water). Geotechnical information available from such sources typically include soil gradation and plasticity data, soil moisture contents and depths to the water table, soil strength and its consolidation characteristics, and the suitability of the underlying material for use in embankments.

The scanning of maintenance records of existing roads and railways in the locale may well provide information regarding troublesome areas (e.g. due to frost heave, seepage water, embankment settlements, or landslides) that will enable the engineer to make better decisions regarding a road's location and design.

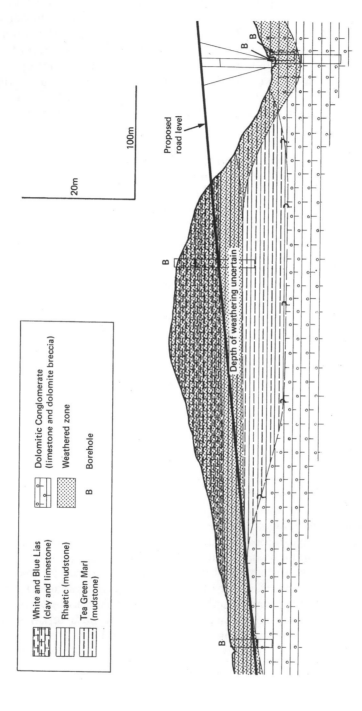

Fig. 2.1 Interpolation of strata between boreholes aided by information from geological survey map (reported in reference 5)

Pedological maps and their associated memoirs have been published at various scales across the UK. Soils information for England and Wales is held by the Soil Survey and Land Research Centre (SSLRC) at Silsoe, the Macaulay Land Use Research Institute at Aberdeen holds the data for Scotland, and the Department of Agriculture for Northern Ireland, holds the Northern Ireland information. These data can be very helpful to ground investigation studies in non-built-up areas. Prepared for agricultural and forestry purposes – hence, their concerns are mainly limited to profile depths of about 1.2 m, albeit peats are described to depths of over 10 m – the Soil Survey information provided for each soil describes such conditions as parent material, texture, subsoil characteristics, and drainage class; the accompanying report provides more detailed information about the physical and chemical properties of each soil. For best use, the Soil Survey maps should be compared with complementary geological maps, e.g. it may be found that the soil maps give a better picture of drift deposits, especially if the geological mapping is old.

Various types of engineering information can be deduced from the above pedological maps and reports. For example, the occurrence of compressible materials, shrinkable clays, shallow depth to rock, unconsolidated sands and degrees of natural soil wetness and drainage can all be assessed from pedological soil maps. The lithology of the upper 1.2 m provides information about the particle-size distribution, stoniness and plasticity of the material. Drainage class (or wetness) is a guide to the duration and depth to waterlogging. Permeability can be gauged from soil porosity data. Compressible soft soils, such as alluvial and peat soils, are well defined and described, e.g. special maps showing peat distribution (as well as groundwater vulnerability to pollution and the risk of erosion by water) are also available. Soils that are likely to attack buried concrete or ferrous metal structures can be identified from recorded pH values.

As discussed previously (Chapter 1) relevant stereoscopic pairs of aerial photographs may already be available, most usually in conventional black-and-white photography. Infra-red, false colour infra-red, and multispectral photography are techniques that have been attempted to ascertain the positions of mine shafts, adits, underground cavities, unstable ground and other geological features. Whilst colour infra-red techniques have been helpful in locating water sources, in general these techniques have not revealed anything more than could be obtained from the conventional black-and-white photography[6].

A thorough *visual inspection* of each proposed route should always be made during the preliminary investigation (see Appendix C of reference 1 for a listing of the procedures for, and methods of carrying out, a *walk-over survey*). As well as answering particular questions and confirming, amplifying and/or supplementing information regarding already identified features, this reconnaissance inspection will often indicate subsurface conditions that require special consideration when locating a roadway, e.g. if sink-holes are likely. Sink-holes or swallow-holes are ground depressions caused by material subsiding into cavities formed because of the solution of the underlying rock. Sink-holes, which are usually found where limestone, chalk or gypsum is close to the ground surface, are not normally marked on geological maps and are best detected from air photographs and/or visual inspection.

Geological faults may be indicated during the walk-over survey by topographical features such as a step in the line of an escarpment or an unexpected depression where the shattered rock has been more rapidly eroded than the surrounding material. In mining areas, steps in the ground are likely the result of mining subsidence; compression and tensile damage to brickwork, buildings, and roads, structures that are out of plumb, and interference with drainage patterns, can also be evidence of mining subsidence.

Potential rock-fall and soil-fall sites can often be anticipated as a result of the field inspection. Broken and terraced ground on hill slopes may be due to land-slips, whilst small steps and inclined tree trunks can be evidence of soil creep. Changes in a crop's appearance may suggest filled pits and quarries or other past man-made interferences with subsoil conditions. Lush growths of water-loving grasses on higher ground suggest that the water table is near the surface, whilst bracken and gorse growths indicate that the soil is well drained. Low-lying flat areas in hill country may be the sites of former lakes, and may suggest the presence of soft silty soils and peat.

Examination of nearby road and railway cuttings, and old pits, during the visual inspection can often indicate soil types and their stability characteristics. However, caution may need to be exercised in interpreting some old excavations that were carried out slowly by hand and/or machine, as their long-term performance may be different from current excavations made by modern equipment.

The visual inspection may also confirm the need for boreholes to be sunk at critical locations, e.g. at the likely sites of major cuttings or embankments. However, the number of these is normally kept to a minimum as they are expensive and many alternative routes may have to be examined.

When all of the preliminary information has been gathered together a preliminary report is normally prepared showing provisional geotechnical maps and longitudinal geological sections, and block diagrams showing the characteristics of the more important engineering situations. Often, sufficient indications regarding the soil and geological conditions may be available at this stage for the various alternative routes to be compared, thereby enabling appropriate costing and engineering geology information to be input into the decision-making regarding the preferred route. Care should be taken at this time to ensure that the provisional geotechnical maps show unfavourable ground conditions[7] such as (a) unstable slopes, (b) materials that require extra-flat slopes in cuttings, (c) unrippable rock above the road's formation level, (d) soft material requiring excavation below the formation level, (e) areas with materials above formation level that are unsuitable as fill, (f) areas with saturated sands and silts above formation level, (g) areas liable to flooding and springs, and areas requiring special drainage measures, (h) locations liable to snow drifting, fog, icing, or excessive winds, (i) mine workings and shafts, swallow-holes, cavernous ground, and areas liable to subsidence, and (j) mineral deposits.

If necessary, appropriate adjustments to the line of the preferred route may also be recommended at the preliminary stage of the subsurface or ground investigation. A proposal for the main ground investigation of the preferred route can then be developed, recognizing that its planning will need to be flexible and subject to change as new information is obtained and assessed.

2.2.2 Main investigation

The ground over and through which a road is to be constructed normally contains soils and rocks that vary considerably, both laterally and vertically, as well as having other hidden features of engineering importance. Some data regarding the underlying materials, and their properties, will already be available from the preliminary study which resulted in the selection of the preferred route. The purpose of the main ground investigation is, therefore, to obtain sufficient additional information from a limited number of subsurface explorations, most usually using boreholes, to determine the strength and frost susceptibility of possible subgrades, and the drainage conditions, so that the road and its associated structures can be properly designed, costed and constructed.

There are no hard and fast rules as to where, how often, or how deep, subsurface explorations should be carried out, albeit guidance can usually be obtained from the preliminary report. The main influencing factors are the nature of the route being examined, the anticipated design of the proposed road, and the locations of its structures. As far as possible the explorations should be carried out in order of importance so that the maximum information is obtained for the budget available. Cheaper methods of investigation should also be carried out first, if they are likely to provide the important information; these will also suggest where more expensive exploration methods need to be employed. Some factors which affect the selection and priority of particular exploration points are:

1. *The need to avoid commitment to an unsatisfactory route.*   Generally, any explorations that could lead to a major relocation of the line of the road, and/or the road's subsequent redesign, should be carried out as soon as possible to minimize the wastage of resources, e.g. the preliminary report may have drawn attention to areas with possible unstable slopes, poor foundations for structures, or limits on the depths of cut or heights of fill.
2. *The need to clarify the geological interpretation of the preferred route.*   Generally, any geological explorations that might affect the design of the main investigation itself should also be carried out as early as possible.
3. *Bridges and other structures.*   Where the sites of structures have been identified with reasonable accuracy, enough exploratory holes should be sunk to enable the detailed geological structure to be plotted and the engineering properties established.
4. *Cuttings and embankments.*   The sites of deep cuttings and high fills normally justify exploration points to obtain information regarding the areas, depths, and properties of (a) soil materials to be excavated from cuttings that might be used/ not used as fill, (b) rock excavations in cuttings, and (c) soft compressible soils, or other adverse ground conditions, that are beneath proposed embankments or below pavement formations in cuttings. When embankment material is to be borrowed from pits alongside the route exploratory holes should also be sunk (to the estimated depths of borrow) at the proposed pit sites.
5. *Soil variations.*   Enough explorations should be carried out to determine the extent of each significant soil type along the route. Critical areas, such as soft ground and the locations of large structures, require special examination. If the soil profile is relatively uniform, and not critical to design, wide spacings may be used between boreholes; however, if the profile changes intermediate explorations

will need to be carried out. e.g. borings for major roads have been carried out at about 300 m intervals in uniform soils and at about 16 m intervals in quickly-changing glacial deposits.

Sufficient exploration holes should be offset from the road centreline to ensure a good transverse coverage of the proposed site, so that lateral variations in ground conditions can be revealed and material obtained for testing for fill suitability. The locations of underground services should be determined beforehand, to ensure that they are not interfered with during the subsurface investigation. The depth to which a particular exploration hole is sunk depends upon the geological conditions encountered and the type of construction anticipated at the location in question. As a basic guide in relation to cuttings, embankments and foundations, the profiles should be determined to depths below the proposed formation at which the stresses imposed by the anticipated design are within the strength capabilities of the ground materials at those depths, with an acceptable amount of consolidation. In many instances an exploration depth of 2 m to 3 m below the formation is adequate, provided that this depth is below deposits that are unsuitable for foundation purposes, e.g. weak compressible soils or manmade ground.

In the case of cuttings this often mean that boreholes only need to be sunk to a depth of 2 m to 3 m below the proposed formation level. However, if the cutting contains rock, a greater exploration depth may be needed to establish whether bedrock or a boulder is involved; if it is bedrock, an accurate outline of the rock contact should be determined and, where necessary, samples obtained to prove its nature and uniformity.

High (7.5–9 m) embankments produce appreciable bearing pressures so that soil borings may need to be extended to a depth of at least 2.5 times the fill height. With low (<3 m) embankments, explorations to a depth of between 3 m and 4.5 m may well be sufficient.

The exact locations and depths of the exploration holes required for structures will need to be agreed with those responsible for their technical approval, before the borings are carried out. A general rule-of-thumb is that, for foundations near the surface, e.g. a conventional footing or a foundation raft, the depth of exploration below the foundation base should be at least 1.5 times the width of the loaded area unless the ground is obviously strong (as in strong rock). In all cases, the planned depths of penetration should be subject to review, depending upon the conditions found at the site during the exploration.

The gathering of data on *groundwater regimes* is essential for drainage design, temporary works, assessing working methods and the use of plant equipment, as well as for the design of earthworks[6]. Thus, groundwater levels in boreholes should always be recorded at the times that they are encountered. If no trace of moisture is found in a test hole in a coarse-grained soil it is reasonable to assume that water table is below the depth of penetration. However, if the borehole is sunk in a soil of low permeability, e.g. a heavy clay, and it appears dry when completed, it should not be assumed that the water table has not been penetrated; in this case the borehole should be protected against the entry of surface water, but not backfilled, for at least 1 or 2 days to allow a check to be made as to whether groundwater is seeping into the hole. If water seepage is detected in a soil of relatively low permeability, the determination of the true phreatic surface may require the insertion of perforated standpipes or sophisticated piezometers at different levels in the hole prior to it being backfilled, and the monitoring of these over a long period of time; this may be necessary at locations where cuttings are proposed or where embankments are be placed on soft alluvial deposits.

It might be noted also here that information about groundwater that is obtained during the subsequent construction should also be accurately recorded, for comparison with the data obtained during the investigation.

2.3 Subsurface exploration methods

Many methods have been devised for securing information regarding the ground subsurface. Whilst each has its particular advantages, the method used in any given instance will vary according to the location and geotechnical features encountered. In fact, not only is it usually desirable to use more than one exploration method on a particular site, but in many instances different techniques may be used at a given location within a site in order to obtain complementary information about the underlying strata.

In general, the ground investigation methods can be divided into the following four main groups[3]: (1) exposures; (2) accessible excavations; (3) boring and probing; and (4) geophysical profiling.

2.3.1 Exposures

Valuable information on ground conditions and materials can often be easily gained by direct observation through natural or existing man-made subsurface 'windows', e.g. outcrops, stream beds, quarries and surface deposits. However, care needs to be taken to ensure that these exposures are typical of the locale as, otherwise, misleading interpretations may be made.

2.3.2 Accessible excavations

Accessible excavations are an excellent way of examining and sampling soft rocks and non-cohesive soils containing boulders, cobbles or gravels. As well as supplementing less-detailed borehole information, there are occasions when test pits or trenches provide the only sensible way of providing needed information on rock bedding, fracturing, and slip planes.

Shallow test pits and trenches, most usually dug by a hydraulic back-hoe excavator, enable a direct visual examination and a photographic record to be made of the in-situ soil conditions above the water table, or the orientation of any in-ground discontinuities. These excavations are very useful also when large disturbed samples of soil are needed for laboratory testing; undisturbed samples, e.g. for triaxial shear and for consolidation tests are, however, more difficult and expensive to obtain (except in stiff clays).

Whilst trenches and pits are easily dug to depths of 4 to 5 m in non-flowing soils, safety considerations generally necessitate the use of shoring supports for excavations deeper than about 1.2 m if they are to be entered by personnel, e.g. for sampling purposes. Fences are required for the safety of the general public if pits are left open to, for example, allow surfaces to dry out and better expose fissures and fabric.

Hand-excavated deep trial pits, or shafts (about 1 m diameter) bored by large power-driven augers, may be economically justified for particular inspection, sampling or in-situ testing reasons. Safety is a governing criterion here, with temporary wall liners being required for the safety of personnel; normally the water table is the maximum depth for which this approach is suitable.

Headings (adits) driven laterally into sloping ground above the water table can be used for in-situ examination of strata, existing foundations and underground constructions, as well as for special sampling and in-situ testing.

2.3.3 Probings and borings

Probings, also known as *soundings*, involve pushing or driving sections of steel rod down through the soil. The simplest type typically consists of a 25 mm dia, pointed, steel rod that is driven into the ground with the aid of a sledge hammer. More elaborate equipment has a pipe surrounding the rod, to eliminate side-friction, and measures the resistance to penetration by recording the number of blows or force of a jack required to advance it a given distance; attempts can then made to interpret the results by comparing the readings with results from known profiles obtained with these penetrometers.

Practically, the results obtainable from probings are very limited. When used in a non-calibrated situation they ensure that rock does not exist above the depth to refusal – what is not known, however, is whether the refusal is due to bedrock or a boulder or, even, a gravel bed. Probing is best used when another ground investigation method indicates a relatively thin layer of soft soil overlying a harder one, and the thickness of the soft stratum needs to be checked, quickly and economically, over a wide area, e.g. to determine the depth of peat deposits.

The main types of *boring equipment* are soil augers, shell and auger rigs, wash borings, and rock drills. Several factors influence the decision as to which of these pieces of equipment to use at a particular site, e.g. site accessibility, equipment availability, soil type, likelihood of rock, sampling requirements, and/or groundwater conditions.

Auger borings are employed at sites where fast testing is required and the use of disturbed samples is satisfactory, e.g. for soil classification purposes along the proposed centreline of a road. Augers operate best in loose, moderately cohesive soils[8]. They are not well suited to use in coarse-grained soils or at sites with high water tables; unless protected with a casing the soil wall will usually collapse into the borehole, and excavated material will fall from the auger as it is being withdrawn.

The simplest hand auger has a helical-shaped bit fastened to a length of piping with a T-handle at the top. The auger is rotated into the ground as far as possible, after which it is withdrawn and the disturbed soil retained on the helical flight is examined and sampled as necessary. Hand-augered boreholes can be as deep as 5 m in suitable soils, i.e. ones with no hard obstructions or gravel-sized particles, and are an easy way of determining the depth to the water table.

A mechanical auger operates in the same fashion as a hand auger except that, of course, the bit is rotated, turned and lifted by a powered drill-stem. Continuous flight augers with hollow stems are powered helical augers mounted on a heavy vehicle that are used for deep (30–50 m) borings in cohesive soils that do not need wall casings to support the borehole. By continuous flight is meant that the lead bit, which is equipped with changeable cutting teeth, can be lengthened by the addition of extension flight pieces (1.5–2.0 m lengths) without having to withdraw the auger when drilling. When augering, the hollow stem (typically, 75 or 125 mm internal dia) is closed at its lower end by a plug to which is attached a centre rod; when the rod and plug are removed, a sampler can be inserted through the hollow stem and an undisturbed sample obtained from the soil immediately below the bit. Also, when rock is met, boring can be extended by core-drilling through the hollow stem.

An exploration method that is rarely used in the UK, but is quite commonly used in North America to investigate sands, silts and clays (but not gravels) is known as *wash boring*. The rig – which comprises a simple winch and tripod – normally uses 65 mm dia. borehole tools and borehole casing. The casing is driven into the ground, one length at a time, by a drop-hammer action similar to that used in pile driving. The soil at the bottom of an appropriate depth of borehole is then broken by the percussive action of a chisel bit and the loosened soil is washed to the surface by water that is forced down, under pressure, through the hollow drill rods.

The soil brought to the surface by the returning wash water is not, of course, representative of the character and consistency of the strata being penetrated. However, an experienced operator can 'feel' a significant change in material and, when this occurs, the churning is stopped and the casing washed clean of loosened soil. After the wash rods are raised from the hole a standard penetration test may be carried out at the top of a stratum or, in the case of a cohesive soil, open tube or piston samples may be taken. Upon completion of this phase of the exploration, the chopping bit is replaced and drilling is continued until another material change is detected.

Soil investigations to depths of 60 m are commonly carried out using *light cable percussion borings* that utilize a mobile rig that is specially designed for ground investigation work. The rigs – which, typically, have a derrick that is around 6 m high and a winch of 1 to 2 t capacity that is driven by a diesel engine – are variously described as 'cable and tool' or 'shell and auger' rigs, with the tools and precise method used depending on the conditions encountered[9].

Shell-and-auger boring is carried out by repeatedly raising and dropping a wire rope that carries the casing sections and boring tools. Except in stiff clays, flush screw-threaded steel casings (typically, 150 or 200 mm dia) are used to line the borehole as it is advanced progressively downward. The borehole is advanced by continuous sampling using drill tools such as a 'clay cutter', a 'shell' or 'bailer', and various chisels. The clay cutter is an open-ended tube, with a heavy weight attached, that is dropped down the casing in damp or dry boreholes in cohesive soils; the tube is removed for examination and appropriate testing, when it is full with disturbed soil. The shell or bailer is an open-ended tube that is fitted with a cutting shoe; it has a leather- or steel-hinged flap close to the shoe that is designed to retain sand and gravel in the presence of water. Chisels, which are dropped down the borehole to break up hard material, are used to facilitate the retrieval of soil by a shell.

Boring in rock, particularly when cores are required, is most usually carried out using *rock drilling*. Rock drilling usually involves either of two processes, 'open-hole' drilling or 'rotary-core' drilling. In either case the drill bit is rotated at high speed at the bottom of the borehole and a drilling fluid (usually clean water or compressed air) is forced into the hole through the hollow steel rods; the fluid serves a dual purpose, i.e. it cools and lubricates the bit and brings the drill cuttings to the ground surface.

Open-hole rotary drilling, also known as 'shot drilling', is a wash boring process in that a plain bit is used, and chilled shot that is fed down through the rods with the lubricating medium acts as the cutting agent and cuts all of the material within the diameter of the borehole. Open-hole drilling is mainly used for quick penetration of strata that is of limited interest to the investigation, as its structure is almost totally destroyed; typically it might be used with weak rocks or soils to install test instruments or as a probing technique to test for voids such as old mine workings or solution cavities.

With conventional rotary-core drilling a ring bit studded with industrial diamonds (for hard rocks) or tungsten carbide inserts (for soft rocks) is screwed onto the bottom of the outer barrel of two concentric cylindrical barrels. The outer barrel is fastened to, and rotated by, extension rod sections of hollow steel tubing. As the bit cuts into the rock, the isolated core feeds into the inner, non-rotating barrel, and this is raised at regular intervals to the surface so that the rock core can be extracted. A spring-steel 'core catcher' is located just above the core bit to prevent the rock from dropping out of the core barrel. The drilling fluid is added to the core bit through the annulus between the two barrels.

2.3.4 Geophysical profiling

Geophysical profiling is a form of subsurface investigation whereby specially developed instruments are used to obtain measurements, at the ground surface, of variations in the physical properties of subsurface materials. By correlating these measurements with known subsurface conditions – if the required correlation data are not already available from previous study, they are usually obtained from test pits or conventional boreholes – the geophysical data can then be interpreted to provide information regarding the underlying materials.

The great advantage of the geophysical approach lies in the speed and relative cheapness with which particular subsurface information can be obtained for a large site with reasonably uniform overburden conditions. The digging of pits or the sinking of boreholes can then be restricted to control points where correlation testing is required, and to locations where anomalous readings are obtained.

The four main civil engineering purposes for which geophysical surveys are carried out are as follows[1] (see also reference 10):

Geological investigation: Mapping geological boundaries between layers; determining the thickness of superficial deposits and depth to 'engineering rockhead'; establishing weathering profiles; and the study of particular erosional and structural features, e.g. the location of buried channels, faults, dykes, etc.

Resource assessment: Location of aquifers and the determination of water quality; exploration of sand and gravel deposits and rocks for aggregate; identification of clay deposits, etc.;

Hazard assessment: Detection of voids and buried artefacts; location of buried mineshafts and adits, natural cavities, old foundations, pipelines, etc.; detection of leaks in barriers; pollution plumes on landfill sites.

Engineering properties assessment: Determination of such properties as dynamic elastic moduli, rock rippability and quality; soil corrosivity for pipeline protection studies, etc.

There are many different types of geophysical survey techniques, each based on different geophysical principles, e.g. seismic refraction and (more recently) seismic reflection, electric resistivity, electromagnetic, magnetic, and gravity methods. The techniques which have tended to be most used in road location/design studies are the seismic refraction and electric resistivity methods; the physical properties that these measure are shock-wave velocity and electrical resistivity, respectively. However, it should be noted that whilst these methods have been extensively used overseas, especially in North America, they have not been much used in the UK on the basis that the usefulness of geophysical methods of site investigation for road works has not been demonstrated except in limited applications[6].

It cannot be emphasized too strongly that the proper implementation and interpretation of a geophysical survey requires the use of specialist personnel to ensure that (a) a cost-effective method is selected that is appropriate to the exploration challenge, and (b) the method provides the degree of precision required of the results. Hence, this type of survey should normally be entrusted to an organization that is expert in this work, if poor results are to be avoided.

With the *seismic refraction* technique, shock waves are generated just below the ground surface by a hammer or falling weight striking a steel plate, or a small explosive charge, and the times taken by the energy to travel from the shot point to vibration detectors – these are termed 'geophones' – located at known distances along a line are measured and transferred instantly to a master recorder. Knowing the distance and travel times, the velocities at which the seismic waves move through the underlying strata can be calculated and the depth to the layer beneath can be determined from a mathematical relationship.

Figure 2.2(a), for example, shows the advance of a wavefront in a simple two-layer system composed of an overburden above bedrock. Assume that the wave travels at velocity V_o in the overburden so that, say, 0.016 s after the explosion it reaches the rock at point A. From A a wavelet then starts into the rock at a velocity of V_r so that after, say, 0.020 s it reaches the distance noted by that circle, not only vertically below A but also in all directions within the rock; thus the 0.020 s wavefront reaches the point B along the overburden–rock interface. Along the interface, however, the wave in the rock starts a new series of wavelets back into the overburden, toward the ground surface. The ones at B and C are but two of many such wavelets, and their envelope is the new refracted wave DE as shown 0.025 s after the shot. Figure 2.2(b) graphs the continued advance of the wavefront through the overburden until more than 0.040 s after the explosion; this travels to the nearer geophones at velocity V_o. However, at some distance along the line of

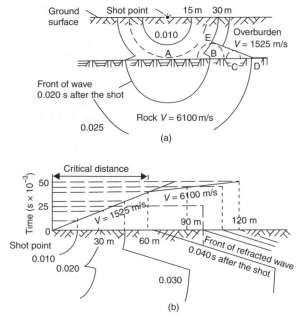

Fig. 2.2 Illustration of the use of the seismic method to find the depth to bedrock

geophones a refracted wave will travel the shorter distance to the surface and be picked up before the original wavefront reaches the same point through the over-burden; beyond this 'critical distance' the refracted wave is always picked up first by the geophones.

The plot of the arrival times at the geophones versus distance from the shot point (Fig. 2.2(b)) indicates that the closer geophones record the overburden veloc-ity and that the plotted points fall along a straight line of slope V_o, whilst the farther geophones record values along a line of slope V_r. The bedrock depth at the shot point is then estimated from

$$H=0.5D[(V_r-V_o)/(V_r+V_o)] \tag{2.1}$$

where H=depth from surface to bedrock (m), D=critical distance (m), V_o=velo-city of the soundwave in the overburden (m/s), and V_r=velocity of the soundwave in the rock (m/s).

Seismic profiling is best used to obtain results along a line where the geolo-gical conditions are uniform and simple, with clear-cut contrasts in the velocities of the sound waves transmitted through overlying formations; in this context, note that velocities generally increase with increasing density of the transmitting medium (Fig. 2.3)[11]. The method is commonly used to determine the depth to bedrock along the line of a road. The results obtained can also be used to assess the rippability of the rock mass, by comparing them with data in published tables[12].

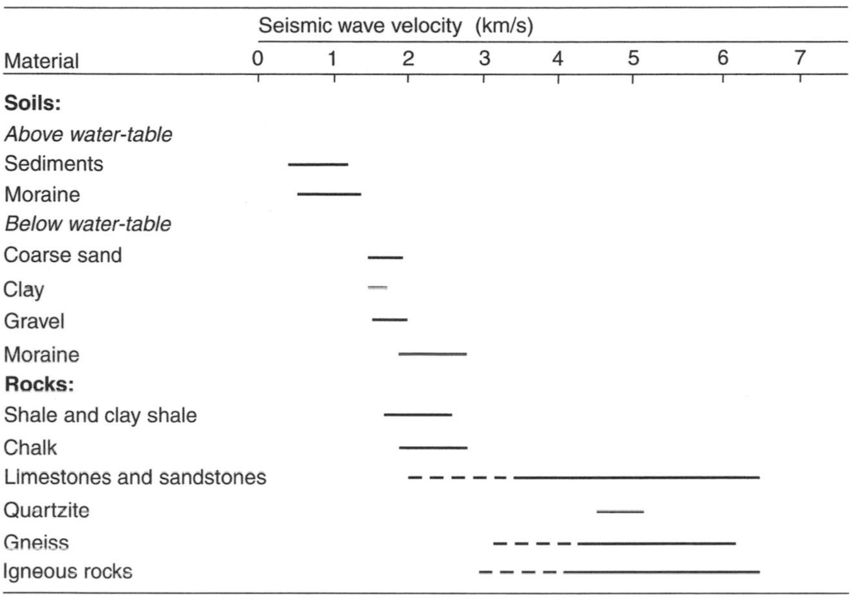

Material	Seismic wave velocity (km/s)							
	0	1	2	3	4	5	6	7

Soils:

Above water-table

Sediments

Moraine

Below water-table

Coarse sand

Clay

Gravel

Moraine

Rocks:

Shale and clay shale

Chalk

Limestones and sandstones

Quartzite

Gneiss

Igneous rocks

Fig. 2.3 Seismic values for natural materials (reported in reference 11)

It must be emphasized again, however, that predicting a material from its velocity requires considerable skill, and correlation tests over known subsurface formations are essential for the successful interpretation of the data gathered.

The *electrical resistivity* of a substance is the resistance offered by a unit cube of the substance to the flow of an electric current perpendicular to one of its faces. With the exception of certain metallic minerals that are good conductors, the constituent minerals of the earth are more or less insulators, and the resistivities of the various subsurface formations are almost entirely dependent upon the salinity of the moisture contained in them. Thus, as Fig. 2.4 suggests, by determining the resistivities of different materials at a site and comparing these with resistivities from known subsurface formations, interpretations can be made regarding the subsurface strata.

To carry out a resistivity survey an electric current is passed into the ground through two metal electrodes (Fig. 2.5(a)). Two potential electrodes are next inserted in the ground so that the four electrodes are in a straight line and spaced at equal intervals. To determine the resistivity of the ground the current I flowing through the ground between the current electrodes is measured on a milliammeter and, at the same time, the voltage drop V between the potential electrodes is measured on a potentiometer. The resistivity of the underlying material is then given by an equation of the form:

$$\rho = 2\pi AV/I \qquad (2.2)$$

where ρ = resistivity (ohm-cm), A = distance between any two adjacent electrodes (cm), V = potential drop (volts), and I = current (amperes).

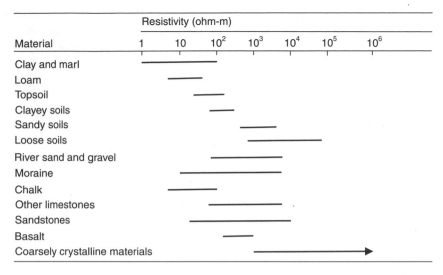

Fig. 2.4 Resistivity values for natural materials (reported in reference 13)

When the electrode spacing is small compared with the thickness of, say, a clay overburden, most of the current will flow through the uppermost depth and the resistivity measured will approximate the true resistivity of that layer. As the electrode spacing is increased the depth of current penetration also increases, and the resistivity measured is an average for all material within the depth A below the surface. As the spacings are progressively expanded, the current flow lines eventually encounter the underlying rock formation and this material, which has a higher resistivity, begins to affect the average values. As the spacings are further increased, the rock has a proportionally greater effect until, eventually, the resistivity measured approaches that of the underlying rock material.

When the resistivity is plotted against the electrode spacing A, a curve is obtained that is similar to the 'individual' curve in Fig. 2.5(b). When the resistivity values are

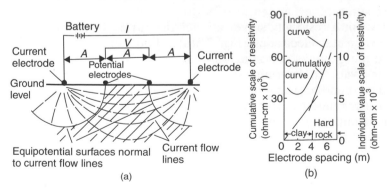

Fig. 2.5 Illustration of the use of the electrical resistivity method[14]

plotted on a cumulative curve, a straight line (or one of gentle curvature) is obtained as long as the effective current flow remains within the overburden layer; however, as the spacing is increased the cumulative curve develops an increasingly upward curvature, reflecting the influence of the higher resistivity of the rock. If straight lines are drawn through as many points as possible on the cumulative curve, their intersection in the region of the increasing curvature gives a good approximation of the thickness of the upper layer, i.e. the overburden.

2.4 Preparing the main report

Subsurface investigations are of little value unless accurate and comprehensive records are kept and reported. It should never be forgotten that within a relatively short time of the report being prepared it will be the only record of what was found, as samples are destroyed or rendered unrepresentative, and memories grow dim. Considerable care and attention should therefore always be paid to the proper preparation of the report.

Commonly, the investigation report for significant projects, is prepared in two parts: the first part is normally a descriptive report containing details of surface conditions at, and locations of, all sites where exploration took place, the equipment, procedures and testing utilized, and the data gathered including (in graphical, photographic, tabular and written forms, as appropriate) borehole logs, geological profiles, field and laboratory test data, etc., whilst the second part has the analysis, conclusions and recommendations.

The reader is referred to section 7 of reference 1 for excellent advice (including examples) regarding the preparation of subsurface investigation reports.

A topographical map showing the line of the proposed road, and the related locations of all boreholes, trial pits, geotechnical tests, and geophysical measurements is an essential component of the report; geological information and surface and borehole observations may be recorded on the same map or separately. A longitudinal section showing the surface topography, and the borehole details located at their correct positions and levels, is also essential. A section is normally prepared for each trial pit to show the encountered strata, the positions of any samples and in-situ tests, water levels, and structure details such as dip, faults, joints, fissures, cleavage, shear planes and slip planes. Additional sections, including some at right angles to the centreline, are normally provided at the sites of important structures, extensive works for side roads, and major cuttings and embankments. The locations and levels of springs, seepages, streams, natural ponds, wells, piezometer installations, artificial drainage systems, and flooded pits and quarries are also shown on maps and sections. Geological plans with the route superimposed are usually prepared at a scale of 1:25 000, whilst longitudinal-sections and cross-sections often have horizontal and vertical scales of 1:2500 and 1:250, respectively.

Depending upon the requirements of the investigation the borehole information may be presented simply as vertical columns showing the geological formations and groundwater levels, the lithology of the strata, and the soil classifications and appropriate test results, and it is then left to the recipient of the report to assess the subsurface conditions between boreholes. Alternatively, this information may be combined with other data from the site, e.g. from geological maps, memoirs, mining records and surface observations, to give continuous, interpolated, geological sections. Whilst the latter form of presentation is more informative, it should not be attempted unless the interpolation can be made with confidence[5].

Details of the boring technique and progress, observations and in-situ tests made during each boring, water levels (including when measured) and the types and locations of soil and rock samples taken are recorded in the finished borehole logs. The results of laboratory tests on samples taken from the field are not normally recorded on the borehole log, but are tabulated on separate sheets.

2.5 References

1. BS 5930: 1999, *Code of Practice for Site Investigations*. London: The British Standards Institution, 1999.
2. Site Investigation Steering Group, *Site Investigation in Construction: Part 2 – Planning Procurement and Quality Management of Site Investigation*. London: Thomas Telford, 1993.
3. Site Investigation Steering Group, *Site Investigation in Construction: Part 3 – Specification for Ground Investigation*. London: Thomas Telford, 1993.
4. National Audit Office, *Quality Control of Road and Bridge Construction*. London: HMSO, 1989.
5. Dumbleton, M.J. and West, G., *Guidance on Planning, Directing and Reporting Site Investigations*, TRRL Report LR 625. Crowthorne, Berkshire: The Transport and Road Research Laboratory, 1974.
6. *Ground Investigation Procedure*, DoT Advice Note HA 34/87. London: The Department of Transport, 1987.
7. Perry, J. and West, G., *Sources of Information for Site Investigations in Britain*, TRL Report 192 (Revision of TRL Report LR 403). Crowthorne, Berkshire: The Transport and Road Research Laboratory, 1996.
8. AASHTO Highway Subcommittee on Bridges and Structures, *Manual on Foundation Investigations*. Washington, DC: American Association of State Highway and Transportation Officials, 1978.
9. Robb, A.D., *Site Investigation*. London: Thomas Telford, 1982.
10. Engineering Group Working Party, Report on engineering geophysicals, *Quarterly Journal of Engineering Geology*, 1988, **21**, pp. 207–27.
11. Stewart, M and Beaven, P.J., *Seismic Refraction Surveys for Highway Engineering Purposes*, TRRL Report LR 920. Crowthorne, Berkshire: The Transport and Road Research Laboratory, 1980.
12. *Handbook of Ripping*. Peoria, Illinois: Caterpillar Tractor Co., Inc., 1988.
13. *Manual of Applied Geology for Engineers*. London: The Institution of Civil Engineers, 1977.
14. Moore, R.W., Applications of electrical resistivity measurements to subsurface investigations, *Public Roads*, 1957, **29**, No. 7, pp. 163–9.

CHAPTER 3

Plans, specifications and contracts

A. Boyle

This chapter primarily deals with the basic details of how a road improvement scheme is executed, paying particular attention to documentation-related issues associated with the construction process. The descriptions are based largely on experience with the road industry in the UK, but they also touch on work in other countries.

Before discussing documentation issues, however, it is useful to provide an overview of the whole road development process so as to put these in context.

3.1 Classical steps in preparing a programme for a major road: an overview

Figure 3.1 shows, in a simplified way, the major steps involved in the preparation and implementation of a major road proposal in the UK. Similar processes are, in general, carried out in other countries.

3.1.1 Scheme identification, feasibility, preferred route, and preliminary design studies

If road authorities, which are normally the owners and operators of road networks, identify a capacity- or safety-related problem at any particular point or section on the network, they will normally request that a study be carried out. This *scheme identification study*, which is likely to be carried out either in-house or by external consultants, is primarily about the quantification of precisely what the problem is and identifying whether there is a potential improvement scheme that would resolve or mitigate it in some way. This may result in a scheme as simple as the use of additional road signs or markings, or it may be a proposal for a complex urban motorway costing many millions of pounds.

Nowadays, and particularly in the case of major road proposals, the client authority also expects other options to be considered which would minimize the use of the road by private cars, e.g. possible alternatives using forms of public transport. Only if the road-based solution is found to be the optimum one will it be allowed to go ahead. This is in line with the international agreement on sustainable

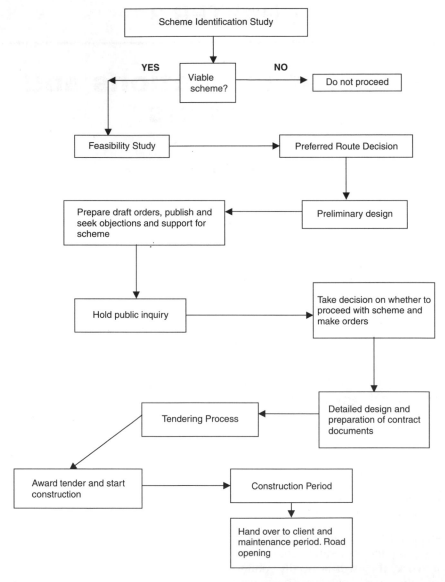

Fig. 3.1 Simplified steps in the preparation of a programme for a new road

development that was reached at Rio in 1999, which is being adhered to by the UK government and by much of the developed world. (Note, however, that the governments of many developing and Eastern Europe countries, i.e. those with relatively poor road infrastructure, may well take the view that their greater need is to improve their road infrastructure, for economic and/or environmental reasons.)

In the UK it is now also necessary to carry out a full risk analysis and assessment and develop a process to manage those risks that have thus been identified, together with a Value Management/Value Engineering study throughout the life of the scheme.

The procedures for so doing are set down in the Highways Agency's *Value for Money Manual*[1].

When the scheme identification study has been completed, the client authority is then likely to request that a *feasibility study* be carried out to identify for public consideration the possible alternative solutions to the problem(s) but using the scheme or schemes that have been identified through the *scheme identification study*. The best of these are put to the public for their consideration at what, in the UK, has hitherto been called a 'preferred route consultation'. In the future, it is likely that this will increasingly be done through a planning conference attended by the relevant planning authorities, government departments, public bodies, action groups, and others who feel that they might have an interest in the outcome and decisions. The alternatives, including the public transport options for solving the problem, will be put to the public for its consideration at this conference.

Having carried out a feasibility study and considered the response of the public to the alternative scheme, the client authority then takes the *preferred route decision* (in the case of a road-oriented solution), and this is published. The published route is then subject to planning controls, and these can be used to restrict other developments which are likely have a detrimental effect on the proposed new road.

A designer is then commissioned to carry out sufficient design to enable a decision to be made as to exactly where the road will be located within the preferred route envelope. In order for this to be done, preliminary topographical and geotechnical surveys will have to be carried out so that the designer has enough detail regarding such physical features as the contours of the land and the composition of the soils along the feasible alignments, to enable the preparation of a *preliminary design* of the proposed road. This design will need to be sufficiently robust to stand up to detailed scrutiny, and to enable options to be discarded that are likely to be too costly or too difficult to build. The impacts that the proposed horizontal and vertical alignments will have on side roads, accesses to private property, footpaths, etc., that are affected by the new road alignment will need to be considered at this stage also. Additionally, an accurate assessment can be made of the need for land purchases, including any Compulsory Purchase Orders that will have to be made under the *Compulsory Purchase of Land Regulations*[2].

3.1.2 Statutory governmental processes

The preliminary design is used to obtain planning permission for the proposed new road. In the case of trunk roads in the UK, this requires the *publication of Orders* under Section 10 (for conventional trunk roads) or Section 16 (for Special Roads, i.e. motorways) of the *Highways Act 1980*. If the proposal requires the alteration and possible closure of other affected roads, Orders will need to be published for these also; these, which are known as Side Road Orders for trunk roads, are published under Section 14 of the Act.

After the Orders have been published, interested parties, e.g. people who for various reasons do not wish the road to be built on the proposed alignment (or not at all), as well as people who are directly affected by the road, are given a period of time in which to register their objections to, support for, or comments on, the Orders. They may also suggest alternatives, including those that were considered but rejected at the public consultation stage. The authority proposing the new road will need to consider these concerns, and seek to resolve them. If, however, some objectors are so adamantly against the proposal that their objections cannot be met, or they have a valid argument with the proposal, it may need to be discussed and argued at a *Public Inquiry*. If the objections that the proposing authority cannot resolve are statutory ones, a public inquiry must be held to consider these (and other non-statutory) objections. A statutory objection is one that involves someone who is directly affected by the scheme.

If a public inquiry is decided upon, an eminent person is appointed by the Lord Chancellor as the Inspector to hear all the objectors, and the proposer's responses to them. This can be a long and complex process, particularly when the proposal relates to roads in urban areas.

In the case of trunk road and motorway schemes, the Secretary of State of the Department of Transport, local Government and the Regions, can take the decision regarding the *confirmation of the Orders*. This is done by either confirming the Orders as published, or with modifications agreed between the parties at the public inquiry or subsequently recommended by the Inspector. The Secretary of State is required to set out the reasons for accepting or rejecting the Inspector's recommendations in the decision letter. A further opportunity is then available for objectors to challenge the decision before the Orders are finally made; this challenge may be taken into the High Court, the Court of Appeal, the House of Lords and, if so required, at the European Court of Human Rights.

As far as local authority roads are concerned the powers to build these are slightly different in that the local authority has the power to grant itself planning permission. However, in most cases the granting of such permission is referred to the Secretary of State for confirmation, and therefore also be subject to a Public Inquiry.

3.1.3 Design and tender processes

Once a road proposal has gained planning permission in whatever form is required by the laws and conventions of the country concerned, the preparation of *detailed design documentation* can begin. Whilst some indication of the topography and geological state of the route of the new road will have been obtained at the preliminary design stage as indicated in subsection 3.1.1 above, further topographical, geological and geotechnical surveys will normally need to be carried out to confirm the earlier surveys and provide the greater level of detailed information required for the final design. The detailed design of bridges and other structures in accordance with known standards is normally initiated at this stage also, alongside the drainage and road pavement designs

Having completed the statutory and detailed design processes, the client authority then seeks interested contractors to bid on the documented scheme via a *tendering process* (see subsection 3.2.4). If too many contractors express an interest in tendering for major road projects in the UK the Highways Agency may reduce the list to a manageable number, e.g. six, typically, who are judged to have the capability to carry out the project. If short-listing such as this is not done the client authority may not only have a very considerable number of tenders to evaluate but, more importantly, much time and money will be wasted by a large number of tenderers in preparing tenders that, for various capability reasons, stand little chance of being accepted. (It might be noted here that, in the European Community (EC), public sector clients have to comply with the Public Procurement Directives of the EC; these require all schemes over a threshold of about 0.5 m pounds sterling to be advertised in the *Official Journal of the European Union*.)

Once the tenders have been received the technical advisers to the client authority consider them, advise the client on the most appropriate one to accept and the client will then award the contract to the successful bidder. Historically, in the UK, the successful bidder has been the one who put in the lowest price; however, this is not necessarily the case now as quality assessment is increasingly included in the selection process.

3.1.4 Construction contracts

Following the award of the contract to the successful tenderer, the construction documents, amended as necessary after appropriate negotiations with the tenderers, are issued in the form of Contract Documents. These documents form the basis of the construction of the new road.

The award of the contract does not mean that the client authority can forget about the project until construction is completed and the road is opened. Almost inevitably there will be alterations and variations during the construction period, and it is in this area that most on-site disputes arise. However, it has to be recognized that the variability of the materials that are encountered on site, as well as those used in civil engineering works in general, are such that the construction process may well necessitate changes that cannot be predicted in advance. The road designers should therefore seek to do all that lies within their power to minimize these changes and disputes, using the risk management techniques set down in reference 1. It is also essential that the form of contract chosen should be one that minimizes the areas for dispute and maximizes opportunities for the client and the contractor to work together in partnership.

In many ways, the key to a successful construction contract lies in the hands of the senior staff employed to carry out the works. If the contractor's senior staff and those employed to supervise the works are highly experienced, the project is more likely to be carried out economically, efficiently, and safely.

3.1.5 Maintenance period

Once the contract is completed the new road is handed over to the client authority for use by traffic. There may then be an opening ceremony, but these are less frequent now in countries where environmental issues have a high profile.

Following the hand-over there is a maintenance period during which the contractor remains responsible for the road structure. Almost inevitably, there will be a list of minor problems identified at the final inspection before the hand-over of the road is accepted, but which are insufficiently serious to warrant their rectification before the road is opened. These will need to be put right at the contractor's expense, along with anything else that comes to light during the early life of the new road. This period is known as the maintenance period, and is usually no longer than 12 months. The 'repair' work carried out by the contractor during the maintenance period is separate from the normal road maintenance activities carried out by the road authority's maintenance workers or contractors.

3.2 Documentation of a major road improvement project

The documentation required for a major road improvement construction contract is much the same as that used for any major construction contract. Its generic components are as follows: (1) Conditions of Contract; (2) Specifications; (3) Methods of measurement; (4) Bills of quantities; and (5) Tender and contract drawings.

3.2.1 Conditions of contract

The conditions of contract set out the general requirements of the contract. In the UK the conditions that are most commonly used for road schemes are those produced by the Institution of Civil Engineers (ICE). The 7th Edition of the ICE conditions[3] have just been brought into use at the time of writing; however, some schemes may still be using the previous (6th and 5th) editions[4] because of the long gestation period associated with some projects. In addition the suite of conditions that form the New Engineering Contract are increasingly being used, as well as the Design and Build option. Internationally, the contract conditions[5] developed by the Federation Internationale des Ingenieurs-Conseils (FIDIC), which are not dissimilar to the ICE conditions, are commonly used.

The ICE conditions clearly define the respective roles of the contractor, the supervising engineer and the client, and provide arbitration methods to resolve disputes. However, some participants in the contractual process have concerns that these conditions also tend to produce cost overruns and excess claims and, consequently, a confrontation-stance between the participating groups. To overcome these problems attempts have been made to encourage the concept of 'partnering', whereby contractual structures are put in place which encourage the development of a common intent by the participating parties to make the construction process a joint exercise that leads to a successful outcome. The need to finance road schemes that cannot be provided from the public purse, has also led to other conditions and contract forms being tried out. These include the Design, Build, Finance and Operate form used in the UK.

3.2.2 Specifications

The specification documentation is that section of the contract that deals with how the road is to be constructed. Its purpose is to ensure that client authorities get what they want when they contract to have a road built.

In the UK specifications may also refer to British Standards, design standards and advice notes. A British Standard contains the agreed and correct composition for a product that is used throughout industry in the UK; some standards, termed *Codes of Practice*, also set out the correct way to carry out construction processes. Increasingly, these standards are derived from the equivalent European Standard, in some cases the British Standard becoming the European Standard and in other cases, standards that are agreed through the European standardization process become British Standards as well (see Section 3.3). A *design standard is* a document, most commonly produced by a governmental agency that helps designers to prepare acceptable and safe designs. An *advice note* is a document that helps designers on how best to apply design standards, i.e. it contains information on how a design standard is to be interpreted, as well as other data which may not be directly related to the design standard.

Generally, there are two main ways of specifying work in order to achieve the desired result. These involve using either 'method' or 'end-product' or 'performance' specifications.

With *method specifications* reliance is placed on research results which show that if certain methods are used to complete particular works, they will meet the requirements of the client. This 'recipe' approach is acceptable provided that the contractor complies with the exact requirements of the specification, and the works are carried out in the scheme exactly as they were carried out in the full-scale trials that resulted in the specification's development. As the latter requirement is not always feasible, method specifications normally permit a certain latitude to allow for deviations. If the specification is faulty in any way, the contractor can claim that the fault was not his if the work is deemed unsatisfactory.

With *end-product* or *performance specifications* the client specifies the desired outcome from the construction, e.g. that the completed works or components of the works meet certain requirements for durability and strength. This form of specification requires the client to make clear to the contractor what is required from the scheme when it is completed. Contractors generally like end-product specifications because they provide opportunities for innovative and lower-cost solutions that meet the performance requirements of the client.

The road specification document most used in the UK is the Highways Agency's *Specification for Highway Works*[6]. This comprises a large number of clauses which are grouped into various series, with each series dealing with different aspects of road construction, e.g. site clearance, fencing, safety barriers, drainage, earthworks, road pavements (all types), signs, electrical works, steelwork, concrete structures, and bridge works. The clauses provide detailed requirements for the manufacture and testing of components and materials used in the construction works, including details on how to erect sections of the works, how to excavate, transport and lay particular materials, and how to complete the works. (Full lists of series and clauses are given in reference 6.)

It might also be noted here that there is much interest now in the UK in applying the tenets of quality management to construction work (as well as to design). Thus, contractors are now often required to develop *quality plans* for construction projects. Typically, these plans include such information as how they propose to carry out the works and what plant they plan to use, together with criteria for checking and measurement so that it can be verified that their targets are being met, as well as advice regarding the site and management personnel proposed.

3.2.3 Bill of quantities

The 'bill of quantities', or BoQ as it is often abbreviated, is that part of the contract documentation that contains the pricing information regarding the items that the designer expects will be needed to form the constructed road. Its purpose is to provide a structure whereby the tenderers can put their prices against these items so that, when these are added up, the total tender price is obtained.

As part of the normal course of design, the client's consultant engineer produces a complete set of design drawings, and these are included in the contract documentation for the road. Copies of this set, which in the main become the *contract drawings* (subject to any changes identified and agreed during the tender period), are then used by tenderers to prepare their tender submissions. It might also be noted here that, upon final completion of the project, the contract drawings are amended as necessary to produce *as-built drawings* which show exactly what was built, and where it was built, for record purposes and for the future information of the maintenance authority.

The client's designer, having completed the design drawings for the road, uses these to estimate the quantities of all of the various types and classes of material required by the work. These are put together in the BoQ in the form of lists of items of work that, in the UK, are mostly compiled from a library of standard bill items in the Highways Agency's *Method of Measurement* publication[7]. These lists contain a quantity column (with measurements in, for example, linear metres, cubic metres, or tonnes), a column for the tenderer's rate to be inserted against each item, and a column for the total price against each item.

Nowadays, bills of quantities are often sent to the tenderers on floppy disk so that, with the aid of various forms of proprietary software, they can use them to complete their tenders. Whilst the floppy disk, with the completed tender, can then be sent back to the client, a hard copy is also required for legal reasons.

On some projects there may be items in the BoQ about which the designer is unclear. For example, it may be known that the soil in an area to be excavated is unsuitable for use as a fill material and will therefore have to be wasted, but the exact volumes of waste soil cannot be determined until the construction is actually carried out. In such instances, it is usual for the designer to include what is known as a *provisional item* in the bill of quantities, and to seek from the successful tenderer information about the rate to be charged for this work to be carried out, assuming that an estimated volume of material has to be moved.

The use of provisional items in bills of quantities should be minimized, because of the potential dangers that it poses for the client. In the case of the waste soil example noted above, for example, if the estimated volume of poor material is small, the tenderer may attach a high unit price hoping that the actual quantity to be moved will prove to be much larger. On the other hand, if the estimated volume of poor material is large and the actual amount moved is much lower, the contractor may subsequently claim a higher rate than was initially quoted, on the grounds that he was provided with misleading information and that, as the actual amount of poor soil was much less than expected, the construction programme was disrupted, hired equipment could not be used, and he was caused to incur a heavy loss.

3.2.4 The tendering process

As noted above (subsection 3.2.3), tendering is the process whereby a select number of contractors are invited to price the bill of quantities for a particular project, taking into account all the information and constraints that are included in the proposed contract. With this process the contractors are provided with a complete set of tender documents, together with any instructions to assist in their correct completion and submission. The contractors take these away and the task of compiling their tenders typically begins with the tender documents being broken into sections and invitations to quote for parts of the works being sent out by each tendering contractor to various suppliers and subcontractors. Simultaneously, each contractor's planning engineers get together with the intended project manager, to start the process of preparing a *programme of work* to build the scheme in the most cost-effective and sensible way.

Programming, which is essentially concerned with the smooth running of the construction work, is an integral part of the tendering process. The programme of work is developed very early in the preparation of the tender, so as to enable the allocation of enough people, plant and other resources to individual parts of the scheme to ensure that each part can be completed within a scheduled time. The programme produced is vital to the task of estimating the cost of the works and the bringing together of the final tender proposal. Issues such as its 'buildability', the potential for claims, and any other considerations that might affect the final price submitted, are also considered in the course of developing the programme, until a final price and timetable is established for submission to the client's representative.

Under the provisions of the Conditions of Contract, the successful tenderer is also required to produce a programme to the client, within a specified period of time after the award of the contract, which will detail how the work is to be carried out and in what time-scale. This programme must be kept up-to-date during the work, so as to reflect the changes that result from coping with the uncertainties of a construction site.

During the tender preparation process it is not unusual for a contractor to notice anomalies in the documents, and to notify the client regarding these. If these include reasonable suggestions, a tender amendment letter is sent out to all tenderers and, if the change is important, the tender period may be extended to allow further time for its consideration.

Some tenderers may consider that if some aspect(s) of the design or specification could be changed they would be able to produce a cheaper tender, because the alternative would better fit their construction processes or be more easily built. They are allowed to submit such alternatives, and to seek to get them accepted, provided that a *conforming tender* is also submitted, i.e. a conforming tender is one that meets, in every respect, the unchanged requirements of the tender documents.

The completed tenders must be submitted, as published, to a certain place nominated by the client authority or its agents by a certain time. When received the tenders are normally stamped to avoid subsequent controversy regarding whether or not the deadline for their receipt was met.

A *Tender Board* is normally established to open the sealed tenders and decide on a ranking for the tenderers based on tender evaluation guidelines. This Board also judges whether the tenders conform to the requirements, and whether to accept a contractor's alternative should one be proposed. The Board's report is submitted to the client authority, which then gets its design consultants to fully assess the tenders. Finally, a consolidated report on the tenders, and a recommendation regarding the successful tenderer, is submitted to whomever the client has nominated as responsible for taking the decision regarding the awarding of the contract.

3.2.5 Contracts

Until relatively recently the standard method of major road procurement in the UK involved the client authority employing an engineering consultant to design the scheme and assist in the statutory planning process, develop the contract documents, and seek tenders as described in the previous section of this chapter. Once the decision was made regarding to whom the contract should be awarded using this approach, the *standard contract* would consist of the conditions of contract, the specification, the priced bill of quantities, the design drawings, and a formal exchange of letters between the client and the successful contractor. If there was any dispute the conditions of contract were used to deal with contractual issues, the specification with the technical matters, and the bill of quantities with any pricing concerns. Contractual variations and claims (and disputes) are not uncommon with this standard approach.

A *variation* is a change to the agreed design that occurs after the contract has been awarded when the contractor finds something on site that was not allowed for in the contracted tender price or which, if included in the contract, cannot be built as intended for some reason. Alternatively, the variation may result from the supervising engineer or the client finding something in the design that has to be changed.

Claims occur when the contractor considers that a variation is affecting his ability to construct the scheme as defined by the contract, and that this is costing him money. Because of construction expediency the claim costs of variations are not always agreed and/or documented at the onset of change and, consequently, a lot of time may have to be subsequently spent on arguments regarding their validity. Not uncommonly, either party to the contract, or both, may resort to the arbitration clauses in the conditions of contract in order to reach a decision on a claim. Often these arguments are not finally resolved until after the scheme has been completed and opened to traffic.

When all the claims are settled and the final measurement is made, the client and the contractor settle the final accounts, and most of the remaining money owed to the contractor is then paid. The only payment that normally remains outstanding after this settlement of the accounts is the 'retention money'; this is a part of the tender sum that is kept by the client for, say, 12 months to pay for the rectification of possible defects in the completed work (and ensure that they are carried out).

In recent years there has been a great deal of interest in the UK regarding the use of other forms of contract in lieu of the standard contract. Examples of these are design and build contracts.

With a *design and build contract* the successful contractor is chosen at an early stage of the development of a scheme; this is usually, but not exclusively, after the planning process has been completed. As its name implies, the concept underlying this contract form is to bring together, in partnership, the parties responsible for the design and the construction of the road scheme. This approach is seen as bringing into the design process the knowledge and skills of the contracting industry so as to make the design more 'buildable' and, thus, to reduce costs whilst, at the same time, providing an opportunity for the minimization of disputes over the design and specification of the work. The contract still has a client side, albeit the contracting side includes not only a contractor (or group of contractors for a very large scheme) but also an engineering designer nominated by the client. With this form of contract, the agent appointed by the client to supervise the implementation of the contract is likely to have a relatively reduced role.

Pressures on the public purse in recent years have led to an increasing consideration of the use of private instead of governmental finance for the procurement of new road schemes. Many countries are now seriously looking at involving non-governmental financial institutions in the provision of essential road (and rail) infrastructure improvements, as well as hospitals, schools, sewerage projects, indeed any construction and operational project which might have been funded from the public purse. These can be done in a number of ways but most usually takes the form of a contractual arrangement whereby a company is given a concession to *design, build, finance and operate (DBFO)* a length of road, or a portion of the road network in a geographical area, for a fixed period, e.g. 20, 30 or 50 years. Whilst the client road authority retains the basic legal responsibility for the road with this approach, the DBFO company agrees both to design, build and pay for the scheme and to maintain and operate the road(s) to agreed standards throughout the period of the concession.

In return for its investment the DBFO company may be given the right to impose tolls, under agreed conditions, on vehicles using the road during the concession period. Alternatively, if the direct payment of a toll is not acceptable (usually for political reasons), the client authority may pay for the vehicle usage via 'shadow tolls' whereby the DBFO company is paid a variable rate, agreed at the onset of the contract, by the commissioning road authority, e.g. as traffic using the road increases with time the annual level of payment may also be increased until a cut-off point is reached beyond which no more increases (except for inflationary ones) are paid.

3.2.6 Supervision of construction contracts

With standard contracts the contractor is responsible for the construction of the works to time, to specification, and to the agreed cost. With these contracts the contractors normally have their own engineering staff whose tasks are to make sure that the schemes are built in accordance with the contractual requirements – and to maximize the profits from them.

To ensure that its requirements are met in a standard contract a client authority will normally employ someone to supervise the contract on its behalf, i.e. to check on what the contractor is doing and to finally agree that the work has been carried out in an acceptable manner. The supervision of construction is an important part of the standard ICE contract, and its conditions of contract lay down a role for a person, i.e. the *Resident Engineer (RE)*, to do this.

Most usually the client will utilize the services of the consulting engineer who designed the scheme to supervise the works, but occasionally a separate firm experienced in construction supervision may be used. As the supervision of a major road construction contract can be fraught with variations and disputes, the appointment of a careful, experienced, resident engineer is critical to ensuring that cost escalation and excessive disputes are avoided, especially in complex projects.

3.3 The European dimension

The European scene is dominated by the actions of the European Union and its Executive, the European Commission, which is based in Brussels. Like most governments, this Executive is split up into a large number of departments known as *Directorates General (DGs)*.

In the fields of construction and transport there are a number of major players as far as the Directorates General of the Commission are concerned. Thus, DG 3 – Industry deals with all standardization issues, whilst DG 7 – Transport is responsible for transport policy, the funding of transport projects, and research into transport issues. DG 13 – Informatics deals with issues of transport telematics from the research aspect, whilst DG7 deals with implementation alongside DG13. DG 15 – Public Administration has to do with Public Procurement Directives, which control the ways in which the public sector lets contracts. Recently, these DGs have been shaken up and the duties outlined above may not now be exactly the same, but they give an indication of the way Brussels works.

The European Union has decided that to facilitate the operation of its single market the specifications for all construction products should, wherever possible, be harmonized. These requirements have been enshrined in the Construction Products Directive (CPD) – the operation of which is managed by DG3 – which is resulting in considerable change and flux in the road construction industry. Member states, and some of the trade and standards' bodies, have their views on harmonization heard through DG3's Standing Committee on Construction (SCC), which meets regularly to consider, debate, and vote on proposals from the Commission for mandates to the Comité Européen de Normalisation (CEN) which is the body that is mainly responsible for developing, through technical committees, European Standards (EN). These mandates include details on what levels of overall quality, and of attestation of conformity, should be included within the harmonized portions of the EN standards. By 'attestation of conformity' is meant the level of checking to be carried out to ensure that the product does what the specification states it can do; the attestation level may range from a simple manufacturer's declaration to a full third-party assessment. The harmonized EN standards are gradually replacing well-established national standards, e.g. BSI in the UK, DIN in Germany and AFNOR in France; however, it is likely to be many years before the full transition is completed. Where developed and agreed the European standards are mandatory for public purchasers and, consequently, when dealing with roadworks.

Whilst this discussion is mainly concerned with trends within Europe, it is appropriate to mention here that there is another important standard-making body that is also influential; this is the International Standards Organisation (ISO). The ISO is the body that sets standards at world level. However, use of ISO standards is not mandatory unless national bodies adopt and use them in their own national series.

Another European body that is of importance in road construction is the European Organisation for Technical Approvals (EOTA), which gives technical approval to proprietary products. If, for example, a manufacturer develops a product such as a bridge deck-waterproofing system, bearing or parapet, for example, then it would submit it to EOTA for testing and certification that it met the general requirements for such a product, and that the manufacturer's claims can be substantiated. The current equivalent body in the UK is the British Board of Agrement (BBA).

Another requirement of the European Union is that member states must recognize as equivalent those products which meet the requirements of the manufacturing country's own national standards even though there may not be, as yet, a harmonized European standard. This can cause problems where there are clear differences between the requirements of different national standards and, in particular, the level of attestation of conformity.

In recent years a number of the erstwhile Eastern Bloc countries have applied for membership of the European Union and it is not unlikely that a number of these will join within the near future. The road and rail transport infrastructure in many of these countries is not up to the standards of those in Western Europe. The EU is committed to funding major improvements to underdeveloped road and rail links so as to create a 'level playing field' for industry and, to that end, is currently funding the development of modern specifications on the western model for these countries. It can be expected that there will be a major expansion in infrastructure building contracts in due course in Eastern Europe.

3.4 References

1. Highways Agency, *Value for Money Manual*. London: HMSO, 1996.
2. *Compulsory Purchase of Land Regulations*, Cmd. 2145. London: HMSO, 1996.
3. Institution of Civil Engineers, *Conditions of Contract*, 7th edition. London: Thomas Telford, 1991.
4. *ibid*, 5th edition, 1973 and the 6th edition, 1991.
5. Federation International des Ingenieurs-Conseils, Conditions of Contract for Works of Civil Engineering Construction, Lausanne: FIDIC, 1987.
6. Highways Agency, *Manual of Contract Documents; Vol. 1 – Specification for Highway Works*. London: HMSO, 1998.
7. Highways Agency, *Manual of Contract Documents; Vol. 3 – Methods of Measurement*. London: HMSO, 1998.

CHAPTER 4

Soils for roadworks

C.A. O'Flaherty

4.1 Soil formation and types of soil

The earth's crust is composed of in-place rock and weathered unconsolidated sediments. The sediments were derived from rock as a result of physical disintegration and chemical decomposition processes, and deposited by gravity, or through ice, water, or wind action. Given that the mechanical forces of nature have been instrumental in moving most soils from place to place, it is convenient to describe them according to the primary means by which they were naturally transported to and/or deposited in their present locations.

4.1.1 Residual soils

Residual soils are inorganic soils formed by the in-situ weathering of the underlying bedrock, and which were never transported. The climate, e.g. temperature and rainfall, mainly determined the type of residual soil formed. Mechanical weathering, i.e. disintegration, dominated the soil forming process in northern cold climates and in arid regions, whilst chemical weathering dominated the process in tropical regions with year-round high temperatures and high rainfall. Residual soils are rarely found in glaciated areas; they have either been moved or they are buried beneath other glaciated soils. By contrast, in-situ lateritic soils are found to great depths in warm humid areas.

4.1.2 Gravity-transported soils

Colluvial soils are formed from accumulations of rock debris, scree or talus which became detached from the heights above and were carried down the slopes by gravity. These materials are usually composed of coarse angular particles that are poorly sorted, and the upper and lower surfaces of the deposits are rarely horizontal. Some deposits are very thin and can be mistaken for topsoil; however, thicknesses in excess of 15 m may be found in non-glaciated valley bottoms and in concave parts of hill slopes. One of the most widespread of colluvial soils is the clay-with-flint overburden that caps many of the chalk areas of South East England.

Organic soils, e.g. peats and mucks, that were formed in-place as a result of the accumulation of chemically-decomposed plant residues in shallow ponded areas, are termed *cumulose soils*. Both peat and muck are highly organic; the fibres of unoxidized plant remains are still visible in peat whereas they are mostly oxidized in the older muck soils. Whilst easily identified on the surface these soils are often encountered at depths below the surface in glaciated areas.

4.1.3 Glacial soils

Glacial soils were formed from materials transported and deposited by the great glaciers of the Ice Ages (spanning ca 2 million years) which ended about 10 000 years ago. These glaciers, often several kilometres thick, covered vast areas of the northern hemisphere and carried immense quantities of boulders, rock fragments, gravel and sand, as well as finer ground-up debris. When the ice-sheet melted and retreated from an area the carried material, termed *glacial drift*, was left behind. The drift deposits of most interest to the road engineer can be divided into two groups: (a) those laid down directly by the retreating ice, e.g. glacial tills (especially boulder clays) and drumlins, and (b) glacial melt-water deposits, e.g. moraines, kames, eskers, glacial outwash fans, and varved clays.

The term *boulder clay* is used to describe the unstratified *glacial till* that was formed underneath the ice-sheet and deposited as an irregular layer of varying depth over the ground surface. This material can be very variable and range from being entirely of clay to having pockets of sand and gravel and/or concentrations of stones and boulders. The 'clay' is mainly ground rock-flour resulting from abrasion of the rocks over which the glacier has passed.

Drumlins are smooth oval-shaped hills that are rich in boulder clay and oriented with their long axes parallel to the direction of ice movement. They are believed to have been formed as a result of irregular accumulations of till being overridden by moving ice. Drumlins usually occur in groups, and can be up to 2 km long, 400 m wide, and 100 m high (although 30–35 m is more common).

Generally, it can be said that the engineering properties of boulder clays vary considerably and they require careful investigation before usage.

Moraines are irregular hummocky hills that contain assorted glacial materials. Four types of moraines are generally recognized, viz. terminal, lateral, recessional, and ground moraines.

Found roughly perpendicular to the direction of glacial movement, *terminal moraines* are long low hills, perhaps 1.5–3 km wide and up to 30 m high, that mark the southermost limit of the glacial advance. As the front of the ice sheet encountered a warmer climate the material carried was dropped by the melting ice whilst, at the same time, the ice front was continually forced forward by the pressures from colder regions so that the hummocks became greater as more dropped material was added to them.

Lateral moraines are similar to terminal ones except that they were formed along the lateral edges of the melting ice sheet and are roughly parallel to the direction of ice movement. These moraines are most clearly defined along the edges of valleys in mountainous terrain.

As the Earth warmed up the ice front melted at a faster rate than it advanced and the ice mass began to recede. The recession was constantly interrupted and, consequently, minor crescent-shaped 'terminal' moraines were formed at irregular intervals; these are known as *recessional moraines*. If a topography is characterized by recessional moraines interspersed with many depressions with small lakes it is often termed a 'knob-and kettle' topography.

When the ice sheet melted rapidly material was continually dropped at the edges of the receding ice sheet. This material is referred to as *ground moraine*.

During the Ice Age, just as today, the weather was warmer at particular times of the year, and melt-water flowed out from the ice mass carrying both coarse and fine particles and spread over the uncovered land with the finer materials being carried further. In narrow valleys the waters flowed quickly and freely and the materials deposited in those valleys tended to be coarse-grained, whilst those left on plains were often extremely varied. Sometimes the outwash waters fed into enclosed valleys and formed lakes within which settlement took place. Streams of water flowed in tunnels at the base of the ice, often with great force, especially during periods of rapid ice melting. Kames, eskers, outwash fans and varves are distinctive landforms deposited at these times, which are now of particular interest to the road engineer.

Kames are hummocky-shaped mounds that were formed parallel to the ice-front by glacial streams which emerged during pauses in the ice retreat. They mostly contain poorly-stratified accumulations of sand and gravel, usually with some clay, that are generally suitable for use in road construction.

Eskers are relatively low (typically <18 m high) sinusoidal ridges, often several kilometres long, which contain highly stratified gravel and sand that is eminently suitable for use in road construction. Usually with flat tops and steep sides, eskers are the end-products of accumulations of glacial material that were deposited by subglacial streams at roughly right angles to the retreating ice front. The lack of fine-grained particles is explained by the fact that they were washed away by streams flowing rapidly from the melting ice.

Good construction material is also found in *outwash fans*. These were formed where the streams emerged from the ice-front and their speeds were checked so that the transported gravel and sand was dropped in a fan-shape.

Varves are fine-grained laminated clays that were deposited in glacial lakes or in still waters impounded in front of a retreating glacier. The different rates of settlement of the silty and muddy materials in the waters meant that the more quickly-settling materials lie below the finer ones and each pair of coarse and fine layers represents a season's melting of the ice. The thickness of each varve can vary from being infinitesimally small to perhaps 10 mm thick, and deep accumulations have been found (e.g. 24 m in an engineering excavation in a Swedish lake). Varved soils give rise to troublesome foundation problems and should be avoided when locating a roadway.

4.1.4 Alluvial soils

This is a general term used to describe soils formed from sediments deposited by river waters that were no longer moving fast enough to keep their loads in suspension or to trundle larger materials along the bottom of the stream.

An *alluvial cone* is often the source of good road construction materials. This landform is found at the change of gradient where a stream emerged suddenly from a mountain valley onto a plain. As the water had a higher velocity as it entered the plain, cone material is generally coarse grained and well drained, and contains few fines.

When a stream carried an excess of water, e.g. after a heavy rainfall, it often broke its banks and spread out over the adjacent land, dropping coarser material adjacent to the stream and finer material further out. Thus natural *levees* of gravel and sand may be found beside the upper reaches of rivers in flood plains. These levee accumulations are often usable for road construction purposes, albeit they normally have to be processed beforehand to remove excess fines. In the case of the lower reaches of mature streams the levee materials are finer and poor-draining swampy land is found behind them.

Mature rivers tend to be on gently-inclined beds and to use up their energies in horizontal meandering rather than in cutting deeper channels. During these meanders coarse-grained material was often deposited at the inside of the curves whilst the moving waters cut into the opposite bank. The deposits at the insides of the bends can be very valuable construction materials, and may influence the location of a road as well as its construction.

Over time mature rivers continually changed their path and old meanders were cut off, leaving behind half-moon-shaped depressions now known as *oxbows*. With the passage of time, run-off waters carrying fine-grained materials flowed into the oxbows so that, eventually, they became filled with silty and clayey sediments. Since they are not continuous but rather occur as pockets of clay-like materials, oxbows should be avoided when locating a road as they are liable to cause differential heaving and/or settlement of a pavement.

In many instances, for geological reasons, an old meandering river suddenly became 'rejuvenated', and having excised a new channel in the old alluvium, left raised *terraces* on either side. Valuable deposits of gravel and sand can be found in these terraces. Also, terraces often provide good routes for roads through valleys, as they reduce the need for costly excavation into the valley sides and the pavement is located above the floodwaters of today's river.

Whilst much of the sediment carried by a river was deposited over its floodplains, a considerable amount of finer-grained material still reached its mouth where it was discharged into the ocean or a lake. Over time roughly triangular-shaped accumulations of these materials, known as *deltas*, were built forward into the water and some eventually became grassed over to form small promontories. Deltaic deposits can be very varied and considerable investigation may be required to determine if they can be used in road construction.

4.1.5 Marine soils

In addition to the material carried to the sea by rivers, the ocean itself is continually eroding the shoreline at one location and depositing the eroded material at another. As oceans receded from old land areas, great areas of new land surface composed of a variety of deposited materials became exposed to the weathering elements. These processes have been going on for millions of years.

It is difficult, in the limited space available here, to make definitive comments about marine soils, except that their usage in roadworks should be treated with considerable care, especially if they contain fine-grained material.

4.1.6 Eolian soils

Eolian soils are those that were formed from material that was transported and laid down by the wind. Of particular interest are dune sands and loess.

Dune sands are found adjacent to large bodies of water where large quantities of beach sand were deposited. When the wind blows consistently from one direction, the coarse sand particles move by 'saltation' across open ground with little vegetative cover, until eventually they lodge together and dunes begin to form. Saltation is a skipping action resulting from sand particles being lifted by the impacts of other wind-blown grains. The dunes formed have relatively gentle slopes on their windward side whilst their leeward slopes are at much steeper angles of repose.

Because of their tendency to move, sand dunes can be a problem. When the dune is sufficiently formed that the amount of sand deposited on it by the wind balances the amount taken from the leeward side by the same wind, the dune will tend to migrate as sand particles are blown up and over the top. Roads that pass through migrating dune areas that are not stabilized by vegetation can be covered by sands and, particularly when the roads are low cost and unsurfaced, rendered impassable.

Loesses are porous, low density eolian silts of uniform grading that are found away from the beaches from which they draw resources. They mostly consist of silt-sized particles that were picked up by the wind and carried through the air to their final destinations. Deposits at particular locations are composed of very uniform particles; the particles sizes get smaller and smaller the further the distance from the beach source. Loesses are relatively unknown in Britain, albeit they occur extensively in Eastern Europe, China and the USA.

Because loess deposits are both free draining and rich in calcium carbonate, it is possible to excavate road cuttings with near vertical sides through them. However, the cementing action of the $CaCo_3$ is lost with manipulation of the loess, so that embankments formed with this material must have side-slopes similar to those formed with other soils.

4.2 Soil profiles

As the sediments described above were moved and reworked, more weathering, abrasion, mixing with organic materials and soluble minerals, and leaching took place until eventually soil profiles were formed. The profile found in any given locale is mainly, therefore, the historic product of five major soil-forming factors; climate, vegetation, parent material, time, and topography.

A soil profile (Fig. 4.1) comprises a natural succession of soil layers which represent alterations to the original sediment that were brought about by the soil-forming process. In easily drained soils with well-developed profiles there are normally three distinct layers known as the A-, B- and C-horizons.

The A-horizon is often called the *zone of eluviation*, as much of the original ultrafine colloidal material and the soluble mineral salts have been leached out by downward-percolating water. This horizon is normally rich in the humus and organic plant residues that are vital to good crop growth, but bad for road construction purposes, because the vegetative matter eventually rots and leaves voids, whilst the layer as a whole often exhibits high compressibility and elasticity, high resistance to compaction, and variable plasticity.

The B-horizon is also known as the *zone of illuviation* as material washed down from the A-horizon accumulates in this layer. The fact that this horizon is usually more compact than those above and below it, contains more fine-grained particles, is less permeable, and is usually more chemically-active and unstable, makes the B-horizon important in road construction. For example, the extra accumulation of fine particles in the B-horizon of an active fine-grained soil on a gentle slope may make the layer so unsuitable as the subgrade for a pavement that it has to be removed and wasted; by contrast, the excavated B-horizon material from a sandy soil may possibly be used to improve the gradation of its A-horizon and improve its foundation capabilities.

The layer in which the road engineer is normally most interested is the C-horizon. This contains the unchanged material from which the A- and B-horizons were originally developed, and it is in exactly the same physical and chemical state as

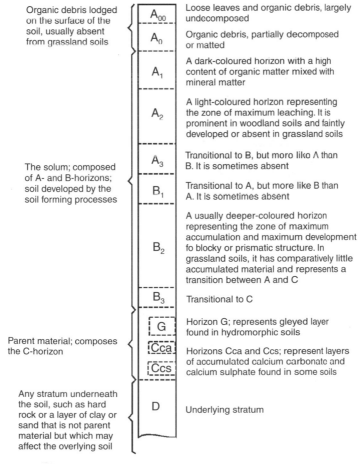

Fig. 4.1 A hypothetical soil profile

when it was deposited in the geological cycle. It is this material which is normally used as fill when building an embankment or, if it is good enough, a road pavement.

Although not a normal occurrence, there may be a further horizon, i.e. the D-horizon, beneath the C-horizon. This underlying stratum can have a significant effect upon the development of the characteristics of the overlying soil profile if it is within, say, 1 to 1.5 m of the surface.

As is shown in Fig. 4.1, the A-, B- and C-horizons may also be subdivided into a number of sub-horizons. More often than not, however, many of these sub-horizons, and sometimes full horizons, may be missing in situ, depending upon the influencing soil-forming factors and erosional features.

4.3 Soil particles

Depending upon its location in the profile, a soil may be described as a porous mixture of inorganic solids and (mainly in the A-horizon) decaying organic matter, interspersed with pore spaces which may/may not contain water. The soil solids component of this assemblage consist of particles of varying size, ranging from boulders to colloids. The coarseness or fineness of a soil is reflected in terms of the sizes of the component particles that are present; those most commonly considered are gravel (see subsection 5.9.3), sand, silt and clay.

Before discussing these solids in detail it is worth noting that whilst all international organizations now accept that the most convenient way to define gravel, sand, silt and clay is on the basis of particle size, various organizations have assigned different values to the relevant sizes. It is important to be aware of this when reading and interpreting the technical literature on soils.

In 1908 the Swedish soil scientist, Atterberg, published what is probably the most important early attempt to put on a scientific basis the limiting sizes of the various soil fractions. Atterberg classified gravel particles as being between 20 and 2 mm in size; he said that these were the limits within which no water is held in the pore spaces between particles and where water is weakly held in the pores. Sand was described as being between 2 and 0.2 mm in size; the lower limit was set at the point where water is held in the pores by capillary action. Atterberg visualized silt as being the soil component which ranges in size from where sand begins to assume clay-like features to the upper limit of clay itself, i.e. between 0.2 and 0.002 mm. The choice of 0.002 mm as the upper limit of the clay fraction was based on the premise that particles smaller than this exhibited Brownian movement when in aqueous suspension.

The *British Soil Classification System for Engineering Purposes*[1,2] has particle size limits that are different from, and more comprehensive than, those devised by Atterberg: boulders are sized as being >200 mm; cobbles are 200–60 mm; coarse gravel is 60–20 mm, medium gravel is 20–6 mm, and fine gravel is 6–2 mm; coarse sand is 2–0.6 mm, medium sand is 0.6–0.2 mm, and fine sand is 0.2–0.06 mm; coarse silt is 0.06–0.02 mm, medium silt is 0.02–0.006 mm, and fine silt is 0.006–0.002 mm; and clay is <0.002 mm.

For completeness it might also be noted that, with the British system, the term 'soil fines' is also used to describe soil material passing a 63 micron sieve. Thus, if the percentage of fines by mass is equal to or greater than 35 per cent the soil is described as a *fine soil*, whereas a soil with less than 35 per cent passing this sieve is termed a *coarse soil*.

4.3.1 Sand

Sand particles (see also subsection 5.9.3) are quite inactive chemically. They are generally bulky in shape, albeit individual grains may be described as angular, sub-angular, rounded or sub-rounded, depending upon the degree of abrasion received prior to their final deposition. Residual sands are usually angular whilst river and beach sands are generally rounded. Wind-blown sands are usually very fine and well-rounded whereas ice-worn sand particles can have flat faces.

Clean sand particles do not exhibit any cohesive properties and are therefore little influenced by changes in moisture content. The pores between the particles are relatively large; hence, sandy soils are very permeable and well-drained, and consolidation effects are small.

Table 4.1 shows that the stability of a sandy soil is greatly influenced by its degree of compaction, gradation, and particle shape. Note that the shearing resistance increases with compaction, and with angularity and size of particle.

4.3.2 Silt

Physically, silts are generally similar to sands in that they derive much of their stability from the mechanical interaction between particles. Coarse silt particles are essentially miniature sand particles and thus they tend to have similar bulky shapes and the same dominant (quartz) mineral.

Unlike sands, silts also possess a limited amount of cohesion due to interparticle water films. Whilst silts are classed as permeable, water can only move through the (small) pore spaces relatively slowly. Where the smaller-sized particles predominate, silts exhibit clay-like tendencies and may undergo shrinkage and expansion when exposed to variations in moisture content.

Table 4.1 Some approximate shearing resistance results for coarse-grained soils[3]

Grain-size	Degree of compaction	Angle of internal friction (deg):	
		Rounded grains, uniform gradation	Angular grains, well-graded
Medium sand	Very loose	28–30	32–34
	Moderately dense	32–34	36–40
	Very dense	35–38	44–46
Sand and gravel			
65% G + 35% S	Loose	–	39
	Moderately dense	37	41
80% G + 20% S	Dense	–	45
	Loose	34	–

Note: By comparison, blasted rock fragments have an internal friction angle of 40–55 deg.

4.3.3 Clay

Clay differs from sand and silt in respect of both physical properties and chemical make-up. It is very important for the road engineer to understand what constitute clay particles.

Physically, clay particles are lamellar, i.e. flat and elongated, and thus have a much larger surface area per unit mass than the bulky-shaped silts and sands. A measure of the differences in surface area of various soil fractions can be gained by assuming that the particles are spherical in shape (see Table 4.2). As the intensity of the physico-chemical phenomena associated with a soil fraction is a function of its exposed surface area, this table suggests why the clay fraction has an influence on a soil's behaviour which can appear to be out of proportion to its mass or volume in the soil.

Any analysis of the clay fraction is to a large extent a study of its colloidal component. In theory, a colloid is any particle that exhibits Brownian movement in an aqueous solution; in practice, however, the term is normally applied to particles smaller than 1 micron in size. *Clay colloids* are primarily responsible for the cohesiveness of a plastic soil, its shrinking and swelling characteristics, and its ability to solidify into a hard mass upon drying. Also, the drainage characteristics of a soil are considerably influenced by the amount and form of its colloidal content.

The importance attached to the colloidal fraction is associated with the electrical charges which the colloids carry on their surface. Figure 4.2 illustrates what is termed the Helmholtz double-layer concept of the make-up of a colloidal particle. Note that the inner part comprises an insoluble nucleus or micelle surrounded by a swarm of positively-charged cations; the inner sheath of negative charges is part of the wall of the nucleus. These positively-charged cations are in equilibrium at different but infinitesimally small distances from the colloid surface. A clay with sodium as the main adsorbed ion is called a sodium clay, whilst a calcium clay mostly has adsorbed calcium ions.

If a clay with adsorbed ions of a particular type is brought into contact with ions of a different type, some of the first type of ions may be released and some of the second ones adsorbed in their place. This exchange of positively-charged ions, termed

Table 4.2 Some physical characteristics of soil separates[19]

Name	Diameter (mm)	Number of particles/g	Surface area of 1 g of each separate (cm^2)
Fine gravel	1.00–2.00	90	11.3
Coarse sand	0.50–1.00	722	22.7
Medium sand	0.25–0.50	5777	45.2
Fine sand	0.10–0.25	46213	90.7
Very fine sand	0.05–0.10	722074	226.9
Silt	0.002–0.05	5776674	453.7
Clay	<0.002	90260853860	11343.5

Note: Each particle is assumed to be a sphere having the maximum diameter of each group.

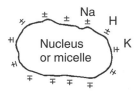

Fig. 4.2 Schematic representation of a colloidal particle

cation exchange, forms the basis of the stabilization of soils with certain chemicals (see Chapter 6).

The manner in which cations are exchanged is most easily explained by considering colloids that are in a solution. Due to heat movement and Brownian motion, the adsorbed ions continually move back and forth (within a limited range) from the surfaces of the particles. If electrolytes are added to the solution, cations are set in random motion because of the Brownian effect and some of them slip between the negative wall of the nucleus and the adsorbed/oscillating ions. These electrolytic cations then become preferentially adsorbed and some previously-oscillating surface ions are released and remain in the solution as exchanged ions. Obviously, the more loosely the surface ions are held the greater is the average distance of oscillation and, hence, the greater is the likelihood of ion adsorption and/or replacement.

Overall, the efficiency with which ions can replace each other in a clay soil is dependent upon the following factors: (1) relative concentration or number of ions; (2) number of charges on the ions; (3) speed of movement or activity of the different ions; and (4) type of clay mineral present. The ease with which cations can be exchanged and adsorbed is expressed in terms of the *cation exchange capacity* (cec) of the soil; this is the number of milli-equivalents (meq) of ions that 100 g of soil can adsorb. By definition, 1 meq is 1 mg of hydrogen or the amount of any other ion that will combine with or displace it; thus, if the cec of a soil is 1 meq it means that every 100 g of dry soil is capable of adsorbing or holding 1 mg of hydrogen or its equivalent.

As noted previously, the affinity of a soil for water is very much influenced by the fineness of its clay particles and upon the clay mineral(s) present. In this respect it should be appreciated that clay particles are essentially composed of minute flakes and in these flakes, as with all crystalline substances, the atoms are arranged in a series of units to form *clay minerals*. The atomic structure of most clay minerals consists of two fundamental building blocks composed of tetrahedrons of silica and octahedrons of alumina. Each of the main clay mineral groups was formed from the bonding together of two or more of these molecular sheets. The clay minerals that are of greatest interest in relation to roadworks are kaolinite, montmorillonite and illite.

The kaolinite structural unit is composed of an aluminium octahedral layer with a parallel superimposed silica tetrahedral layer intergrown in such a way that the tips of the silica layer and one of the layers of the alumina unit form a common sheet. The kaolinite mineral is composed of a stacking of these units (Fig.4.3(a)); its structure can be considered akin to a book in which each leaf is 0.7 nm thick.

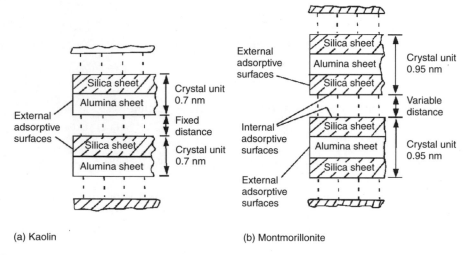

Fig. 4.3 Schematic diagrams of the structure of typical kaolin and mont-morillonite crystals

Successive sheets are held together by hydrogen bonds which allow the mineral to cleave into very thin platelets.

Kaolinitic clays are very stable. The hydrogen bonds between the elemental sheets are sufficiently strong to prevent water molecules and other ions from penetrating; hence, the lattice is considered to be non-expanding. The kaolinite platelets carry negative electrical charges on their surfaces which attract thick layers of adsorbed water; however, as the lattice is non-expanding the effective surface area to which water molecules can be attracted is limited to the outer faces and, consequently, the plasticity, cohesion, and shrinkage and swelling properties of kaolinitic clays are very low when compared with other silicate clays.

Figure 4.3(b) shows that a montmorillonite crystal unit comprises two tetrahedral sheets separated by an octahedral sheet (with the tips of each tetrahedral sheet and a hydroxyl layer of the octahedral sheet intergrown to form a single layer). The minimum thickness of each crystal unit is about 0.95 nm and the dimensions in the other two directions are indefinite. These sheets are stacked one above the other, like leaves in a book. There is very little bonding between successive crystal units and, consequently, water molecules and cations can readily enter between the sheets. In the presence of an abundance of water the clay mineral can be split into individual unit layers.

The ease with which water can enter between the crystal units makes mont-morillonitic clays very difficult to work with. The very large areas of charged sur-face that are exposed mean that hydrated ions/water are easily attracted to them. Note that each thin platelet of montmorillonite has the power to attract a layer of adsorbed water of up to 20 nm thick to each flat surface so that, assuming zero pressure between the surfaces, the platelets can be separated by 40 nm and still be 'joined'. This capability gives to these clays their high plasticity and cohesion, marked shrinkage on drying, and a ready dispersion of their fine flaky particles, e.g. typical

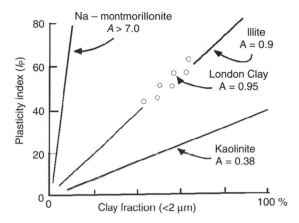

Fig. 4.4 Relationships between the plasticity index (I_P) and the clay fraction (%) for kaolinitic, illitic and montmorillonitic soil samples[5]. (Note: Activity (A) = I_P ÷ % clay.)

plastic index (I_P)-values for Na- and Ca-montmorillonitic clays (cec=80–100 meq/100 g) are 603 and 114, respectively, vis-à-vis 26 and 37 for Na- and Ca-kaolinitic clays (cec=5–15 me q/100 g) respectively[4].

The thickness of each illite crystal unit is 1 nm. Like montmorillonite, illite has a 2:1 lattice structure. It differs from montmorillonite, however, in that there is always a substantial (ca 20 per cent) replacement of silicons by aluminums in the tetrahedral layers. The valencies vacated by this substitution are satisfied by positively-charged potassium ions which lie between the structural units and tie them together. The strength of this potassium bond is intermediate between the hydrogen bond of kaolinite and the water linkage of montmorillonite and the net result is that illites have properties that are intermediate between those of the other two clay minerals.

The predominant clay mineral present in a soil can be easily and inexpensively indicated (see Fig. 4.4) from its *activity* (A). The activity of a soil is the ratio of its plasticity index (I_P) to the percentage by mass of soil particles within it that are less than 2 micron. Inactive (good) and active (difficult) soils have A-values of less than 0.75 and more than 1.25, respectively.

4.4 Soil water

Water from precipitation, which does not evaporate or flow away in the form of surface run-off, penetrates the ground and percolates downward under the action of gravity until a depth is reached below which the soil pores are completed saturated. The upper surface of this saturated zone, which is most easily determined in coarser-grained soils, is variously termed the *water table*, the *groundwater surface*, or the *phreatic surface*. The zone between the ground and the groundwater surfaces is called the *zone of aeration*: its thickness can range from >100 m (exceptionally, in very dry climates) to <1 m (commonly in the UK).

The more coarsely grained the soil in the zone of aeration the more quickly the water infiltrates down. Whilst fine-grained soils have considerable porosity, gravitational water moves more slowly through them due to the resistance to flow (associated with the narrow percolating passages and high sidewall frictions) encountered by the water as it attempts to infiltrate downward. Sometimes the zone of aeration contains a coarse-grained soil which lies on top of an inclined layer of clayey soil, e.g. in glacial till, so that the water is able to infiltrate more quickly laterally than vertically; then the percolating water may appear as seepage water in, for example, road cuttings following precipitation and special drainage may have to be provided to cope with this.

4.4.1 Groundwater

Below the groundwater surface the soil pores are completely saturated. The water table is not a horizontal plane, but is as constantly changing as the topography above it. Thus, the groundwater is rarely at rest as elevation differences in its surface cause it to flow laterally until it emerges as ground seepage water or to feed lakes, streams or swamps. Consequently, in dry weather the water table can be considerably lowered, whereas in wet weather it will rise as precipitation adds to the moisture in the soil and lakes and streams are replenished. During soil survey work, therefore, it is important to note the date when any measurement is taken to the water table, so that a seasonal correcion can be applied later, if necessary.

By definition, the *porewater pressure* (u_w) at any depth in the saturated soil is the excess pressure in the porewater above that at the groundwater surface, where it is equal to the atmospheric pressure. Thus, if $u_w = 0$ kPa at the groundwater surface in, say, a gravelly soil, and it increases linearly with depth, then at a depth z m (assuming ρ_w = density of water, Mg/m^3)

$$u_w = + \rho_w z \qquad (4.1)$$

4.4.2 Held water

If gravity was the only force acting on the water in the aeration zone the soil pores above the water table would be dry at all times except when precipitation water is infiltrating downward. However, if a cross-section is taken through the zone, it will be found that there is a layer above the groundwater surface within which the pores are wholly or partially filled with moisture. This layer is termed the *capillary fringe*. It is typically 2 to 5 cm thick in coarse soils, 12–35 cm in medium sands, 35–70 cm in fine sands, 70–150 cm in silts and 200–400 cm or more (after long periods of time) in clays.

Water that is continuous through the pores within the capillary fringe is held by surface tension and adsorption forces, which impart to it a negative pressure or suction with respect to atmospheric pressure[6].

With the coarser types of fine-grained soils, e.g. fine sand and silts, the greater proportion of the held water can be attributed to capillarity in the pores between the particles, if it is assumed that the pores in such soils form an interrelated mass of irregular 'capillary tubes'. If the lower ends of a number of different-sized capillary tubes are immersed in water it will be found that (i) the water level in each tube will rise above the free water level until the mass of the water column is just supported by the surface tension force in operation at the interface of the water meniscus at the top of the column and the inside surface of the tube, and (ii) the height of this capillary rise will increase as the radius of the tube decreases. Similarly, the height of the capillary fringe will increase as the sizes of the pores in such soils are decreased, i.e. the more finely grained the soil the greater the potential for water to rise through capillary action. The porewater pressure (u_w) at any height z above the groundwater surface within the fringe is less than atmospheric pressure, and equals $-\rho_w z$.

The mass of water supported by the surface tension tends to pull the particles together and compress the soil. However, as the fine sand and coarse silt particles are already in physical contact with each other little volume change occurs as a result of this capillary rise.

Below coarse silt size, adsorbed moisture becomes more important in terms of its contribution to the amount and distribution of water within the capillary fringe. It is very important when the soil is an active (e.g. montmorillonitic) clay, as changes in moisture content within the fringe will affect soil density, volume, compressibility and stability.

The moisture content and its distribution with depth within a clayey soil at any given time are in equilibrium and reflect the environmental conditions, including seasonal precipitation and evaporation. If the ground is covered with a pavement so that it is protected from precipitation and evaporation, a new moisture equilibrium is eventually reached with respect to the position of the groundwater surface. In the process, pressure from the overburden and gravity seek to expel the moisture whilst the suction of the soil seeks to retain it, i.e.

$$s = \alpha\sigma - u_w \qquad (4.2)$$

where s = soil suction, u_w = porewater pressure, σ = vertical pressure from overburden, and α = compressibility factor, i.e. the proportion of overburden pressure acting on the porewater. Whilst in coarse-grained soils the overburden pressure is carried by the intergranular contacts and $\sigma = 0$, it is entirely carried by the porewater in active clay soils and $\sigma = 1$. With intermediate soils the extent to which the overburden pressure is carried by the porewater depends upon the plasticity characteristics of the soil in question, i.e.

$$\alpha = 0.027 I_P - 0.12 \qquad (4.3)$$

where I_P = plasticity index of the soil (between 5 and 40).

A test is available[7] to measure the soil suction, which is the pressure that has to be applied to the porewater to overcome the soil's capacity to retain the water. For a given soil there is an increase in soil suction with decreasing moisture content, and this relationship is continuous over its entire moisture range, e.g. the suction value can range from up to $10^7 \, \text{N/m}^2$ for oven-dry soils to zero for saturated soils that will take up no more water. Because of this large variation the soil suction is usually expressed as the common logarithm of the length in centimetres of an equivalent suspended water column; this is termed the pF-value of the soil moisture. Thus, for example, a soil with a pF of 1 equals $10^1 \, \text{cm}$ water ($97.9 \times 10^1 \, \text{N/m}^2$), pF3 = $10^3 \, \text{cm}$ ($97.9 \times 10^3 \, \text{N/m}^2$), pF5 = $10^5 \, \text{cm}$ ($97.9 \times 10^5 \, \text{N/m}^2$), and pF7 = $10^7 \, \text{cm}$

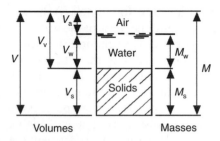

Fig 4.5 Diagrammatic illustration of the three phases of a soil

$(97.9 \times 10^7 \text{ N/m}^2)$. Note that, because of the logarithmic nature of the scale, $pF=0$ does not exactly relate to zero suction. Oven-dryness corresponds closely to $pF=7$.

The density of wet soil is close to twice that of water, so that if u_w and s are also expressed in centimetres of water, Equation 4.2 can be used with an appropriate soil suction curve to deduce the equilibrium moisture content above the groundwater surface (see reference 7 for an excellent treatise on this).

4.5 Soil phase relationships

As noted previously, soil is an assemblage of mineral particles interspersed with pore spaces that may/may not be filled with water. It is convenient to visualize the different soil phases, and the relationships between them, by representing, in graphical form, a soil sample in which the solid, liquid and gaseous phases are segregated. Figure 4.5 shows such a sample that is placed in an equivolume cylinder of unit cross-section; this arrangement makes it possible to consider the component volumes as represented by their heights, so that it is a simple matter to develop some important relationships.

4.5.1 Moisture content

The moisture content of a soil is its mass of water expressed as a percentage of the mass of dry solids in the soil. Thus

$$w = (M_w/M_s) \times 100 \qquad (4.4)$$

where $w =$ moisture content (%), $M_w =$ mass of water, and $M_s =$ mass of solids.

4.5.2 Voids ratio

The voids ratio is the ratio of the volume of voids to the volume of solids in the soil. It pays no attention to the proportions of water, air or other gases which may constitute the pore spaces. Thus

$$e = V_v/V_s \qquad (4.5)$$

where $e =$ voids ratio, $V_v =$ volume of voids, and $V_s =$ volume of solids. Also

$$V_v = e/(1+e) \qquad (4.6)$$

and

$$V_s = 1/(1+e) \tag{4.7}$$

In a saturated soil, the voids ratio is directly proportional to the moisture content. In this case the volume of voids, V_v, is equal to volume of water, V_w. Then

$$e_s = V_w/V_s = (M_w/M_s)(\rho_s/\rho_w) = w\,(\rho_s/\rho_w) \tag{4.8}$$

where e_s = voids ratio when the soil is saturated, ρ_s = density of solid particles, and ρ_w = density of water.

4.5.3 Porosity and percentage air and water voids

The *porosity* of a soil is the ratio of the volume of the voids to the total volume of the soil, expressed as a percentage value. As with the voids ratio, porosity pays no regard to the constituent volumes of the pore spaces. Thus

$$n = (V_v/V) \times 100 \tag{4.9}$$

where n = porosity (%) and V = total volume of the soil mass. Porosity can also be expressed by the formula

$$n = [e/(1+e)] \times 100 \tag{4.10}$$

The percentages of the total volume of the soil that are occupied by the air in the voids and by the water in the voids are referred to as the *percentage air voids* and *percentage water voids*, respectively. Thus

$$V_a = (V_v - V_w) \tag{4.11}$$

$$n_a = 100(V_a/V) \tag{4.12}$$

and

$$n_w = 100(V_w/V) \tag{4.13}$$

where n_a = air voids (%), n_w = water voids (%), V_a = volume of air voids, V_w = volume of water voids, and V = total volume of the soil mass. Also, the sum of the percentages of air and water voids equals the porosity, i.e.

$$n = n_a + n_w \tag{4.14}$$

4.5.4 Degree of saturation

The extent to which the voids in a soil are filled with water is termed the degree of saturation, S_r (%). It is the ratio of the volume of water to the volume of voids, expressed as a percentage. Thus

$$S_r = 100\,(V_w/V_v) \tag{4.15}$$

4.5.5 Bulk density and dry density

The mass of the wet solid particles plus the water contained in the pore spaces of a soil per unit volume is called the *bulk (or wet) density*. Thus

$$\rho = M/V = (M_s + M_w)/(V_s + V_w + V_a) \tag{4.16}$$

where ρ = bulk (or wet) density of the soil (Mg/m^3).

The mass of the dry solids per unit volume of soil is the *dry density*. Thus

$$\rho_d = M_s/V = M_s/(V_s + V_w + V_a) = \rho/(1 + w) \tag{4.17}$$

where ρ_d = dry density of the soil (Mg/m^3).

4.5.6 Relationships between dry density, bulk density, moisture content, and percentage air voids

The relationship between the bulk, ρ, and dry, ρ_d, densities of a soil follow from their definitions. Thus

$$\rho/\rho_d = (M/V)(V/M_s) = (M_s + M_w)/M_s = 1 + w/100 \tag{4.18}$$

and

$$\rho_d = 100\rho/(100 + w) \tag{4.19}$$

The relationship between dry density, ρ_d, moisture content, w, and the percentage air voids, n_a, is deduced as follows:

$$\rho_d = M_s/(V_s + V_w + V_a) \tag{4.20}$$

and

$$1/\rho_d = (V_s/M_s) + (V_w/M_s) + (V_a/M_s)$$
$$= (1/\rho_s) + (w/100\rho_w) + (V_a/V\rho_d) \tag{4.21}$$

hence

$$(1/\rho_d)[100 - (100V_a/V)] = (100/\rho_s) + (w/\rho_w) \tag{4.22}$$

$$(1/\rho_d)(100 - n_a) = (100/\rho_s) + (w/\rho_w) \tag{4.23}$$

and

$$\rho_d = \rho_w[1 - (n_a/100)]/[(\rho_w/\rho_s) + (w/100)]$$
$$= \rho_w[1 - (n_a/100)]/[(1/r_s) + (w/100)] \tag{4.24}$$

where r_s = specific gravity of the soil particles and ρ_w = density of water.

Equation 4.24 shows that if any two of dry density, moisture content, and percentage air voids are known for a soil of a particular specific gravity, then the third can be easily established.

4.5.7 Specific gravity

Also known as *relative density*, the specific gravity is the ratio of the mass of the soil particles to the mass of the same (absolute) volume of water. Thus

$$r_s = M_s/V_s \, \rho_w = \rho_s/\rho_w \tag{4.25}$$

where r_s = specific gravity, ρ_w = density of water (Mg/m^3) and ρ_s = particle density (Mg/m^3). Note that particle density, whose units are Mg/m^3, is the average mass per unit volume of the solid particles (where the volume includes any sealed voids contained within the solid particles); it is numerically equal to the dimensionless specific gravity.

4.6 Frost action in soils

Frost action can be defined as any action resulting from freezing and thawing which alters the moisture content, porosity or structure of a soil, or alters its capacity to support loads. The two aspects of frost action that are of major interest to the road engineer are frost heaving and thawing.

4.6.1 Frost heaving

By frost heaving is normally meant the process whereby a portion of a frost-susceptible soil, and the road pavement above it, are raised as a result of ice formation in the soil. If the heaving occurs non-uniformly, pavement cracking can occur as a result of differential heaving pressures. Differential heaving is likely at sites where there are non-uniform conditions of soil (e.g. where subgrades change from clean sands to silty materials) or of water availability (e.g. differences in subgrade moisture between the edges and mid-portion of a road pavement). In Britain, significant frost heaving has been associated with severe winters having 40 or more days of continuous frost[9].

Whilst the volume increase of soil water upon freezing (ca 9 per cent) obviously contributes to frost heaving, its major cause by far is the formation of ice lenses. For ice lenses to develop three conditions must be met: (i) sufficiently cold temperatures to freeze some of the soil water, (ii) a supply of water available to the freezing zone, and (iii) a frost-susceptible soil. Ice lensing does not occur when any one of these three conditions is missing. Figure 4.6 illustrates generally

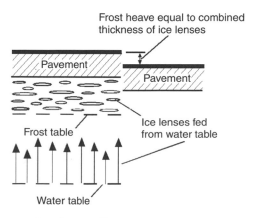

Fig. 4.6 Ice lens formation in a soil

the manner in which layers of segregated ice are formed in a soil. The ice lenses, which normally grow parallel to the ground surface and perpendicular to the direction of the heat flow, can vary from being of hairline thickness to many centimetres thick.

The frost heave mechanism is most simply explained as follows. When the *freezing isotherm* penetrates into a fine-grained soil some of the water in the soil pores becomes frozen. As a consequence, the unfrozen moisture content decreases and the suction associated with its surface tension and adsorption forces rises rapidly and causes moisture from the water source in the unfrozen soil to move to the freezing zone and form ice lenses; this process continues until a suction value is reached which inhibits further freezing at that temperature and a new state of equilibrium is established. As the temperature drops and the freezing isotherm penetrates further into the soil the equilibrium is disturbed and the suction process initiates another cycle of ice lens formation at a different depth in the soil; and so on. The accumulated thickness of the ice lenses then accounts for the frost heave.

If the freezing penetration of the soil occurs rapidly, the water sucked up is frozen into many thin lenses; if it occurs slowly the lenses are thick. From a road maintenance viewpoint a rapid early freeze is preferable to a slow one as it results in the critical upper layers of the soil having relatively less ice so that, when the thaw begins, there is less moisture accumulation in the soil at a given instance and this benefits pavement stability at that vulnerable time.

Trouble from frost heave can be expected when the water table in a frost-susceptible soil is close to the ground surface and just below the freezing zone. The rough rule-of-thumb for road-making in Britain is that a potentially troublesome *water source* for ice lens formation is present when the highest water table level at any time of the year is within, say, 1–1.5 m of the formation or of the bottom of any frost-susceptible material in the pavement. When the depth to the water source is more than 3 m throughout the year, substantial ice segregation is usually unlikely; however, this does not mean that no ice lenses will be formed but, rather, that those which do occur are likely to be within tolerable limits in normal pavement applications.

Investigators of *frost susceptible soils* have long found that certain soils are more vulnerable to frost action than others, e.g. Casagrande[8] stated in 1932 that, under natural freezing conditions and with a sufficient water supply, considerable ice lensing could be expected in any non-uniform soil with more than 3 per cent of its particles smaller than 0.02 mm and in very uniform soils with more than 10 per cent smaller than 0.02 mm.

British experience[9] is that cohesive soils can be regarded as non-frost-susceptible when the plasticity index, I_P, is greater than 15 per cent for a well-drained soil, or >20 per cent when the soil is poorly drained (i.e. when the groundwater surface is within about 0.6 m of the bottom of the pavement resting upon it). The pore sizes in these cohesive soils are generally too small to allow significant migration of water to the freezing front during the relatively short periods of freezing normally associated with British winters. The liability of these soils to frost heave also decreases with decreasing state of compaction.

Non-cohesive soils (except limestone gravels) can be regarded as non-frost-susceptible if the mass of material passing the 75 micron sieve is less than or equal to 10 per cent. Limestone gravels with an average particle saturation moisture content in excess of 2 per cent are normally considered as potentially susceptible to frost action.

Soils falling between the above two groups tend to be considered frost susceptible. All chalk soils (and crushed chalk) are susceptible to frost heave and their use should be avoided in road construction in locales subject to freezing temperatures. Degree of compaction has little influence upon the frost susceptibility of these chalk materials.

In the Scottish Highlands the freezing isotherm regularly penetrates about 450 mm into a soil; in the south west and western areas of England penetrations are normally less than 200 mm. Table 4.3 (column 7) summarizes experience with respect to the vulnerability of British soils to frost heave.

A test procedure[10] has been developed in Britain for assessing the frost susceptibility of soils (see Subsection 4.9.9).

4.6.2 Thawing impacts on a road pavement

When thawing begins, the ice melts primarily from the top down. If the thawing occurs at a faster rate than the melt water can escape into underdrains or the more pervious layers of the pavement system, or be reabsorbed into adjacent drier areas, then heavy traffic loads will result in the generation of excess pore-water pressures which cause a decrease in the load-carrying capacity of the pavement and can seriously shorten its service life.

If thawing does not occur at the same rate over all parts of the pavement, non-uniform subsidence of the heaved surface will result. Differential thaw is most commonly associated with: (a) different thermal properties of adjacent pavement sections, caused by non-uniformity of subsoil strata and soil conditions, (b) non-uniform exposure to the sun's rays and differing angles of incidence, (c) shading of portions of the pavement due to deep cuts, trees, overpasses, or buildings, (d) proximity to surface and surface drainage, and/or (e) differing pavement colours.

Another undesirable consequence of thawing is the subsidence of coarse open-graded subbase materials into thaw-weakened silt and clay subgrade soils, as the latter flow into the large pore spaces in the coarse material. There have been many instances where a pavement has become frost-susceptible as a result of the upward inpregnation of a silt subgrade into a previously non-frost-susceptible subbase during the thawing process.

If the melt water content is sufficiently high and the traffic conditions are sufficiently heavy to cause a reworking of the subsoil, a free-flowing mud may be formed which is forced out at the edges of, or breaks in, the pavement. This action, termed *frost boil*, is most commonly associated with rigid pavements.

Initial thaw damage to flexible pavements is usually reflected in the form of a close network of cracks accompanied by distortion of the carriageway.

Table 4.3 Field characteristics of soils and other materials used in earthworks[11]

Material	Major divisions	Subgroups	BSCS group symbol	Casagrande group symbol	Drainage characteristics	Potential frost action	Shrinkage or swelling properties	Value as a road foundation when not subject to frost action	Bulk density before excavation (mg/m³)		Coefficient of bulking (%)
									Dry or moist	Submerged	
(1)	(2)	(3)	(4)	(5)	(6)	(7)	(8)	(9)	(10)	(11)	(12)
Coarse soils and other materials	Boulders and cobbles	Boulder gravels	–	–	Good	None to very slight	Almost none	Good to excellent	–	–	–
	Other materials	Hard: Hard broken rock, hardcore, etc	–	–	Excellent	None to slight	Almost none	Very good to excellent	–	–	20–60
		Soft: Chalk, soft rocks, rubble	–	–	Fair to practically impervious	Medium to high	Almost none to slight	Good to excellent	1.10–2.00	0.65–1.25	40
	Gravels and gravelly soils	Well-graded gravel and gravel–sand mixtures, little or no fines	GW	GW	Excellent	None to very slight	Almost none	Excellent	1.90–2.10	1.15–1.30	10–20
		Well-graded gravel–sand mixtures with excellent clay binder	GWC	GC	Practically impervious	Medium	Very slight	Excellent	2.00–2.25	1.00–1.35	
		Uniform gravel with little or no fines	GPu	GU	Excellent	None	Almost none	Good	1.60–1.80	1.00–1.11	
		Gap-graded gravel and gravel–sand mixtures, little or no fines	GPg	GP	Excellent	None to very slight	Almost none	Good to excellent	1.60–2.00	0.90–1.25	
		Gravel with fines, silty gravel, clayey gravel, poorly graded gravel–sand–clay mixtures	GM/GC	GF	Fair to practically impervious	Slight to medium	Almost none to slight	Good to excellent	1.80–2.10	1.10–1.30	

		Description									
Sands and sandy soils		Well-graded sands and gravelly sands, little or no fines	SW	SW	Excellent	None to very slight	Almost none	Excellent to good	1.80–2.10	1.05–1.25	5–15
		Well-graded sand with excellent clay binder	SWC	SC	Practically impervious	Medium	Very slight	Excellent to good	1.90–2.10	1.15–1.30	
		Uniform sands with little or no fines	SPu	SU	Excellent	None to very slight	Almost none	Fair	1.65–1.85	1.00–1.15	
		Gap-graded sands, little or no fines	SPg	SP	Excellent	None to very slight	Almost none	Fair to good	1.45–1.70	0.90–1.00	
		Sands with fines, silty sands, clayey sands, poorly graded sand–clay mixtures	SM/SC	SF	Fair to practically impervious	Slight to high	Almost none to medium	Fair to good	1.70–1.90	1.00–1.15	
Fine Soils	Soils having low compressibility	Silts (inorganic) and very fine sands, rock floor, silty or clayey fine sands with low plasticity	ML/SCL MS/CLS	ML	Fair to poor	Medium to very high	Slight to medium	Fair to poor	1.70–1.90	1.00–1.15	20–40
		Clay of low plasticity (inorganic)	CL	CL	Practically impervious	Medium to high	Medium	Fair to poor	1.60–1.30	1.00–1.11	
		Organic silts of low plasticity	MLO	OL	Poor	Medium to high	Medium to high	Poor	1.45–1.70	0.90–1.00	
	Soils having medium compressibility	Silt and sandy clays (inorganic) of intermediate plasticity	CIS	MI	Fair to poor	Medium	Medium to high	Fair to poor	1.55–1.30	0.95–1.11	–
		Clays (inorganic) of intermediate plasticity	CI	CI	Fair to practically impervious	Slight	High	Fair to poor	1.60–2.00	1.00–1.10	
		Organic clays of intermediate plasticity	CIO	OI	Fair to practically impervious	Slight	High	Poor	1.50	0.50	

(Continued on p. 88)

Table 4.3 (continued)

Material (1)	Major divisions (2)	Subgroups (3)	BSCS group symbol (4)	Casa-grande group symbol (5)	Drainage character-istics (6)	Potential frost action (7)	Shrinkage or swelling properties (8)	Value as a road foundation when not subject to frost action (9)	Bulk density before excavation (mg/m³) Dry or moist (10)	Submerged (11)	Coefficient of bulking (%) (12)
	Soils having high compressibility	Micaceous or diatomaceous fine sandy and silty soils, elastic silts	–	MH	Poor	Medium to high	High	Poor	1.75	1.00	–
		Clays (inorganic) of high plasticity, fat clays	CH	CH	Practically impervious	Very slight	High	Poor to very poor	1.70	0.70	
		Organic clays of high plasticity	CHO	OH	Practically impervious	Very slight	High	Very poor	1.50	0.50	
Fibrous organic soils with very high compressibility		Peat and other highly organic swamp soils	Pt	Pt	Fair to poor	Slight	Very high	Extremely poor	1.40	0.40	–

4.7 Identification and description of soils in the field

Soil (and rock) descriptions are important aspects of a ground investigation. A full description provides detailed information about a soil as it occurs in situ and, consequently, very few soils have identical descriptions. Soils are commonly identified and described in the field on the basis of data recovered from boreholes and excavations, and undisturbed materials seen in excavations and cuttings.

A soil's description is based on the particle-size grading of the coarser particles and on the plasticity of the finer particles; this is because of the roles that these characteristics play in determining engineering properties of the soil. Section 6 of reference 12 describes, in detail, the process for describing undisturbed soils that is recommended for use in the UK. Figure 4.7 summarizes, in flow chart form, the descriptive process used whilst Table 4.4 provides details of the criteria involved.

In relation to Table 4.4 it might be noted that the main characteristics of a soil are normally described in the field in the following order:

1. *Mass characteristics*: Density/compactness/field strength (Col. 2); discontinuities (Col. 3); and bedding (Col. 4).
2. *Material characteristics*: Colour (Col. 5); particle shape and composition (Col. 7) and size (Col. 8); principal soil type (Col. 9);
3. *Geological formation, age and type of deposit*: Stratum name (Col. 12).
4. *Classification*: field (rapid) assessment of soil group symbol (optional).

The term 'mass characteristics' as used above refers to descriptions of those soil characteristics that depend on structure and, therefore, can only be observed in the field or in some undisturbed samples. 'Material characteristics' refers to those characteristics that can be described from a visual and manual examination of either disturbed or undisturbed samples.

With this system, a fine-grained soil might be described as a

firm closely-fissured yellowish-brown CLAY (LONDON CLAY FORMATION). Loose brown sub-angular fine and medium flint GRAVEL (TERRACE GRAVELS).

Note that, in the above example, additional minor information is given at the end of the main description after a full stop, so as to keep the standard main description concise.

As another example, soil materials in interstratified beds might be described as

thinly interbedded dense yellow fine SAND and soft grey CLAY (ALLUVIUM).

Note: As with soils, rock identifications and descriptions are also obtained in the field on the basis of samples recovered from excavations and boreholes, and from direct examination of in-situ materials in natural outcrops, cuttings, quarries, etc. In this case the rocks are described in the following sequence: (a) material characteristics such as (i) strength, (ii) structure, (iii) colour, (iv) texture, (v) grain size, and (vi) rock name, in capitals, e.g. GRANITE; (b) general information, such as (i) additional information and minor constituents, and (ii) geological formation; and (c) mass characteristics such as (i) state of weathering, (ii) discontinuities, and (iii) fracture state. For space reasons rock identifications and descriptions are not further discussed here, and the reader is referred to reference 12 for discussion regarding the detailed criteria employed in these processes.

Table 4.4 Criteria involved in the identification and description of in-situ soils[12]

(1) Soil group	(2) Density/compactness/strength		(3) Discontinuities		(4) Bedding		(5) Colour	(6) Composite soil types (mixtures of basic soil types)		(7) Particle shape	(8) Particle size	(9) PRINCIPAL SOIL TYPE
		Field test	Scale of spacing of discontinuities		Scale of bedding thickness			For mixtures involving very coarse soils, see 41.4.4.3 in ref 12		Angular	— 200	BOULDERS
Very coarse soils — Loose		By inspection of voids and particle packing	Term	Mean spacing (mm)	Term	Mean thickness(mm)	Red Orange Yellow	Term	Approx % secondary[c]	Sub-angular	— 60	COBBLES
Very coarse soils — Dense												
Coarse soils (over about 65% sand and gravel sizes) — Borehole with standard penetration test N-value (blows / 300 mm pen)			Very widely	Over 2000	Very thickly bedded	Over 2000	Brown Green	Slightly (sandy[d])	< 5	Sub-rounded	Coarse — 20	GRAVEL
Very loose	0 - 4		Widely	2000 to 600	Thickly bedded	2000 to 600	Blue White			Rounded	Medium — 6	
Loose	4 - 10		Medium	600 to 200	Medium bedded	600 to 200	Cream	(sandy[d])	5 to 20[b]	Flat	Fine	
Medium dense	10 - 30		Closely	200 to 60	Thinly bedded	200 to 60	Grey Black			Tabular	— 2	
Dense	30 - 50		Very closely	60 to 20	Very thinly bedded	60 to 20	etc.	Very (sandy[d])	>20[b]	Elongated	Coarse — 0.6	SAND
Very dense	>50		Extremely closely	Under 20	Thickly laminated	20 to 6				Minor constituent type	Medium — 0.2	
Slightly cemented	Visual examination: pick removes soil in lumps which can be abraded		Fissured	Breaks into blocks along unpolished discontinuities	Thinly laminated	Under 6		SAND AND GRAVEL	about 50[b]	Calcareous, shelly, glauconitic, micaceous etc. using terms such as	Fine — 0.06	
Fine soils (over about 35% silt and clay sizes) — Un-compact	Easily moulded or crushed in the fingers		Sheared	Breaks into blocks along polished discontinuities	Inter-bedded	Alternating layers of different types prequalified by thickness term if in equal proportions. Otherwise thickness of, and spacing between, subordinate layers defined	Light Dark Mottled	Term	Approx % secondary[c]	Slightly calcareous	Coarse — 0.02	SILT
Compact	Can be moulded or crushed by strong pressure in the fingers							Slightly (sandy[e])	<35	Calcareous Very calcareous	Medium — 0.006 Fine	
Very soft 0-20	Finger easily pushed in up to 25 mm				Inter-laminated					% defined on a site or material-specific basis or subjective	— 0.002	CLAY/ SILT
Soft 20-40	Finger pushed in up to 10 mm		Spacing terms also used for distance between partings, isolation beds or laminae, dessication cracks, rootlets etc,					(sandy[e])	35 to 65[a]			
Firm 40-75	Thumb makes impression easily											CLAY
Stiff 75-150	Can be indented slightly by thumb											
Very stiff 150-300	Can be indented by thumb nail							Very (sandy[f])	>65[a]			
Hard (or very weak mudstone) Cu > 300 kPa	Can be scratched by thumbnail (see 41.2.2 in ref 12)											

Organic soils						Transported mixtures	Colour	
Firm	Fibres already compressed together	Fibrous	Plant remains recognizable and retains some strength	Slightly organic clay or silt / Slightly organic sand	Grey as mineral	Contains finely divided or discrete particles of organic matter, often with distinctive smell, may oxidise rapidly. Describe as for inorganic soils using terminology above.
				Organic clay or silt / Organic sand	Dark grey / Dark grey	
Spongy	Very compressible and open structure	Pseudo-fibrous	Plant remains recognizable, strength lost	Very organic clay or silt / Very organic sand	Black / Black	
				Accumulated in situ		Predominately plant remains, usually dark brown or black in colour, distinctive smell, low bulk density. Can contain disseminated or discrete mineral soils.
Plastic	Can be moulded in hand and smears fingers	Amor-phous	Recognizable plant remains absent	Peat		

(9) PRINCIPAL SOIL TYPE	(10) Visual identification	(11) Minor constituents	(12) Stratum name	(13) Example descriptions
BOULDERS	Only seen complete in pits or exposures			
COBBLES	Often difficult to recover whole from boreholes	Shell fragments, pockets of peat, gypsum crystals, flint gravel, fragments of brick, rootlets, plastic bags etc.		Loose brown very sandy sub-angular fine to coarse flint GRAVEL with small pockets (up to 30 mm) of clay. (TERRACE GRAVELS)
GRAVEL	Easily visible to naked eye; particle shape can be described, grading can be described.	using terms such as; with rare with occasional with abundant/frequent/numerous	RECENT DEPOSITS, ALLUVIUM, WEATHERED BRACKLESHAM CLAY,	Medium dense light brown gravelly clayey fine SAND. Gravel is fine (GLACIAL DEPOSITS)
SAND	Visible to naked eye; no cohesion when dry; grading can be described.	% defined on a site or material specific basis or subjective	LIAS CLAY, EMBANKMENT FILL,	Stiff very closely sheared orange mottled brown slightly gravelly CLAY. Gravel is fine and medium of rounded quartzite. (REWORKED WEATHERED LONDON CLAY)
SILT	Only coarse silt visible with hand lens; exhibits little plasticity and marked dilatancy; slightly granular or silky to the touch; disintegrates in water; lumps dry quickly; possesses cohesion but can be powdered easily between fingers		TOPSOIL, MADE GROUND OR GLACIAL DEPOSITS ? etc.	
CLAY/ SILT	Intermediate in behaviour between clay and silt. Slightly dilatant			Firm thinly laminated grey CLAY with closely spaced thick laminae of sand. (ALLUVIUM)
CLAY	Dry lumps can be broken but not powdered between the fingers; they also disintegrate under water but more slowly than silt; smooth to the touch; exhibits plasticity but no dilatancy; sticks to the fingers and dries slowly; shrinks appreciably on drying usually showing cracks.			Plastic brown clayey amorphous PEAT. (RECENT DEPOSITS)

Notes: [a]Or described soil depending on mass behaviour; [b]Or described as fine soil depending on mass behaviour; [c]%coarse or fine soil type assessed excluding cobbles and boulders; [d]Gravelly or sandy and/or silty or clayey; [e]Gravelly and /or sandy; [f]Gravelly or sandy.

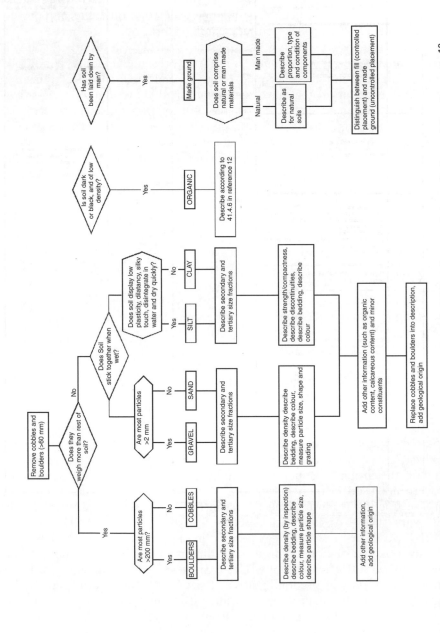

Fig. 4.7 Flow chart of the general process used in the identification and description of in-situ soils[12]

4.8 Soil classification

In contrast with soil identification and description, soil classification is concerned with placing a soil within a limited number of soil groups on the basis of the grading and plasticity characteristics determined from soil tests that are carried out on disturbed soil samples (usually in the laboratory). In this case, the characteristics measured are independent of the particular condition in which the soil occurs and pay no regard to the influence of the soil structure, including fabric, of the soil mass.

Grading and plasticity provide very useful guides as to how a disturbed soil will behave when used as a construction material in earthworks, under various conditions of moisture content. Thus, for example, Table 4.3 indicates some relationships (based on British experience) that have been established between a soil's classification (col. 4) and its drainage characteristics (col. 6), frost action potential (col. 7), shrinkage or swelling properties (col. 8), and value as a road foundation (col. 9).

Internationally, the soil engineering classification systems that are in widest use today are based on the Airfield Classification System[13] devised by Professor Arthur Casagrande during World War II. The British Soil Classification System (BSCS)[2], which is an outgrowth of the Casagrande system, is primarily used in the UK.

Under the BSCS, any soil can be placed in one of a number of main groups on the basis of the grading of its constituent particles, and the plasticity of the soil fraction that passes the 425 micron BS sieve; the grading and plasticity characteristics can be estimated in the field (rapid method) or, for detailed classification, tests must be carried out in the laboratory to enable the main groups to be divided into subgroups.

Classification is carried out on gravel, sand, silt and/or clay material that is nominally finer than 60 mm, i.e. passing the 63 mm sieve. Thus, if the soil sample contains any particles greater than 60 mm, they must first be picked out and their proportions by mass of the whole sample recorded as cobbles (60 to 200 mm) or boulders (>200 mm).

Table 4.5 shows that the grading and plasticity characteristics are divided into a number of clearly defined ranges, each of which is referred to by a descriptive name and a descriptive letter. Table 4.6 shows how the soil groups are formed from combinations of the ranges of characteristics, and gives both the names of the groups and the symbols that may be used with them. Note that the name of the soil group is always given when describing a soil, supplemented (if required) by the group symbol; the group or subgroup symbols are put in brackets if laboratory methods were not used in their derivation, e.g. (GC). The letter describing the dominant size fraction is always placed first in the group symbol: e.g. CS, sandy CLAY; SC, very clayey SAND; or S-C (spoken 'S dash C'), clayey SAND. The designation FINE SOIL or FINES, F, may be used in place of SILT, M, or CLAY, C, when it is not possible or not required to distinguish between them. Use of the alternative term M-SOIL, instead of SILT, is suggested as a means of avoiding confusion with materials of predominantly silt size, which only form part of the group. Any group may be qualified as 'organic', if organic matter is a significant constituent, by suffixing the letter O to the group symbol: e.g. CHO, organic CLAY of high plasticity, or CHSO, organic sandy CLAY of high plasticity. A plus sign is used to form a symbol for mixtures of soil with boulder-sized or cobble-sized

Table 4.5 Names and descriptive letters for grading and plasticity characteristics, used with the British Soil Classification System

Components	Terms	Descriptive name	Letter
Coarse	Main	GRAVEL	G
		SAND	S
	Qualifying	Well graded	W
		Poorly graded	P
		Uniform	Pu
		Gap graded	Pg
Fine	Main	FINE SOIL, FINES may be differentiated into M or C	F

		SILT (M-SOIL)
		    plots below A-line of plasticity
		    chart* (of restricted plastic range)
			M

		CLAY plots above A-line* (fully plastic)	C
	Qualifying	Of low plasticity	L
		Of intermediate plasticity	I
		Of high plasticity	H
		Of very high plasticity	V
		Of extremely high plasticity	E
		Of upper plasticity range** incorporating groups I, H, V and E	U
Organic	Main	PEAT	Pt
	Qualifying	Organic may be suffixed to any group	O

*See Fig. 4.9, ** This term is a useful guide when it is not possible or not required to designate the range of liquid limit more closely, e.g. during the rapid assessment of soils

particles: e.g. BOULDERS with CLAY of high plasticity, B+CH; or GRAVEL with COBBLES, G+Cb.

Coarse soils are distinguished by the fact that they have less than 35 per cent of material nominally finer than 0.06 mm, i.e. passing a 63 micron test sieve, and their classification groupings are primarily determined on the basis of their particle-size distributions. Plotting the distribution on a grading chart will assist in designating a soil as well or poorly graded and, if poorly graded, whether uniform or gap-graded; Fig. 4.8 shows typical examples of these materials. Figure 4.8 also shows particle-size distributions for a fine soil (London Clay) and a soil showing a very wide range of sizes (till).

Fine soils have more than 35 per cent by mass of silt- and clay-sized particles. If plasticity measurements are made on material passing a 425 micron sieve then, within the fine soils group, silts and clays may be differentiated according to where their liquid limit and plasticity index plot on the plasticity chart shown at Fig.4.9. Thus, CLAY, C, plots above the A-line, and is fully plastic in relation to its liquid limit. SILT (M-SOIL), M, which plots below the A-line, has a restricted plastic range in relation to its liquid limit, and relatively low cohesion; fine soils of this type include clean silt-sized materials and rock-flour, micaceous and diatomaceous soils, pumice and volcanic soils, and soils containing halloysite.

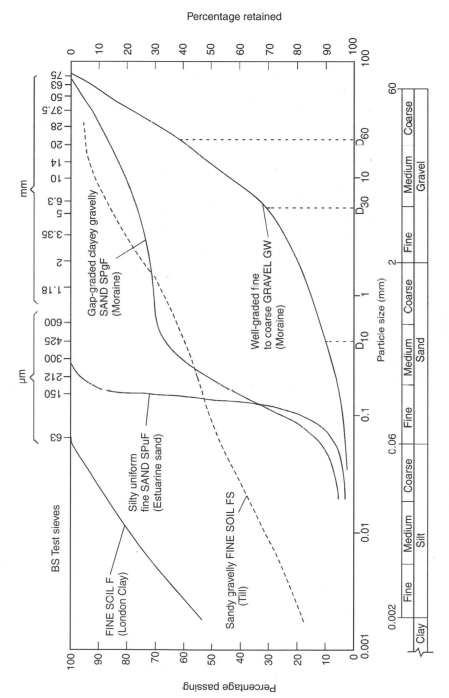

Fig. 4.8 BSCS grading chart for soils, with grading curves of selected soil types

Table 4.6 The British Soil Classification System for Engineering Purposes

Soil groups		Subgroups and laboratory identification				
GRAVEL and SAND may be qualified Sandy GRAVEL and Gravelly SAND, etc. where appropriate		Group symbol	Subgroup symbol		Fines (% less than 0.06 mm)	Liquid limit %	Name

COARSE SOILS (Less than 35% of the material is finer than 0.06mm) — GRAVELS (More than 50% of coarse material is of gravel size (coarser than 2 mm))						
Soil description	Group symbol	Subgroup symbol	Fines (% less than 0.06 mm)	Liquid limit %	Name
Slightly silty or clayey GRAVEL	G	GW		0 to 5		Well-graded GRAVEL
		GP	GPu GPg			Poorly graded/Uniform/Gap-graded GRAVEL
Silty GRAVEL	G-F	G-M	GWM GPM	5 to 15		Well-graded/Poorly graded silty GRAVEL
Clayey GRAVEL		G-C	GWC GPC			Well-graded/Poorly graded clayey GRAVEL
Very silty GRAVEL	GF	GM	GML etc	15 to 35		Very silty GRAVEL; subdivide as for GC
Very clayey GRAVEL		GC	GCL GCI GCH GCV GCE			Very clayey GRAVEL (clay of low, of intermediate, of high, of very high, of extremely high plasticity)
Slightly silty or clayey SAND	S	SW		0 to 5		Well-graded SAND
		SP	SPu SPg			Poorly graded/Uniform/Gap-graded SAND

Group	Soil name	Group symbol	Sub-group symbols	% fines	Description
SANDS (More than 50% of coarse material is of sand size (finer than 2mm))	Silty SAND	S-F, S-M	SWM, SPM	5 to 15	Well-graded/Poorly graded silty SAND
	Clayey SAND	S-C	SWC, SPC		Well-graded/Poorly graded clayey SAND
	Very silty SAND	SF, SM	SML etc		Very silty SAND; subdivided as for SC
	Very clayey SAND	SC	SCL, SCI, SCH, SCV, SCE	15 to 35	Very clayey SAND (clay of low, of intermediate, of high, of very high, of extremely high plasticity)
Gravelly or sandy SILTS and CLAYS (35% to 65% fines)	Gravelly SILT	FG, MG	MLG etc		Gravelly SILT; subdivided as for CG
	Gravelly CLAY	CG	CLG, CIG, CHG, CVG, CEG	< 35 / 35 to 50 / 50 to 70 / 70 to 90 / > 90	Gravelly CLAY of low plasticity / of intermediate plasticity / of high plasticity / of very high plasticity / of extremely high plasticity
SILTS and CLAYS (65% to 100% fines)	Sandy SILT	FS, MS	MLS etc		Sandy SILT; subdivide as for CG
	Sandy CLAY	CS	CLS etc		Sandy CLAY; subdivide as for CG
FINE SOILS (More than 35% of the material is finer than 0.06 mm)	SILT (M-SOIL)	F, M	ML etc		SILT; subdivide as for C
	CLAY	C	CL, CI, CH, CV, CE	< 35 / 35 to 50 / 50 to 70 / 70 to 90 / > 90	CLAY of low plasticity / of intermediate plasticity / of high plasticity / of very high plasticity / of extremely high plasticity
ORGANIC SOILS	Descriptive letter 'O' suffixed to any group or sub-group symbol.				Organic matter suspected to be a significant constituent. Example MHO: - Organic SILT of high plasticity.
PEAT	Pt				Peat soils consist predominantly of plant remains which may be fibrous or amorphous

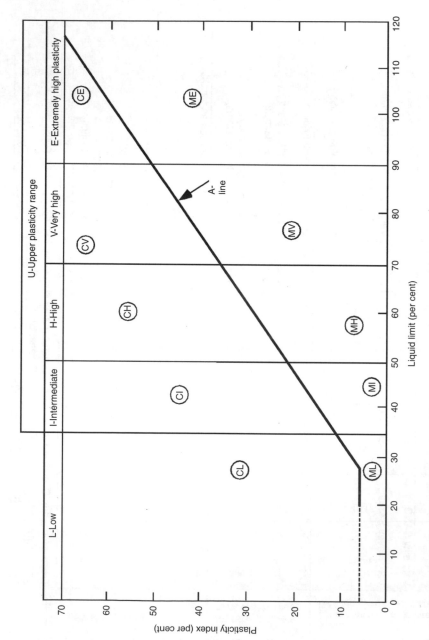

Fig. 4.9 BSCS plasticity chart for the classification of fine soils and the finer part of coarse soils. (*Note*: The letter O is added to the symbol for any material containing a significant proportion of organic matter, e.g. MHO)

The most important group of organic soils also usually plot below the A-line on the plasticity chart. These soils, which are designated ORGANIC SILT, MO, include most M-SOILS of high liquid limit and above. (Organic soils are normally unsuitable as road construction or foundation materials because of their compressibility characteristics.)

4.9 Basic soil tests

The following are summary descriptions and discussions of some of the more common physical tests carried out on soils in connection with roadworks, which have not previously been discussed in detail. The emphasis in each discussion is upon the purpose for which each test is carried out, and detailed information regarding test procedures is readily available in the literature, e.g. see references 1 and 14.

Before discussing the individual tests, however, it must be emphasized that the samples to be sent to the laboratory must be of sufficient quality for the test(s) in question and, when they arrive at the laboratory, they should be handled with the minimum amount of shock and disturbance and stored at constant temperature and humidity. In this context, the importance of having an adequate sample of material to work with in the laboratory must be also stressed. When the approximate number and types of test are known, it is a simple matter to estimate the total amount of soil that has to be obtained from the field. When the programme of laboratory tests is uncertain, Table 4.7 can be used to provide guidance on desirable amounts of soil for various series of tests.

Also, before the soil samples are passed to the laboratory, the possibility that they may be contaminated with harmful substances should be assessed, so that laboratory personnel can be advised of the need for any special procedures that may be necessary to ensure their safety.

Laboratory tests should be carried out at a constant temperature – preferably in an air-conditioned laboratory – and the test conditions utilized should be representative of those in the field at the time being considered in the design. Much practical experience underlies soil testing and, when the tests are properly carried out, reliable predictions usually result if the derived data are used with skill and experience.

Table 4.7 Mass of soil sample required for various laboratory tests[12]

Purpose of sample	Soil types	Mass of sample required (kg)
Soil identification, including Atterberg limits; sieve analysis; moisture content and sulphate content tests	Clay, silt, sand Fine and medium gravel Coarse gravel	1 5 30
Compaction tests	All	25–60
Comprehensive examination of construction materials, including soil stabilization	Clay, silt, sand Fine and medium gravel Coarse gravel	100 130 160

4.9.1 Determination of particle-size distribution (grading)

Gradation testing involves determining the percentages of the various grain sizes present in a soil from two test processes: sieving and sedimentation.Two methods of sieving are specified, dry sieving and, if the sample contains silt or clay, wet sieving. The relative proportions of silt and clay can only be determined from sedimentation testing.

Dry sieving is appropriate only for soils containing insignificant quantities of silt and clay, i.e. essentially cohesionless soils.

Wet sieving covers the quantitive determination of the particle-size distribution down to fine sand size: this requires the preparation of a measured quantity of soil and the washing of all silt and clay through a 63 micron sieve. The retained coarser material is then dry sieved through a series of successively smaller sieves. The mass retained on each sieve, as well as the material passing the 63 micron sieve, is then expressed as a percentage of the total sample.

Sedimentation testing can be used to determine a soil's particle-size distribution, from coarse sand size (2 mm) to clay size (0.002 mm). If significant quantities of both coarse- and fine-grained particles are present in a soil, sedimentation testing is only carried out on the material passing the 63 micron sieve, and the results of both the wet sieving and sedimentation tests are combined to give the overall particle-size distribution for the soil.

All sedimentation testing is based on the assumptions that soil particles of different size (i) are spherical in shape, (ii) can be dispersed uniformly through a constant-temperature liquid without being close enough to influence each other, and (iii) have settling velocities in accordance with Stoke's Law (which states that the terminal velocity of a spherical particle settling in a liquid is proportional to the square of its diameter). Whilst soil particles, especially clay particles, are not spherical, the results obtained are sufficiently realistic for practical soil testing purposes.

In a typical sedimentation test, a suspension of a known mass of fine particles is made up in a known volume of water. The mixture is shaken thoroughly and the particles are then allowed to settle under gravity. With the *pipette* method the mixture is sampled at a fixed depth after predetermined periods of time and the distribution of particle sizes is determined by mass differences. With the *hydrometer* method the density of the soil–water mixture is determined at fixed time intervals using a relative density hydrometer of special design, and the distribution of particle sizes is then determined by formula calculation or nomograph.

The results of a particle-size analysis are presented either (a) as a table which lists the percentage of the total sample that passes each sieve size or is smaller than a specified particle diameter, or (b) as a plot of the sieve or particle size versus percentage passing each sieve or smaller than each diameter, with the sieve or particle size on a logarithmic scale and the percentages finer on an arithmetic scale. Some classic particle-size distributions for British soils are shown in Fig. 4.8.

Note that a *well-graded soil*, i.e. one that compacts to a dense mass, has a wide range of particle sizes and a smooth and upward-concave size distribution with no deficient or excess sizes in any size range. A soil is said to be well graded if it has a *coefficient of uniformity*, C_u, of 5 or more, and a *coefficient of curvature*, C_c, between 1 and 3, as determined from

$$C_u = D_{60}/D_{10} \qquad (4.26)$$

and

$$C_c = (D_{30})^2/(D_{10}D_{60}) \qquad (4.27)$$

where D_{10}, D_{30}, and D_{60} (see Fig. 4.8) are the particle diameters at which 10, 30 and 60 per cent, respectively, by mass of the soil are finer.

If the range of sizes is small, the gradation curve will be steep, and the soil is described as a *uniform soil* (see the estuarine sand in Fig. 4.8). A uniform soil is poorly graded and has a C_u-value of 2 or less.

A poorly graded soil may have a near-horizontal 'hump' or 'step' in its gradation curve, indicating that it is missing some intermediate sizes. This is a *gap graded soil* (e.g. see the poorly graded moraine soil in Fig. 4.8).

The results of particle-size distribution tests are of considerable value when used for soil classification purposes. Specifications for roadbases and subbase construction use particle-size analysis in the quality control of soil materials. Use is frequently made of particle-size analyses for mix-design or control purposes, when stabilizing soils by mechanical or chemical means.

The use of these analyses to predict engineering behaviour should be treated with care, however, and limited to situations where detailed studies of performance or experience have allowed strong empirical relationships to be established, e.g. as previously discussed (Table 4.3) relationships based on particle size have been used to estimate the frost-susceptibilities of soils.

4.9.2 Atterberg consistency tests

Consistency is that property of a soil which is manifested by its resistance to flow. As such it is a reflection of the cohesive resistance properties of the finer fractions of a soil rather than of the intergranular fractions.

As discussed previously, the properties of the finer fractions are considerably affected by moisture content, e.g. as water is continously added to a dry clay it changes from a solid state to a semi-solid crumbly state to a plastic state to a viscous liquid state. The moisture contents, expressed as a percentage by mass of the oven-dry soil samples, at the empirically-determined boundaries between these states, are termed the Atterberg limits after the Swedish scientist who was the first (in 1910) to attempt to define them. The Atterberg limits are determined for soil particles smaller than 425 microns.

The *shrinkage limit* (abbreviated w_s or *SL*) is the moisture content at which a soil on being dried ceases to shrink; in other words, it is the moisture content at which an initially dry soil sample is just saturated, without any change in its total volume. The test is carried out by immersing a sample of soil in a mercury bath and measuring its volume as its moisture content is decreased from the initial value, plotting the shrinkage relationship (volume/100 g of dry soil vs moisture content, %), and reading the shrinkage limit from this graph.

The *plastic limit* (abbreviated w_P or *PL*) is the moisture content at which a soil becomes too dry to be in a plastic condition, and becomes friable and crumbly. This boundary was originally defined by Atterberg as the moisture content at which the sample begins to crumble when rolled into a thread under the palm of the hand;

this has now been standardized for test purposes as the moisture content at which the soil–water mixture has been rolled into an unbroken thread 3 mm in diameter and it begins to crumble at this size.

Pure sands cannot be rolled into a thread and are therefore reported as 'non-plastic'. The w_P-values for silts and clays normally range from 5 to 30 per cent, with the silty soils having the lower plastic limits.

The moisture content at which a soil passes from the liquid to the plastic state is termed the *liquid limit* (w_L or *LL*). Above the liquid limit the soil behaves as a viscous liquid whilst below w_L it acts as a plastic solid. The liquid limit test carried out in the UK involves measuring the depths of penetration, after 5 s, of an 80 g standardized cone into a soil pat, at various moisture contents. The moisture content corresponding to a cone penetration of 20 mm, as determined from a moisture content versus depth of penetration relationship (Fig. 4.10), is taken as w_L.

Sandy soils do not have liquid limits and are therefore reported as non-plastic. For silty soils, w_L-values of 25 to 50 per cent can be expected. Liquid limits of 40 to 60 per cent and above are typical for clay soils; montmorillonitic clays having even higher values, e.g. Gault clays, are reported[7] as having w_L-values ranging from 70 to 121 per cent (and w_P-values of 25–32 per cent).

Values commonly used in association with the liquid and plastic limits are the *plasticity index*, (I_P or *PI*) and the *liquidity index*, I_L. The plasticity index is the percentage moisture range over which the soil is in a plastic state, whilst the liquidity index is the ratio of the difference between the in-situ moisture content and the plastic limit to the plasticity index. Thus

$$I_P = w_L - w_P \tag{4.28}$$

and

$$I_L = (w - w_P)/I_P \tag{4.29}$$

where w = in-situ moisture content (%) and w_P = plastic limit (%).

Note that if $I_L > 1$ then the in-situ moisture content, w, is more than w_L, which indicates that the soil is extremely weak in its natural state. If $I_L < 0$, i.e. is negative, it indicates that, because the in-situ moisture content is less than the plastic limit, the soil is like a brittle solid and will crumble when remoulded. If the liquidity

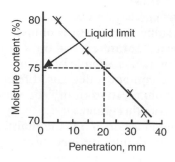

Fig. 4.10 Determining the liquid limit

Table 4.8 Some overall characteristics indicated by the consistency tests

Characteristic	Comparing soils of equal w_L with I_p increasing	Comparing soils of equal I_p with w_L increasing
Compressibility	About the same	Increases
Permeability	Decreases	Increases
Rate of volume change	Increases	–
Dry strength	Increases	Decreases

index is between 1 and zero, it indicates that the in-situ moisture content is at or below the liquid limit.

The most common use of the plasticity test results is in the classification of fine-grained soils and the fine fractions of mixed soils. Some overall soil characteristics that are directly indicated by the consistency tests are given in Table 4.8.

Both w_L and I_p can be used as a quality-measuring device for pavement materials, to exclude granular materials with too many fine-grained particles. If the plasticity index of a soil is known, it can be used to give a rough approximation of its clay content, e.g. soils with I_p-values of 13 and 52 per cent can be expected to have clay contents of 20 and 60 per cent, respectively (see Table 10.2 of reference 7). The plasticity index has also been used to estimate the strength of a cohesive soil, as reflected by its California Bearing Ratio (CBR) value[15].

4.9.3 Particle density and specific gravity tests

The two methods of test that are most used to determine soil particle density are the gas jar and small pyknometer methods. The *gas jar method* is suitable for all soils, including gravels, provided that not more than 10 per cent material is retained on a 37.5 mm test sieve. The *small pyknometer method* is the definitive method (in the UK) for soils composed of sand, silt and clay-sized particles. Both methods require the soil particles to be oven-dried at 105°C and then placed in a container for weighing with and without being topped up with water. The particle density is determined from the following equation:

$$\rho_s = (m_2 - m_1)/[(m_4 - m_1) - (m_3 - m_2)] \tag{4.30}$$

where m_1 = mass of container (g), m_2 = mass of container and soil (g), m_3 = mass of container, soil and water (g), and m_4 = mass of container and water (g).

These tests measure the weighted average of the densities of all the mineral particles present in the soil sample. As about 1000 minerals have been identified in rocks (mostly with densities between 2 and 7 Mg/m^3), and as soils are derived from rocks, the results obtained can be expected to vary according to the original rock sources and the amount of mixing inherent in the soil's derivation. However, because of the natural preponderance of quartz and quartz-like minerals the densities of soil particles in the UK are most often between 2.55 and 2.75 with the lower values generally indicating organic matter and the higher values metallic material. In practice, a value of 2.65 is commonly used, unless experience suggests otherwise.

Particle density is numerically equal to the dimensionless *specific gravity (relative density)*. This is used in computations involving many tests on soils, e.g. determining voids ratios and particle-size analyses (sedimentation testing).

4.9.4 Moisture-density test

Compaction is the process whereby the solid particles in a soil are packed more closely together, usually by mechanical means, to increase its dry density.

In 1933, Ralph Proctor of the Los Angeles Bureau of Waterworks and Supply devised a laboratory test, now generally known as the *Proctor test*, to ensure that dry densities were attained under compaction that gave desired impermeabilities and stabilities to earth dams. After compacting over 200 different soils at various moisture contents, and using laboratory compacting energies then considered equivalent to those produced by field compaction, Proctor demonstrated that water acts as a lubricant which enables soil particles to slide over each other during compaction and that, as a soil's moisture content was increased, its compacted dry density increased to a maximum value, after which it decreased. Only a small amount of air remains trapped in a compacted soil at the point of maximum dry density, and any increase in moisture content above what Proctor termed its 'optimum moisture content' results in soil particles being replaced by water with a consequent reduction in dry density.

Over the course of time standardized laboratory procedures were developed in many countries to evaluate the moisture–density relationship. The British version of this basic test is now carried out by compacting a prepared air-dried soil sample in a specified way into either a metal cylindrical mould with an internal volume of 1 litre and an internal diameter of 105 mm, or a California Bearing Ratio (CBR) mould with an internal diameter of 152 mm and depth of 127 mm, depending upon the maximum particle size, i.e. the test restricts the maximum particle size used in the 1 litre mould to 20 mm and in the CBR mould to 37.5 mm. With the *standard Proctor test* a 50 mm diameter, 2.5 kg metal rammer is dropped through 300 mm for 27 blows (1 litre mould) or 62 blows (CBR mould) onto each of three, approximately equally thick, layers of soil in the mould. In the *modified Proctor test* a greater compactive effort is employed and a 50 mm diameter, 4.5 kg rammer is allowed to fall through 450 mm, for the same numbers of blows as with the standard test, onto each of five layers in the same-sized moulds. Following compaction, surplus compacted soil is struck off flush with the top of the mould, and the mass of the soil is then determined and its bulk or wet density is calculated by dividing the mass by the known volume of the mould. A sample is then taken of the soil used in the mould and this is dried in an oven at 110°C to determine its moisture content.

After the compacted material has been removed from the mould a new air-dried sample of soil is prepared, a higher increment of water is added, and a new compacted sample is prepared using the same procedure. The preparation of samples is continued until the mass of a compacted sample is less than that of the preceding measurement. Knowing the bulk density and moisture content of each sample, their dry densities are then calculated from the formula

$$\rho_d = 100\rho/(100+w) \tag{4.31}$$

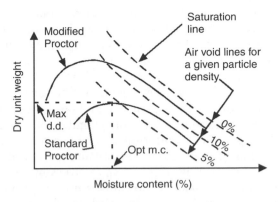

Fig. 4.11 Moisture–density relationships at differing compactive efforts

where ρ_d = dry density (Mg/m³), ρ = wet density (Mg/m³), and w = moisture content (%). If the dry densities are plotted against their corresponding moisture contents, and a smooth curve is drawn through each set of data points, a relationship similar to those shown in Fig. 4.11 is obtained, and the optimum moisture content and the maximum dry density is then read from the appropriate moisture–density curve. Also shown in this figure are the curves corresponding to 0, 5 and 10 per cent air voids, calculated from

$$\rho_d = [1 - (n_a/100)]/[(1/\rho_s) + (w/100\,\rho_w)] \qquad (4.32)$$

where ρ_d = dry density (Mg/m³), ρ_s = particle density (Mg/m³), ρ_w = density of water (Mg/m³) assumed equal to 1, n_a = volume of air voids expressed as a percentage of the total voids in the soil, and w = moisture content (%).

Table 4.9 shows some typical values that were obtained with various soils using the standard and modified Proctor tests. Note that the greater compaction applied in the modified test had the greatest effect upon the heavy and silty clays, with large increases in the maximum dry densities and large decreases in the optimum moisture contents, i.e. when the air voids contents are large the effect of increasing the compaction is significant whereas it is negligible when the air voids content is small. Table 4.9 also shows that the maximum dry density increases and optimum

Table 4.9 Comparison of standard and modified Proctor test results for various soils

Type of soil	Standard Proctor test:		Modified Proctor test:	
	Max. dry density (Mg/m³)	O.m.c. (%)	Max. dry density (Mg/m³)	O.m.c. (%)
Heavy clay	1.555	28	+0.320	−10
Silty clay	1.670	21	+0.275	−9
Sandy clay	1.845	14	+0.210	−3
Sand	1.940	11	+0.145	−2
Gravel-sand-clay	2.070	9	+0.130	−1

moisture content decreases as the soil becomes less plastic and more granular. With uniformly-graded fine sands, however, which have significant voids between their single-sized particles, the dry density increase at the optimum moisture content is fairly small; further, the optimum moisture content is fairly difficult to determine because of the free-draining characteristics of these soils.

Whilst the Proctor tests provide useful guides to the range of moisture contents suitable for compaction with most soils, they are relatively poor guides for specifications relating to the compaction of permeable granular soils with little silt and clay, e.g. fine-grained clean gravels or uniformly-graded coarse clean sands, and so the *vibrating hammer test* has been developed in Britain for use with coarse soils (maximum particle size = 37.5 mm). With this test an electric vibrating hammer operating with a power consumption of 600 to 750 watts is used to compact three layers of moist soil (each layer is vibrated for 60 s) in a CBR mould; the total downward force during compaction is maintained at between 300 and 400 N. The hammer's tamping foot has a circular base that almost completely covers the area of the mould and produces a flat surface; the depth of the specimen after compaction must lie between 127 and 133 mm. The wet or bulk density, ρ, is then calculated from

$$\rho = M/18.1h \tag{4.33}$$

where M = mass of the wet compacted soil (g), and h = compacted specimen height (mm), and the dry density is calculated as for the Proctor tests (Equation 4.31). By varying the soil moisture content, a moisture-content – dry-density relationship is obtained from which the optimum moisture content and maximum dry density are determined.

Theoretically, the maximum dry density achievable with a soil having a specific gravity of 2.70 is 2.70 Mg/m³; this could only occur, however, if all the particles could be fitted against each other exactly. As soil particles come in various shapes and sizes, there are always air voids and so this theoretical density can never be achieved. Well-graded dense gravel mixtures can be compacted to maximum dry densities, at their optimum moisture contents, with 5–10 per cent air voids, whilst uniformly graded sands and gravels may have 10–15 per cent air voids. Soils often have in-situ dry densities of 50–60 per cent of their theoretical maximum values whilst peats and mucks may be around 15 per cent.

The moisture–density test is designed to assist in the field compaction of earthworks. It assumes that the shear strength of a soil increases with increasing dry density, and (controversially) that the compactive efforts used in the laboratory tests are similar in effect to those achieved by particular construction equipment in the field. Consequently, specifications often state the dry density required for earthworks as a percentage of that achieved in the laboratory when compacted at the optimum moisture content for maximum dry density for the specified compactive effort, e.g. a relative compaction of 90 or 95 per cent of modified Proctor dry density for granular-type soils and 90 or 95 per cent of the standard Proctor value for fine-grained silts and clays.

Generally, the optimum moisture contents derived from the Proctor tests on a given soil are a useful guide to the moisture content range that is suitable for that soil's compaction. Generally, for UK climatic conditions, well-graded or uniformly graded soils are normally suitable for compaction if their in-situ moisture contents are not more than 0.5 to 1.5 per cent above the optimum moisture content for maximum dry density as determined from the standard Proctor test. In the case of cohesive soils the upper in-situ moisture content limit is typically 1.2 times the plastic limit, for compaction in the field.

4.9.5 Moisture condition value (MCV) test

As will have been gathered from the discussion in relation to the Proctor tests, there is a relationship between compaction effort, moisture content, and density. Figure 4.11 shows that at a fixed compactive effort (i.e. number of rammer blows) there is no increase in dry (or bulk) density once the optimum moisture content is present in the soil. Table 4.9 and Fig. 4.11 show that, as the compactive effort is increased, the optimum moisture content decreases and the maximum density increases. Similarly, it can be shown that, if the moisture content of a given soil is held constant, there is a unique compactive effort (i.e. an 'optimum' number of rammer blows) which produces a maximum dry density, and increasing the number of rammer blows beyond this optimum will not result in any density increase; also, the higher the moisture content of the soil the lower the compactive effort required to achieve the maximum dry density.

The *moisture condition value (MCV) test* involves testing a soil at a fixed moisture content and, by incrementally increasing the number of blows of a rammer, determining the compactive effort beyond which no further increase in density occurs. As the MCV test is relatively new – it was only developed in the late 1970s (see reference 16) – the test and its interpretation will be described in some detail here, as it is now in wide use.

Figure 4.12(a) shows the apparatus used to determine the moisture condition value for a soil sample. The apparatus has a 100 mm by 200 mm high mould which sits on a detachable heavy base, a free-falling 97 mm diameter cylindrical rammer with a mass of 7 kg, and an automatic release mechanism that is adjustable to maintain a constant drop height of 250 mm onto the surface of the soil sample. The base has a mass of 31 kg, to allow the apparatus to be used in the field on surfaces of varying stiffness (including on soft soil). When testing, 1.5 kg of soil passing the 20 mm sieve is placed loosely in the mould and covered with a 99 mm diameter lightweight rigid fibre disc. The rammer is then gently lowered to the surface of the disc and allowed to penetrate into the mould, under its own weight, until it comes to rest; the automatic release mechanism is then adjusted to ensure a drop of 250 mm, using the Vernier scales attached to the guide rods and the rammer. The rammer is then raised until it is released by the automatic catch, dropped, and its penetration into the mould (or the length protruding above the mould) is measured using the scale attached to the rammer, and noted (Table 4.10). This process is carried out repeatedly, with the release mechanism being adjusted to ensure a constant drop of 250 mm, with penetration (or protrusion) measurements being taken after selected accumulated numbers of blows, until there is no additional rammer penetration, i.e. no additional densification. A sample of soil is then taken from the mould and its moisture content determined in the normal way.

The penetration of the rammer at any given number of blows, n, is then compared with that for $4n$ blows and the difference, which is a measure of density change, is calculated and noted. If the change (on a linear scale) is plotted against

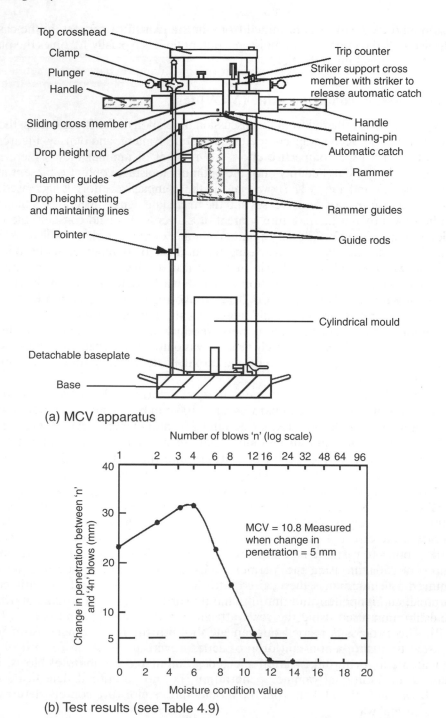

(a) MCV apparatus

(b) Test results (see Table 4.9)

Fig. 4.12 Moisture condition apparatus and graph of test results for a heavy clay

Table 4.10 Results of a determination of the moisture condition value for a heavy clay[20]

Number of blows of rammer '*n*'	Protrusion of rammer above mould (mm)	Change in penetration between '*n*' and '*4n*' blows of rammer (mm)
1	106.3	23.5
2	96.4	29.2
3	89.0	31.8
4	83.8	31.7
6	74.7	23.1
8	67.2	15.6
12	57.2	5.6
16	52.1	0.5
24	51.6	0
32	51.6	
48	51.6	

the initial cumulative number of blows, *n* (on a logarithmic scale), in each instance, a curve such as that shown in Fig.4.12(b) is obtained. Note that the steepest possible straight line is drawn through the points immediately before or passing though the 5 mm 'change in penetration' value, which has been arbitrarily selected as indicating the point beyond which no significant change in density occurs. The moisture condition value (MCV) for the sample is then defined as

$$MCV = 10 \log B \qquad (4.34)$$

where B = number of blows at which the change in penetration equals 5 mm, as read from the straight line. (Alternatively, the MCV can be read from the lower horizontal axis of a previously prepared chart.) For the example in Table 4.10 and Fig. 4.12(b), MCV = 10.8 and the measured moisture content = 31.5 per cent.

The test, which takes about half an hour to complete, is repeated on a number of samples of the same soil, each at a different moisture content, and a different MCV is obtained for each sample. When the moisture condition values are plotted against their respective moisture contents, a straight line relationship is established for the soil, which can then be used as a calibration line for subsequent determination of the moisture content in the field. Thus, if the moisture content in the field is required quickly at any given time, its MCV can be determined easily using the portable test apparatus, and the relevant moisture content can then be read from the calibration chart.

As a general construction guide, an MCV of 8.5 is recommended[17] as the lower limit of acceptability of a soil for compaction at its natural moisture content, and difficulties in earthworking can be expected when the MCV drops significantly below this value.

4.9.6 California Bearing Ratio (CBR) test

The California Bearing Ratio (CBR) test was originally devised by O.J. Porter of the California State Highways Department, following an extensive investigation into flexible pavement failures in California in the nine years prior to 1938. It was taken up by the US Corps of Engineers during World War II and adapted for airport pavement design purposes. After the war other organizations became interested in the test as a means of empirically measuring soil strength for pavement design purposes, and the original CBR test and design procedures were subsequently adapted to meet particular needs in various countries.

The CBR test is normally carried out in the laboratory. The principle underlying the test involves determining the relationship between force and penetration when a cylindrical plunger of a standard cross-sectional area is made to penetrate a compacted soil sample at a given rate. At certain values of penetration the ratio of the applied force to a standard force, expressed as a percentage, is defined as the California Bearing Ratio (CBR) for the soil.

The test is carried out in the UK on soil having a maximum particle size not exceeding 20 mm. It involves placing a predetermined mass of soil in a steel mould with an attached collar that is 152 mm in diameter by 178 mm high, and compacting it either statically or dynamically until a 127 mm high specimen is obtained that is at the moisture and dry density conditions required by the design. Following compaction the specimen, still in its mould, is covered with annular surcharge weights approximately equivalent to the estimated mass of flexible pavement expected above the soil layer in the field; it is then placed in a testing machine and a plunger with an end diameter of 49.6 mm is caused to penetrate the compacted soil at a rate of 1 mm per minute. The penetration of the plunger is measured by a dial gauge, and readings of the applied force are read at intervals of 0.25 mm to a total penetration not exceeding 7.5 mm.

After the test a load versus penetration curve is drawn for the data obtained, as ideally illustrated by the convex-upward curve A in Fig. 4.13. In practice, an initial seating load is applied to the plunger before the loading and penetration gauges are set to zero and, sometimes, if the plunger is not perfectly bedded, a load–penetration curve like that at curve B may be obtained. When this happens, the curve has to be corrected by drawing a tangent at the point of greatest slope and then projecting it until it cuts the abscissa; this is the corrected origin, and a new penetration scale is (conceptually) devised to the right of this point. The CBR is then obtained by reading from the corrected load–penetration curve the forces that cause penetrations of 2.5 and 5 mm and expressing these as percentages of the loads which produce the same penetration in a standard crushed stone mix (also shown in Fig. 4.13); the forces corresponding to this curve are 11.5 kN at 2 mm penetration, 17.6 kN at 4 mm, 22.2 kN at 6 mm, 26.3 kN at 8 mm, 30.3 kN at 10 mm, and 33.5 kN at 12 mm. The higher of the two percentages is then taken as the CBR-value for the soil; for example, the CBRs obtained from curves A and B are 54 and 43 per cent, respectively.

The CBR of a soil is an indefinable index of its strength which, for a given soil, is dependent upon the condition of the material at the time of testing. This means that the soil needs to be tested in a condition that is critical to its design.

At any given moisture content, the CBR of a soil will increase if its dry density is increased, i.e. if the air content of the soil is decreased. Thus, a design dry density should be selected which corresponds to the minimum state of compaction expected in the field at the time of construction.

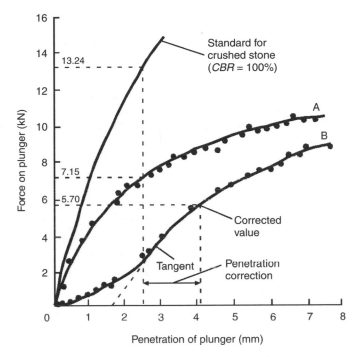

Fig. 4.13 Example California Bearing Ratio load–penetration curves

At a given dry density the CBR also varies according to its moisture content. Hence, the design CBR-value should be determined for the highest moisture content that the soil is likely to have subsequent to the completion of earthworks. The selection of this moisture content requires an understanding of the manner of moisture fluctuations in a soil.

Before a soil is covered by a road pavement, its moisture content will fluctuate seasonally. For instance, a gradient of moisture within the soil profile that is determined during the summer might look like curve *A* in Fig. 4.14, whilst another taken in the winter might look like curve *B*. Some time after construction of the pavement the soil moisture conditions will become stabilized and subject only to changes associated with fluctuations in the level of the water table; when this happens the moisture gradient might look like curve *C*.

If the pavement is to be built during the winter, the soil should be tested at the moisture content at point *b*; this high content should be used as the road may be opened to traffic before the moisture content can decrease to that at point *c*. If the road is to be built during the summer, the soil should be tested at the moisture content at *c* as, ultimately, this will be its stable moisture content.

The determination of the *equilibrium moisture content* (point *c*) to use in CBR testing is best determined from laboratory suction tests on soil samples taken from the proposed formation level, using a suction corresponding to the most adverse anticipated water table conditions. If facilities for conducting suction tests are not available, this moisture content can be estimated at a depth below the zone of seasonal fluctuation but at least 0.3 m above the water table, i.e. at or below point *d*. In

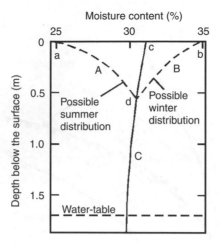

Fig. 4.14 Diagrammatic illustration of seasonal variations in the moisture content of a heavy soil

cohesive soils in Britain, especially those supporting dense vegetation, point *d* is typically at a depth of 1–1.25 m; in sandy soils, which do not normally have deep-rooted grasses, the moisture content at a depth of 0.3–0.6 m will usually provide a good estimate of the moisture content at point *d*. If the cohesive soil is to be used in an embankment the moisture content at a depth of 1–1.25 m should be increased by about 2 per cent to allow for the disturbance to the natural soil structure; if silty and sandy soils are to be used in embankments the moisture content at point *d* need only be increased by about 1 per cent.

An *in-situ CBR test* can also be carried out in the field, using a suitably-rigged vehicle. In-situ CBR results can be significantly different from laboratory-derived results (except for heavy clays having an air voids content of 5 per cent or more) and, consequently, the in-situ CBR test is not normally used as a quality assurance measure to ensure specification implementation.

One further point which might be noted here relates to the use of the *Plate Bearing Test* to determine the stiffness modulus, E (MN/m²), of a soil. This test, which is described in detail in Part 9 of reference 1, is very laborious to set up and carry out, and requires a truck or excavator to provide reaction. Also, the speed of loading is slow, giving poor simulation of traffic loading. In practice, therefore, the following relationship, which has been derived empirically for typical UK soils, is widely used to determine the modulus:

$$E = 17.6(CBR)^{0.64} \tag{4.35}$$

This equation is approximately valid for CBR-values between 2 and 12 per cent.

4.9.7 Unconfined compressive strength (UCS) test

The unconfined compressive strength (UCS) test can only be carried out on saturated non-fissured fine-grained soils, or soils that are stabilized with additives that bind the particles together. With fine-grained soils the test, which is a rapid substitute for the undrained triaxial test, is carried out most commonly on 38 mm diameter cylindrical specimens having a height to diameter ratio of 2:1. In the UK stabilized soils are normally tested using 150 mm cubes for coarse-grained and medium-grained materials; for fine-grained (passing the 5 mm sieve) soils 100 mm diameter by 200 mm high or 50 mm diameter by 100 mm high cylindrical samples are recommended. Before testing, stabilized samples are compacted to a predetermined dry density (usually the maximum value obtained during the moisture–density test) either by static (fine- and medium-grained soils) or dynamic (medium- and coarse-grained soils) compaction.

During the compression test for stabilized cubes the load is applied at a uniform rate of increase of stress of 3.5 MN/m^2 per minute until specimen failure; in the case of cylindrical specimens the load is applied at a uniform rate of deformation of 1 mm/min. With natural cohesive soils the load is applied at about 8 mm/min, the aim being to achieve a test time of around 2 min for a specimen failing at 20 per cent strain. Whilst a load versus deflection curve can be plotted, it is the UCS at failure that is usually used in road engineering, especially soil stabilization, work.

For soil stabilization work, the test is used to determine the suitability of a soil for treatment with a given additive and to specify the additive content to be used in construction; it is also used as a quality control mechanism in the field. Unconfined compressive strength results are considerably influenced by such factors as the amount and type of additive, the method and length of curing of the test specimen, and whether or not the specimens are saturated before testing.

4.9.8 Shear tests

Laboratory shear tests seek to evaluate representative soil specimens in a way that is similar to that anticipated under field conditions. Hence, test programmes are devised to approximate the expected loading conditions of the soil, so that the results can be translated into reliable predictions of soil responses in terms of the parameters needed for the design analysis contemplated.

Direct shear box tests are carried out only on coarse-grained soils, usually in square boxes with 60 mm or 100 mm sides. Samples of coarse-grained soils are more easily prepared for shear box testing than for triaxial testing. However, drainage conditions cannot be controlled nor pore pressures determined with this form of testing, and the plane of shear is fixed by the nature of the test.

General practice with respect to laboratory shear testing is to carry out *triaxial compression tests* on most soils. Triaxial tests are normally carried out on 70, 100 or 150 mm dia samples with a height to diameter ratio of 2:1. The tests can be unconsolidated undrained or drained tests or, if consolidated in the apparatus prior to testing, consolidated undrained or drained tests; any undrained or drained tests in which pore pressures are measured are normally consolidated before shearing.

The shearing resistance of a soil is important in relation to the design of earth slopes for road cuttings and embankments. It is derived from

$$s = c + \sigma_n \tan \phi \qquad (4.36)$$

where s = shear strength (kPa), c = cohesion (kPa), σ_n = stress normal to the shear plane (kPa), and ϕ = angle of shearing resistance (deg). The tests are designed to derive the angle of internal friction and the cohesion, which are considered constant for both laboratory and field conditions.

Generally, sandy soils develop their shearing resistance through friction, with little or no cohesion. Internal friction is mainly affected by the shape of the coarse particles, and is little affected by moisture content; however, it increases rapidly with increasing dry density. The bulk of a clay's shear resistance comes from cohesion associated with water bonds between the particles. Cohesion is greatly influenced by moisture content, i.e. it decreases with increasing moisture to reach a low level at the plastic limit and almost zero at the liquid limit. When an unconfined compressive strength test is carried out on a natural clay soil sample, the internal friction can be assumed equal to zero, and the c-value is then one-half of the compressive strength.

Typical ranges for friction and cohesion are as follows: (i) for sandy soils, $\phi=28\text{--}45$ deg and $c=0\text{--}2.06\,\text{MN}/\text{m}^2$, and (ii) for clay soils, $\phi=0\text{--}15$ deg and $c=0.7\text{--}13.8\,\text{MN}/\text{m}^2$.

The *vane test* is a field shear test whereby a cruciform vane on the end of a solid rod is forced into a clay soil and then rotated. The test is most applicable to uniform clay soils as the presence of coarse particles, rootlets, or thin layers of sand or silt will distort the torque. The shear strength is calculated from

$$\tau_v = 1000M/[3.142D^2(0.5H+0.167D)] \tag{4.37}$$

where τ_v=vane shear strength (kPa), M=applied torque (N mm) obtained by multiplying the maximum angular rotation of the torsion spring (deg) by the calibration factor (N mm/deg), D=overall vane width (mm), and H=length of vane (mm).

4.9.8 Consolidation tests

When an embankment is placed on a saturated soil mass the immediate tendency is for the particles in the foundation material to be pushed closer together. However, the water in the soil, being incompressible, must initially carry part of the applied load and this results in the production of an initial pressure, i.e. the 'pore water pressure', which continues as water drains from the soil. During the drainage period, which can take many years to finish, the soil particles are continually forced closer, thereby producing the volume change termed 'settlement'. One-dimensional consolidation tests attempt to estimate in an accelerated manner both the rate and total amount of settlement of a soil layer under an applied load.

The standard (incremental loading) dead-weight oedometer test is restricted to representative samples of saturated clays, fine silts, and other soils of low permeability. The test involves cutting and trimming a 50, 75 or 100 mm diameter by 20 mm thick soil sample which fits into a special metal ring used in the test and, after ceramic porous discs have been placed above and below the sample, placing the assembly in a loading unit. In the case of a stiff clay, for example, a careful compressive load–unload and reload sequence is then applied using small increments and decrements, and the changes in thickness of the sample are read at set time intervals. With the continuous loading oedometer test stresses, strains or pore pressures are varied continuously instead of, as in the standard test, applying the loads in discrete increments.

Major parameters derived from the consolidation test are the compressible index, the coefficient of consolidation, and the degree of consolidation. The *compressibility index*, C_c, is used in the analysis of total settlement; it is a dimensionless factor which normally ranges from 0.1 to 0.3 for silty clays, and from 0.2 to 1.0 for clays. The *coefficient of consolidation*, c_v, ranges from 0.2 to 2.0 cm²/s for silty clays, and 0.02 to 0.10 cm²/s for clays; this coefficient is used to estimate the settlement for a given period of time under a given increment of load. The *degree of consolidation*, U, is the ratio of the settlement for the time period to the total settlement, expressed as a percentage.

In practice, actual settlements can take place more rapidly than may be predicted by the one-dimensional consolidation test. Also, drains can be used in the field to accelerate settlement by providing more readily accessible passageways for the water to escape.

4.9.9 Frost heave test

There are some 22 different frost-heave tests in use throughout the world[18] where test specimens in the laboratory are exposed to environments that are supposed to simulate those in the field. As a consequence of the many approaches, various countries have different frost-susceptibility criteria. Practically, therefore, it should be appreciated that the test (BS 812) used in the UK is, on the whole, related to British environmental conditions.

The laboratory test[10] for assessing the susceptibility of compacted soil to frost heave involves the preparation of nine 102 mm diameter by 152 mm high specimens for testing in a special freezing cabinet. The specimens are usually compacted at the optimum moisture contents and maximum dry densities derived from the standard Proctor test; with some clay soils, however, the soils may be compacted to 5 per cent air voids at the natural (as-dug) moisture content. Each test specimen is prepared in a specified manner and placed on a porous disc in water so that the water level is just in contact with the bottom of the specimen. The water and air temperatures in the freezing cabinet are maintained at +4°C and −17°C, respectively, for 250 h, and the heave–time relationship is plotted for each sample.

With this test, soils with mean heaves of 13 mm or less during the 250 h period are regarded as non-frost susceptible, those with heaves between 13 and 18 mm are considered marginal, and those with heaves in excess of 18 mm are classified as frost-susceptible[7].

4.10 Soil testing: a comment on the European dimension

Soil testing in most countries is normally carried out according to procedures and criteria decreed by national standards, e.g. the British Standards Institution (BSI) is charged with the preparation and publication of standard test methods in the UK. Many of the standard tests adopted in different countries are very similar, whereas many others are very different, having been adapted and modified to meet local influencing needs. As a consequence, there is no agreed 'European Standard' at this time relating to soil testing.

This situation is now in the process of changing, however. Within the EEC the Comité Européen de Normalisation (CEN) has been given the responsibility for 'normalizing' current national standards and preparing agreed European ones. The

CEN members at this time are the national standards bodies of Austria, Belgium, Czech Republic, Denmark, Finland, France, Germany, Greece, Iceland, Ireland, Italy, Luxembourg, Netherlands, Norway, Portugal, Spain, Sweden, Switzerland, and United Kingdom, and committees comprising members from these countries have been set up to derive agreed approaches and standards. It is important to appreciate that, until the developed documents have been agreed and approved, it is permissible to keep conflicting national standards in each country as a basis for testing and design.

The official English language version[21] of a provisional European Prestandard (ENV) for soil testing for geotechnical design purposes was issued in the UK in April 2000, and comments regarding its acceptability are currently being sought. A review of this publication will be initiated two years after its publication so that a decision can be taken regarding its status at the end of its 3-year life. Depending upon the replies received at the end of this review period, the responsible BSI Committee will decide whether to support its conversion into a European Standard (EN), to extend the life of the provisional standard, or to withdraw it.

4.11 References

1. BS 1377: Parts 1–9, *Methods of Test for Soils for Civil Engineering Purposes*. London: British Standards Institution, 1990.
2. Dumbleton, M.J., *The British Soil Classification System for Engineering Purposes: Its Development and Relation to Other Comparable Systems*, TRRL Report LR1030. Crowthorne, Berkshire: The Transport and Road Research Laboratory, 1981.
3. Leonards, G.A., Engineering properties of soils, Chapter 3 in Leonards, G.A., *Foundation Engineering*. London: McGraw-Hill, 1962.
4. *Manual of Applied Geology for Engineers*. London: The Institution of Civil Engineers, 1976.
5. Skempton, A.W., Soil mechanics in relation to geology, *Proceeding of the Yorkshire Geological Society*. 1950, **29**, pp. 33–62. Reported in Blyth, F.T.H. and de Freitas, M.H., *A Geology for Engineers,* 7th edn. London: Arnold, 1984.
6. Croney, D., Coleman, J.D. and Black, W.P.M., Movement and distribution of water in soil in relation to highway design and performance, *Highway Research Board Special Report 40*, 1958, pp. 226–52.
7. Croney, D. and Croney, P., *Design and Performance of Road Pavements*, 3rd edition. London: McGraw-Hill, 1998.
8. Casagrande, A., Discussion on frost heaving, *Highway Research Board Proceedings*, 1932, **11**, Part 1, pp. 167–172.
9. Jones, R.H., Frost heave of roads, *Quarterly Journal of Engineering Geology*, 1980, **13**, No. 2, pp. 77–86.
10. Roe, P.G. and Webster, D.C., *Specification for the TRRL Frost-heave Test*, TRRL Report SR829. Crowthorne, Berkshine: The Transport and Road Research Laboratory, 1984.
11. BS 6031, *Code of Practice for Earthworks*. London: British Standards Institution, 1981.
12. BS 5930: 1999, *Code of Practice for Site Investigations*. London: British Standards Institution, 1999.

13. Casagrande, A., Classification and identification of soils, *Proceedings of the American Society of Civil Engineers*, Part 1, 1947, **73**, No. 6, pp. 783–810.
14. BS 1924, *Methods of Test for Stabilized Soil*. London: British Standards Institution.
15. Black, W.P.M., A method of estimating the California Bearing Ratio of cohesive soils from plasticity data, *Geotechnique, London*, 1962, **12**, No. 4, pp. 271–282.
16. Parsons, A.W., *Compaction of Soils and Granular Materials: A Review of Research Performed at the Transport Research Laboratory*. London: HMSO, 1992.
17. *The Use and Application of the Moisture Condition Apparatus in Testing Soil Suitability for Earthworking*, SDD Applications Guide No. 1. London: The Scottish Development Department, 1989.
18. Technical Committee Report on Testing of Road Materials, *Proceedings of the XVII World Road Congress of the Permanent International Association Road Congresses*, 1983, pp. 76–8.
19. Millar, C.E., Turk, L.M. and Foth, H.D., *Fundamentals of Soil Science*. New York: John Wiley, 1962.
20. Parsons, A.W. and Toombs, A.F., *The Precision of the Moisture Condition Test*, Research Report 90. Crowthorne, Berkshire: The Transport and Road Research Laboratory, 1987.
21. DD ENV 1997–2:2000, *Eurocode 7: Geotechnical design – Part 2: Design assisted by laboratory testing*. London: British Standards Institution, 2000.

CHAPTER 5

Materials used in road pavements

M.J. Brennan and C.A. O'Flaherty

All road pavements require the efficient use of locally available materials if economically constructed roads are to be built. This requires the design engineer to have a thorough understanding of not only the soil and aggregate properties that affect pavement stability and durability but also the properties of the binding materials that may be added to these.

The most important pavement materials are bitumen and tar, cement and lime, soil (see Chapter 4), and rock, gravel and slag aggregates. In more recent years, for economic and environmental reasons, renewed attention has been given to the use of 'waste' materials in lieu of conventional aggregates in pavements so these are discussed here also.

5.1 Penetration-grade refinery bitumens

Bitumens used for road construction are viscous liquid or semi-solid materials, consisting essentially of hydrocarbons and their derivatives, which are soluble in trichloroethylene[1]. Whilst bitumens occur naturally (e.g. in lake asphalts containing mineral materials), the predominant majority of those used in roadworks are the penetration-grade products of the fractional distillation of petroleum at refineries.

It might be noted here that the term 'asphalt' is used in the American technical literature to describe what is termed 'bitumen' in the UK. In the UK (and in Europe, following agreement by the Comité Européen de Normalisation) the term 'asphalt' is reserved for materials containing a mixture of bitumen and mineral matter, e.g. lake asphalt or hot rolled asphalt.

Bitumens that are produced artificially from petroleum crudes (usually napthenic- and asphaltic-base crudes) are known as refinery bitumens. Not all petroleum crudes are suitable for the production of road bitumen; those used in the UK are mainly derived from Middle East and South American sources, as most of the North Sea crudes contain little or no bitumen.

Bitumen is obtained by a refinery distillation process, which involves condensation in a fractionating column. The first distillation is normally carried out at oilfield refineries where the crude is heated, at atmospheric pressure, to not greater than 350°C to remove naphtha, gasoline and kerosene fractions. The 'topped' oil is then

Table 5.1 Penetration-grade bitumens commonly used for road purposes[1]

Property	Grade of bitumen:				
	15 pen	**40 pen HD**	**50 pen**	**100 pen**	**200 pen**
Penetration at 25°C	15±5	40±10	50±10	100±20	200±30
Softening point (°C)					
minimum	63	58	47	41	33
maximum	76	68	58	51	42
Loss on heating for 5 h at 163°C (%, maximum)					
loss by mass	0.1	0.2	0.2	0.5	0.5
drop in penetration	20	20	20	20	20
Solubility in trichloroethylene (% by mass, minimum)	99.5	99.5	99.5	99.5	99.5
Relative permitivity at 25°C and 1592 Hz	–	2.650	2.650	–	–

shipped to a destination refinery where it is heated, at reduced pressure, to collect the heavier diesel and lubricating oils; without the reduced pressure, chemical changes, i.e. cracking, would impart undesirable properties to the bitumen residue. The residue is then treated to produce a wide range of penetration-grade bitumens that, depending upon the amount of distillate removed, range in consistency from semi-solid to semi-fluid at room temperature. In practice, it is common for a refinery to prepare and stock large quantities of bitumen at two extremes and to blend these to obtain intermediate grades.

Penetration-grade refinery bitumens are primarily designated by the number of 0.1 mm units that a special needle penetrates the bitumen under standard loading conditions, with lower penetration depths being associated with harder bitumens. Ten grades of bitumen, from 15 pen (hardest) to 450 pen (softest), are used in pavement materials in the UK. The harder grades (15–25 pen) are used in mastic asphalts, the medium grades (35–70 pen) in hot rolled asphalts, and the softer grades (100–450 pen) in macadams. All are black or brown in colour, possess waterproofing and adhesive properties, and soften gradually when heated. Properties of a number of penetration-grade bitumens are listed in Table 5.1.

At 25°C the density of a bitumen typically varies from 1 to 1.04 g/cm³, whilst its coefficient of thermal expansion is about 0.00061 per °C. It might be noted that the 40 pen heavy duty (HD 40) bitumen in Table 5.1 is 'blown' by passing air through it; this makes it less susceptible to temperature change than the other bitumens. HD 40 bitumen is used to provide a stiffer asphalt which deforms less under traffic; however, its mixing, laying and rolling must be carried out at higher temperatures than are used with conventional bitumens of the same grade.

5.1.1 Bitumen tests and their significance

Specifications with regard to the design and construction of a bituminous pavement are of little value if the properties of the binder used are not adequately controlled. To aid in ensuring that the bitumen has the desired qualities, various tests have been devised to measure these properties for particular purposes. As detailed information on practice in carrying out these tests is readily available, they are only briefly described here.

Bitumens are termed 'visco-elastic' materials in that (a) at temperatures above about 100°C they exhibit the properties of a viscous material, (b) at temperatures below about minus 10°C they behave as an elastic material, and (c) at temperatures in between they behave as a material with viscous and elastic properties, with the predominating property at any given time depending upon the temperature and rate of load application. *Viscosity*, which is the property of a fluid that retards its ability to flow, is of particular interest at the high temperatures needed to pump bitumen, mix it with aggregate, and lay and compact the mixed materials on site. For example, if the viscosity is too low at mixing, the aggregate will be easily coated but the binder may drain off whilst being transported; if the viscosity is too high, the mixture may be unworkable by the time it reaches the site. Also, if too low a viscosity is used in a surface dressing, the result may be 'bleeding' or a loss of chippings under traffic.

The measurement of a bitumen's absolute viscosity requires the careful use, at a standard temperature, of relatively sophisticated laboratory equipment such as a sliding plate viscometer[2]. The determination of the absolute viscosity of a bitumen using this viscometer is explained as follows.

If the space between two parallel flat plates 5 to 50 microns apart is filled with a bitumen and one of the plates is moved parallel to the other, the particles attached to the moving surface will move at a speed that depends on the distance between the plates and the viscosity of the binder. The viscosity, η, in Pascal seconds (Pa.s) of the bitumen is the ratio of the shear stress to the rate of strain:

$$\eta = (F/A)/(v/d) \tag{5.1}$$

where F = applied shear force, N; A = surface area of each plate, m^2; v = velocity of one surface relative to the other, m/s; and d = distance between the plates, m.

The viscosity of a penetration-grade bitumen during hot mixing is, ideally, 0.2 Pa.s. During compaction with a roller, the viscosity depends upon the workability of the mix; typically, however, it ranges from 1 to 10 Pa.s for hot rolled asphalt and 2 to 20 Pa.s for dense bitumen macadam. In service in the pavement, a bitumen is a semi-solid with a viscosity of the order of 10^9 Pa.s. (By comparison, water has a viscosity of 0.001 Pa.s.)

In industrial practice, where simplicity and rapidity are of the essence, viscosity is more likely to be measured indirectly using the long-established penetration and softening point tests as empirical proxies.

As shown in Fig. 5.1(a), the *penetration test*[3] measures the depth to which a standard needle will penetrate a bitumen under standard conditions of temperature (25°C), load (100 g) and time (5 s). The result obtained is expressed in penetration units, where one unit equals 1 dmm (0.1 mm).

The penetration test on its own is simply a classification test, and is not directly related to binder quality. Nonetheless, the penetration grade of a bitumen is inextricably linked to the composition and use of a bituminous material. For example, higher penetration bitumens are preferred for use in pavements in colder climates (to reduce cracking problems) whilst lower penetration ones have preference in

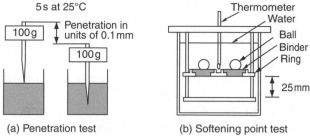

Fig. 5.1 Penetration and softening point tests for bituminous binders

hot climates. When used to bind a well-graded aggregate with high internal friction, a medium-grade bitumen will enhance the workability of the mix. A low-penetration bitumen will provide stability to a mastic-type of mix with a high sand content, for use at locations where traffic stresses are very high, e.g. at bus stops.

The *softening point test*[4] determines the temperature at which a bitumen changes from semi-solid to fluid. The softening point is an equi-viscous temperature in that it is the temperature at which all refinery bitumens have the same viscosity, i.e. about 1200 Pa.s. There is a rule-of-thumb to the effect that the mixing temperature of a penetration-grade bitumen is 110°C above its softening point; thus, for example, nominal 50 pen and 100 pen bitumens are often mixed at about 162°C and 156°C, respectively.

As indicated in Fig. 5.1(b), the test involves placing a 3.5 g steel ball on a disc of bitumen that is supported by a brass ring, and immersing it in water. The water is heated uniformly at the rate of 5°C/min until the disc is sufficiently soft for the ball, enveloped in bitumen, to fall 25 mm through the ring onto a base plate. The water temperature at which the binder touches the base plate is recorded as the softening point of the bitumen.

A bituminous binder should never reach its softening point under traffic. Many bitumens, but not all, have a penetration of 800 pen at their softening point. Below the softening point temperature the relationship between the logarithm of the penetration (dmm) and temperature (°C) is linear for non-blown bitumens, i.e.

$$\log pen = \text{constant} + \alpha t \tag{5.2}$$

where, for a bitumen of a given origin, the slope α is a measure of its temperature susceptibility, and is obtained by measuring the penetration at two temperatures. It is well established that values of α vary from 0.015 to 0.06; however, other than showing that there are considerable variations in temperature susceptibilities, these numbers mean little to the road engineer. For ease of interpretation, therefore, another measure of temperature susceptibility, i.e. the *penetration index (PI)*, which assumes that a 'normal' 200 pen Mexican bitumen has a $PI = 0$, was devised[5] and other road binders are then rated against this standard. The *PI* is obtained from

$$PI = \{20(1 - 25\alpha)\}/(1 + 50\alpha) \tag{5.3}$$

The value of α is derived from penetration measurements at two temperatures, t_1 and t_2, and using Equation 5.4.

$$\alpha = \frac{(\log pen \text{ at } t_1 - \log pen \text{ at } t_2)}{(t_1 - t_2)} \qquad (5.4)$$

Initially, a bitumen with a *PI* < 0 was regarded as being inferior in terms of temperature susceptibility; nowadays, most road bitumens are considered acceptable if they have a *PI* between −1 and +1. The *PI* is also used to estimate the stiffness of a bitumen for analytical pavement design purposes[6].

The *Fraas breaking point test*[7], which measures the (low) temperature at which a bitumen reaches a critical stiffness and cracks, is used as a control for binders used in very cold climates. The Fraass breaking point is the temperature at which a thin film of bitumen attached to a metal plate cracks as it is slowly flexed and released whilst being cooled at the rate of 1°C/min. If the penetration and softening point of a bitumen are known, the breaking point can be predicted as it is equivalent to the temperature at which the bitumen has a penetration of 1.25.

As the loading times for penetration, softening point and the Fraass breaking point tests are similar for a given penetration-grade bitumen, a process has been devised that enables these and viscosity test data to be plotted against temperature on a *bitumen test data chart*[8,9]. The temperature scale in this chart (Fig. 5.2) is linear, the penetration scale is logarithmic, and the viscosity scale has been devised so that low-wax content bitumens with 'normal' temperature susceptibility or penetration indices give straight line relationships (i.e. curve S, for Straight, in Fig. 5.2); consequently, only the penetration and softening point of the binder need be known to predict the temperature/viscosity characteristics of a penetration-grade bitumen. Blown (B) bitumens can be represented by two intersecting straight lines, where

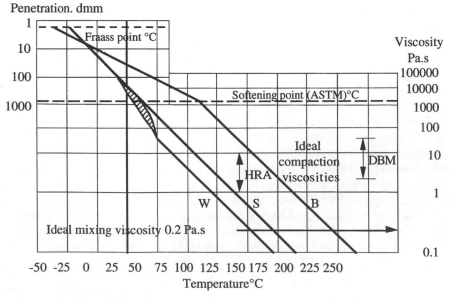

Fig. 5.2 A bitumen test data chart

the slope of the line in the high-temperature range is approximately the same as that for the unblown (S) bitumen. Four test values are required to characterize the bitumen. Waxy (W) bitumens also give intersecting lines of nearly equal slopes, but are not aligned. Between the two lines, there is a transition range in which the test data are scattered.

A chart such as in Fig. 5.2 has two main uses: (a) to provide a fingerprint of the rheological behaviour of the binder, and (b) to determine the ideal mixing and compaction temperatures for the bitumen. A viscosity of 0.2 Pa.s is generally used for mixing, whilst the viscosity adopted for rolling, which depends on the workability of the mix, ranges from 1–10 Pa.s for a hot rolled asphalt (HRA) to 2–20 Pa.s for a dense bitumen macadam (DBM)[10].

The *loss-on-heating test*[11] is essentially an accelerated volatilization test that is designed to ensure that excessive hardening of the bitumen does not occur, through loss of volatile oils, during the storage, transport and application of a binder. The test involves placing a 50 g sample of the binder in a small container and leaving it for 5 h in a revolving-shelf oven that is maintained at 163°C. The percentage loss in mass after the sample has cooled, and the percentage loss in penetration, are then recorded.

A solubility requirement of 99.5 per cent is found in all UK specifications for refinery bitumens. The *solubility test*[12] is used to ensure that the amounts of impurities picked up during the storage of a bitumen, and the amounts of salt that may not have been removed during the refining process, do not exceed the allowable limit. With this test, a specified amount of binder is dissolved in trichloroethylene, a very toxic solvent; after filtering the solution through a fine-porosity filter, the percentage material that is insoluble is obtained by difference.

For the purpose of the *permittivity test*[13] the permittivity or dielectric constant of a bitumen – this is a dimensionless value – is defined as the ratio of the capacitance of a capacitor with bitumen as the dielectric to the capacitance of the same capacitor with air as the dielectric. The susceptibility of a bitumen to weathering has been correlated with this parameter.

With the *rolling thin film oven (RTFO) test*[14] eight bottles, each containing 35 g bitumen, are arranged horizontally in openings within a vertical carriage in an oven that is heated to 163°C. The carriage is rotated at 15 rev/min and every 3 or 4 seconds a heated jet of air is blown onto the moving film of bitumen in each bottle, for 75 min. At the end of the test the loss in mass, penetration, and softening point of the bitumen are determined. The RTFO test measures hardening or 'ageing' of the bitumen by both oxidation and evaporation, and the results obtained have been found to correlate well with the ageing that occurs during mixing with hot aggregate.

The *flash point test* is carried out by heating a sample of bitumen at a uniform rate whilst periodically passing a small flame across the material. The temperature at which the vapours first burn with a brief flash is the flash point of the binder, whilst that at which the vapours continue to burn for at least 5 s is its fire point. The flash point is the more important of the two values as it indicates the maximum temperature to which the binder can be safely heated. The flash points of most penetration-grade bitumens lie in the range 245–335°C. (By contrast, medium-curing cutback bitumens usually flash between 52 and 99°C.)

CEN specifications: a comment

Historically, the testing of bituminous materials (including bitumens, tars and aggregates) in the UK has been governed by British Standards. This is now in the process of change as a consequence of Britain's involvement in the European Community (EC), and it can be expected that the traditional BS test publications will eventually be replaced by harmonized EN standards produced by the Comité Européen de Normalisation (CEN) and its technical committees.

The strategy underlying the preparation of CEN specifications for bituminous binders is to control quality by maximizing the use of traditional commonly-used test methods. To satisfy a wide range of national practices, economic interests, technological requirements and climatic conditions, it is envisaged that a set of mandatory tests will be applied; however, countries will also be able to choose from different additional optional tests in their national preface. Thus, for example, it can be expected that the mandatory tests will include standard tests for penetration, softening point, solubility, and viscosity, that the rolling thin film oven (RTFO) test will be used to assess resistance to hardening, and that the flash point test will be included for safety purposes. Optional tests may include wax content (mainly of interest in France and Germany), Fraass breaking point (to indicate cold weather behaviour), penetration index (to provide information regarding thermal cracking properties for, especially, Mediterranean countries), and changes in the softening point and Fraass breaking point after RTFO test hardening[15].

It might also be noted that CEN has now decreed that, following harmonization, bitumens will be graded by the following penetration ranges: 20–30, 30–45, 35–50, 40–60, 50–70, 70–100, 100–150, 160–220, and 250–330 pen.

5.1.2 Bitumen composition

As the chemical composition of a binder ultimately determines its performance, it is useful to very briefly consider the main components of a bitumen. Bitumen, which has a very complex structure, is generally regarded as being a colloidal system of high molecular weight asphaltene micelles dispersed in a lower molecular weight maltene medium. As indicated in Fig. 5.3, the micelles are considered to be asphaltenes together with an absorbed sheath of high molecular weight aromatic resins which act as a stabilizing solvating layer[16]. Further away from the centre of the micelles, there is a gradual transition to less aromatic resins and these, in turn, extend into the less aromatic oily maltene dispersion medium.

The *asphaltenes* are brown to black, highly polar, amorphous solids containing, in addition to carbon and hydrogen, some nitrogen, sulphur and oxygen. The asphaltene content significantly influences the rheological characteristics of a bitumen, e.g. increasing the content gives a lower-penetration harder bitumen with a higher softening point and, consequently, a higher viscosity. The *resins* are dark brown semi-solids or solids. They are very polar and this makes them strongly adhesive. The *aromatics* are dark brown viscous liquids; they comprise 40 to 60 per cent of a bitumen. The *saturates* are non-polar viscous oils that are straw or white in colour; they comprise 5 to 20 per cent of a bitumen.

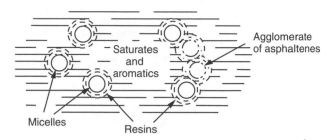

Fig. 5.3 The main chemical components of bitumen[17]

Blowing air through a bitumen during its manufacture increases the asphaltene content and decreases the aromatic content, thereby decreasing the temperature susceptibility of the binder. Mixing (especially), storage, transport, placement and compaction, as well as in-service use, also have the ageing effect of increasing the asphaltene content and decreasing the aromatic content.

5.1.3 Engineering properties of bitumen

The stress–strain relationship for bitumen is more complex than for conventional civil engineering materials such as steel and cement concrete. At low temperatures and short loading times, bitumen behaves almost as an elastic (brittle) solid, whereas at high temperatures and long loading times, its behaviour is almost that of a viscous fluid. At intermediate temperatures and loading times (e.g. such as are experienced at ambient air temperatures and under the spectrum of pulse loadings caused by moving traffic), bitumen's response is in the visco-elastic range. This visco-elastic hallmark is stamped on the performance of all bituminous materials.

A sinusoidal loading form is used to simulate the pulsed loading that occurs under vehicular traffic, when measuring the stress–strain relationship for bitumen. When a visco-elastic material is subjected to a repeated sinusoidal stress of the form $\sigma_0 \sin \varpi t$, the resulting strain $\varepsilon_0 \sin(\varpi t - \phi)$, where ϕ is the phase lag, lags behind because of the viscous component of the response. The strain can be expanded trignometrically as $\varepsilon_0 \{\cos \phi \sin \varpi t - \sin \phi \cos \varpi t\}$ showing that the response is composed of one component in phase with the stress ($\varepsilon_0 \cos \phi$) and another 90 degrees out of phase ($\varepsilon_0 \sin \phi$). The relationship between stress and strain is evidently complex. A physical interpretation of this is provided in the rotating vector diagram in Fig. 5.4, with the variations in stress and strain being provided by the horizontal or vertical projections of the vectors $\tilde{\sigma}(t)$ and $\tilde{\varepsilon}(t)$. By establishing real and imaginary axes, stress and strain are expressed as complex numbers:

$$\tilde{\sigma}(t) = \sigma_0 (\cos \varpi t + i \sin \varpi t) = \sigma_0 e^{i\varpi t} \tag{5.5}$$

$$\tilde{\varepsilon}(t) = \varepsilon_0 (\cos(\varpi t - \phi) + i \sin(\varpi t - \phi)) = \varepsilon_0 e^{i(\varpi t - \phi)} \tag{5.6}$$

Accordingly, the complex modulus, E^*, which is the equivalent of Young's modulus for an elastic material, is defined as a complex number

$$E^* = \frac{\sigma}{\varepsilon} = \frac{\sigma_0}{\varepsilon_0} e^{i\phi} = \frac{\sigma_0}{\varepsilon_0} \cos \phi + i \frac{\sigma_0}{\varepsilon_0} \sin \phi = E_1 + iE_2 \tag{5.7}$$

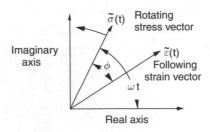

Fig. 5.4 Rotating stress and strain vectors

The real component of the modulus, E_1, is in phase with the stress and represents that part of the stored energy in the material that can be restored; this is called the storage modulus. The imaginary component of the modulus, E_2, gives an indication of the amount of energy that is dissipated in the material by internal friction; this is the loss modulus. The absolute value of the complex modulus, $|E^*|$, is termed the *stiffness modulus*,

$$|E^*| = \frac{\sigma_O}{\varepsilon_O} = \sqrt{(E_1^2 + E_2^2)} \tag{5.8}$$

For linear and homogeneous materials, the relationship between E^* and the complex shear modulus, G^*, is given by

$$E^* = 2G^*(1 + \nu^*) \tag{5.9}$$

where $\nu^* =$ complex Poisson's ratio.

The complex shear modulus, G^*, and the visco-elastic phase angle, δ, can be measured directly using the sliding plate rheometer[18]. The dissipated energy $G^* \sin \delta$ is used to characterize fatigue performance, whilst the characteristic value $G^* \div \sin \delta$ is used to assess rutting potential.

The stiffness modulus of a bitumen, S_b (MN/m^2), under a given loading time, t (s), at a temperature, T (°C), can be estimated from the well-known Van der Poel nomograph[19], using the American Society for Testing and Materials (ASTM) softening point for which the water is not stirred during the test, SP, and the penetration index, PI. For repeated numerical calculations, the following equation, which is derived[20] from the nomograph, can be used to estimate S_b:

$$S_b = 1.157 \times 10^{-7} t^{-0.368} e^{-PI} (SP - T)^5 \tag{5.10}$$

provided that $0.01s < t < 0.1s$, $-1 < PI < +1$, and $10° < (SP - T) < 70°C$.

Once the stiffness modulus of the bitumen has been estimated it can, in turn, be used to estimate the stiffness modulus of the material in which it is being used as a binder. Allowing for hardening during mixing and laying to about two-thirds of the original penetration, typical stiffnesses for original 50 pen and 100 pen bitumens at 15°C are 110 MPa and 55 MPa, respectively, for traffic moving at 50 km/h (i.e. $t = 0.02$ s).

5.2 Natural asphalts

The largest natural deposit of *lake asphalt* occurs on the Island of Trinidad off the north-west coast of South America. The main lake of asphalt covers an area of roughly 35 ha, has a depth of about 90 m, and is estimated to contain 10 to 15 million tonnes of material. Following excavation, the asphalt is heated to 160 °C to drive out gases and moisture, and then run through strainers to remove vegetable debris before being poured into wooden barrels for export under the name 'Trinidad Épuré' or 'Refined Trinidad Lake Asphalt'.

The refined lake asphalt product typically contains about 55 per cent bitumen, 35 per cent mineral matter, and 10 per cent organic matter. Following the first commercial shipment to England (in 1840), it was widely used in road construction until the introduction of pitch-bitumen in the 1960s, i.e. a blend of 70–80 per cent bitumen with 20–25 per cent coal-tar pitch, which had similar qualities.

Épuré is renowned for its ability to weather well, i.e. by oxidizing, it helps to maintain a rough surface texture and good skid resistance with ongoing abrasion by vehicle wheels. Nowadays, it is usually blended with 100–300 pen refinery bitumens; this is done by melting the hard Épuré (ca 1 dmm pen) and adding appropriate proportions of refinery bitumen to produce 35, 50 and 70 pen binder blends. For rolled asphalt surfacings, a 50:50 mixture of Épuré and a 200 pen bitumen tends to give near optimum results for most normal roads, bus stops, and footpaths[21].

Natural rock asphalts are mainly limestones and sandstones that are impregnated with, typically, 5–15 per cent of natural bitumen. Historically, the natural rock asphalt used in the UK was imported from the Val de Travers region in Switzerland and from the Gard region in France. Whilst their usage is covered by a British Standard[22], natural rock asphalts are rarely employed in road construction in the UK today.

5.3 Cutback bitumens

The consistencies of penetration-grade bitumens vary from very viscous to semi-solid at ambient air temperatures, so they are normally heated to quite high temperatures (typically 140–180°C) for use in road pavements. In many instances, however, it is neither necessary nor desirable to use a penetration-grade binder, and instead use is made of a cutback bitumen that is capable of being applied at ambient temperatures with little or no heating.

Various types of cutback bitumens are in use throughout the world. Broadly, these can be classified as slow-curing, medium-curing, and rapid-curing, depending upon the nature of the volatile solvent used in their preparation. A medium-curing cutback is used in the UK, most usually for surface dressing or maintenance patching purposes, but also in open-textured bitumen macadams that are porous enough for the solvent to evaporate fairly quickly. This cutback is produced by blending kerosene or creosote with a 100, 200 or 300 pen bitumen. After application the solvent dissipates into the atmosphere and leaves the cementitious bitumen behind.

In practice, an adhesion agent such as tar acids or an amine salt is usually included in the formulation of cutback bitumen to assist the 'wetting' of the aggregate and minimize the risk of the early stripping of bitumen from the stone in damp or wet weather, whilst the flux oil is evaporating.

As the viscosity of the cutback bitumen is low, it is specified by the time, in seconds, required for 50 ml to flow through a 10 mm diameter orifice in the Standard Tar Viscometer (STV)[23]. If required, its absolute viscosity, in Pa.s, can be determined by multiplying the STV flow time, (s) by the density of the cutback bitumen (g/ml) and by a constant (=0.400). Three cutback bitumen grades are specified[1]: 50 s, 100 s and 200 s containing 8 to 14, 6 to 12, and 4 to 10 per cent distillate by volume at 360°C, respectively. The 100 s grade is that which is most commonly used.

The flash point of a cutback bitumen can be as low as 65°C, due to the volatile oils. Consequently, these binders require care to ensure their safe handling during storage and at working temperatures.

5.4 Bitumen emulsions

As bitumen and water are immiscible, another way of reducing the viscosity of the bitumen is to add it to water and create an emulsion. An emulsion is a fine dispersion of one fluid (in the form of minute droplets) in another liquid in which it is not soluble. Bitumen emulsions mostly used in the UK are a stable suspension of 100, 200 or 300 pen bitumen globules, each typically less than 20 micron in diameter, in a continuum of water containing an emulsifying agent; the function of the emulsifier, which is adsorbed on the surfaces of the globules, is to prevent them from coalescing.

Many factors affect the production, storage, use and performance of a bitumen emulsion, e.g: the chemical properties, hardness, and quantity of the base bitumen; the bitumen globule size; the type, properties, and concentration of the emulsifying agent; and the manufacturing equipment and the conditions under which it is used, e.g. temperature, pressure, shear.

Most emulsions are produced by heating the base bitumen to ensure a viscosity of less than 0.2 Pa.s (the ideal mixing viscosity) and shredding it in a colloidal mill in the presence of a solution of hot water and emulsifier. The temperature of the emulsifying solution is adjusted so that the temperature of the finished emulsion is less than the boiling point of water; for safety reasons it is normally less than 90°C. An emulsifier consists of a long-chain hydrocarbon terminating with a polar molecule which ionizes in the water; the non-polar paraffin portion is lipophilic with an affinity for bitumen, whilst the polar portion is hydrophilic with an affinity for water. A *cationic emulsion* is produced with the aid of an emulsifying solution prepared by dissolving a long chain of fatty amine in hydrochloric or acetic acid:

$$CH_3 - CH_2 \ldots \ldots NH_2 + HCl \rightarrow CH_3 - CH_2 \ldots \ldots NH_3^+ + Cl^- \qquad (5.11)$$

An *anionic emulsion* is produced with an emulsifier prepared by dissolving a fatty acid in sodium hydroxide:

$$CH_3 - CH_2 \ldots \ldots COOH + NaOH \rightarrow CH_3 - CH_2 \ldots \ldots COO^- + Na^+ + H_2O \qquad (5.12)$$

When the concurrent streams of hot bitumen and hot water and emulsifier are forced through the minute clearance in the colloidal mill the shearing stresses created cause the bitumen to disintegrate into globules, many of the lipophilic chains are adsorbed by these globules, and positively charged NH_3^+ ions are located at the bitumen surface in the case of a cationic emulsion, and negatively charged COO⁻ ions at the bitumen surface in the case of an anionic emulsion. In either case, the like-charged globules repel each other when they come into contact, and this prevents their coalescence.

It is vitally important that the cationic and anionic types of bitumen emulsions are never mixed, as breakdown of both emulsions will then occur.

When a bitumen emulsion – it is brown in colour – comes in contact with a mineral aggregate, the emulsifier ions on the globules and in solution are attracted to opposite charges on the aggregate surfaces; this causes a reduction of the charges on the globule surfaces and the initiation of the *breaking* of the emulsion. The first step in the breaking process is termed 'flocculation'; it results in the electrostatically-reduced globules forming a loose network of globule flocs. The greater the amount of emulsifier present the slower the rate of globule flocculation, i.e. a small amount of emulsifier results in fast flocculation whilst a large amount is reflected in a slow break. Bitumen coalescence takes place rapidly when the charges on the globule surfaces become depleted and the aggregate surfaces become covered with hydrocarbon chains. The breaking process is completed when continuous black bitumen films are formed that adhere to the aggregate particles and the residual water in the system has evaporated. The evaporation of the water can be fairly rapid in favourable weather, but it is slowed by low temperatures, high humidities, and low wind velocities (including calms).

The aggregate grading is important to the breaking process, with fine aggregates (i.e. with large surface areas) breaking more quickly. Dirty aggregates accelerate the breaking process. The more porous and dry the road surface and/or the aggregate being coated, the more quickly water from the emulsion will be removed by capillary action, and the more rapidly breaking will be completed under the prevailing atmospheric conditions.

Cationic emulsions are generally considered to be more effective than anionic ones in coating some aggregates, under all weather conditions. With granite, for example, the positively-charged cationic emulsifier attached to the globules is attracted to the negative charge on the surface of the aggregate, whereas there is no attraction between the negatively-charged anionic emulsifier and the negatively-charged surface of the idealized granite. In practice, however, this is not very important as the negative charge on the aggregate is rapidly neutralized, and further breaking of the emulsion depends on evaporation[24].

5.4.1 Emulsion specifications

Bitumen emulsions used in the UK are classified[25] according to a three-part classification code.

The first part of the code designates the emulsion as being either A (for anionic) or K (for cationic).

The second part of the code is a number ranging from 1 to 4 that indicates the stability or breaking rate of the emulsion. Thus, the higher the number the more stable the emulsion. Class A1, A2 and A3 anionic emulsions are also described as being labile, semi-stable and stable, respectively, whilst class K1, K2 and K3 cationic emulsions are described as rapid acting, medium acting, and slow acting, respectively. A special class A4 emulsion has also been formulated for use in slurry seal work.

The ease with which *class A1 emulsions* break make them unsuitable for mixing with aggregates, and they are mostly used for surface dressing, tack coating, patching, formation and subbase sealing, and concrete curing. Labile emulsions (and semi-stable ones) cannot be stored out of doors in very cold weather as they will usually break upon freezing. *Class A2 emulsions* have enough emulsifier to permit mixing with aggregates provided that the percentage passing the 75 micron fraction is well below 5 per cent. *Class A3 emulsions* can be cold-mixed with aggregates, including those with large proportions of fines or chemically-active materials such as lime or cement. They can also be stored out-of-doors in cold climates without breaking and, in warm climates, they can be admixed with soil in soil stabilization works. When *class A4 emulsions* are mixed with aggregates specified for use in slurry seals, the mixes form free-flowing slurries that are sufficiently stable to be sustained throughout the laying procedure. Also, their setting times can be varied.

Class K1 cationic emulsions are characterized by a rapid deposition of bitumen on aggregates and road surfaces, with consequent early resistance to rain. Whilst normally fluid enough to be applied at atmospheric temperatures, the high bitumen content K1 grades must be applied hot. Whilst these emulsions are generally unsuitable for mixing with aggregates, they are commonly used for surface dressing, grouting, patching, and formation and subbase sealing. With the *class K2 emulsions* the rate of deposition of the bitumen is sufficiently delayed to allow mixing with some clean coarse aggregates. Applied cold, these emulsions are mostly used to prepare coated materials that can be stockpiled, e.g. for remedial patching. With the *class K3 emulsions*, which are also applied cold, the rate of deposition of the bitumen is sufficiently delayed for them to be mixed with certain fine aggregates. Cationic slurry seals are also formed with these emulsions.

The third part of the emulsion classification code is a number that specifies the bitumen content of the emulsion. Thus, for example, a K1-70 emulsion means that it is a cationic emulsion, rapid-acting, with 70 per cent by mass of residual bitumen.

The *viscosity* of an emulsion is of considerable importance as most of these binders are applied through nozzles on a spray bar. It is measured using either an Engler viscometer or a Redwood II viscometer[25]. In the case of the Engler apparatus, it is the time, in seconds, for 200 ml emulsion at 20°C to flow under gravity through a standard orifice; with the Redwood II test, it is the time, in seconds, for 50 ml emulsion to flow from the viscometer at 85°C.

5.5 Road tars and tar-bitumens

Tar is a viscous liquid, with adhesive properties, that is obtained by the destructive distillation of coal, wood, shale, etc. By destructive distillation is meant that the raw material is subjected to heat alone, without access to air.

The crude tars that now form the basis of road tars are a by-product of the distillation of coal at about 1000°C in the production of metallurgical or domestic coke (i.e. high-temperature coke oven tar) or smokeless solid fuel at 600–750°C (i.e. low-temperature tar). In either distillation process light oils, e.g. benzene, toluene and xylene are first collected, and then middle and heavy oils, e.g. creosote oils and naphthalene, are extracted, leaving a residual material that is termed 'pitch'. The pitch is subsequently fluxed back with light tar oils to obtain road tars of a desired standard. The dividing line between tar and pitch is essentially one of viscosity, so that a material that is below 60°C evt (equivalent to an R and B softening point of 42°C) is called tar, whilst anything more viscous is called pitch.

Road tars used to be heavily used in road construction in the UK prior to the discovery and exploitation of North Sea gas in the mid-1960s. Since then the crude tars produced as a by-product of the manufacture of town gas have essentially disappeared and, as a consequence, the crude tars now available for the production of road tars amount to only a small fraction of those previously produced. Crude tars are now mainly produced by carbonization ovens associated with the iron, steel and coal industries.

Two types of road tar are specified in a British Standard[26], types S and C. Each type is graded by viscosity with the measure being the tar's nominal equi-viscous temperature as determined using the Standard Tar Viscometer[23], i.e. this is the temperature in degrees Celsius at which the time of flow of 50 ml tar through a 10 mm diameter orifice is 50 s. The type S group, which is used for surface dressings, has four grades ranging from 34°C evt to 46°C evt in 4 degree increments, S34, S38, S42, and S46. The type C group has eight grades, also in 4 degree increments, C30 to C58; its tars are used in coated macadams, with the higher number (more viscous) tars being used in warm weather and in mixes in heavily-trafficked pavements requiring a higher stiffness modulus.

Tar–bitumen blends seek to combine the best qualities of both binders whilst minimizing their individual weaknesses. Blends used in surface dressing[27] normally fall within the composition ranges of 35 to 55 per cent of road tar and 45 to 70 per cent of refinery bitumen. The tar normally has not less than 25 per cent pitch of softening point 80°C (R and B). Two tar–bitumen grades are specified for use with surface dressings; they have viscosities (STV) at 40°C of 100 s and 200 s. The main features of these blends are their good weathering characteristics; they also have less tendency to strip from aggregates (as compared with pure bitumens), and are less brittle (as compared with pure tars).

Tar–bitumen mixtures used in coated macadam surfacings have penetration values of 100 or 200 pen, and they normally contain 10 per cent tar and 90 per cent bitumen.

Routine tests required by the British Standard, in addition to viscosity, involve determining the distillation residues after certain temperatures, R and B softening point, and the density at 20°C. The SEGAS or Beckton tray tests may be carried out, depending upon the tar type or tar-bitumen blend.

The *SEGAS test* is applied only to road tars and tar–bitumen blends used in surface dressings. With the standard test, a film of tar on a flat plate is exposed, at 45°C, to a controlled current of air for 43 h; the increase in equiviscous temperature is then measured and compared with an allowable increase. There is a close correlation between SEGAS test results and the equiviscous temperature increases of tar binders in service after a year. The brittle points of these tars and tar–bitumen blends are about 65°C and 75°C below their evt values, respectively, and they occur at road temperatures between −15 and −5°C and −20 and −10°C, respectively.

The SEGAS result can thus be used to predict whether a loss of surface dressing chippings is likely in that first year.

The *Becton tray test*, which is applied to tars and tar–bitumen blends used in coated macadams, measures the rise in equiviscous temperature that a thin film of binder undergoes in 1 h in a heated oven. The temperature of the oven varies according to the binder; 80°C for C30 to C42 tars and 100 or 200 s (at 40°) tar–bitumens, 100°C for C46 to C54 tars, and 140°C for C58 tar. The measured evt rise is then compared with recommended limits for the tar binder under consideration.

During the mixing of coated material, binder is usually spread as a 0.1 mm thick film on aggregate particles. Whilst all binders undergo hardening when mixing occurs at elevated temperatures, those containing volatile fluxing oils will harden more because of evaporation of the oils. The application of the recommended Becton tray test limits to the tar binders ensures that excessive hardening will not occur during coating, provided that the recommended aggregate coating temperatures are not exceeded.

5.5.1 Tars vs bitumens

The addition of tar to a bitumen generally improves the wetting of an aggregate during mixing, thereby improving, for example, surface-dressing adhesion. However, the main advantage that tar has is that it is much less vulnerable than bitumen to the dissolving action of petroleum solvents or distillates; thus, tar surfacings usually have longer lives than conventional bitumen ones when used in locations where spillages of petrol and oil are likely, e.g. parking areas.

Historically, it is well established that people engaged in specific occupations in the carbonizing and coal-tar byproduct industries have shown a high incidence of skin cancer. However, it is reported[28] that the carcinogenic hydrocarbon concentrations emitted during tar-paving operations are sufficiently low as to not represent a health hazard.

Tar is more temperature-susceptible than bitumen and has a narrower working temperature range. At high ambient temperatures mixtures containing tar are less resistant to plastic flow; however, at low temperatures, tar is more brittle. Surfacing materials containing tar must be very well compacted to minimize the intrusion of air and water, i.e. if not they will oxidize and harden more rapidly than mixes containing bitumen and early failure from cracking is more likely.

5.6 Modified bitumens

The past 25 years, in particular, has seen a continuing, underlying, demand for higher-quality bituminous binders. As a consequence, many modifying additives have been investigated with the aim of improving the properties of the penetration-grade bitumens used in road pavements. Some of these modifiers are proprietary products; others are in the public domain. Some have been commercially successful; others have not. Whatever, the fact that there has been such a multiplicity of modifiers can create confusion as to their relative roles and values.

When evaluating a particular modifier, it is useful to have in mind that (a) it is the predominantly viscous response of bitumen at high temperatures, i.e. its ability to flow, that makes it workable for contractors to produce and lay bituminous materials, and (b) it is the elasticity of bitumen, which predominates at relatively lower pavement temperatures, that gives the bituminous material its structural integrity. At ambient road temperatures, however, a small viscous component of the bitumen still remains and it is the accumulation of the viscous response to millions of axle load applications that abets rutting (especially at high ambient temperatures). At low temperatures, bitumen is susceptible to brittleness and cracking. These weaknesses can be considerably alleviated by increasing the overall stiffness of the bitumen used in locales with high ambient road temperatures (to reduce rutting) and/or by decreasing the stiffness of the binder used at locales with low road temperatures (to reduce cracking). These are the main reasons why, in the UK, small quantities of (relatively expensive) modifiers are added to bitumens used in wearing courses at difficult road sites, e.g. at roundabouts, traffic lights or pedestrian crossings on busy urban streets with large volumes of slow-moving vehicles in locales that are subject to extremes of temperature.

Modifiers have other advantages also, and it can be expected that their use in bituminous surfacings will grow in future years.

For discussion purposes, the main bitumen modifiers in current use can be divided into adhesion agents, thermoplastic crystalline polymers, thermoplastic rubbers, thermosetting polymers, and chemical modifiers.

5.6.1 Adhesion agents

The coating of an aggregate with bitumen is facilitated by using a low-viscosity high-penetration bitumen in the mix. However, if damp weather conditions prevail during or immediately after the laying, water may be preferentially attracted to the aggregate surfaces so that stripping of the bitumen takes place; however, the more viscous the binder the better its adhesion to the stone and the less likely it is that stripping will occur. Better adhesion to the stone and less stripping is likely when using more viscous binders. As a consequence, the choice of bitumen to use in a particular circumstance is very often a trade-off between ease of mixing, i.e. the workability of the mix, and the desire for better adhesion so as to minimize the risk of stripping.

One of the roles of a conventional filler material in a bituminous mix is to increase the viscosity of the binder, thereby lessening the risk of stripping. It is now well established that, if a small amount of *hydrated lime or cement* (say, 1 to 2 per cent by mass of the aggregate in the mix) is included as a replacement for some of the conventional filler material, a chemical action will take place between either additive and the bitumen that results in the formation of compounds that are adsorbed on negatively charged aggregate surfaces, and this has the effect of improving adhesion and rendering the bitumen less vulnerable to stripping.

5.6.2 Thermoplastic crystalline polymers

Also known as *thermoplastic plastomers*, these polymers include ethylene vinyl acetate (EVA), polyethylene (PE), polypropylene (PP), polystyrene, and polyvinylchloride (PVC). Of these the EVA polymer, which results from the copolymerization of ethylene and vinyl acetate, is probably the thermoplastic copolymer that is most widely used to modify bitumens for roadworks. The properties of an EVA copolymer are controlled by its molecular weight and vinyl acetate content. The lower the molecular weight the lower the viscosity and, hence, the stiffness. The greater the vinyl acetate content the more 'rubbery', i.e. flexible, the material. Various copolymer combinations are now available for use with (usually) more workable, lower viscosity, bitumens. All blend well with bitumen and are thermally stable at normal asphalt mixing temperatures.

A 70 pen bitumen with 5 per cent EVA is now promoted as an alternative to using a 50 pen bitumen in hot rolled asphalt wearing courses; tests have shown that, as well as being more workable, this combination provides increased stiffness and greater rutting resistance at ambient temperatures, albeit the bitumen's elastic properties are little affected. The addition of EVA to a wearing course also aids compaction during laying at low temperatures by improving the workability of the mix, and by providing more time for the embedment of coated chippings[29].

5.6.3 Thermoplastic rubbers

Also known as *thermoplastic elastomers*, these modifiers include polymers such as natural rubber, vulcanized rubber, styrene-butadiene-styrene block copolymer (SBS), styrene-ethylene-butadiene-styrene block copolymer (SEBS), styrene-isoprene-styrene block copolymer (SIS), and polybutadiene. As thermoplastics they soften on heating and harden when cooled to ambient temperatures.

Experiments on the use of rubber in bitumen or tar have been carried out for well over 100 years, and a wealth of literature exists on the subject. One review of the literature[30], which included the use of natural and vulcanized rubber in bituminous mixtures, concluded that rubber additives enhance the elastic responses of bitumen at higher ambient temperatures, and result in a material that has a marked increase in resistance to deformation simultaneously with reduced brittleness at low temperatures. However, when crumb rubber is added to a bitumen, only a proportion of the rubber particles disperse in the binder and the remainder acts as a soft 'aggregate' or cushion between stones in the total aggregate–binder structure. Later work[31] has shown that if crumb rubber recycled from used tyres is used with a catalyst that makes it soluble in the bitumen in dense surfacings, the result is a low noise-emission material.

The SBS and SIS styrenic block copolymers are now regarded[16] as the elastomers with the greatest potential for admixing with bitumen. It is critical that their admixing be with bitumens with which the copolymers are compatible; if they are not, the viscosity (stiffness) of the modified bitumen may not be changed in the direction required.

A polymer is a long molecule consisting of many small units, i.e. monomers, that are joined end-to-end, and copolymers are composed of two or more different monomers. By commencing polymerization with styrene monomer, then changing to a butadiene feedstock, and finally reverting to styrene, it is possible to make a butadiene rubber that is tipped at each end with polystyrene; this product is called styrene-butadiene-styrene (SBS). The polystyrene end blocks give strength to the polymer whilst the mid-block butadiene rubber gives it its elasticity. At mixing and compaction temperatures, i.e. above 100°C, the polystyrene will soften and allow the material to flow; on cooling to ambient temperatures, the copolymer will regain its stiffness and elasticity. At high in-situ temperatures of 60°C, SBS-modified bitumens are significantly stiffer and more resistant to permanent deformation than equivalent unmodified bitumens; at very low temperatures, the viscosity is reduced and the modified binders exhibit greater flexibility and provide greater resistance to cracking[32].

5.6.4 Thermosetting polymers

Polymers such as the acrylic, epoxy, or polyurethane resins result from the blending of a liquid resin and a liquid hardener which react chemically with each other. They are described as thermosetting because unlike thermoplastic polymers, their flow properties are not reversible with a change in temperature, once the material has cooled down to the ambient temperature for the first time.

When the two-component polymers are mixed with bitumen, the resulting binder displays the properties of modified thermosetting resins rather than of conventional thermoplastic bitumens[31]: the binder (a) becomes an elastic material that does not exhibit viscous flow when cured, (b) is more resistant than bitumen to attacks by solvents, and (c) is less temperature-susceptible than bitumen and is essentially unaffected by ambient pavement temperature changes.

The dynamic stiffness of a conventional asphalt is very considerably lower than that of an equivalent thermosetting polymer-modified asphalt, especially at high ambient temperatures.

5.6.5 Chemical modifiers

Sulphur and manganese are the two most well-known chemical modifiers considered for use with bitumen. The research work on these modifiers has mostly been carried out in North America where they are readily available.

Sulphur has an interesting temperature–viscosity relationship. Whilst a yellow solid material at standard conditions of temperature and pressure, it melts at about 119°C and exists as a low viscosity liquid (compared with bitumen) between 120°C and 153°C – its viscosity is lowest at 150°C – and then, at about 154°C, it starts to become rapidly, and very, viscous. Above 150°C the molten sulphur reacts very vigorously with hot bitumen to release hydrogen sulphide (H_2S), which is a toxic gas and a dangerous health hazard; also, the higher the temperature the more rapid the rate of H_2S evolution.

Depending upon the composition of the bitumen, at least 15 to 18 per cent by mass of sulphur is normally added to the bitumen prior to admixing with the aggregate. Some of the sulphur reacts chemically with the bitumen, whilst the remainder remains in suspension as a separate phase provided that the temperature does not exceed 150°C. As the mix cools, the excess sulphur in the void spaces slowly re-crystallizes, keying in the coated particles and increasing interparticle friction, and imparting high mechanical stability to the mix. Because of sulphur's fluidity when heated, its addition to the bitumen reduces the viscosity and increases the workability of the binder–aggregate mix; because of the sulphur's beneficial effect on the stiffness modulus after cooling, higher-penetration bitumens can be used, thereby rendering the mix even more workable.

Because of the relative fluidity of the mix, sulphur–bitumen materials are very adaptable to thin-layer work. For example, there is a proprietary material marketed in the UK that is promoted for repairing surface damage to roads; the heated material is simply poured, levelled and then, when it has cooled, vehicles can travel over it, i.e. no roller compaction is required.

Oil-soluble *organo-manganese compounds* have also been proposed as additives to bitumen on the grounds that they improve the temperature susceptibility of the binder, thereby improving the dynamic stiffness and resistance to deformation of the bituminous material. These compounds must be premixed with a carrier oil if they are to be dispersed quickly in a bitumen.

5.7 Cements

Technically, a cement is any material which, if added in a suitable form to a non-coherent assemblage of particles, will subsequently harden by physical or chemical means and bind the particles into a coherent mass. This broad definition allows such diverse materials as bitumen, tar and lime to be termed 'cements'. In general, however, the term is most often associated with Portland, slag, pozzolanic, and high alumina cements, all of which are finely-ground powders which, in the presence of water, have a chemical reaction (hydration) and produce, after setting and hardening, a very strong and durable binding material.

Setting is the change in the cement paste that occurs when its fluidity begins to disappear; the start of this stiffening process is the 'initial set' and its completion is the 'final set'. *Hardening*, which is the development of strength, does not begin until setting is complete. The setting time and rate of setting are considerations of vital importance affecting the use of cement in road pavements. Gypsum (i.e. calcium sulphate) retards setting and for this reason a small amount is added to cement during its production.

The literature on cement is legion and easily available, as are the methods of testing cement. Hence, the following is only a very brief and general description of the more important cements used in roadworks.

5.7.1 Portland cement

Portland cement, first used in 1824, is named after the natural limestone found on the Portland Bill in the English Channel, which it resembles after hydration. The raw materials used in the preparation of cement are calcium carbonate, found in the form of limestone or chalk, and alumina, silica, and iron oxide, found combined in clay or shale. Marl, which is a calcareous mudstone, can also be used in the manufacture of cement.

The cement is prepared by grinding and mixing proportioned amounts of the raw materials, and feeding the intimate mix through (most usually) a coal-fired high temperature kiln. When the temperature of the material in the kiln is at about

Table 5.2 Effects of the main Portland cement compounds

Principal constituent compounds	Rate of chemical reaction and heat generation	Most active period	Contribution toward final strength
C_3S	Moderate	2nd to 7th day	Large
C_2S	Slow	7th day onwards	Moderate
C_3A	Fast	1st day	Small

1450°C, incipient fusion takes place and the components of the lime and the clay combine to form clinkers composed of tricalcium silicate ($3CaO.SiO_2$, abbreviated C_3S), dicalcium silicate ($2CaO.SiO_2$, or C_2S), tricalcium aluminate ($3CaO.Al_2O_3$ or C_3A), and calcium aluminoferrite ($4CaO.Al_2O_3.Fe_2O_3$ or C_4AF). (*Note*: It is customary in cement chemistry to denote the individual clinker minerals by short symbols such as $CaO = C$, $SiO_2 - S$, $Al_2O_3 - A$ and $Fe_2O_3 = F$.) The burnt clinkers are allowed to cool and taken to ball-and-tube mills where they are ground to a fine powder. Gypsum (typically 1–5 per cent) is then added during the grinding process.

Table 5.2 compares the contributions to final strength made by the main cement compounds.

British Standard Portland cement[33] with the suffix N (known as *ordinary Portland cement*) is the workhorse of the cement industry; having a medium rate of hardening, it is this cement that is most commonly used in road pavements.

Portland cement with the suffix R (*rapid-hardening Portland cement*) is similar to the ordinary cement except that the final cement powder is more finely ground; consequently, a much greater surface area is available for hydration and it is able to harden more rapidly. Rapid-hardening cement is more expensive than ordinary cement because of the extra grinding of the clinker that is required to achieve the fine powder. Its higher rate of strength development lends itself to use in situations where an accelerated road-opening to traffic, or a reduction in the time when the pavement is vulnerable to frost action, is essential.

Sulfate-resisting Portland cement[34] is similar to ordinary cement except for its special capabilities in resisting chemical attack from sulfate (sulphur trioxide, SO_3) present in seawater, some groundwaters, gypsum-bearing strata, certain clay soils in hot countries, and/or some industrial wastes. This cement is now rarely used in rigid pavement construction in the UK because the design of the subbase and the placement of a membrane between the slab and the subbase are usually sufficient to separate the concrete from the disintegrating effects of sulfate in the soil.

5.7.2 Portland blastfurnace cement

This cement[35] is a mixture of Portland cement and up to 65 per cent of ground, granulated, blastfurnace slag. Granulated slag used in cement[36] is a by-product of the manufacture of iron and, by itself, is a relatively inert material. However, its important characteristic is that it is pozzolanic, i.e. it will react with lime in the presence of water to form cementitious products. Thus, the hardening of blast-furnace cement is characterized by two processes: (a) the cement clinker hydrates when water is added, and (b) as this hydration occurs, there is a release of free calcium hydroxide which reacts with the ground slag as it hydrates. A Portland blastfurnace cement that has a low slag content behaves like an ordinary Portland cement whilst one with a high content reflects the influence of the slag ingredients.

In general, blastfurnace slag cement hardens more slowly than ordinary cement, but in the long run there is little difference in the final strength achieved by either. The use of slag cement is most justified in locales where this cement is economically produced and where high early pavement strength is not an essential requirement.

5.7.3 Pozzolanic cement

This cement is made by grinding together an intimate mixture of Portland cement clinker and a pozzolanic material such as natural volcanic ash or finely-divided pulverized fuel ash. Pulverized fuel ash (known as 'pfa' in the UK and 'fly ash' in the USA) is a waste product (see Subsection 5.10.3) obtained from the burning of coal in the generation of electricity. Pfa used in cement in Britain must meet specified requirements (e.g. see references 37 and 38).

As with blastfurnace slag cement, the hardening of a pozzolanic cement is characterized by the hydration of the cement powder and the release of free calcium hydroxide which reacts with the pfa as it hydrates. It is a slow-hardening low-heat cement, and its utilization in a pavement is dependent upon high early strength not being important. If a pozzolanic cement is included in a concrete mix, it permits lower water-to-cement ratios for a desired workability, thereby providing a denser concrete of lower permeability and greater durability. The cement is best used in locales where it is economically produced and/or it is desired to impart a degree of resistance to chemical attack from sulfate and weak acid.

5.8 Limes

Lime is calcium oxide (CaO), and is most often produced by calcining (burning) crushed limestone in either shaft (vertical) or rotary (near-horizontal) kilns. If the limestone is a pure or near-pure calcium carbonate ($CaCO_3$), the limes produced are termed *calcitic* or *high-calcium limes*. If the limestone is a dolomitic limestone containing a high proportion of magnesium carbonate ($MgCO_3$), the lime products are termed *dolomitic* or *magnesian limes*. Calcitic and dolomitic limes are used in road pavements in both quicklime and hydrated lime forms.

At atmospheric pressure and a temperature of 900°C, the product of the following reaction is termed *calcitic* or *high-calcium quicklime*:

$$CaCO_3 + heat \rightarrow CaO + CO_2 \qquad (5.13)$$

If a dolomitic limestone is calcined to 900°C at atmospheric pressure, the decomposition product is a mixture of calcium oxide (CaO) and magnesium oxide (MgO), which is *dolomitic quicklime*.

'Slaking' is a general term used to refer to the combining of quicklime and an excess amount of water to produce a slaked lime slurry of varying degrees of consistency. When just sufficient water is added to a high-calcium quicklime to satisfy its chemical affinity for moisture under the hydration conditions, all of the calcium oxide will be converted to calcium hydroxide, $Ca(OH)_2$, with the evolution of heat, and the product is called *calcitic* or *high-calcium hydrated lime*. The MgO component of dolomitic quicklime does not hydrate so readily at the temperatures, atmospheric pressure and short retention times used in the normal hydration process; in this case the product is $Ca(OH)_2 + MgO$, which is termed *dolomitic monohydrate lime*.

Hydrated lime is produced as a fine dry powder containing, typically, about 30 per cent water; this makes it more expensive than quicklime, especially if large quantities and long haulage distances are involved. Both hydrated limes and quick-limes release heat upon contact with water; the heat given off by quicklime, however, is much greater because of the highly exothermic hydration reaction, and this makes it much more dangerous for construction workers to use, especially in windy weather conditions.

5.9 Conventional aggregates

The main aggregates used in road pavements on their own or in combination with a cementitious material are either natural rock materials, gravels and sands, or slag aggregates.

5.9.1 Natural rock aggregates

If the glacial drift overburden could be removed and the underlying rocks exposed, the geological map of Great Britain and Ireland would appear as simplified in Fig. 5.5. Solid rock suitable for the manufacture of road aggregates is almost exclusively quarried from formations of the Palaeozoic and pre-Palaeozoic geological ages[39,40] as few of the later formations have been sufficiently strengthened by heat and/or pressure processes to be of value for this purpose – which is why there are few quarries in the south east of Britain.

Geologists have classified rocks into three main groups, based on their method of origin; igneous, sedimentary, and metamorphic.

Igneous rocks were formed at (extrusive rocks) or below (intrusive rocks) the earth's surface by the cooling of molten material, called magma, which erupted from, or was trapped beneath, the earth's crust. Extrusive magma cooled rapidly and the rocks formed are very often glassy or vitreous (without crystals) or partly vitreous and partly crystalline with very small grain sizes. An extrusive rock may also contain cavities that give it a vesicular texture. By contrast, the intrusive rocks are entirely crystalline, due to the magma cooling slowly, and the crystals may be sufficiently large to be visible to the naked eye.

The best igneous roadstones normally contain medium grain sizes. Particles with coarse grains (>1.250 mm) are liable to be brittle and to break down under the crushing action of a compacting roller. If the grains are too fine (<0.125 mm), especially if the rock is vesicular, the aggregates are also liable to be brittle and splintery.

When an igneous rock contains more than about 66 per cent silica (SiO_2), it is described as 'acidic', and as 'basic' if it has less than 55 per cent. Rocks with 55–66 per cent total SiO_2 are termed 'intermediate'. Acidic rocks tend to be negatively

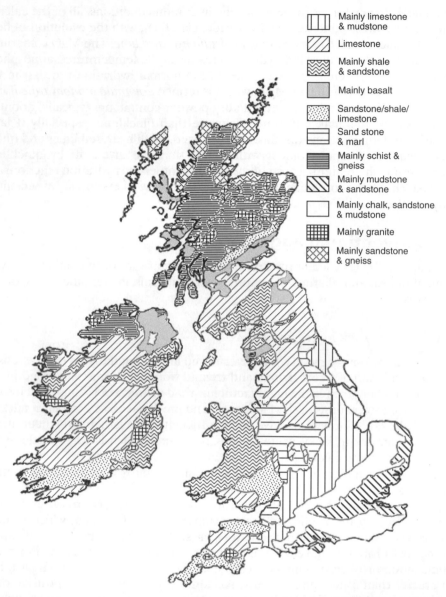

Fig. 5.5 Simplified geological map of Great Britain and Ireland

charged and aggregates containing large amounts of feldspar and quartz in large crystals do not bind well with bitumen (which also has a slight negative charge), whereas aggregates that are rich in ferromagnesian minerals, e.g. basalt and gabbro, bind well with bitumen.

As many rock aggregates differ little from each other in respect of practical road-making abilities, it is convenient to combine them into groups with common characteristics, e.g. the important igneous rock aggregates belong to the basalt, gabbro, granite and porphyry groups. The main rocks in the basalt group are basalt, dolerite, basic porphyrite and andesite; they are mostly basic and intermediate rocks of medium and fine grain size. The gabbro group is composed primarily of basic igneous rocks, gabbro, basic diorite and basic gneiss. Members of the granite group are mostly acidic and intermediate rocks of coarse grain size; the heavily used members of this group are granite, quartz-diorite, gneiss and syenite. Porphyry group members are acid or intermediate igneous rocks of fine grain size; typical examples are porphyry, granophyre, microgranite and felsite.

Basalt aggregates are strong and many have a high resistance to polishing; however, those containing olivine which has decomposed to clay (e.g. some in Scotland) have high drying-shrinkage characteristics that can lead to problems in concrete[39]. Granites are strong and their resistance to polishing is usually good; however, being acidic, more attention may have to be paid to anti-stripping treatment if they are to be used with bitumen. The porphyries are generally considered to be good all-round roadstones.

Sedimentary rocks were formed when the products of disintegration and/or decomposition of any older rock were transported by wind or water, redeposited as sediment, and then consolidated or cemented into a new rock type, e.g. siliceous rocks. Some rocks were also formed as a result of the chemical deposition of organic remains in water, e.g. calcareous rocks.

Argillaceous siliceous rocks were formed when fine-grained particles were deposited as clays or muds and then consolidated by pressure from overlying deposits. These rocks are very fine-grained, highly laminated, and very often are easily crushed into splinters. Consequently, they are rarely used as pavement aggregates and never in bituminous surfacings.

Arenaceous siliceous rocks were formed from deposits of sand and silt that became lithified as a result of pressure from overlying strata, or by the deposition of cementing material between the grains. Some of these rocks are brittle, whilst others are quite hard. The predominant mineral is either quartz or chalcedony (both SiO_2), which tends to make good adhesion between these arenaceous aggregates and bitumen more difficult.

Calcareous rocks resulted from great thicknesses of the remains of marine animals being deposited on the ocean floor. The predominant mineral is calcite ($CaCO_3$), which renders the rocks basic. Some calcareous rocks are too porous to be used as roadstones and tests[41] have shown that all crushed chalks, and oolitic and magnesian (dolomitic) limestones having an average saturation moisture content greater than 3 per cent, are frost susceptible and, hence, they should not be used in road pavements in Britain.

With sedimentary rocks, the most important road-making aggregate groups are the gritstones and limestones. Acceptable gritstones are abrasive and highly polish-resistant, e.g. greywackes, tuffs, breccias, fine-grained well-cemented sandstones, siltstones or flagstones. Unacceptable gritstones are mostly coarse-grained sandstones that are deficient in 'cement'. Limestone aggregate is widely used for all construction purposes. However, as they have a high susceptibility to polishing, most limestone aggregates are not used in bituminous wearing courses.

Metamorphic rocks are those which, as a result of great heat (thermal metamorphism) or great heat and pressure (regional metamorphism), were transformed into new rocks by the recrystallization of their constituents. Thermal metamorphic rocks, which are almost all harder than the rocks from which they were transformed, are generally in demand as road aggregates. Regional metamorphic rocks are relatively coarse-grained and some are highly foliated, e.g. schist (from igneous material) and slate (from shale). Aggregates from these foliated rocks are generally not desirable in pavements as they can be quite fissile and are liable to be crushed when compacted with rollers.

The main metamorphic aggregate groups of importance in road-making are hornfels, quartzite and, to a lesser extent, schists. Apart from having poor resistance to polishing, hornfels aggregates are very hard and make excellent pavement materials. Other than having a tendency to strip because of poor adhesion to bitumen, metamorphic quartzites make good road aggregates; they also have good resistance to polishing.

The simplified geological map at Fig. 5.5 is adapted from the Geological Survey Map of the main sources of rock in Great Britain and Ireland. This diagram shows very graphically that many of the sources of good aggregate are well away from where they are needed for developmental purposes. This is particularly clear in respect of the south-east of England where the in-situ materials are mainly chalk, clays and sands. As a consequence, the primary aggregates required for construction purposes (including road-making) in the heavily-populated south-east region have to be imported, mainly by road transport from the south-west.

5.9.2 Natural rock aggregate production

The production of rock aggregate consists of three processes; extraction, crushing and screening.

Extraction is usually carried out using explosives to 'win' the stone from the source rock in quarries. Very large pieces of won rock have to be reduced in size before they can be taken to the crusher; this is done by either secondary blasting ('pop-blasting'), dropping a heavy ball from an overhead jib ('drop-balling'), and/or using a hydraulic hammer.

Crushing involves continually reducing the size of the extracted stone to the sizes and shapes required by the pavement specifications, using either compression or impact crushers. Compression crushers squeeze the rocks between a fixed plate and a moving member that advances and recedes from the fixed one, as the stone moves through the crusher until it is small enough to pass out of the crushing chamber. Impact crushers subject the rock to repeated hammer blows as it passes through the reducing chamber until the particles are small enough to exit the chamber (often through a grid at the outlet).

The predominant size of the multi-sized aggregate produced by a crusher approximates its setting, where the setting is the smallest gap between the fixed and moving parts. However, the exact setting required to produce a predominant size of chipping is best determined by trial and error.

With a compression crusher, the dominant factor affecting the shape of the aggregate produced is the crusher reduction ratio; this is the ratio of the size of the feed opening to its setting. The lower the ratio the better the aggregate shape. Also, the best shapes are obtained with aggregate sizes at or about the setting.

Overall, impact crushers produce better-shaped aggregates than compression ones[39]. However, for economic reasons, i.e. to minimize wear of crusher parts, impact crushers are mostly employed to produce aggregates from more easily-crushed rock, e.g. limestone.

The production of aggregates of varying sizes involves a *screening* process in conjunction with the use of a number of crushing stages. These crushing stages commonly involve a primary crusher, a secondary reducing crusher, and two or three tertiary reduction crushers, e.g. cone crushers, which successively reduce the sizes of the stones produced at each stage. Before each crushing stage, material smaller than the next crusher setting is allowed to fall onto vibrating screens; this reduces the amounts of material passing through the crusher. Usually double- or triple-deck wire mesh screens are used to sort the various aggregate sizes, except at the primary stage when heavy-duty perforated plate screens are used to remove waste material or 'scalpings'.

The final screening, after the tertiary reduction stage, requires the crushed stones to be fed onto a series of screens to produce commercially saleable single-sized aggregates (typically 40, 28, 20, 14, 10, 6 mm and 3 mm dust) that are held in storage bins until required.

5.9.3 Gravels and sands

Gravel and sand are unconsolidated, natural, coarse-grained rock particles that have been transported by wind, water or glacial ice and deposited when movement slowed or stopped (see also Chapter 4). The individual particles are hard and usually rounded or irregular rather than angular, depending upon the amount of abrasion encountered during the prior movement.

Sands and gravels are commonly used in concrete pavements as their roundness results in good workability and their 'as dug' gradings often do not require additional processing. Hydrated lime or an anti-stripping agent may need to be added to some gravels, e.g. those containing flint, if they are to be used in bituminous mixes, to prevent the binder from stripping from the particles. Some gravels used in roadworks have to be crushed during processing to make them more angular.

Sands and/or gravels are most usually found on land in aeolian deposits such as sand dunes, in alluvial stream deposits such as valley terraces or alluvial fans, and in glacial deposits such as moraines, eskers and kames. They may be 'won' from dry pits using excavation equipment such as drag lines, scrapers, loading shovels, dozers, etc., or from wet pits using suction dredgers, floating cranes and grab or drag line excavators.

Sands and gravels are also extracted from offshore using suction dredgers equipped with hydraulic pumps capable of sucking sand, gravel and water from depths of up to 36 m, through rigid pipes that are dropped to the sea bed. As sand and gravel accumulates in the holds of the dredgers, the sucked water flows overboard. Processing of the materials won from the seabed is carried out on land.

The processing of sand and gravel first involves washing the extracted materials into separate sand and gravel fractions. The washing also removes silt and clay from the fractions and, in the case of marine materials, reduces the chlorides present from the sea water. The gravel fraction (i.e. the plus 5 mm size material) is then separated into various sizes using vibratory screens; if necessary, crushing is also carried out. The sand fraction is further washed and, usually, separated into coarse and fine fractions by hydraulic means.

5.9.4 Slag aggregate

The main 'secondary' aggregate used in road pavements in the 1950s and 1960s was air-cooled dense *blastfurnace slag*. Large quantities of this by-product of the smelting of iron ore were accumulated in waste tips in Britain in the latter half of the 19th and first half of the 20th centuries. Blastfurnace slag, which was initially used as bulk fill, became highly regarded as an aggregate and British Standard specifications were developed in relation to its use (e.g. see reference 42). Most of these early slag heaps have been used up and this, combined with changes in the method of producing steel (without intermediate pig iron) mean that the locations where significant quantities of this premium material are available are now much reduced.

Air-cooled blastfurnace slag has very good anti-skid properties and, hence, it is highly regarded as a surface dressing aggregate. The high angularity and irregular shapes of the slag particles mean that pavements incorporating this aggregate have high internal friction. Bituminous surfacings using slag are normally very stable; however, an additional amount of binder is normally required (vis-à-vis rock aggregates) to compensate for the binder content absorbed by the slag pores.

Steel slag produced from the current steel-making process has good polish-resistant properties. However, it is not normally used as bulk fill or selected granular fill because of the possible presence of free lime (CaO) and free magnesia and the consequent risk of expansion when hydration occurs.

5.9.5 Aggregate properties

The most important engineering properties of the aggregates used in road pavements are: cleanliness; size and gradation; shape and surface texture; hardness and toughness; durability; and relative density.

A clean aggregate is one that has its individual particles free from adherent silt-size and clay-size material. Aggregate *cleanliness* is normally ensured by inclusion in the pavement specifications of criteria relating to the maximum allowable amounts of adherent deleterious materials present in the coarse and fine aggregate fractions, as these reduce the bonding capabilities of cements and bituminous binders in mixes.

The aggregate *size and gradation*, i.e. the maximum particle size and the blend of sizes in an aggregate mix, affect the strength, density and cost of a pavement. When particles are to be bound together by a Portland cement or a bituminous binder, a variation in the gradation will change the amount (and consequently the cost) of binder needed to produce a mix of given stability and quality. Aggregate size and gradation have a major influence upon the strength and stiffness characteristics of a bituminous mix, as well as its permeability, workability, and skid resistance. For example, the aggregate grading in a dense bituminous surfacing (which depends upon being well graded for its denseness and consequent stability) is more critical than the grading used in a bituminous macadam (in which stability is primarily dependent upon the interlocking of the coarse particles).

Care should be taken to minimize the amount of handling and transporting to which well-graded aggregates are subjected, as particle-size segregation can easily occur during these operations, and this can be expensive to fix. Well-graded aggregates (e.g. Equation 5.14) have gradation curves resembling a parabola, i.e. a combination of a curve approaching an ellipse for the fines' portion and a tangential straight line for the coarse portion.

Particle *shape and surface texture* are used to describe aggregates (Table 5.3) and to provide guidance regarding their internal friction properties, i.e. those which (by means of the interlocking of particles and the surface friction between adjacent

Table 5.3 Descriptive evaluations of mineral aggregates

(a) Particle shape

Classification	Description	Examples
Rounded	Fully water-worn or completely shaped by attrition	River or seashore gravel; desert, seashore and wind-blown sand
Irregular	Naturally irregular, or partly shaped by attrition and having rounded edges	Other gravels; land or dug flint
Flaky	The thickness is small relative to the other two dimensions	Laminated rock
Angular	Possessing well-defined edges formed at the intersection of roughly planar faces	Crushed rock of all types; talus; crushed slag
Elongated	Usually angular, in which the length is considerably larger than the other two dimensions	–
Flaky and elongated	The length considerably larger than the thickness	–

(b) Surface texture

Surface texture	Characteristics	Examples
Glassy	Conchoidal fracture	Black flint, vitreous slag
Smooth	Water-worn, or smooth due to fracture of laminated or fine-grained rock	Gravels, chert, slate, marble, some rhyolites
Granular	Fracture showing more or less uniform round grains	Sandstone, oolite
Rough	Rough fracture of fine- or medium-grained rock containing no easily visible crystalline constituents	Basalt, felsite, porphyry, limestone
Crystalline	Containing easily visible crystalline constituents	Granite, gabbro, gneiss
Honeycombed and porous	With visible pores and cavities	Brick, pumice, foamed slag, clinker, expanded clay

surfaces) resist the movement of aggregates past each other under the action of an imposed load.

Crushed basalt, for example, is generally considered to be an excellent pavement aggregate because of the high internal friction associated with the angular shapes and rough surface texture of the particles. By contrast, rounded smooth gravels have relatively low internal friction as particle interlock and surface friction are poor; hence, many specifications require that gravel aggregate be crushed to produce jagged edges and rougher surfaces before being used in a pavement. Flat and flaky or long and thin aggregate particles produce poor interlock and, consequently, poor-quality pavement layers; when held with bitumen in surface dressings at carriageway level, they tend to be crushed by a roller and the result is poor embedment.

Particle shape has an appreciable effect upon the physical properties of bituminous mixes; it is critical with respect to the properties of open-graded mixes[43]. Rough-textured aggregates (both coarse and fine) contribute much more to the stability of a pavement than equivalent-sized aggregates with smooth surface textures.

Hard aggregates have the ability to resist the abrasive effects of traffic over a long time. The carriageway macrotexture that facilitates the rapid drainage of water from the surface whilst it is in contact with vehicle tyres (as well as utilizing the hysteresis effects in the tyre-tread rubber to absorb some of the kinetic energy of the vehicle) is dependent for its continuance upon the resistance of its ingredient materials to deformation and abrasion under traffic. The basic mechanism by which a road surface becomes slippery is the loss of aggregate microtexture brought about by the wearing and smoothing, i.e. polishing, of exposed particle surfaces by the tyres of vehicles. Also, the abrasion resistance of aggregates used in all pavement layers should be such as to ensure that the risk of fine material being produced by the movement of particles is kept at an acceptable level.

Tough aggregates are those which are better able to resist fracture under applied loads during construction and under traffic. The aggregates in each pavement layer must be tough enough not to break down under the crushing weight of the rollers during construction or the repeated impact and crushing actions of loaded commercial vehicles.

Durable or *sound* aggregates are those that are able to resist the disintegrating actions of repeated cycles of wetting and drying, freezing and thawing, or changes in temperature. Aggregates with high water absorptions (i.e. >2 per cent) are often considered to be vulnerable to frost action if they are placed in a pavement within 450 mm of the road surface. Aggregates coated with a bituminous binder are normally considered not to be vulnerable to frost problems in Britain[44].

The *relative density on an oven-dry basis* (RD_{od}) of an aggregate particle is the ratio of its mass in air (including the particle voids which are permeable to water intrusion as well as those which are impermeable) to the mass in air of an equal volume of distilled water.

Road aggregates are normally proportioned by mass and, hence, the relative density on an oven-dry basis is of vital importance in determining the proper particle-size blend. For example, gradation specifications are only valid if the particles in the coarse and fine fractions have approximately the same relative densities.

Thus, if the mean relative density of all the particles in the fine fraction is much greater than that for the coarse, the result may be a mixture which, because of a lack of fines, is too harsh; if the mean of the coarse fraction is the greater, the resultant mix may be too rich in fines.

The *saturated surface-dry relative density* (RD_{ssd}) is the ratio of the saturated surface-dry particle mass to the mass in air of an equal volume of water. It is used when determining aggregate abrasion values.

The *apparent relative density* (RD_{app}) of an aggregate particle is the ratio of its mass in air (including the particle's impermeable voids but excluding the permeable ones) to the mass in air of an equal volume of distilled water.

By definition, an aggregate particle's apparent relative density is greater than its relative density on an oven-dry basis, and consequently this leads to different estimates of the voids content in bituminous mix design. For this reason it is preferable to use an effective relative density that reflects the extent to which the binder in the mix penetrates the permeable voids in the aggregate. Alternatively, if it were assumed that the bitumen did not penetrate the permeable voids, the relative density on an oven-dry basis would be appropriate; the alternative assumption that the bitumen did penetrate the permeable voids would justify using the apparent relative density.

5.9.6 Aggregate tests

About 30 per cent of the aggregates produced in Great Britain are used in road construction, in all pavement layers from subgrade improvement (capping) layers to skid-resistant wearing courses[44]. Aggregate properties can be expected to vary from source to source, and the suitability of a given aggregate for a particular project will depend upon the pavement layer and the construction material used in that layer. The diversity of the variations and needs has resulted in the development of tests (which must be carried out on representative samples[45]) to assist the engineer in decision-making as to the suitability of particular aggregates.

The following are brief descriptions of some of the more common road aggregate tests used in the UK. Table 5.4 shows test results obtained for roadstones belonging to various aggregate groups.

Gradation tests are used by aggregate producers for quality control purposes and by users for checking compliance with specifications. They provide the quantities, expressed in percentages by mass, of the various particle sizes of which a sample of aggregate is composed.

In the case of coarse (retained on a 3.35 mm sieve) and fine (mainly passing the 3.35 mm and retained on the 75 micron sieves) particles, these quantities are determined[46] by sieve analysis, i.e. by allowing the aggregate sample to fall through a stack of standard sieves of diminishing mesh size that separate the particles into portions retained on each sieve. Depending upon need, the British Standard sieves that have historically been used in a stack have included all or some of the following sizes: 75, 63, 50, 37.5, 28, 20, 14, 10, 6.3, 5, 3.35, 2.36, 1.70 and 1.18 mm, and 850, 600, 425, 300, 212, 75 and (sometimes) 63 micron; following CEN harmonization the possible sieve sizes in a stack will be 125, 63, 40, 31.5, 20, 16, 14, 10, 8, 6, 4, 2, and 1 mm, and 500, 250, 125, and 63 micron. If the fines content is high, wet sieving

Table 5.4 Summary of means and ranges of test values for roadstones in various aggregate groups. (The ranges are noted in brackets)

Group	Aggregate crushing value	Aggregate abrasion value	Water absorption	Relative density	Polished-stone value
Basalt	14	8	0.7	2.71	61
	(15–39)	(3–15)	(0.2–1.8)	(2.6–3.4)	(37–64)
Granite	20	5	0.4	2.69	55
	(9–35)	(3–9)	(0.2–2.9)	(2.6–3.0)	(47–72)
Gritstone	17	7	0.6	2.69	74
	(7–29)	(2–16)	(0.1–1.6)	(2.6–2.9)	(62–84)
Limestone	24	14	1.0	2.66	45
	(11–37)	(7–26)	(0.2–2.9)	(2.5–2.8)	(32–77)
Porphyry	14	4	0.6	2.73	58
	(9–29)	(2–9)	(0.4–1.1)	(2.6–2.9)	(45–73)
Quartzite	16	3	0.7	2.62	60
	(9–25)	(2–6)	(0.3–1.3)	(2.6–2.7)	(47–69)
Gravel	20	7	1.5	2.65	50
	(18–25)	(5–10)	(0.9–2.0)	(2.6–2.9)	(45–58)
Slag	28	8	0.7	2.71	61
	15–39	3–15	0.2–2.6	2.6–3.2	37–74

should be carried out in preference to dry sieving, as otherwise an incorrect grading is likely to be obtained, i.e. the amount of very fine material will be underestimated.

The resultant particle-size distribution by mass is expressed either as the total percentages passing or retained on each sieve of the stack, or as percentages retained between successive sieves (see Table 5.5). The cumulative percentage-passing-by-mass method is very convenient for the graphical representation of a grading and is used in most aggregate specifications. The percentages-retained-on-particular-sieves method is preferred for single-sized aggregate specifications.

An aggregate is said to be 'single-sized' if its gradation is such that most of the particles are retained on a given sieve after having passed the next larger sieve. A

Table 5.5 Laboratory determination of the gradation of an aggregate (sample = 2000 g)

BS sieve size	Individual mass retained on sieve		Cumulative mass retained on sieve		Total mass passing sieve
(mm)	(g)	(%)	(g)	(%)	(%)
28	0	0	0	0	100
14	160	8	160	8	92
3.35	240	12	400	20	80
2	200	10	600	30	70
1.18	300	15	900	45	55
0.600	180	9	1080	54	46
0.300	180	9	1260	63	37
0.150	340	17	1600	80	20
0.063	220	11	1820	91	9

'gap-graded' aggregate is one where significant components of the particles are retained on a number of larger sieves and then the remainder are mainly retained on a number of smaller sieves, with only small amounts being retained on the in-between sieves. A 'well-graded' or 'continuously-graded' aggregate has sufficient particles retained on each sieve of the nest of sieves to ensure that a dense material is obtained when the aggregate is compacted; the ideal gradation is given by Fuller's curve:

$$P = 100(d/D)^n \qquad (5.14)$$

where P = total percentage by mass passing sieve size d, D = maximum size of aggregate particle in the sample, and n = typically 0.5 (without binder) to 0.45 (with binder).

The proportion(s) of aggregate material that comprises the filler content, i.e. historically, the material finer than 75 micron (or 63 micron, under the harmonized CEN standards), is determined by a gravimetric method[17], which applies Stoke's Law to the rate of fall of the particles in suspension in water.

The cleanliness of an aggregate, which is reflected in the surface-active 'dust' component of the filler that is smaller than clay-size, i.e. <2 micron, can be measured by the *methylene blue (MB) test* for the assessment of fines[48,49]. The basic stain test, the aim of which is to measure the ionic adsorption potential of at least 200 g of 0/2 mm particle size, involves admixing a series of 5 ml injections of a standard dye solution to a suspension of fines in 500 ml of distilled water. After each dye injection, a drop of the suspension is taken with a standard glass rod and deposited on a filter paper to form a stain that is composed of a generally-solid, blue colour, central deposit of material surrounded by a colourless wet zone. The amount of the drop must be such that the diameter of the deposit is between 8 and 12 mm. The test is deemed to be positive if, in the wet zone, a halo consisting of a persistent light blue ring of about 1 mm is formed around the central deposit. The methylene blue value is then recorded from the following equation:

$$MB = 10(V_1/M_1) \qquad (5.15)$$

where M_1 = mass of the 0/2 mm test portion (g), V_1 = total volume of dye solution injected (ml), and MB = methylene blue value (g/kg).

The measure of particle shape that is most often included in road aggregate specifications is the flakiness index.

The *flakiness test*[50], which is not applicable to particle sizes greater than 63 mm or smaller than 6.3 mm, is carried out by first separating the aggregate sample into the amounts retained on various sieve sizes and then using a special thickness gauge with slots to determine the mass of flaky material in each amount. Particles are classified as flaky if they have a thickness (smallest dimension) of less than 0.6 of their mean sieve size, i.e. this size being the mean of the limiting size apertures used for determining the size fraction in which the particles occur. The *flakiness index* is reported as the sum of the masses of aggregate passing through the various thickness gauges, expressed as a percentage of the total mass of the sample that is gauged. The lower the index the more cubical the aggregate.

Allowable flakiness index levels vary according to purpose; pre-coated chippings used in Great Britain have a maximum allowable index of 25, a bituminous mix with a very open grading may also only be allowed an index of up to 25, while other bituminous mixes may be permitted up to 50.

Many *hardness* tests have been developed to evaluate the ease (or difficulty) with which aggregate particles are likely to wear under attrition from traffic. The two most widely used in the UK are the aggregate abrasion test and the accelerated polishing test.

The *aggregate abrasion value (AAV)* test[51] is carried out on two samples of at least 24 aggregate particles, all passing the 14 mm sieve and retained on the 20 to 14 mm flake-sorting gauge with a slot width of 10.2 mm. The particles in each sample are mounted in, but project above, the surface of a resin compound contained in a small shallow tray. For testing, each tray is up-ended and the aggregate is pressed against a 600 mm diameter steel grinding lap that is rotating at 28–30 revolutions per minute. As the disc rotates, contact is maintained with the aid of a 2 kg load as a standard (Leighton Buzzard) abrasive sand is fed continuously to its surface just in front of the inverted tray. On completion of 500 revolutions, each test sample is removed, the aggregate loss in mass after abrasion is calculated and the *AAV* for each sample is determined from the following equation:

$$AAV = 3(M_1 - M_2)\, RD_{ssd} \qquad (5.16)$$

where M_1 and M_2 are the initial and final sample masses, respectively, and RD_{ssd} is their saturated surface-dry relative density. The *AAV* for the aggregate is then taken as the mean of the two values.

The lower the *AAV* the greater the aggregate resistance to abrasion. Thus, for example, aggregates with values greater than 12 are not allowed in coated macadam wearing courses carrying more than 2500 commercial vehicles per lane per day, whereas values of 16 are allowed in courses carrying less than 1000 cv/lane/day[52].

The *polished-stone value (PSV)* test[53] is a two-part skid-resistance test that is carried out on aggregate particles subjected to accelerated attrition by a polishing machine in the laboratory (see Figs 18.10 and 18.11). *PSV* is the only parameter relating to the microtexture of an aggregate which can be measured in a standardized manner and which has been related to traffic and site conditions in the UK[54].

In the first part of this test particles meeting specified size and shape criteria are set in resin and clamped onto the flat periphery of a 406 mm diameter 'road wheel' so as to form a 46 mm wide continuous surface of stone particles; this wheel is then rotated at 320 revolutions per minute. A 200 mm diameter solid rubber-tyred wheel with a width of 38 mm is next brought to bear on the aggregate surfaces with a force of 725 N and corn emery is fed continuously to the interacting tyre-aggregate surfaces for 3 h; at the end of this process a new wheel is brought to bear on the aggregate for another 3 h, only this time the feeding material is emery flour.

In the second part of the PSV-test, at the end of the 6 h of attrition, the state of polish of the aggregate sample is measured (in terms of the coefficient of friction between its wet surface and the rubber slider of a standard pendulum-type portable skid-resistance tester, see Fig. 18.7) and compared with the results from specimens from a control aggregate, to calculate the polished stone value. With the TRL tester, a falling pendulum swings further past the vertical when testing highly-polished aggregates compared with less polished ones. The tester gives results that represent the skidding resistance of a patterned tyre skidding at 50 km/h.

The *PSV* for the aggregate sample is determined from

$$PSV = S + 52.5 - C \qquad (5.17)$$

where S and C are means of test and control specimen values, respectively.

Aggregates with PSV-values of 68 or more are now recommended in new bituminous wearing courses at difficult sites in the UK, e.g. at and on approaches to traffic signals, pedestrian crossings, roundabouts, railway level crossings, or on gradients longer than 50 m and greater than 10 per cent. *PSVs* of 55 to 65 are recommended for most busy main roads (including motorways). Aggregates with *PSVs* < 50 are not recommended for use in any wearing courses, even if wet skidding accidents are unlikely.

Unfortunately, many high-PSV natural aggregates have low abrasion resistances.

The tests in common use that are reflective of *toughness* are the aggregate crushing, aggregate impact, and ten per cent fines tests.

The *aggregate crushing value (ACV)* test[55], which is a measure of the resistance of an aggregate to crushing under a gradually applied compressive load, is carried out on aggregate passing the 14 mm sieve and retained on the 10 mm sieve. The aggregate is placed in a standard mould using a specified procedure and then loaded uniformly so that a compressive force of 400 kN is slowly applied over 10 min. This load is then released and the mass of material passing the 2.36 mm sieve, expressed as a percentage of the total sample mass, is termed the *ACV*.

Note that the lower the *ACV* the tougher the aggregate, and that those with *ACVs* above 25 are rarely used in road pavements.

The *aggregate impact value (AIV)* test[56] is carried out by subjecting aggregate passing the 14 mm sieve and retained on the 10 mm sieve to 15 blows of a 14 kg hammer falling through a height of 380 mm. Following completion of the series of blows the material passing the 2.36 mm sieve is expressed as a percentage of the sample and recorded as the *AIV*.

The AIV test provides a relative measure of the resistance of an aggregate to sudden shock, e.g. as might occur under a vibratory roller. As with the *ACV*, the lower the *AIV* the stronger the aggregate. Aggregates with *AIV* > 25 are normally regarded as being too weak and brittle to use in a pavement.

A major advantage of the AIV test is that it uses equipment that is portable; this is in contrast with the ACV test which requires the use of a heavy compression testing machine that is not portable.

The *ten per cent fines value (TFV)* test[57], which uses the same equipment as the ACV test, is also carried out on dry aggregate passing the 14 mm sieve and retained on the 10 mm sieve. The aggregate is placed in a standard mould using a specified procedure and then loaded uniformly for 10 min. This action causes a degree of crushing which is reflected in the mass of fines (i.e. passing the 2.36 mm sieve) in the sample at the end of the test period. The mass of fines is weighed and expressed as a percentage of the total sample, whilst the force causing this degree of crushing is also noted. The procedure is repeated a number of times and a simple formula is then applied to the results to determine the load (in kN) that will cause 10 per cent fines. This force is the ten per cent fines value.

The TFV test tends to be used to evaluate weaker aggregates, i.e. where application of the 400 kN force required in the ACV test would result in an excessive amount of fines being produced so that the consequent cushioning effect could give rise to misleading *ACV* results. Note that the lower the *TFV* the less tough the aggregate. Aggregates with *TFV* < ca 8 kN are not normally used in road pavements.

A measure of the *soundness* of an aggregate, i.e. the extent to which it is susceptible to weakening by water, is obtained by carrying out the *soaked ten per cent fines* test[57]. The procedure is similar to that for the TFV test except that each aggregate sample is immersed in water for 24 h prior to the application of the crushing load (after the particles have been surface-dried), and the crushed particles are oven-dried before sieving.

The soaked TFV test is most often carried out on aggregates proposed for sub-base and capping layers in pavements. If the soaked *TFV* is significantly lower than the dry *TFV*, the aggregate may be judged unsound.

The three main *relative density* tests for aggregates[58] are a wire basket method for aggregate particles larger than 10 mm, a gas jar method for aggregates between 40 mm and 5 mm, and a pyknometer or gas jar method for aggregates 10 mm nominal size and smaller. The method for filler materials involves the use of a density bottle.

The principles underlying relative density determinations are most easily explained in relation to the wire basket method. Basically, this test involves immersing the prepared sample in water and after 24 h weighing it in the water, following a prescribed procedure. The saturated aggregate is then removed from the water and the individual particles are surface-dried with a towel before being weighed in the air; they are then placed in an oven set at 105°C. After 24 h the particles are removed from the oven and immediately weighed again in the air.

Formulae used to calculate the various relative densities (Mg/m^3) are:

$$\text{relative density on an oven-dry basis } (RD_{od}) = \frac{D}{A - (B - C)} \qquad (5.18)$$

$$\text{relative density on a saturated and surface-dry basis } (RD_{ssd}) = \frac{A}{A - (B - C)} \qquad (5.19)$$

$$\text{apparent relative density } (RD_{app}) = \frac{D}{D - (B - C)} \qquad (5.20)$$

where A = mass of the saturated surface-dry aggregate in air (g), B = apparent mass in water of the basket plus the saturated aggregate (g), C = apparent mass in water of the empty basket (g), and D = mass of the oven-dried aggregate in air (g).

The average relative density of an aggregate composed of fractions of different relative densities can be calculated from the individual values:

$$RD_{ave} = 100/\{(M_1/R_1) + (M_2/R_2) + (M_3/R_3)\} \qquad (5.21)$$

where RD_{ave} = average relative density of the final 'blended' aggregate, and M_1 and R_1, M_2 and R_2, and M_3 and R_3 are the mass percentages and relative densities of the individual fractions.

The *water absorption (WA)*, as a percentage of the dry mass of the aggregate, is also normally determined at the same time as the relative densities:

$$WA\ (\%) = 100\ (A - D)/D \qquad (5.22)$$

where *A* and *D* are as defined for relative density. The greater the water absorption percentage the more likely it is that an aggregate will have frost susceptibility problems in cold climes.

Once the water absorption and saturated surface-dry relative density are known, the porosity of an aggregate can be determined from

$$\text{Porosity } (\%) = 100 \ (WA)(RD_{ssd})/(WA + 100) \tag{5.23}$$

CEN specifications: a comment As noted previously in respect of bitumen testing, the testing of aggregates in the UK is currently governed by British Standards. As with other countries in the EC, these Standards are currently under review by the Comité Européen de Normalization (CEN) and its technical committees, and it can be expected that many of them will be changed soon. For example, as noted above, a decision has already been taken as to the sieve sizes to be used under the normalization process, and the reader is referred to reference 59 for detailed advice regarding this.

5.10 Secondary aggregates

The current and projected high demand for conventional quarried rock aggregates and the increasing difficulty of obtaining planning consent for their extraction, combined with a greater awareness of the considerable quantities of 'waste aggregates' that are both stockpiled and currently arising from the mineral extraction industries, the construction/demolition industry, and industrial processes and municipal incinerators, have stimulated greater interest in the use of secondary aggregates in road construction in Britain in recent years[60]. Benefits arising from the use of these low-grade aggregates include reductions in the demand for conventional materials, less energy and environmental costs associated with the extraction and transport of natural aggregates, and lower environmental and economic costs related to the storage and dumping of waste. Also, secondary aggregates often cost less than conventional materials as they are normally priced into the industrial processes that created them.

The normal uses of most secondary aggregates are as ordinary fill in embankments and as capping and subbase materials in pavements (see reference 61). Problems associated with their use include the (relatively poor) quality of the materials being considered, the variability between and within dumps and, with some materials, the subsequent release of contaminants through leaching from roadworks. Current requirements for secondary aggregates used in Britain are given in various Series of the *Specification for Highway Works*[62], viz. Series 600 (earthworks and capping), Series 800 (unbound sub-bases), Series 900 (bitumenbound layers), and Series 1000 (cement-bound roadbases and subbases, and pavement-quality concrete).

The main secondary aggregates considered for use in roadworks are blastfurnace and steel slags (see Subsection 5.9.4), colliery shale, spent oil shale, pulverised fuel ash and furnace bottom ash, china clay waste, slate waste, incinerated refuse and demolition and construction waste.

5.10.1 Colliery shale

Colliery shale, large quantities of which are located in the British Midlands, the North of England, and South Wales and Central Scotland, is the waste product of coal mining which was either removed to gain access to the coalface, or unavoidably brought from the pit with the coal and had to be separated out at the coal-cleaning plant. It is available in two forms in coal-mining areas: *unburnt shale* (now called *minestone*), and *burnt shale* (*burnt minestone*) which resulted from the spontaneous combustion of coal particles within the spoil heaps.

Shales, which are mainly composed of silica, alumina and iron, can differ considerably between and within heaps[63]. Nonetheless, hard well-burnt shale, i.e. shale that is free from ashy refuse and rubbish that may soften when wet, has a history of successful use as a fill material in embankments, and as a capping and subbase material, in roads that are relatively close to spoil tips. Normally, material of less than 76 mm maximum size is preferred for ease of laying and compaction.

Unburnt shale has been less used, mainly because of fears about spontaneous combustion. However, experience has shown that the risk of fire is heavily influenced by the ease with which oxygen can penetrate into and through the material, and that spontaneous ignition does not occur if the unburnt shale is well compacted as it is being laid. Consequently, increasing amounts of unburnt shale are now being used as bulk fill material.

Excess sulphate present in some shales can cause damage if it leaches into concrete and cement-bound materials. Thus, if a shale is to be used within 0.5 m of a bridge abutment, or a concrete pavement or pipe, the sulphate content of a 1:1 shale–water extract should not exceed 2.5 g sulphate (as SO_3) per litre. Also, unburnt shale should not be used in reinforced earth structures as sulphuric acid, which is formed from iron sulphide (FeS_2) reacting with oxygen and water from the atmosphere, could corrode the metal components of the reinforcing straps.

Most unburnt shales, and some burnt ones, are frost susceptible and are, therefore, not normally used in pavements in Britain unless they have at least 450 mm of cover or they are stabilized with cement.

5.10.2 Spent oil shale

The commercial exploitation of oil-producing shale in Europe has tended to be concentrated in Scotland, Italy and Spain. At its peak (in 1913) the West Lothian area of Scotland produced about 3.3 million tonnes of shale oil per annum, and petrol stations throughout Scotland supplied significant quantities of refined shale oil (known as 'Scotch') to motorists. However, this became uneconomic over time and production was finally stopped in 1962.

During production various oils were removed from the mined oil shale. The spent oil shale, and other waste shales brought to the surface, were dumped on land close to the mines and refineries. Spontaneous combustion sometimes occurred in the heaps, causing more changes to them.

This shale is most appropriately used as bulk fill in embankments and as a capping material; its loss-on-ignition is normally low, and there is no risk of spontaneous combustion if the material is properly compacted as it is being placed. The sulphate content can be high and care needs to be taken when using spent oil shale close to bridge abutments, concrete pipes, etc.

Water absorption tests give an average value of about 15 per cent for spent oil shale. The shale particles are relatively soft and tend to break down under compaction. Tests indicate that spent oil shale normally exhibits frost heaves that are well in excess of 18 mm after 250 h; as this is the maximum permitted in Scotland, this material is normally not used within 450 mm of a road surface, unless it is cement-stabilized.

5.10.3 Pulverized fuel ash and furnace bottom ash

Pulverized fuel ash (pfa) is the solid fine ash carried out in the flue gases of power station boilers that are fired with pulverized coal in the generation of electricity. This ash (also called *fly ash*) is a fine powder that is removed from the flue gases with the aid of mechanical precipitators and initially collected in hoppers; this dry white powder is called 'hopper ash'. If the hopper ash is passed through a mixer-conveyor plant where a measured amount of water is added before stock-piling, it is termed 'conditioned pfa'. At some power stations, the pfa is mixed to a slurry and hydraulically transported to storage ponds; this product is known as 'lagoon ash'.

The coarser component of the residual ash from the burnt pulverized coal is not carried over with the flue gases but instead falls to the bottom of the furnace; this material is termed *furnace bottom ash (fba)*. At some power stations the bottom ash is sluiced out to the ponds in conjunction with the hopper ash; separation of the coarse from the fine particles of the mixed material then takes place during settlement so that the lagoon pfa at these stations is both coarser and more variable than elsewhere.

The main pfa ingredients are finely-divided, glassy spheres of silica, SiO_2 (45 to 51 per cent) and alumina, Al_2O_3 (24 to 32 per cent). Magnetic and non-magnetic iron compounds, some alkali, and limited amounts (2 to 3 per cent) of water soluble materials are also present. Calcium oxide derived from the burning of limestone present in the original coal may also occur alone or in combination with other ingredients in the fly ash. Residual unburnt carbon is also normally present, together with inclusions of unfused ash.

Pulverized fuel ashes are pozzolans, i.e. in the presence of lime and water a chemical reaction takes place which results in the formation of hydrous calcium aluminates and silicates that are similar to the reaction products of hydrated cement. Thus, fly ashes containing significant amounts of water-soluble lime (CaO and MgO) and calcium sulphate can become involved in pozzolanic reactions and will self-harden.

The particle-size compositions of British hopper and lagoon ashes can vary considerably from power station to power station. Overall, the hopper ashes in particular are mainly silt-size uniformly-graded materials[64]; as such they are liable to be frost susceptible if located within 450 mm of a road surface. Generally, however, ashes with less than 40 per cent of particles passing the 75 micron sieve are unlikely to be frost susceptible in Britain.

Analysis of the different size-fractions indicates that the chemically-reactive non-combustible materials, i.e. silica, alumina and haematite, are mainly found in the finer fractions of the pfa. A reactive fly ash tends to have at least 80 per cent by mass of its particles smaller than 42 micron.

The quantity of unburnt carbon present in a pfa (determined as the loss-on-ignition) depends on the efficiency with which the pulverized coal is burnt in the furnace. Thus, older power stations can have carbon contents in excess of 10 per cent whilst more modern stations have fly ashes with less than 2 per cent. As the carbon exists as irregular porous particles, a high carbon content increases the moisture content requirement of any pfa mixture; it also results in lower dry densities, reduces the proportion of reactive surface area available to enter into pozzolanic reactions, and physically limits the contacts of cementitous materials. Thus, a pfa with a high carbon content is normally a lesser-quality construction material.

Sometimes a hopper pfa will retain latent electrical charges which cause the individual ash particles to repel each other, thereby causing bulking of the massed material. These fly ashes should be rejected for use in road construction due to the difficulty in achieving adequate compaction.

Furnace bottom ash is much coarser than pulverized fuel ash. It varies in particle size from fine sand to coarse gravel. Its main use in Great Britain is as an embankment and capping material.

The main uses of pfa in road construction in Britain are as bulk fills for embankments, in capping layers, and in cement-bound bases. The self-hardening properties of some fly ashes are especially useful when selected fill is required behind bridge abutments, as their settlement is less than that for most conventional fill materials. Compacted pfa has a lower dry density (typically $1.28\,Mg/m^3$) than most other materials used in embankments, and this lightweight property is very advantageous when mass filling is required on highly-compressible soils.

Overseas, in warm climates, pfa is often used with lime and soil, or cement and soil, to produce stabilized subbases and roadbases (see Chapter 6).

5.10.4 China clay wastes

The mining of china clay (kaolin) involves the use of high-pressure jets of water to extract the kaolinized granite from steep-sided open pits. The broken-up rock is first collected in a slurry and processed to remove the sand waste; the residual slurry is then de-watered and a second process is carried out to remove the fine clayey sand and mica residue. Nearly 9 t of waste results from the production of 1 t of china clay, namely 3.7 t of coarse sand, 2 t of waste rock, 2 t of overburden, and 0.9 t of micaceous residue. The wastes are normally tipped onto land adjacent to the pits, in mounds up to 45 m high.

Except for the micaceous residue, most china clay wastes – especially the sand waste – have potential for use in road-making. Many china clay sands are not frost susceptible and can be used in cement-bound bases and pavement-quality concrete, as well as in embankments and capping layers. When used in concrete, they require more cement (15–20% by mass) than river sands to achieve the same strength; this is probably because of the angularity or harshness of the waste sand mixes.

A synthetic surface dressing aggregate utilizing china clay sand has been developed which is highly resistant to wear and traffic polishing. Other uses for china clay sand include back-filling for pipe trenches and French drains, and as permeable backings to earth-retaining structures.

A feature of china clay wastes is the presence of mica. All micas – muscovite ($K_2O.3Al_2O_3.6SiO_2.2H_2O$) is the most common – have a layered structure in which successive sheets are able to part easily in the plane parallel to their larger surfaces, and to form thin flakes. These flakes are extremely resilient when subjected to pressure on their larger surfaces, and behave somewhat like leaves of a leaf-spring. However, studies have shown that the proportion of mica normally present in china clay sand is too low to cause problems in road construction; the main detrimental effect is that its resilience reduces the degree of compaction achievable for a given compactive effort, by about 0.007 and 0.012 Mg/m^3 for each 1 per cent of fine (<75 micron) and coarse (mostly 212–600 micron) mica, respectively.

The china clay industry is concentrated in Devon and Cornwall and, unfortunately, transport costs are often too high for the waste heaps to be used as an embankment material at far away locales.

5.10.5 Slate waste

For every 1 t of slate produced in the UK an additional 20 t of waste material is produced as a by-product. Consequently, very considerable quantities of slate waste are available in dumps adjacent to quarries in the Lake District, in Devon and Cornwall, and, especially, in North Wales. However, its use in roadworks is relatively low, primarily because of the high cost of transporting the waste to construction sites. Also, the relative remoteness of the accumulations has meant that limited research has been carried out to maximize their suitability for road-making.

Whilst waste slate is a crushed rock, its nature varies with its origin. Thus, cherts (which are sometimes interbedded with the slate) and igneous rocks are often found in slate waste accumulations. By contrast, 'mill waste' consists mainly of slate blocks and chippings from the dressings of the slate.

Slate waste is used in cement-bound bases and pavement-quality concrete, as well as in capping layers and as a bulk fill material. The flaky nature of the waste particles causes problems in compaction, and grid rollers have been found to be most useful in overcoming this as they break the longer needle-shaped pieces of slate into short pieces.

5.10.6 Incinerated refuse

Incinerated refuse residues consist mainly of clinker, glass, ceramics, metal and residual unburnt matter. The unburnt residual material may be paper, rag, or putrescible substances, with the amounts depending upon the type of furnace, the temperature of the firebed or the length of time that the materials are contained within it, and the composition of the raw refuse.

Many incinerator ashes can be used in embankments, with some being better than others. Whilst the ash obtained at one incinerator can vary considerably from that obtained at another, there is reasonable consistency in the material obtained at a given source. As concrete and cemented products are attacked by sulphates dissolved in water, incinerated refuse in which the soluble sulphate content exceeds 2 g/l should not be used within 0.5 m of a concrete structure, unless the structure is protected through the use of, say, supersulphated cement or sulphate-resisting Portland cement.

5.10.7 Demolition debris

Demolition wastes are often proposed for road construction, particularly as they are mostly available in urban ares where there is often a shortage of conventional aggregates. Unfortunately, much demolition debris is produced in relatively small quantities and spasmodically, so that advanced planning for its use can be difficult. Also, it is very variable and the major components (brick, building stone and concrete) are very often intermixed with plaster, wood, glass, etc., so that the resultant mixture can be so heterogeneous that it is unsuitable for use even as a bulk fill.

Debris that is relatively free from contamination, i.e. that is mainly composed of brick and non-reinforced crushed concrete, has considerable potential for road construction, and is sometimes included in specifications as clean rock-like 'hard-core' for use in embankments, capping layers, and cement-bound bases.

5.10.8 Chalk

Chalk accounts for about 15 per cent of the major geological formations in England as a whole and more than 50 per cent in south-east England. It is a soft white porous limestone that was laid down as an ooze on the sea-bed when most of north-western Europe was covered by sea. Thin layers of very hard chalk are found at intervals in the chalk mass. Where the ooze layers became mixed with clay muds, a greyish or buff-coloured chalk marl was formed that is sometimes found as separate layers or homogeneously distributed throughout the chalk in varying proportions, e.g. non-chalk constituents as high as 40 per cent have been measured.

Chalk is a low-grade material that is generally frost-susceptible. Because of this it is primarily used in embankment construction. When stabilized with cement it may also be used in pavement bases and subbases.

A problem associated with using soft fresh chalk in embankments is that its natural rock structure is partly broken down and some of the contained water is released during the excavation and construction processes. The natural moisture contents of fresh chalk are normally at or about their saturation contents; these typically range from 8 to 36 per cent. A fresh fill normally consists of a mixture of chalk lumps and fines, and the fines can form a slurry ('putty chalk') at high moisture contents. If the proportion of putty chalk in the fill is sufficient to control the behaviour of the whole, an unstable embankment results and work may have to be stopped because compaction plant cannot operate efficiently on the spread material.

As the stability of a freshly-placed chalk fill depends upon its moisture content and the degree of crushing it has experienced (i.e. its fines content), a classification system based on these two parameters has been developed[65] to assist in the use of this material in embankments.

5.11 References

1. BS 3690; Part 1, *Bitumens for Building and Civil Engineering: Specification for Bitumens for Roads and Other Paved Areas*. London: British Standards Institution, 1990.
2. Szatkowski, W., Determination of the elastic recovery of binder/polymer mixtures using a modified sliding plate microviscometer, LR14, Crowthorne, Berkshire: Road Research Laboratory, 1967.
3. BS 2000: Part 49, *Penetration of Bituminous Materials*. London: British Standards Institution, 1983.
4. BS 2000: Part 58, *Softening Point of Bitumen (Ring and Ball)*. London: British Standards Institution, 1983.
5. Pfeiffer, J.Ph. and van Doormaal, P.M., Rheological properties of asphaltic bitumen, *Journal of the Institute of Petroleum*, 1936, **22**, No. 154, pp. 414–40.
6. Heukelom, W., Improved nomographs for bitumen, *Shell Bitumen Review*, 1973, **43**, pp. 8–9.
7. Fraass, A., Test methods for bitumen and bituminous mixture with specific reference to low temperature, *Bitumen*, 1937, **7**, pp. 152–5.
8. Heukelom,W., A bitumen test data chart for showing the effect of temperature on the mechanical behaviour of asphaltic bitumens, *Journal of the Institute of Petroleum*, 1969, **55**, pp. 404–17.
9. Heukelom, W., An improved method of characterizing asphaltic bitumens with the aid of their mechanical properties, *Proceedings of the Association of Asphalt Paving Technologists*, 1973, **42**, 67–98.
10. Nicholls, J.C. and Daines, M.E., Laying conditions for bituminous materials, In *The Asphalt Yearbook 1994*. Stanwell, Middlesex: Institute of Asphalt Technology, 1994, pp. 94–8.
11. BS 2000: Part 45, *Loss of Heating of Bitumen and Flux Oil*. London: British Standards Institution, 1982.
12. BS 2000: Part 47, *Solubility of Bituminous Binders*. London: British Standards Institution, 1983.
13. BS 2000: Part 357, *Permittivity of Bitumen*. London: British Standards Institution, 1983.
14. Hveem, F.N., Zube, E. and Skog, J., Proposed new test and specifications for paving grade asphalt, *Proceedings of the Association of Asphalt Paving Technologists*, 1963, **32**, pp. 271–352.
15. Gubler, R., Hugener, M. and Partl, M.N., Comparison of different approaches in standardization and characterization of bituminous binders. In H. di Benedetto and L. Francken (eds), *Mechanical Tests for Bituminous Binders: Proceedings of the Fifth International RILEM Symposium*, pp. 53–60. Rotterdam: A.A. Balkema, 1997.
16. Whiteoak, D., *The Shell Bitumen Handbook*. Chertsey, Surrey: Shell Bitumen UK, 1990.
17. Ramond, G. and Such, C., Bitumen et bitumes modifiés – relations structures, propriétés, composition, *Bulletin de Liaison des Laboratoires des Ponts et Chaussées, Thématique, Bitumen et Enrobés Bitumineux*, 1992, pp. 23–45.

18. Carswell, J., Claxton, M.J. and Green, P.J., Dynamic shear rheometers: Making accurate measurements on bitumens. In *The Asphalt Yearbook 1997*. Stanwell, Middlesex: Institute of Asphalt Technology, 1997, pp. 79–84.
19. Van der Poel, C., A general system describing the visco-elastic properties of bitumen and its relation to routine test data, *Journal of Applied Chemistry*, 1954, **4**, pp. 221–36.
20. Ullidtz, P.A., Fundamental method for prediction of roughness, rutting, and cracking of pavements, *Proceedings of the Association of Asphalt Paving Technologists*, 1979, **48**, pp. 557–86.
21. The modern use of Trinidad Lake Asphalt, *Highways and Road Construction*, 1975, **43**, No. 1786, pp. 18–20.
22. BS 1446, *Mastic Asphalt (Natural Rock Asphalt) for Roads and Footways*. London: British Standards Institution, 1973.
23. BS 2000: Part 72, *Determination of Viscosity of Cutback Bitumen*. London: British Standards Institution, 1993.
24. Jackson, N. and Dhir, R.K., *Civil Engineering Materials*, 4th edn. London: Macmillan, 1988.
25. BS 434: Parts 1 and 2, *Bitumen Road Emulsions (Anionic and Cationic)*. London: British Standards Institution, 1984.
26. BS 76, *Tars for Road Purposes*. London: British Standards Institution, 1974.
27. Transport Research Laboratory, *Design Guide for Road Surface Dressing*, 3rd edn. Crowthorne, Berkshire: The Transport Research Laboratory, 1992.
28. Jamieson, I.L., The *Carcinogenic Potency of Tar in Road Construction*, Technical Report RB/2/78. South Africa: National Institute for Transport and Road Research, 1978.
29. Nicholls, J.C., EVATECH H polymer-modified bitumen, PR109. Crowthorne, Berkshire: Transport Research Laboratory, 1994.
30. Dickinson, E.J., A critical review of the use of rubbers and polymers in bitumen bound pavement surfacing materials, *Australian Road Research*, 1977, **7**, No. 2, pp. 45–52.
31. Sainton, A., Dense noiseless asphalt concrete with rubber bitumen. Paper presented at the Eurasphalt and Eurobitume Congress held at Brussels, 1996.
32. Preston, J.N., Cariphalte DM – a binder to meet the needs of the future, In *Bitumen Review 66*. London: Shell International Petroleum Co., May 1992, pp. 16–19.
33. BS 12, *Portland Cement*. London: British Standards Institution, 1996.
34. BS 4027, *Sulfate-resisting Portland Cement*. London: British Standards Institution, 1996.
35. BS 146, *Portland Blastfurnace Cements*. London: British Standards Institution, 1996.
36. BS 6692, *Ground Granulated Blastfurnace Slag for Use with Portland Cement*. London: British Standards Institution, 1992.
37. BS 3892: Part 1, *Pulverized-fuel Ash for Use with Portland Cement*. London: British Standards Institution, 1997.
38. BS 6588, *Pulverized-fuel Ash Cement*. London: British Standards Institution, 1996.
39. Hosking, R., *Road Aggregates and Skidding:* TRL State of the Art Review **4**. London: HMSO, 1992.
40. Blyth, F.G.H., and de Freitas, M.H., *A Geology for Engineers*. London: Arnold, 1984.

41. Croney, D. and Jacobs, J., *The Frost Susceptibility of Soils and Road Materials*, Report LR90, Crowthorne, Berkshire: Road Research Laboratory, 1967.

42. BS 1047, *Air-cooled Blastfurnace Slag Aggregate for Use in Construction*. London: British Standards Institution, 1983.

43. Benson, F.J., *Effects of Aggregate Size Shape, and Surface Texture on the Properties of Bituminous Mixtures – A Literature Survey*, Highway Research Special Report **109**, 1970, pp. 12–21.

44. White, E. (ed.), *Bituminous Mixes and Flexible Pavements*, London: The British Aggregate Construction Materials Industries (BACMI), 1992.

45. BS 812: Part 102, *Methods for Sampling*. London: British Standards Institution, 1989.

46. BS 812: Part 103, Section 103.1, *Method for Determination of Particle Size Distribution: Sieve Test*. London: British Standards Institution, 1989.

47. BS 812: Part 103, Section 103.2, *Method for Determination of Particle Size Distribution. Sedimentation Test*. London: British Standards Institution, 1989.

48. prEN 933, *Tests for Geometrical Properties of Aggregates – Part 9: Assessment of Fines – Draft Methylene Blue Test*. Brussels: European Committee for Standardization, 1996.

49. Brennan, M.J., Kilmartin, T., Lawless A. and Mulry, B., A comparison of different methylene blue methods for assessing surface activity. In Di Benedetto, H. and Francken, L. (eds), *Mechanical Tests for Bituminous Materials; Proceedings of the Fifth International RILEM Symposium*, pp. 517–23. Rotterdam: A.A. Balkema, 1997.

50. BS 812: Part 105, Section 105.1, *Method for Determination of Particle Shape. Flakiness Index*. London: British Standards Institution, 1989.

51. BS 812: Part 113, *Method for Determination of Aggregate Abrasion Value*. London: British Standards Institution, 1990. (and Amendment No.1, 1991)

52. *Specification Requirements for Aggregate Properties and Surface Texture for Bituminous Surfacings for New Roads*, Technical Memorandum H16/76. London: Department of Transport, 1976.

53. BS 812: Part 114, *Method for Determination of the Polished Stone Value*. London: British Standards Institution, 1989.

54. Roe, P.G. and Hartshorne, S.E., *The Polished Stone Value of Aggregates and In-service Skidding Resistance*, TRL Report 322. Crowthorne, Berkshire: The Transport and Road Research Laboratory, 1998.

55. BS 812: Part 110, *Method for Determination of Aggregate Crushing Value (ACV)*. London: British Standards Institution, 1990.

56. BS 812: Part 112, *Method for Determination of Aggregate Impact Value (AIV)*. London: British Standards Institution, 1990.

57. BS 812: Part 111, *Methods for Determination of Ten Per Cent Fines Value (TFV)*. London: British Standards Institution, 1990.

58. BS 812: Part 2, *Methods for Determination of Density*. London: British Standards Institution, 1995.

59. BS EN 933–2, *Tests for Geometrical Properties of Aggregates: Part 2, Determination of Particle Size Distribution – Test Sieves, Nominal Size of Apertures*. London: British Standards Institution, 1996.

60. Nunes, N.C.M., Bridges, M.G. and Dawson, A.R., Assessment of secondary materials for pavement construction: Technical and environmental aspects. *Waste Management*, 1996, **16**, Nos. 1–3, pp. 87–96.
61. *Conservation and the Use of Reclaimed Materials in Road Construction and Maintenance*, Technical Memorandum HD 35/95. London: HMSO, 1995.
62. *Specification for Highway Works*. London: HMSO (undated).
63. Sherwood, P.T., *The Use of Waste and Low-grade Materials in Road Construction: (2) Colliery Shale*. TRRL Report LR649. Crowthorne, Berkshire: The Transport and Road Research Laboratory, 1975.
64. Sherwood, P.T., *The Use of Waste and Low-grade Materials in Road Construction: (3) Pulverized Fuel Ash*. TRRL Report LR686. Crowthorne, Berkshire: The Transport and Road Research Laboratory, 1975.
65. Ingoldby, H.C. and Parsons, A.W., *The Classification of Chalk for Use as a Fill Material*. TRRL Report LR806. Crowthorne, Berkshire: The Transport and Road Research Laboratory, 1977.

CHAPTER 6

Soil-stabilized pavements

C.A. O'Flaherty

6.1 Why stabilize soils?

Soil stabilization is any treatment (including, technically, compaction) applied to a soil to improve its strength and reduce its vulnerability to water; if the treated soil is able to withstand the stresses imposed on it by traffic under all weather conditions without deformation, then it is generally regarded as stable. This definition applies irrespective of whether the treatment is applied to a soil in situ or after the soil has been removed and placed in a pavement or embankment.

The continued demand for aggregates in highly-developed countries has now resulted in shortages of stone and gravel in many localities, e.g. in large urban areas where aggregate availability has traditionally been low and/or where the production of aggregates is prohibited for environmental reasons. In many developing countries the availability of good cheap supplies of aggregates is limited and is not able to keep up with the needs of their burgeoning economies. As a consequence, greater attention is now being paid to the use of locally-available substitute materials such as stabilized soils (and 'waste' materials – see Chapter 5) to meet road construction needs.

Current practice in the UK is to use stabilized soil only in capping and subbase layers in road pavements. In many other countries stabilized soil is also used to form roadbases and surface courses.

In practice, the main methods by which soils are stabilized for road purposes are: (1) mechanical or granular stabilization; (2) cement stabilization; (3) lime and lime-pozzolan stabilization; and (4) bituminous stabilization. An excellent analysis of the technical literature suggests the recommendations in Fig. 6.1 in relation to the particle-size distributions and plasticity indices of soils that are generally suitable for stabilization by these means. In practice, it is often found that a particular soil can be stabilized in a number of ways and that the decision as to which method to use is primarily a financial one, coupled with consideration of the skills, resources and construction equipment available and previous experience with the alternative methods.

6.2 Mechanical stabilization

Improving the gradation of a raw soil by admixing a coarse and/or fine material (usually 10 to 50 per cent), with the aim of achieving a dense homogeneous mass

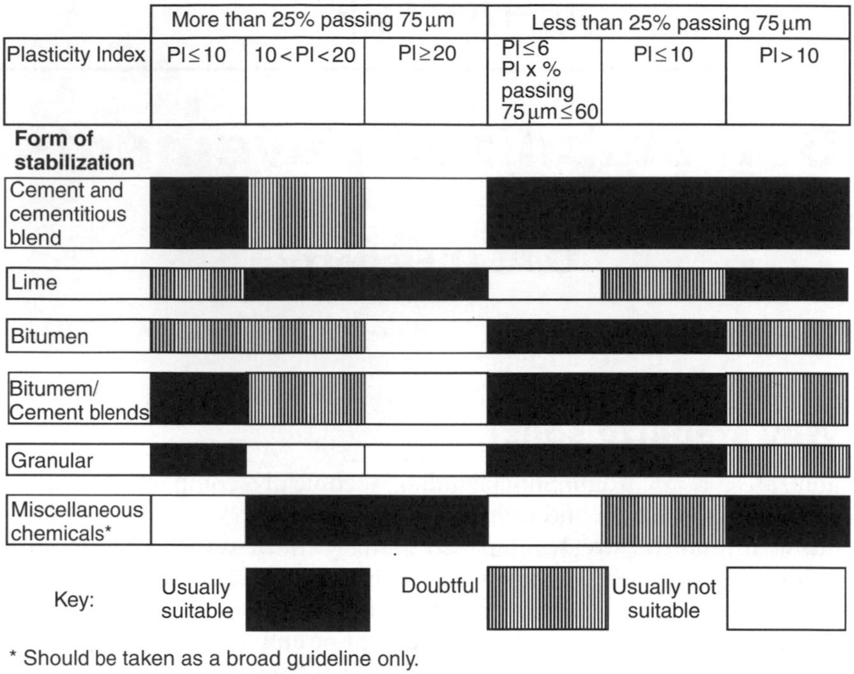

Plasticity Index	More than 25% passing 75 μm			Less than 25% passing 75 μm		
	PI ≤ 10	10 < PI < 20	PI ≥ 20	PI ≤ 6 PI x % passing 75 μm ≤ 60	PI ≤ 10	PI > 10

Form of stabilization

Fig. 6.1 Guide to selecting a method of soil stabilisation[1]

when compacted, constitutes mechanical stabilization. Mechanical stabilization is also known as *granular stabilization* and *soil–aggregate stabilization*; in the latter context the term 'aggregate' refers to the addition of non-fines material greater than 0.06 mm that contributes to the development of internal friction.

The world-wide use of mechanical stabilization in road-making results from its (a) maximization of the use of cheap locally-available poorly-graded materials, e.g. dune- or river-deposited sands, silty sands, sandy clays, and silty clays, and (b) its reliance for stability upon the inherent cohesive and internal friction properties of the controlled proportions of raw soils that are admixed.

The use of *chloride additives* is often associated with granular stabilization. Soils that already have mechanical stability can often be further improved by the addition of chlorides.

Mechanically-stabilized roads range from thin unsealed pavements, carrying limited traffic volumes, that are formed from a subgrade soil with which another soil is admixed in order to meet identified property deficiencies and then compacted, to sealed roads that utilize blends of different soil materials in one or more of a pavement's component layers and satisfactorily carry heavy traffic volumes. In countries with well-developed economies, unsealed granular-stabilized pavements are now primarily used in rural areas, commonly to provide access to farming properties, forests and undeveloped land. When the estimated future traffic demand might justify a higher-quality pavement, but insufficient funds are available at the time, it is quite common for a granular-stabilized pavement to be constructed as part of a stage construction. By *stage construction* is meant the step-by-step improvement of a road pavement as expenditure is justified by additional traffic so that, at the final stage, the initial granular-stabilized pavement may form part or all of the roadbase and/or subbase of a sealed road that is designed to carry the heavier traffic flows.

6.2.1 Some design criteria

The main uses of mechanically-stabilized soils are in the unsealed surface courses, roadbases, and subbases of lightly-trafficked roads, in the subbases and roadbases of (mainly) single-carriageway roads with bituminous surfacings, and in the subbases and capping layers of more heavily-trafficked dual carriageway pavements.

Traffic intensity is the dominant factor influencing the surface wear of unsealed pavements, albeit climatic conditions are also important. Thus, *unsealed surface courses* should only be used in roads which carry less than, say, 200–300 veh/day, as higher volumes result in high maintenance needs.

From the point of view of stability and resistance to abrasion, the materials used in an unsealed surface course should consist of a well-graded mixture of hard, tough, durable fragments of stone or gravel, finely-divided mineral aggregate such as sand, and a small proportion of clayey binding material. The aim is to ensure the formation of a low-dust tightly-bonded layer when it is compacted and shaped.

For ease of grading and compaction it is undesirable to use aggregate particles greater than 20 to 25 mm in surface courses as they are likely to be torn from the surface by a motor grader carrying out maintenance. Best aggregate interlock is obtained by using angular particles; hence, crushed or glacial gravels and sands are preferred to river materials. For resistance to ravelling 20–60 per cent of the surfacing material should be retained on a 2.36 mm sieve. The main function of the clay binder is to help prevent the smaller aggregate pieces from being whipped away under the action of traffic, especially in dry weather. In hot climates the clay also helps to retain, against evaporation, the moisture content necessary for stability. A suitable surface course grading is given in Table 6.1.

Table 6.1 Suggested particle size distribution for unsealed soil-aggregate pavements[1]

Sieve size (mm)	% passing sieve: Wearing course (roadbase)	Subbase**
26.5	100	100
19.0	85–100	70–100
9.5	65–100	50–80
4.75	55–85	32–65
2.36	20–60	25–50
0.425	25–45	15–30
0.075*	10–25	5–15

* The 0.075 mm fraction is the fraction containing the dust particles.
**The maximum particle size for a subbase is often increased to 40 mm.

From the point of view of pavement stability, as little water as possible should penetrate an unsealed soil–aggregate surfacing. Thus, these roads should have A-shaped cambers with crossfalls of 4–5 per cent; as well as being cheaper and easier to construct and maintain than conventional curved ones, these cambers shed rainwater more quickly into roadside ditches. The clay content also helps prevent moisture infiltration, as the particles plug the soil pores during wet weather – provided the clay content is just enough to perform this function, and excessive swelling does not cause dislocation of the granular materials. For stability, and to reduce permeability, the fines-to-sand ratio should lie between 0.20 and 0.60, i.e.

$$0.2 < [(\% \text{ less than } 0.075 \text{ mm})/(\% \text{ less than } 2.36 \text{ mm})] < 0.6$$

Also, the *PI* should lie in the range 4 to 15. The lower end of the *PI* range is appropriate for wetter climates, higher traffic, and materials with lower stone contents, whilst the higher end is applicable to arid climates, lower traffic and high stone content materials[8]. The corresponding linear shrinkage range should be 2 to 8, with the lower end of the range being for wetter climates.

The primary requirement of soil–aggregate *bases* is that they should be composed of a mixture of coarse and fine soils so that, when compacted, they form a highly-stable dense material with high shear strength. Dense close-packed mixtures with minimum voids are obtained when the final particle-size distribution tends toward the gradation given by the modified Fuller's law:

$$p = 100(d/D)^n \qquad (6.1)$$

where p = percentage by mass of the total mixture passing any given sieve of aperture-sized (mm), D = size of the largest particle in the mixture (mm), and n = an exponent between 0.33 and 0.5. Mixtures having a maximum size of 19 mm and gradation curves corresponding to n-values of 0.45 and 0.50 are desirable in that only 8 and 6 per cent fines, respectively, pass the 75 micron sieve. When $n < 0.33$ the fines content may be excessive and lead to the development of positive pore pressures (due to reduced permeability) and consequent instability during compaction and in service; if $n > 0.5$ the mix will be too harsh and more difficult to work. In practice, too great an emphasis should not be placed on achieving the ideal gradation; rather stress should be placed on obtaining a mixture that is sufficiently dense to meet stability needs whilst maximizing the use of low-cost readily available soils.

The clay content of a granular-stabilized base should be very carefully controlled. Whereas in the case of an unsealed surface course some clay is needed to prevent water infiltration and to bind the coarse particles together, the opposite is normally true for bases, especially if they are to be covered with a bituminous surfacing. The presence of a significant quantity of clay in a base will also attract water which, if it cannot escape, will accumulate and eventually cause failure of the pavement.

Soil variability is such that there is no absolute prescription regarding the amounts and nature of fine material that can be used in granular-stabilized pavements; rather, most recipes are based on experience, and different pavement mixtures may perform satisfactorily in different environmental areas. As a general guide, however, mixtures used in the bases of unsealed pavements should have plasticity indices in the range 4–9 and should not have liquid limits greater than 35, whilst those in

Table 6.2 Gradations of granular materials commonly used in the subbases of new roads in the UK[3]

BS sieve size (mm)	Percentage passing by mass:	
	Type 1	Type 2
75	100	100
37.5	85–100	85–100
20	60–100	60–100
10	40–70	40–100
5	25–45	25–85
0.600	8–22	8–45
0.075	0–10	0–10

bases in sealed pavements should not have *LL*- and *PI*-values that are greater than 25 and 6, respectively.

Specifications for soil–aggregate mixtures are described in many publications, e.g. those of the American Association of State Highway and Transportation Officials (AASHTO) and the National Association of Australian State Road Authorities (NAASRA). As a basic premise, however, advice regarding suitable specifications to use in particular geographic areas should be sought in those locales in the first instance. (Reference 2 is a useful guide to the factors underlying the design of soil–aggregate roads for use in developing countries.)

Whilst soil–aggregate pavements are rarely constructed for general use in the UK because of the wet climate, the Department of Transport's *Specification for Highway Works*[3] allows the use of natural sands and gravels in the subbases of major roads (see Table 6.2). The Type 2 subbase material specified in the table can be natural sands, gravels, crushed rock or slag or concrete, or well-burnt non-plastic shale, and the fraction passing the 425 micron sieve must have *PI* < 6. The Type 1 material can be crushed rock or slag or concrete, or well-burnt non-plastic shale, but it may contain up to 12.5 per cent by mass of natural sand that passes the 5 mm BS sieve, and the fraction passing the 425 micron sieve must be non-plastic. Both types of subbase material must not be gap-graded and have a 10 per cent fine value of at least 50 kN.

6.2.2 Use of chlorides in unsealed granular-stabilized pavements

The chlorides used in unsealed granular-stabilized pavements are sodium chloride (NaCl) and, most commonly, calcium chloride ($CaCl_2$). Both chlorides occur naturally in the ocean, in salt lakes, and in brines present in cavities in rock formations. In addition, great quantities of calcium chloride are obtained by refining the $CaCl_2$-rich waste from the soda-ash industry.

Chlorides are not cementing agents and thus a chloride-stabilized pavement must be already mechanically-stable and have low permeability. They are most effectively used when added to well-graded materials and compacted to dense impervious masses; this reduces the ease with which the chlorides are leached downward from the upper parts of unsealed pavements. Open-graded permeable mixtures are not suitable for chloride treatment.

Chlorides have excellent hygroscopic and deliquescent properties. The hygroscopicity of a material is a measure of its ability to absorb moisture from the air; a deliquescent material is a hygroscopic one that will dissolve in moisture. These two properties mean that, in dry non-arid climatic areas, moisture that is evaporated from the roadway during the day can be replaced by moisture from the atmosphere during the night. If the atmospheric humidity is sufficient, chloride-treated gravel-stabilized unsealed roads will remain dust-free until the chlorides are leached out.

Chloride solutions are good lubricants, which means that a higher soil–aggregate dry density can be obtained with a given compactive effort; in this context $CaCl_2$ is a better lubricant than NaCl. Compaction in dry weather can also be carried out over a longer period of time at a given moisture content when chlorides are dissolved in the mixing water; this is because evaporation takes place at a slower rate, due to reduced vapour pressure.

The development of ravelling and corrugation conditions in the surfacings of unsealed gravel-stabilized pavements in warm climates is considerably retarded by the addition of a chloride. Chlorides provide firm hard-packed surface courses with a smoothness and riding quality approaching that of bituminous surfacings. Thus, ravelling is reduced as it is harder for aggregates to be plucked from the surfaces. Corrugations are created as a consequence of the oscillatory pounding actions that are produced when vehicles encounter irregularities on the carriageway; thus, the smoother the surfacing the less likely it is that corrugations will be formed.

The use of $CaCl_2$ in soil–aggregate bases in unsealed pavements is also advantageous. If the surface course contains $CaCl_2$, and the climate is temperate, the calcium chloride leached from a base will be replenished to a certain extent by the downward movement of moisture through the surfacing. Further, $CaCl_2$ present in a subbase or roadbase will reduce frost heave, and consequent deleterious frost action, by lowering the freezing temperature of the moisture.

A satisfactory calcium chloride content for an unsealed pavement is about $1 \, kg/m^2$ per $100 \, mm$ of compacted thickness; if NaCl is used higher contents may be required. Most economical usage, and reduced surface deterioration, is obtained by periodic surface grading and by reapplication of chlorides.

An undesirable feature of chlorine-stabilized roads is the corroding effect that splash-water has upon the steel in motor vehicles.

6.2.3 Blending soils to meet soil–aggregate specifications

In some instances, deposits of naturally-occurring soils will be found which meet the selected soil–aggregate specification. More often than not, however, the development of an acceptable granular-stabilized mixture involves the sampling and testing of locally-available soils in the laboratory for their gradations and plasticity values, carefully proportioning two or three of them to meet the specification criteria, and then making up and testing trial batches to ensure that the preferred proportions are within the specification limits. Methods for determining the blend proportions essentially utilize trial and error testing, based on gradations and plasticity indices. Whilst proportioning to fall within gradation limits is a relatively simple trial-and-error process for two soils whose gradations are known, that for three soils is more effectively initiated using graphical methods (e.g. see reference 4, p. 336); however, space does not permit their illustration here.

A guide to the initial proportions to be blended to give a desired plasticity index for a mixture of two soils can be obtained from the following equations:

$$a = 100S_B (P - P_B)/[S_B (P - P_B) - S_A(P - P_A)] \tag{6.2}$$

and

$$b = 100 - a \tag{6.3}$$

where a = amount of soil A in the blended mix (%), b = amount of soil B in the blended mix (%), P = desired PI of the blended mix, P_A = PI of soil A, P_B = PI of soil B, S_A = amount of soil A passing the 425 micron sieve (%), and S_B = amount of soil B passing the 425 micron sieve (%).

6.2.4 Construction methods

Mechanical stabilization is normally carried out by one of three construction methods using mix-in-place, travelling plant, or stationary plant equipment. Whatever method is used the constituent soils which are intended to form the stabilized product should be carefully proportioned and thoroughly mixed to produce a homogeneous material that can be compacted and shaped as specified.

At its simplest the *mix-in-place construction* of a granular-stabilized subbase, for example, may involve scarifying the in-situ subgrade soil, spreading the required thickness of imported borrow soil(s) on top to a uniform depth, and then admixing the dry constituent materials using conventional blade graders to move them from one side of the road to the other. (A better mix is obtained, however, if specialized rotary mixing machines are used.) Upon completion of the blending process the dry-mixed material is spread in a layer of uniform thickness and water is added via the spray bar of a water tanker to bring it to the desired moisture content for compaction. Further mixing is then carried out to ensure a uniform moisture distribution, after which the damp material is re-spread to a uniform thickness and compacted.

The thickness of the compacted 'lift' depends upon the mix gradation and compaction equipment available at the site, e.g. granular mixtures lend themselves to thick lifts (up to 200 mm compacted thickness) using vibratory rollers, whereas thin lifts of ca 100 mm, using pneumatic rollers, are more common with less-granular materials. Whatever the equipment used in the initial compaction process, the top of the final layer should be shaped to the desired camber, using a blade grader, before it is finally compacted by smooth-wheel or pneumatic-tyred rollers to present a smooth, dense, closely-knit surface.

With *travelling plant construction* the process is similar to that described above, except that a single pass of specialized moving equipment is all that is necessary to dry mix the soils, admix water, and spread the moist material to a uniform depth, prior to compaction.

With *stationary plant construction*, the soils to be blended are brought to a central location which is often at the source of supply, e.g. at a gravel pit, and specialized equipment is then used to proportion and mix the materials and water, after which the moist mixture is brought to the construction site by truck, spread to a uniform depth (often by a self-propelled aggregate spreader) and compacted. Whilst this construction process is more expensive than the two previously described, it generally gives a better proportioned and mixed end-product and there are fewer bad weather delays.

6.3 Cement stabilization

Internationally, cement stabilization is most commonly used in subgrade capping and/or subbase layers in major road pavements, and in subbases and/or roadbases of secondary-type roads. It is never used in surface courses because, as well as having poor resistance to abrasion, it must be protected from moisture entry into the cracks that will inevitably form in the cement-treated material.

Factors which ensure that cement stabilization is very widely used are: (1) cement is available in most countries at a relatively low price; (2) the use of cement usually involves less care and control than many other stabilizers; (3) more technical information is available on cement-treated soil mixtures than on other types of soil stabilization; and (4) most soils (except those with high organic matter or soluble sulphate contents) can be stabilized with cement if enough is used with the right amount of water and proper compaction and curing.

The two main types of cement-treated soil mixtures are the cement-bound and cement-modified materials. It should be clear, however, that there is no clear demarcation between the two cement-treated materials albeit, for reasonably well-graded base materials, values of 80 kPa for indirect tensile strength, 0.8 MPa for 7-day moist-cured unconfined compressive strength, or 700–1500 MPa for resilient modulus, are useful 'transition' guides[1].

A pavement layer that is *cement-bound* is a hardened material formed by curing a mechanically-compacted intimate mixture of pulverized soil, water and cement, i.e. sufficient cement is admixed to harden the soil, and the moisture content of the mixture is adequate for both compaction and cement hydration. Cement-bound natural gravels and crushed rocks typically have elastic moduli in the range 2000–20 000 MPa (versus 200–500 MPa for unbound materials), and 28-day unconfined compressive strengths greater than 4 MPa. Moduli for cement-bound fine-grained soils are usually less.

A combination of high tensile strength, drying shrinkage, and subgrade (or subbase) restraint can result in a regular pattern of widely-spaced (0.5–5 m) open cracks in a cement-bound capping, subbase or roadbase. However, provided that the cracks are not more than about 2 mm wide, and they are sealed, this will not normally lead to structural problems if the layer is protected from the elements. If, however, the cracks are in a cement-bound roadbase and they are not sealed, reflection cracking is likely to occur in the surface course, moisture will find its way into the openings, and the pavement can deteriorate quickly because of the 'pumping' of underlying wet fines under traffic. The early trafficking of cement-bound layers is desirable as this is associated with the development of closely-spaced cracks, and these are more easily protected than the widely-spaced ones that occur when early trafficking is not allowed.

The prevention of fatigue cracking is the main design concern for cement-bound bases, and it is critical that they be properly constructed. The bound layers are relatively brittle and fail in tension under fairly low strain, and the critical strain normally reduces with increasing modulus. Thus, a thin cement-bound layer with a high elastic modulus, that is continuously subjected to heavy traffic volumes, is especially prone to fatigue failure, e.g. as an approximation, a 10 per cent construction reduction in either the design thickness or stiffness can lead to a 90 per cent cut in a base's fatigue life[1].

A *cement-modified* mixture is one in which a smaller amount of cement is admixed with a soil so as to improve its moisture-resisting and stability properties without much increasing its elastic modulus and tensile strength. Whilst the resultant material has weak cementitious bonds, these can be sufficient to greatly reduce moisture-induced shrinkage and swell, even with expansive soils. Modified materials tend to have very closely-spaced networks of fine cracks and to fragment under traffic and thermal stresses; consequently, they are most effectively used in pavements if they are formed from soils with good particle interlock. For design purposes, a cement-modified base material is assumed to have the same elastic modulus as an untreated material performing the same function, and to behave in the pavement as an unbound material that does not have tensile strength and thus is not distressed by fatigue cracking.

6.3.1 Mechanism of cement stabilization

When water is added to a mixture of dry soil and cement the primary hydration reactions result in the formation of calcium silicate and calcium aluminate hydrates, and the release of hydrated lime.

As with concrete, the first two of the above products constitute the major cementitious components that bind the soil particles together, irrespective of whether they are cohesionless sands, silts or highly reactive clays. The hydration reactions proceed rapidly and significant strength gains can be achieved within short periods of time.

The third product, the released hydrated lime, enters into secondary reactions with any reactive elements (e.g. clay or an added pozzolanic material such as fly ash) in the mixture to produce cementitious products similar to those formed during the primary reactions, which contribute further to interparticle bonding. In addition there is an exchange reaction with cations already absorbed on the clay particles which change the plasticity properties of the soil. The cementitious reactions proceed at a slower rate, similar to that in lime stabilization (see Section 6.4.1). The greater the fines content of a soil and the more reactive its clay mineral the more important are the secondary reactions.

6.3.2 Some factors affecting cement–soil properties

A soil's *gradation* and *plasticity* are the most important factors affecting its suitability for cement stabilization. Whilst any soil can theoretically be stabilized with cement, the content required to achieve given design criteria (especially in the case of cement-bound materials) must normally be increased as the fines content is increased, other factors remaining constant. Also, excessive amounts of reactive clay can cause major construction problems when pulverizing, mixing and compacting cement-treated soil mixtures. Thus, for example, it may not be economic to attempt to stabilize heavy clays with cement.

Soils that are already well-graded and mechanically stable are most suitable for either form of cement treatment. By contrast, fine-grained gravels, very clayey gravels, silty sands and other soils without good particle interlock are generally considered not to be suitable for use as cement-modified materials; if used, they should be designed as cement-bound materials. Soils with significant amounts of organic matter are often unsuitable for cement stabilization; if a cement–soil paste containing 10 per cent Portland cement has a pH < 12.1 one hour after adding water the soil should be rejected as this indicates the presence of organic matter capable of hindering the hardening of the cement. A soil usually cannot be economically stabilized if it has a sulphate content in excess of about 0.25 per cent.

Table 6.3 provides a general guide to soils that, from both engineering and economic aspects, are normally likely to be suitable for stabilization with cement. The maximum aggregate size is usually 75 mm, but this depends upon the mixing plant available.

Table 6.4 summarizes the gradation requirements of the cement-bound materials allowed in subbases and roadbases in the UK. CBM 5, CBM 4 and CBM 3 are the highest quality materials and are most usually prepared at a central plant from batched amounts of processed crushed gravel or rock and/or crushed air-cooled blast furnace slag. Note that the gradations for CBM 5, 4 and 3 are contained within those for the lesser-quality CBM 2 and that CBM 1 meets the requirements for CBM 2. The CBM 2 is most usually formed from gravel-sand or crushed rock and/or blastfurnace slag aggregates, whilst CBM 1 materials include raw granular soils; both materials can be mixed in-place or at a central plant.

The UK requirements for cement-stabilized capping layers are less stringent than those for subbases, and permit a range of granular and silty cohesive soils. The main requirement is that any soil used for capping purposes must have a minimum CBR-value of 15 per cent before the addition of cement. Other raw soil criteria are: liquid limit < 45 per cent, plasticity index < 20 per cent, organic content < 2 per cent, and total sulphates < 1 per cent.

Any *cement* can be used to stabilize soils but ordinary Portland cement is commonly used. Well-graded soils need only small amounts of cement to satisfactorily stabilize them, compared with poorly-graded high void-content soils. Overall, the strength of a cement-treated material is directly proportional to the amount of cement admixed. The cement content required in any given instance depends upon the design criteria governing the end-product and these, in turn, are influenced by

Table 6.3 Guide to soil property limits for effective cement stabilization[1]

Sieve size (mm)	Percentage passing	Plasticity limits	
4.75	>50	Liquid limit,%	<40
0.425	>15	Plastic limit,%	<20
0.075	<50	Plastic index	<20
0.002	<30		

Note: Soils at the upper content limit of 2 micron clay may need pretreatment with lime

Table 6.4 Grading limits for cement-bound materials used in roadbase and sub-base construction in the UK[3]

| BS sieve size (mm) | Percentage by mass passing for: | | CBM3, CBM4 and CBM5 Nominal maximum size | |
	CBM1	CBM2	40 mm	20 mm
50	100	100	100	–
37.5	95	95–100	95–100	100
20	45	45–100	45–80	95–100
10	35	35–100	–	–
5	25	25–100	25–50	35–55
2.36	–	15–90	–	–
0.600	8	8–65	8–30	10–35
0.300	5	5–40	–	–
0.150	–	–	0–8*	0–8*
0.075	0	0–10	0–5	0–5

*0–10 for crushed rock fines

environmental as well as traffic load factors, e.g. northern US states typically require cement-bound soils to meet strict freeze–thaw criteria whereas wet–dry durability criteria take precedence in South Africa. In the UK the practice is to specify the desired stabilities of cement-bound materials in terms of the 7 day compressive strengths of 150 mm cubes, e.g. those for the moist-cured plant-mixed CBM 5, 4 and 3 cubes are 20, 15 and 10 N/mm^2, respectively, vis-à-vis 7 and 4.5 N/mm^2 for CBM 2 and 1, respectively.

The strengths of cement-bound materials increase rapidly in the first few days, after which strength gain continues (Fig. 6.2), albeit at a slower rate, provided that proper curing is continued. The elastic modulus also continues to increase.

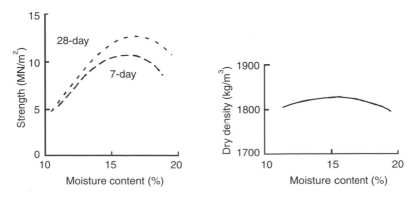

Fig. 6.2 Influence of moisture content on the unconfined compressive strength and dry density of a cement-bound mix composed of 20 per cent cement and a 50:50 mixture of dune sand and montmorillonite clay

Water is required in a cement–soil or cement–aggregate mix to hydrate the cement, improve workability, and facilitate compaction. This water should be potable, have less than 0.5 per cent sulphates, and be free from harmful amounts of organic matter. Ideally, the water used for laboratory testing purposes should be the same as that which is to be used during construction.

Cement-treated soil mixes exhibit similar moisture–density relationships as raw soils, i.e. for a given compactive effort there is one moisture content that will give a maximum dry density. It is important to appreciate, however, that the optimum moisture content for maximum dry density is not necessarily the same as that for strength, and that it can vary according to the manner and duration (see Fig. 6.2) of curing. In general, the optimum moisture content for maximum strength tends to be on the dry side of the optimum for maximum dry density for sandy soils, and on the wet side for clayey soils. For cement-bound clayey soils the optimum moisture content for strength is dependent not only upon the amount of clay present in the soil but also upon the reactivity of the clay mineral.

An even *distribution of cement* throughout the full design depth of a cement-treated layer ensures its most economic use, and the more likely it is that the material will have good stability and durability. In practice, a uniform field distribution requires the admixing of a minimum of 2 per cent cement with an in-situ soil; however, the laboratory-determined cement content may have to be increased in the field (by, typically, 0.5–1 per cent) to allow for mixing inefficiencies. Good distribution is most easily ensured with high-quality central plant-mixed materials; more difficulty is experienced with mix-in-place cement–soil treatments, especially if the soil is cohesive. Generally, cohesive soils are mixed-in-place and, in the process, are pulverized until most clay lumps pass a 19 mm sieve. With highly plastic soils it is common practice to premix a small amount of lime to initiate cation exchange and flocculation/agglomeration reactions which immediately reduce the soil's plasticity, improve its workability, and facilitate the subsequent admixing of (consequently, a reduced amount of) the cement.

Compaction should be completed as soon as possible after the addition of cement to a soil. The chemical reaction between cement and water begins at the time of mixing, and the strength-giving products of hydration begin to form after about 1 h. Thus, for example, in the case of a fixed compactive effort, increasing the mixing time and/or delaying compaction after the completion of mixing generally results in an increase in the optimum moisture content for maximum dry density, and reductions in the dry density, durability and unconfined compressive strength. The extent to which each occurs in any instance is dependent upon the soil type, the cement content, the moist mixing and/or subsequent delay times before compaction is initiated, and the amount of compaction applied. The dry density reduction is due to particle bonding as the cement hydrates, which hinders the ease with which particles can slide over each other during compaction. Consequently, specifications usually place an upper limit on the length of time between when water is added to the mixture and when compaction is completed, e.g. for cement-stabilized capping layers the general practice is for compaction be completed within 2 h of mixing being initiated.

As in the case of concrete, the *curing conditions* applied to a cement-treated soil mixture affects the quality of the product obtained. A cement–soil mixture should always be covered with an impermeable membrane to ensure moist curing during the early stages of its life, i.e. so that sufficient moisture is available to meet its hydration needs, and to reduce the severity of shrinkage cracking as the hydration reactions proceed and the material gains strength. The higher the ambient curing temperature the greater the strength gain. This can be of economic benefit in countries with hot climates as, all other things being equal, higher long-term strengths will be obtained with cement-bonded clayey soil mixes if they are cured at high ambient temperatures and the mixtures are kept moist.

6.4 Lime stabilization

The limes most commonly used for soil stabilization purposes in road construction are hydrated high-calcium lime, $Ca(OH)_2$, and calcitic quicklime, CaO; however, monohydrated dolomitic limes and dolomitic quicklimes (see Chapter 5) are also used. Hydrated limes are available in both powder and slurry forms. Quicklimes are only available as dry granular materials.

There are many similarities between cement and lime stabilization, including that both are used for soil modification and bonding purposes (albeit modification is by far the more common with lime). Major differences, however, relate to (a) the nature and rate of the lime–soil reactions, and (b) the fact that there is no value in using lime with cohesionless or low cohesion sands and gravels unless a pozzolanic material, e.g. pulverized fuel ash (pfa), is added.

Lime is used as a soil treatment for a variety of reasons, e.g. to expedite construction on weak clay subgrades, or to improve the engineering properties of plastic sands, plastic gravels, and reactive clays. Whilst lime is mainly used in subbase and subgrade capping layers in pavements, it is not uncommon for it to be also used overseas to modify soil–aggregate roadbases with high plasticity indices and/or clay contents.

6.4.1 Mechanism of lime stabilization

When lime is admixed with a moist clayey soil a number of reactions take place involving: (1) cation exchange; (2) pozzolanic reaction; and (3) carbonation.

The first of these reactions, that of *cation exchange*, takes place immediately and causes the individual clay particles to change from a state of mutual repulsion to one of mutual attraction, typically due to excess $Ca^{2}+$ replacing dissimilar cations from the exchange complex of the soil. (As a general rule the order of replaceability of the common cations associated with soils follows the lyotropic series $Na^+ < K^+ < Ca^{2+} < Mg^{2+}$, with cations on the right tending to replace cations to the left in the series and monovalent cations being usually replaceable by multivalent cations.) This has the immediate positive effect of promoting flocculation of the particles and a change in soil texture, i.e. with clay particles 'clumping' together into larger-sized aggregates or lumps – thereby improving the soil's gradation, permeability and handling properties

Note that the extent of the change to a soil that is brought about by cation exchange is dependent upon the amount of clay, the clay's reactivity, and the nature of the dominant cation that was originally adsorbed. For example, a sodium-dominated montmorillonitic clay (which has a high cation-exchange capacity) will be much more dramatically affected by the addition of lime than, say, a kaolinitic clay (which has a low cation-exchange capacity); however, the montmorillonitic clay will requires a large addition of lime to achieve calcium saturation and the full flocculation effect, whilst a kaolinitic clay requires only a small amount of lime to achieve its full flocculation potential.

For a *pozzolanic reaction* to take place, the amount of lime added to a clayey soil must be in excess of that required by the soil for cation exchange and flocculation/aggregation. The pozzolanic reaction, which is affected by the clay content, the type of clay mineral, and the curing temperature, occurs more slowly than the cation-exchange reaction, and results in the formation of cementitious products that have long-term effects on the strength, volume stability, and (in colder climates) resistance to frost action of the stabilized soil.

By definition, a pozzolanic material is a siliceous or siliceous and aluminous material which in itself possesses little or no cementitious value, but which (when in a finely-divided form and in the presence of moisture) will chemically react with lime at ordinary temperatures to form compounds possessing cementitious properties. Some components of natural soils, notably clay minerals, are pozzolanic and the lime–soil pozzolanic reaction results in a slow, long-term, cementing together of soil particles at their points of interaction. When the lime is added the pH of the soil is raised, typically to about 12.4, and this highly alkaline environment promotes the dissolution of the clayey particles (particularly at the edges of the clay platelets) and the precipitation of hydrous calcium aluminates and silicates that are broadly similar to the reaction products of hydrated cement.

By *carbonation* is meant the process whereby carbon dioxide from the air, and from rainwater, converts free calcium and magnesium oxides and hydroxides into their respective carbonates. Because of carbonation some lime which would normally take part in the pozzolanic reaction becomes unavailable for this purpose, so that soil–lime (and soil–lime–pfa) mixtures develop lower strengths than might otherwise be expected. In practice, therefore, the carbonation effect means that lime needs to be protected whilst in storage and in shipment prior to field use, and that lime stabilization is best carried out in non-industrial locales. Long intensive mixing and processing during construction should also be avoided if high strength gain is a primary objective, and compaction should occur as soon as possible after the addition of the lime.

6.4.2 Some factors affecting lime–soil properties

Whilst the lime–soil pozzolanic reaction is very important to the long-term stability and durability of a pavement layer, the most striking effect of admixing lime is its immediate impact upon a soil's plasticity properties. Adding lime to a clayey soil normally results in a reduction in the plasticity index, as the liquid limit is usually decreased and the plastic limit is increased (see Fig. 6.3). The resultant agglomerated, more friable, nature of the lime-treated soil means that its *workability* is improved, and this expedites its subsequent manipulation and placement during construction, including during wet weather. The greatest PL increases occur with montmorillonitic soils and the least with kaolinitic ones.

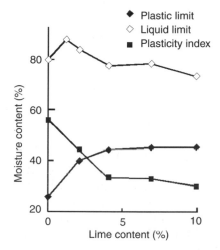

Fig. 6.3 Effect of the addition of lime on the plasticity properties of London Clay (CV: very high plasticity)[5]

The addition of lime also reduces the thickness of the adsorbed water layer on a clay particle and this, combined with the flocculation effect, allows water to flow more easily between particles, thus increasing permeability. Another effect is that there is a decrease in the moisture absorption capability of the clay component, thereby enhancing the soil's volume stability, i.e. there is less swelling when water is added to a lime–soil mix as compared with the raw soil.

The amount of lime required to cause flocculation/aggregation is usually quite small, i.e. typically, between 1 and 3 per cent by mass of the lime–soil mix, and additions of chemically-equivalent (on a molecular weight basis) amounts of hydrated lime and quicklime have the same modifying effect. The lime content which marks the maximum increase in a soil's plasticity limit is termed the *lime fixation point*, and it is the lime that is in excess of this fixation point that becomes involved in pozzolanic reactions and the cementation of particles.

The immediate modifying effect of lime upon soil plasticity has a corresponding effect on *stability*, and this is reflected in an equally-immediate increase in, for example, California Bearing Ratio test values. As curing time progresses, the CBRs increase further as the pozzolanic reactions begin to take effect and tensile and unconfined compressive strength gains occur.

In general, the early (first 7 days) strength gains occur quite rapidly (albeit much more slowly than with cement), and then increase more slowly at a fairly constant rate for some months. For a given curing period, the lime–soil strength will normally increase as the lime content is increased up to a critical amount, beyond which strength either declines or remains constant, i.e. a lime content in excess of that able to react with the pozzolanic material present in the soil during that curing period will not result in any more strength gain. However, the lime–soil reaction may continue for a longer time until all the free lime is used up; for example, if adequate moisture is present a very highly-reactive clay to which sufficient lime is added may require many years before the pozzolanic reaction is completed and equilibrium is finally reached. Most commonly, however, the greater part of the observed increase in strength in lime-treated soils is obtained with about 3 per cent hydrated lime by mass for kaolinitic soils and about 8 per cent for montmorillonitic soils[5].

Soil–lime pozzolanic reactions are very *temperature* dependent (see Fig. 6.4). One reason why lime stabilization is used frequently in warm climes relates to the fact that high ambient temperatures are very beneficial to the rate of strength development. The lime–soil pozzolanic reaction is curtailed at temperatures below about 10°C; however, it will start again once the temperature is raised and if free lime is available.

If a soil has a non-reactive clay mineral and a low clay content, significant pozzolanic strength development will not take place, irrespective of the lime type, content, or length of curing. Normally, there is little value in adding lime to silty, sandy or gravelly soils with *small clay contents* (say, less than about 10 per cent), as the pozzolanic reaction will be minimal. If, however, a second highly-pozzolanic material such as pulverised fuel ash (pfa) can be economically added to these low clay-content soils, the pfa will react preferentially with the free lime and significant strength gains will be obtained. However, the presence of a significant *reactive clay fraction* in lime–soil–pfa mixtures can have a detrimental effect on strength as the lime will then react preferentially with the clay to produce cementitious products that are weaker than the lime–pfa products.

Both hydrated limes and quicklimes (but not slurry limes) are effective in *drying wet soil*s. However, quicklime has a faster drying effect as the chemical reaction between this type of lime and the water in the soil removes free water from the soil and the heat produced by the reaction assists in drying[6].

Moisture–density relationships for lime–soil mixtures are variable, and it is important therefore that the desired relationship be clearly defined in specifications, for field control purposes.

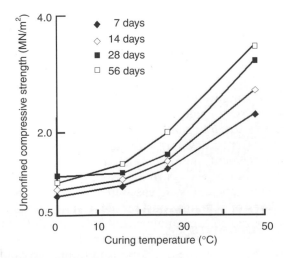

Fig 6.4 Influence of curing temperature and time on the strength of clay (CV: very high plasticity) stabilized with 5 per cent lime[5]

For the same compactive effort lime-treated soils have lower maximum dry densities (typically by 50–80 kg/m^3) and higher optimum moisture contents (typically by 2–4 per cent) than untreated soils; also, the maximum dry density normally decreases further as the lime content is increased. If a lime-treated soil is allowed to gain strength in a loose state prior to compaction, additional reductions in maximum dry density and increases in optimum moisture content are likely to occur; however, the effects are not as severe as is associated with delaying the compaction of a cement-treated soil. Nonetheless, delays in compaction should be avoided once the design amount of lime is added, so that the opportunity for a higher final strength is maximized.

The optimum moisture content for maximum strength of a lime-treated soil is not necessarily the same as that for maximum dry density. With clayey soils the optimum moisture content for strength tends to be on the high side of that for dry density, but with silty soils the opposite may be true. Furthermore, the optimum moisture contents for strength are usually different for different curing periods.

As noted previously, lime-treated (calcium-saturated) clays have a reduced affinity for water. This effect, combined with the formation of the pozzolanic cementitious matrix, means that the *swelling potential* is further reduced when, for example, an expansive clay is treated with a lime content greater than its lime fixation point.

Limes are irritants. Thus, for *safety* reasons, workers should be required to wear protective clothing and masks when spreading and admixing lime.

Hydrated limes can create *dust* problems and should not be used in built-up areas, especially in windy conditions. Dust is normally not a serious problem with quicklimes or, of course, slurry limes.

6.4.3 Selecting the design lime content

As might be expected from the above discussion on lime–soil properties, the design lime content for use with a given soil is normally based on an analysis of the effect of different lime contents on selected engineering properties of the lime–soil mixture. In this respect, it is important to appreciate that different design contents may be determined for the same soil, depending upon the objectives of the lime treatment and upon the design procedure used for testing.

Typically, the basic components of a design procedure are: (1) the method used to prepare the soil–lime mixtures; (2) the procedure used to prepare and cure the test specimens; (3) the test methods used to evaluate the selected properties of the mixtures; and (4) the criteria selected to establish the design lime content. Excellent guidelines relating to mixture design, as well as to the construction of lime–soil pavements, are available in the literature[7].

6.5 Construction of cement- and lime-treated courses in pavements

Cement- and lime-stabilization construction procedures can be divided according to whether they involve mix-in-place equipment or stationary plant. Mix-in-place stabilization is most appropriately employed in the construction of capping and subbase layers, particularly for smaller projects in remote areas where the use of a central plant would be logistically difficult. Stationary plant stabilization requires the establishment of a central mixing plant at which the stabilization ingredients (including the soil) are separately batched and mixed before being transported to the construction site for placement. Whilst a better construction product is normally obtained with central plant stabilization, it is more expensive and, for economic reasons, it is therefore most usually justified in the construction of roadbases and subbases on large road projects.

6.5.1 Mix-in-place construction

The basic steps involved in a conventional mix-in-place process are: (1) the initial preparation of the in-situ soil; (2) pulverization and admixing of the soil, stabilizing agent, and water; (3) compaction and finishing; and (4) curing.

Typically, the *initial preparation* requires excavating the subgrade soil down to the desired formation level within the C-horizon and, if the in-situ soil is to be stabilized, grading the loosened soil to the desired formation cross-section and profile. If subbase borrow material is to be stabilized it is spread on the formation and graded. In either case the soil is scarified thoroughly (and pulverized, if specialized mix-in-place equipment is not available) so as to make it easier to admix the cement or lime.

Cement and quicklime (which is granular) is then spread on top of the loose soil, preferably through a vane feeder on a bulk pneumatic tanker which has an on-board speed-regulated weigh system to control the spread rate. Before mixing, the quicklime will need to be wetted, i.e. 'slaked', until it is totally hydrated. Hydrated lime is a very fine powder and to minimise dust problems on large projects it is best added to the soil in the form of a slurry.

The more uniform the depth of spread of the additive the more likely it is that a uniform distribution will be obtained during the mixing process. If a uniform distribution is not obtained, detrimental cracking may subsequently occur that is associated with localized high and low strengths.

Purpose-designed 'travelling plant' equipment may be used to *mix* the soil, additive and water in one pass of the equipment. Conventional machines are able to handle mixing widths varying from 1.8 to 2.4 m and depths of up to 300 mm. A water tanker coupled to the travelling plant may be used to feed the desired amount of water into the mixing hood of the equipment during the one-pass mixing process; if a slurry is to be used, the lime is injected through the mixing hood at the desired moisture content. As well as ensuring a better distribution, one-pass mixing also has the advantage that it helps to ensure that compaction can commence, and be completed, within a relatively short time after the addition of water. With some heavy soils two passes may be required to ensure the uniform distribution of the cement or lime; in this case the water is usually not added until the second pass, and the first pass is carried out on the 'dry' soil and additive.

If specialized mix-in-place equipment is not available, the water required to ensure that the mixture is at the required moisture content for compaction is added to and mixed with the loose soil, prior to the subsequent admixing of the cement or lime additive using the spreading and mixing equipment available.

It is important that *compaction* be initiated as soon as possible (and no later than 30 min) after the admixing of water to cement-treated mixtures, and that as many passes as possible be carried out within, say, the first 15 min of compaction. Table 6.5 is an abbreviation of a method specification[3] that is recommended[6] for compacting cement-treated soils in the UK. Compaction, including trimming, should be completed within 2 h of the beginning of mixing. Trimming is normally initiated before compaction is finished, to ensure good bonding of any corrected shape; however, care needs to be taken that any thin layer of removed material is wasted and not placed over previously compacted material as this may not bond to the material below and may flake away at a later time under traffic.

Generally, early compaction is also desirable with lime-treated mixtures (especially in hot climatic areas) as this will lead to higher strengths. With lime, however, the time factor is far less critical than with cement. In fact, compaction may be deliberately delayed if a soil is difficult to work, e.g. if it is clayey and its plasticity

Table 6.5 An abbreviated method specification for compacting stabilized soils

Type of compaction plant	Category	Number of passes for compacted layer thicknesses of:		
		110 mm	150 mm	250 mm
Smooth-wheeled roller	*Mass/m-width of roll (kg)*:			
	>2700 up to 5400	16	Unsuitable	Unsuitable
	>5400	8	16	Unsuitable
Pneumatic-tyred roller	*Mass/wheel (kg)*:			
	>4000 up to 6000	12	Unsuitable	Unsuitable
	>6000 up to 8000	12	Unsuitable	Unsuitable
	>8000 up to 12 000	10	16	Unsuitable
	>12 000	8	12	Unsuitable
Vibratory roller	*Mass/m-width of vibratory unit (kg)*:			
	>700 up to 1300	16	Unsuitable	Unsuitable
	>1300 up to 1800	6	16	Unsuitable
	>1800 up to 2300	4	6	12
	>2300 up to 2900	3	5	11
	>2900 up to 3600	3	5	10
	>3600 up to 4300	2	4	8
	>4300 up to 5000	2	4	7
	>5000	2	3	6
Vibrating plate compactor	*Mass/unit area of base plate (kg/m²)*:			
	>1400 up to 1800	8	Unsuitable	Unsuitable
	>1800 up to 2100	5	8	Unsuitable
	>2100	3	6	12
Vibro-tamper	*Mass (kg)*:			
	>50 up to 65	4	8	Unsuitable
	>65 up to 75	3	6	12
	>75	2	4	10
Power rammer	*Mass (kg)*:			
	100 up to 500	5	8	Unsuitable
	>500	5	8	14

needs improving, or if it is too wet and needs to be dried out. In either case, part of the design lime content is first admixed, and the mixture is lightly compacted and allowed to 'condition' for up to 48 h; the rest of the lime is then admixed (with additional moisture, as necessary), and the final compaction is carried out.

Upon completion of compaction the final surface of the cement- or lime-treated layer should appear smooth, dense, tightly packed and free from cracks and compaction planes. Construction traffic (other than that which, of necessity, is involved in the curing process) should be kept off the surface until after the end of the curing period.

The *curing* of a cement- or lime-stabilized material is concerned with promoting strength gain and preventing moisture loss, through evaporation, from the compacted layer for a reasonable length of time (usually at least 7 days). Its purpose is to allow the cementitious hydration reactions to continue, to reduce shrinkage and shrinkage cracking, and to prevent the development of a variable strength profile in the pavement layer. Curing is necessary in all climatic areas but it is particularly important in locales with hot ambient temperatures and/or where windy conditions occur.

The simplest approach to curing is to spray the surface of the stabilized layer intermittently during the first week, making sure that the surface is neither flooded nor allowed to dry out. Nowadays, however, the most common curing methods involve covering the compacted layer with impermeable sheeting (taking care to overlap at joints), or spraying the top of the compacted layer with a bituminous sealing compound or a resin-based aluminous curing compound.

6.5.2 Stationary plant construction

Normally, the stationary plant equipment is located in a quarry or borrow pit close to the construction site. Overall, the stationary plant construction process is similar to that employed with the mix-in-place process, except that the materials to be stabilized – i.e. more than one soil and/or aggregate may be used in a given operation – are stockpiled, proportioned, and mixed with the requisite amounts of additive and water, using conventional equipment, at the central plant. The hydrated lime or cement additives are normally added on the conveyor belt enroute to a pugmill mixer. The mixed product is then transported to the construction site in covered vehicles and spread to a uniform depth, prior to compaction. The more uniform the depth of spread the better the finished product; hence, purpose-built spreader-finishers are preferably used to spread the mixed material.

In the case of cement-treated materials the stationary plant needs to be close enough to the construction site to ensure that mixing, transporting, spreading and compaction can be completed within the specified time period. If the travel time is too long a retarder may need to be added to the mix at the central plant.

6.6 Bituminous stabilization

The mechanisms involved in the stabilization of a soil with a bituminous material (usually either hot bitumen, cutback bitumen, or anionic or cationic bitumen emulsion) are very different from those involved with cement or lime.

With coarse-grained non-plastic soils the main function of the bituminous material is to add cohesive strength. Thus, the stabilization emphasis with granular soils such as gravels and sands, and sandy soils, is upon the thorough admixing of an optimum amount of binder so that particles are thinly coated with binder and held together without loss of particle interlock.

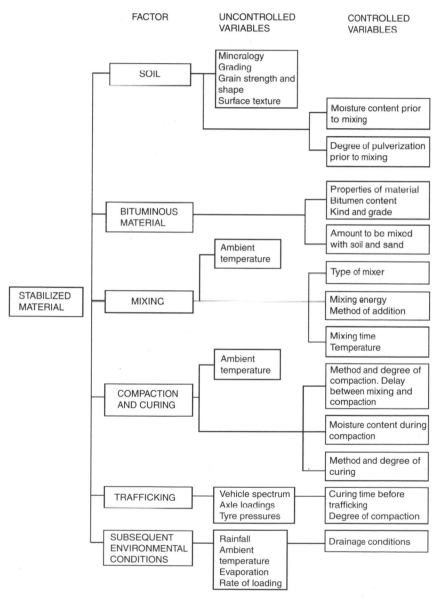

Fig. 6.5 Factors affecting the design and behaviour of bitumen-stabilized materials[1]

In the case of a soil that already has cohesion the bituminous material is admixed in order to waterproof the soil and maintain its existing strength. Here the emphasis is upon impeding the entry of water by adding sufficient bituminous material to (a) wrap soil particles or agglomerates of particles in thin bituminous films or membranes, and (b) plug the soil-void 'channels'.

In practice, of course, a combination of the above mechanisms occurs in most soils that are stabilized with bituminous materials.

The main factors which influence the behaviour and design of soils stabilized with bitumen are summarized in Fig. 6.5. To obtain good results following construction (a) the bituminous material should be thoroughly and uniformly admixed with the soil, (b) soils stabilized with cutback bitumen or bitumen emulsion should be allowed sufficient time to aerate after mixing and before compaction so that excess volatiles and/or moisture can escape, and (c) the soil should be compacted at a uniform fluid content.

Bituminous stabilization is most appropriately used in hot climatic areas where there is normally a need for additional fluid to be added to a soil at the time of construction, to ensure adequate mixing and compaction. Its potential for use in Britain is very limited because of the island's regular rainfalls, i.e. the moisture contents of most soils are normally fairly high throughout the year, and the admixing of additional fluids in the form of bituminous materials could cause loss of strength[6].

6.7 References

1. NAASRA, *Guide to Stabilisation in Roadworks*. Sydney: National Association of Australian Road Authorities, 1998.
2. *Structural Design of Low-volume Roads*, Transportation Technology Support for Developing Countries, Synthesis 4. Washington, DC: The Transportation Research Board, 1982.
3. *Specification for Highway Works*. London: HMSO, 1998.
4. O'Flaherty, C.A., *Highways: Highway Engineering*, Vol. 2, 3rd edn. London: Arnold, 1988.
5. Bell, F.G. and Coulthard, J.M., Stabilisation of clay soils with lime, *Municipal Engineer*, 1990, **7**, No. 3, pp. 125–40.
6. Sherwood, P.T., *Soil Stabilisation with Cement and Lime*. London: HMSO, 1995.
7. TRB Committee on Lime and Lime-Fly Ash Stabilization, *Lime Stabilization: Reactions, Properties, Design, and Construction*. Washington, DC: The Transportation Research Board, 1987.
8. *Unsealed Roads Manual: Guidelines to Good Practice*. Vermont, Vic: Australian Road Research Board, 1993.

CHAPTER 7

Surface drainage for roads

C.A. O'Flaherty

7.1 Importance of surface drainage

Excess water, which is precipitated as rain, hail, snow or sleet, is the enemy of earthwork foundations, pavements – and traffic. Consequently, proper surface drainage design is an essential and integral part of economic road design.

Roads cut across and obstruct natural drainage paths and, hence, there may be locations where flood waters have the power, if given the opportunity, to destroy a roadway by force or to hold up traffic by flooding a carriageway. The barrier effects of new roads can result in blockages on existing local drainage channels which cause redirection and redistribution of stream flows and alter local catchment areas and boundaries. Thus, good drainage design begins with good route location, and roads that avoid poorly-drained areas, unstable foundation soils, frequently flooded areas, and unnecessary stream crossings greatly reduce the costs and dangers associated with these aspects of road drainage. Good location may also make it more economic to relocate stream channels rather than provide bridges, major culverts and/or other expensive drainage features to accommodate them – assuming that the impacts of the proposed channel changes upon the environment are acceptable to the public.

Water on a road always poses potential danger to moving vehicles, due to the longer distances required to stop under wet conditions. The risk of skidding, which is associated with the reduced braking grip between tyres and the wet carriageway, is especially important at locations where braking is required, e.g. at sharp curves, prior to controlled intersections, and at roundabouts and exit lanes on high-speed roads. Water depths in wheel tracks at these locations, and at those subject to long flow paths, should always be kept below 4 mm, and preferably below 2.5 mm. Whilst a relatively rare happening[1], full aqua-planing – which is associated with the complete separation of a tyre from the road surface – can occur at water depths in excess of 4 mm, depending upon tyre condition, vehicle speed and the texture of the carriageway surface; partial aqua-planing, which is associated with reduced tyre friction, starts at around 2.5 mm.

Water on the carriageway is also the source of reduced windscreen visibility as a result of splashing from the tyres of other vehicles, especially commercial vehicles. Road markings and lines, including reflectorized ones, become much more difficult to see, and often become nearly invisible as a result of water accumulating on the carriageway. The positive effects of road lighting are much reduced when there is water on the road surface.

Pedestrians and cyclists are vulnerable to splashing from the tyres of passing vehicles, if there is any significant depth of water on a carriageway.

If the drainage facilities beside or through a road are inadequately designed and/or constructed, water from adjacent land and 'upstream' road areas may encroach on the carriageway and excessively reduce its traffic capacity; if the road surfacing is not impervious, the retained water may seep into and through the pavement and soften the foundation. If a shallow roadside (longitudinal) channel is not made impermeable water may also penetrate the road laterally and soften the foundation. Legal problems can arise as a result of water from inadequate road drains overflowing onto adjacent properties.

The road engineer in the UK is mostly concerned with coping with run-off from rain; it is only in very cold countries subject to rapid (upward) temperature changes that special consideration is normally given to the need for extra drainage facilities for run-off water from melting snow accumulations. Road drainage design is there-fore basically concerned with selecting a design storm, estimating the likely run-off from that storm from the design catchment area, and deciding how to collect and remove the water to a suitable discharge point, i.e. an outfall, so that it can be disposed of safely and economically. The following discussion is concerned with the control of run-off water from relatively small catchment areas, mainly road carriageways, verges and cuttings, so that it can be diverted to suitable disposal points with the minimum detrimental effect upon road users, the road or the environment. Advice on how to deal with flood flows from large catchments is available in the literature (e.g. see reference 2) and is not further discussed here.

7.2 Types and uses of surface drains

Most road carriageways are sloped both laterally and longitudinally so that run-off water from the pavement surface is constantly moving under gravity. The surface run-off may be allowed to flow over the edges of the verges and collected in grassed open ditches cut to, at least, the natural angle of repose of the soil or, more usually in the construction of new roads in the UK, it is collected by specifically-designed channels and carried to suitable outfalls.

7.2.1 Over-the-edge drainage

Over-the-edge verge drainage over embankment slopes (Fig. 7.1(a)), and into open ditches or preformed channel blocks (Fig. 7.1(b)), is mainly used in rural-type locales and should never be employed where a footpath abuts a carriageway. If used on high road embankments, over-the-edge drainage can be the cause of soil erosion, softening of side slopes and, eventually, embankment instability; conse-quently, its use is most applicable to shallow embankments with gentle slopes (to preclude topsoil instability) formed from very stable materials, e.g. rock fill.

Over-the-edge drainage into ditches or channel blocks that carry the surface water to convenient outfalls is normally used in conjunction with, but separate

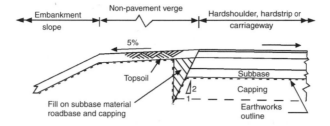

(a) Embankment – Verge draining over embankment slope (Type 2A)

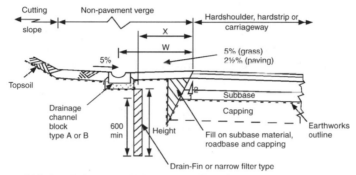

(b) Cutting – Drainage channel blocks and drains (Type 4A)

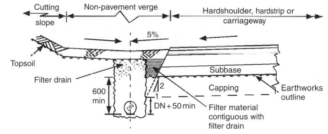

(c) Cuttings – Combined surface and groundwater filter drains (Type 1A)

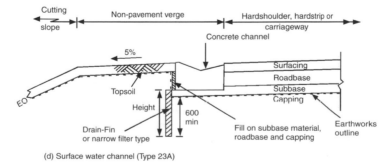

(d) Surface water channel (Type 23A)

Fig. 7.1 Some edge-of-flexible-pavement drainage construction details[3]

from, a subsurface filter drain. The small-capacity channel block system shown in Fig. 7.1(b) is an inexpensive way of handling small volumes of water in, for example, cuttings in impermeable soils. In this case the function of the filter drain is to drain the pavement layers and remove small quantities of groundwater. The distance of the channel block to the edge of the carriageway varies according to the road type, but is never less than 1 m.

7.2.2 Kerbs and gutters

In most streets with footpaths in urban areas, vertical kerbs and sloping gutters are used to form triangular channels that carry the run-off water to inlets in gully pits. The gullies are connected to subsurface pipes that then transport the water to outfalls. The subsurface pipe lengths, which are normally either of concrete or ceramic material, have watertight joints (to prevent leakages that would result in softening of the subgrade) and are laid in trenches that are backfilled with a suitable material, and well compacted, prior to the placement of the pavement.

A *gutter* begins at the bottom of the kerb and extends toward the carriageway. Whilst it usually has the same longitudinal grade as the carriageway, a gutter will often have a steeper crossfall and be composed of a different material from the pavement surface.

A *gulley* is essentially a precast concrete or vitrified clay chamber, the base of which is, typically, up to 900 mm below the invert of an outlet that is connected to the subsurface pipe system. Its design is such that grit and suspended solids can settle in the chamber, for subsequent removal, before the run-off water 'overflows' into the outlet. The gulley outlet is normally fitted with a rodding eye so that it can be cleared from the surface if it should become clogged with floating debris.

Water enters a gulley pit either laterally through an open side-entry inlet that is set into the line of the kerb, or vertically through a grating with bars that is set into the gutter in the line of the flowing water, or through a combination inlet of these. The side-entry inlet is normally the least efficient of these inlets as the small hydraulic head acting at right angle to the direction of flow is insufficient to move much water over the side weir, especially if the gutter is located on a steep slope; its efficiency can, however, be much improved if the gutter approach is shaped just prior to the inlet so that the opening is more directly in the path of the water flow. The other, more efficient, types of inlet can collect up to 95 per cent of the run-off water, provided that the width of flow does not exceed about 1.5 times the grating width[4]. As inlets are vulnerable to clogging, it is good maintenance practice to ensure that carriageways are swept clean, and gulleys are inspected, on a regular basis.

Gulleys in urban areas are normally installed at low points, i.e. sags, on the road, where water naturally accumulates. Other appropriate locations are just prior to bus stops, laybys, taxi ranks, and the upstream radii of corner kerbs, so that the run-off water can be intercepted before it reaches carriageway areas heavily used by pedestrians and/or turning vehicles.

Historical rule-of-thumb guides to gulley spacing were (a) that gulleys on urban streets should not be more than about 40 to 50 m apart, and (b) that the impermeable area of housing estate road draining to a gulley should not be more than

$200 \, \text{m}^2$. These rough guides have now been replaced by a rational approach to gulley spacing design developed by the Transport and Road Research Laboratory[4]. Based on Manning's equation for hydraulic flow, this approach uses tables incorporating data on the efficiencies of different types of inlet when used with varying gutter crossfalls and longitudinal gradients, to determine the drained area of road which, under a rainfall intensity of 50 mm/h, results in kerb-gutter channel flow widths adjacent to the gulley of 0.5 m, 0.75 m, and 1.0 m. Thus, the gulley spacing on a high-speed urban road or on a street with many pedestrians might be chosen so as to result in a channel flow width no greater than 0.5 m at the edge of the carriageway, whilst a design flow width of, say, 1 m might permitted at the side of an unimportant road.

Unless provided for specific road layout reasons, kerbs are not provided for general use on rural trunk roads that do not have adjacent footways[5].

7.2.3 Combined surface and groundwater filter drains

Combined surface and groundwater filter drains beside the carriageway (Fig. 7.1(c)) were a standard feature of major road design in the UK for many years. The open texture of the filter material enables these drains – which are also known as *French drains* – to quickly pass rainwater from the road and verge area. Also, as the pipe diameters are quite large the drains have large groundwater capacity which extends as a cut-off to below the capping layer in the pavement.

Combined systems have, however, many disadvantages. First and foremost, they bring large quantities of surface water into the ground at pavement foundation level, where any malfunctioning of the drainage system is not easily noted albeit the consequent risk to the foundation is greatest. Also, filter drain costs quickly increase with increasing nominal pipe diameter (DN) because of the cost of stone, the filter materials have to be recycled or replaced about every 10 years, and there is a need for regular maintenance to prevent a build up of grass kerb-type barriers. For these reasons these drains are now only used in new construction in exceptional cases, e.g. where large groundwater flows from cuttings have to be dealt with and a combined system shows significant cost savings, or where the road has long lengths of zero longitudinal gradient. Because of the high cost of stone in some regions, and since large diameters mean long drain runs with a greater risk of saturated conditions downstream, pipe diameters in French drains are now[5] normally limited to 250–300 mm.

7.2.3 Surface water channels

Figure 7.2 shows the layout of a classical drainage system developed for a high-quality dual carriageway road in a built-up area in the 1970s, prior to the acceptance of the use of surface drains for new major roads. It consists of longitudinal drains beside the road and in the central reserve, cross-drains, chute and toe-of-slope carrier drains and outfalls. Note that with this particular design, very many of which are still in operation, the surface run-off water is required to enter the subsurface longitudinal pipes either through filter material (over open-jointed or perforated pipes) or gullies in the verges and in the central reserve. Cross-drains

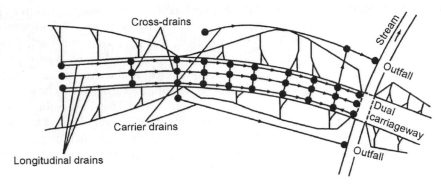

Fig. 7.2 Layout of a 1970s drainage system for a dual carriageway in a built-up area

are provided at appropriate intervals to convey the relatively large volumes of water from the longitudinal pipes to the watertight carrier channels (via a paved chute in the case of the embankment connection) that convey the water to the out-falls. Manholes, to provide access for maintenance, are installed at pipe inter-sections, changes in pipe size or slope, and at intervals along drains subject to maximum spacing restrictions.

The current drainage philosophy in the UK is that French drains are funda-mentally undesirable, and that surface water should be kept on the surface, but clear of the carriageway running surface, for as much of its journey to its ultimate outfall as possible. Surface water channels (Fig. 7.1(d)), which are now the normal means of providing for surface water drainage in new rural road construction unless there are overriding constraints, can carry large volumes of water over long distances, thereby reducing the need for additional carrier pipes. The filter drain has tended to be replaced by a separate subsurface drain, whilst the outlet loca-tions can be more widely spaced and chosen to suit the topography and the avail-ability of natural watercourses into which the water can be discharged. The most economic use of these channels is obtained when the road geometry and the selected outfall locations allow them to be designed to carry water over entire lengths of cuttings and embankments directly to outfalls, thus minimizing the lengths of piped drainage. If carrier pipes are also needed they are used in prefer-ence to carrying water in the filter drain system.

Surface channels are formed from concrete and have either triangular or rec-tangular cross-sections. Triangular channels are usually sited at the edge of the carriageway in front of the safety fence, beside the hardstrip or hard shoulder, and their maximum depth is normally limited to 150 mm. In both verges and cen-tral reserves the channel cross-slopes are normally not steeper than 1:5, except in exceptional circumstances when 1:4 is permitted for either or both faces. Triangu-lar channels deeper than 150 mm, and rectangular channels, are also used provided that safety fencing is installed between the carriageway and the channel, and the channel is not located in the zone behind the safety fence into which the fence might reasonably be expected to deflect during vehicle impact.

7.3 Estimating the surface run-off

Many formulae have been proposed to measure the quantity of run-off from a storm over a given area (see reference 6 for a review of 18 methods). Three methods that have gained particular credibility in the UK to estimate the run-off from carriageway catchments are the Rational/Lloyd-Davies method, the Department of Transport's Advice Note HA37/88 method, and the Transport and Road Research Laboratory (TRRL) Hydrograph method.

The Rational method is described in some detail here because, as well as being well accepted internationally for use in the piped drainage design of rural roads (including shoulders and adjacent roadside areas in cuttings), and in the design of urban stormwater systems, it illustrates particularly well the factors that must be considered in surface run-off calculations. The HA 37/88 method is now recommended for use in the hydraulic design of roadside surface drains for rural trunk roads in the UK. The TRRL method was developed as a simple sewer design tool for use in a large urban area; details of the method are readily available in the literature[7,8] and, for space reasons, it is therefore only briefly discussed here.

Also described here is the TRRL run-off formula used in the determination of culvert sizes for minor streams that must pass under the line of a road.

7.3.1 Rational method

Also known as the Lloyd-Davies method, this uses a run-off formula that is variously attributed to an Irishman[9] in 1850, an American[10] in 1859, and a Welshman[11] in 1906. It relates the peak rate of run-off from a given catchment to rainfall of a given average intensity by means of the following equation:

$$Q = 0.167(60/t_c)Apr \qquad (7.1)$$

where Q = discharge (m³/min), t_c = time of concentration (min), A = catchment area (ha), p = proportion of total rainfall running off after allowing for soakage and evaporation, r = total rainfall during time of concentration (mm), and 0.167 = run-off (m³ per min per ha) for a rainfall of 1 mm/h on a completely impervious surface. If I is the average rainfall intensity (mm/h) for a storm of duration t_c min, then

$$r = (t_c/60)I \qquad (7.2)$$

and

$$Q = 0.167AIp \qquad (7.3)$$

If Q is expressed in litres per second, then Equation 7.3 becomes

$$Q = 2.78AIp \qquad (7.4)$$

A fundamental feature of rainfall is that its *average intensity (I)* throughout a given storm is inversely proportional to the length of storm, i.e. as the duration of a rainfall increases its average intensity decreases (see Table 7.1). Thus, whilst the engineer designing for a large bridge or a complete flood-control system is mainly interested in storms that cover very large areas and last for hours and perhaps days, the road engineer engaged in the design of a culvert beneath a pavement, or a

Table 7.1 Some rainfall intensities (mm/h) at Crowthorne, Berkshire (National Grid Reference 4833E 1633N), as a function of duration and frequency[8]

Duration (min)	Return period (years)				
	1	5	10	50	100
2.0	75.6	120.5	138.3	187	213
2.5	76.5	113.4	130.4	177	202
3.0	66.3	107.2	123.4	168	192
3.5	62.8	101.7	117.3	161	184
4.0	59.6	96.8	111.8	154	176
4.1	59.1	95.9	110.8	152	174
4.3	57.9	94.1	108.8	150	172
4.5	56.9	92.4	106.9	148	169
4.7	55.8	90.8	105.1	145	166
4.9	54.8	89.2	103.4	143	164
5.1	53.9	87.7	101.7	141	162
6.0	50.0	81.6	94.9	132	152

roadside subsurface pipe or surface channel that is carrying run-off from a carriageway, is mainly interested in high-intensity short-duration storms, i.e. those which result in the peak rate of run-off from smaller catchment areas. In this respect it should be noted that, regardless of topography, rainfall intensities during short (<1 h) storms are generally fairly similar at different locations in a relatively small island such as Britain, e.g. the average rainfall expected in 60 min, with a frequency of once in 5 years, is 18 mm plus or minus 3 mm over 85 per cent of the island. However, the variations due to geographical location increase significantly as the storm durations and return periods are increased.

Motorists have difficulty in seeing through their windscreens when driving in storms with rainfall intensities of greater than around 50 mm/h. As shown in Table 7.1 a 50 mm/h intensity rainfall with a duration of 6 min is likely to occur every year at Crowthorne, Berkshire (the site of the Transport Research Laboratory). Whilst it is not shown in this particular table, the same rainfall intensity can be expected to occur at Crowthorne every five years with a duration of 15 min.

The *storm duration* that is usually chosen for road design purposes based on the Rational method of estimating the peak run-off assumes (a) that the maximum discharge at a point in a drainage system occurs when the entire catchment area that is tributary to that discharge point is contributing to the flow, and (b) that the rainfall intensity producing this maximum run-off is the average rate which falls throughout the catchment during the time required for a raindrop to flow to the discharge point from the most time-remote point of the catchment. The 'most time-remote point' is that from which the raindrop's time of flow is the greatest, and the least time taken by that raindrop to make the trip from the most time-remote point to the discharge outlet under design is termed the *time of concentration* (t_c). In other words, the maximum run-off is assumed to occur when the storm duration is equal to the time of concentration; if a storm duration of less than t_c is selected not all of the catchment will contribute to the run-off calculations, whilst the average rainfall intensity will be reduced if a storm duration longer than t_c is chosen.

The time of concentration has two components, the *entry time* (t_e) and the *flow time* (t_f). If, say, the catchment area is the road's carriageway and the point under consideration is an inlet that discharges directly into an adjacent stream, then t_f is zero and t_e, the time for the raindrop to flow off the carriageway and along the gutter to the inlet/outfall point, equals t_c. If, however, the discharge outfall is further down the line within the drainage system, then t_e is the time taken by the raindrop to travel to the gulley plus the time to travel along the connecting pipe from the gulley to the roadside drain, and the flow time, t_f, is the time spent by the raindrop in the roadside drain until it reaches the outfall; the time of concentration, t_c, is then equal to the sum of t_e and t_f.

The entry time varies according to the nature of the catchment, e.g. 2 min is not uncommon for road carriageways whilst large flat areas such as car parks have longer times, i.e. 5 min or more. There is limited value in estimating the entry time to an accuracy greater than, say, 0.5 min, as the effect upon the design intensity is fairly insignificant (see Table 7.1) in comparison with the capacity available within the practical range of pipe sizes used in road drainage design, e.g. one increment change of pipe size from 150 to 230 mm roughly trebles the flow-carrying capacity.

The flow time is dependent upon the size and gradient of the pipe used in the design. In the case of the pipe size an initial assumption has to be made as to its diameter; usually this is either the greater of the smallest acceptable or that immediately upstream from the junction point. As the run-off water usually carries suspended particles that will settle out in the pipe if the velocity of flow is too low, the gradient is normally chosen to have a self-cleansing velocity, i.e. generally more than 0.75 m/s. A practical rule-of-thumb (which includes a safety margin) is that a pipe will be self-cleansing if its gradient is one-half its diameter in millimetres, e.g. if the pipe diameter selected is 150 mm then the gradient should be at least 1 in 75. In the UK the velocity of flow, and hence the flow time, along an underground pipe of a given size and gradient, is often calculated using the Colebrook–White formula[20].

Once the time of concentration has been established, the next step is to determine the design intensity of rainfall for a storm duration equal to the time of concentration. In 1935, E.G. Bilham published[12] an analysis of rainfall data which showed that, for a given rainfall intensity, a *storm's frequency* or, as it is also termed, a *storm's return period*, was directly proportional to its duration. The original relationship developed by Bilham for England and Wales is

$$I = (1/t)[14.14(Nt)^{1/3.55} - 2.54] \qquad (7.5)$$

where I = average rainfall intensity (mm/h), t = rainfall duration (h), and N = storm return period or frequency (years) expressed as a storm of this intensity occurring, on average, once every N years. This formula was subsequently modified and improved, and it is now possible to obtain the design intensity directly from tables similar to, but more detailed than, Table 7.1. These tables, which are available from the Meteorological Office, relate intensity, duration and return period for any geographical area in the UK whose National Grid Reference is quoted.

Note that Equation 7.1 (also Table 7.1) shows that, for a storm of given duration, the average rainfall intensity increases as the return period is lengthened. In other words, the longer the return period the greater is the capital expenditure required to provide a drainage system that is able to fully cope with the larger run-off from a storm that, statistically, is less likely to happen. On the other hand, if too low a return period is used, carriageway flooding, delays and danger to traffic and, possibly, damage to adjacent roadside properties, will occur more often.

The decision as to what storm frequency to adopt in any given drainage design is, in theory, a matter of balancing average annual benefits versus average annual costs, having regard to the standard of protection from flooding that the community insists upon. Generally, however, the factors that mainly influence the selection of the return period are local in character, with the most influential being the locational vulnerability of the road section under consideration. In practice, therefore, the frequency is often determined on the basis of previous local experience, tempered by a check of the implications of the risks at particular locations. For example, a risk factor that would influence the choice is the type of property beside the road, e.g. whether it is industrial, retail commercial, residential or open space, and how it might be affected by flooding if too short a return period is selected (e.g. the flooding of a nursing home for senior citizens might be deemed unacceptable for all but the most extreme conditions). Road locations where traffic volumes and/or speeds are relatively high and there is limited or no escape for flooding water other than to overflow onto the carriageway, e.g. at sags in cuttings, would also require special consideration.

Whilst a 1 year storm frequency is commonly used for major roads in the UK (see reference 5), blind adherence to a recommended guideline when selecting the storm frequency should always be avoided. A practical approach toward selecting the design frequency would be (i) to design the drainage system for a reasonable 'past experience' return period, (ii) check the implications of a range of longer return periods upon this design and, if they are financially great or sensitive, cost the additional works needed to provide acceptable conditions in relation to providing appropriate protection to the adjacent properties, and/or traffic movement and safety,(iii) if the costs determined in step (ii) are significant, redesign the drainage system for a longer return period, and (iv) compare the costs determined at steps (ii) and (iii) and then decide upon the drainage design to use.

A *catchment area (A)*, may be homogeneous or mixed. A homogeneous area would typically be composed of paved surfacing only, e.g. an appropriate homogeneous catchment might be the area of carriageway and sealed shoulders of a rural road, or of an urban or rural motorway, that drains by gravity toward an inlet. A mixed catchment might be composed of a paved area plus adjacent land from which water also flows under gravity to the drainage system.

A catchment area might also be divided into 'averaged' subcatchments, with the scale of the averaging depending upon the design circumstances. Thus, a rural mixed catchment might be divided into a few general subcatchments which vary according to, say, the slope of the land, whilst an urban mixed catchment might be subdivided into paved and sealed areas, paved and unsealed areas, roof areas, garden and cultivated areas, and/or grassed areas[1].

After falling, rainwater either infiltrates into the soil, evaporates back into the atmosphere, is subject to storage (outside the scope of this text), or becomes surface run-off. The *proportion of rainfall running off, (p)*, also termed the *run-off coefficient*, is therefore dependent upon the nature of the surface of the catchment area. Total run-off is relatively rare, even in the case of an apparently impervious, homogeneous, paved surfacing due to, for example, ponding of water in surface irregularities. Factors which affect the run-off from a mixed catchment are the area of paved surface, average slope of the catchment, density of buildings, soil type(s) and vegetative cover, and its surface storage characteristics. Of considerable importance in both urban and rural areas are the catchment slopes, e.g. steep slopes result in a shorter time of concentration and, therefore, a greater peak discharge, than do slopes that are more gentle. Likely future changes in land use should always be taken into account when applying run-off coefficients to mixed catchments, especially in built-up areas.

Some suggested *p*-values for mixed urban and rural catchments are given in Tables 7.2 and 7.3, respectively.

If a mixed catchment is composed of surfacings with run-off qualities that vary significantly from each other, it may be appropriate to develop a weighted *p*-value and apply it to the catchment area as a whole.

The following illustrates how the Rational method is used to design a piped drainage system for a road.

Step 1 Prepare a key plan (Fig. 7.3) of the proposed piped drainage system, based on the contour plan of a mixed catchment. Note that the proposed inlets are

Table 7.2 Impermeability values suggested for urban areas in Great Britain[13]

Type of surface	*p*-value
Urban areas	
with considerable paved areas	1.00
average	0.50–0.70
residential	0.30–0.60
industrial	0.50–0.90
playgrounds, parks, etc	0.10–0.35
Housing development (houses/ha)	
10	0.18–0.20
20	0.25–0.30
30	0.33–0.45
50	0.50–0.70
General development	
paved areas	1.00
roofs	0.75–0.95
lawns, depending upon slope and subsoil	0.35–0.50
heavy clay soils	0.70
average soils	0.50
light sandy soils	0.40
vegetation	0.40
steep slopes	1.00

Table 7.3 Impermeability values suggested for rural areas

Type of surface	p-value
Concrete or bituminous surfacings	0.8–0.9
Gravel or macadam surfacings	0.4–0.7
Bare impervious soils*	0.4–0.7
Impervious soils with turf*	0.3–0.6
Bare slightly pervious soils*	0.2–0.4
Slightly pervious soils with turf*	0.1–0.3
Pervious soils*	0.1–0.2
Wooded areas	0.1–0.2

*These values are applicable to relatively level ground. When the slopes are >2 %, the p-value should be increased by 0.2 (to a maximum of 1.0) for every 2 % increase in slope

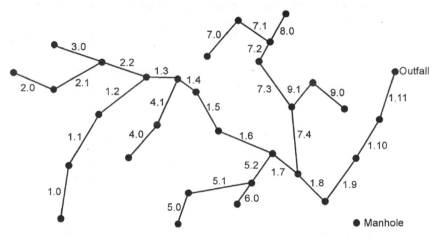

Fig. 7.3 Example key plan of a piped drainage system in an urban area illustrating the pipe-numbering convention

identified in this figure, as are the manholes and pipe lengths in between. Each pipe length is numbered, with the farthest pipe upstream from the outfall identified and numbered 1, and its individual sections numbered 1.0, 1.1, 1.2, 1.3, etc. The manhole in drain 1 at which the first piped branch from adjacent land enters the system is next identified; the longest branch to this point is numbered 2, and individual sections are numbered 2.0, 2.1, 2.2; and so on for branches 3–7. A manhole, whose function is to provide access for maintenance in the event of pipe blockages, is provided where two or more drains meet, where there is a change in pipe direction, gradient, size or type and/or, on long straight sections of pipe, at intervals of not more than about 100 m.

Step 2   Set up a design table (Table 7.4) and complete columns 1–4 for all sections.

Table 7.4 Example design table for Rational method calculations for run-off[8]

1	2	3	4	5	6	7	8	9	10	11	12	13	14
								Impermeable area (ha)					
Pipe length number	Difference in level (m)	Length (m)	Gradient	Velocity (m/s)	Time of flow (min)	Time of concentration (min)	Storm intensity (mm/h)	Roads	Buildings, yards, etc	Total (9+10)	Cumulative	Rate of flow (l/s)	Pipe diameter (mm)
1.0	1.10	63.1	1 in 57	1.33	0.79	2.79	67.9	0.089	0.053	0.142	0.142	26.8	150
1.1	1.12	66.1	1 in 59	1.70	0.65	3.44	62.5	0.077	0.109	0.186	0.328	56.9	225
1.2	0.73	84.7	1 in 116	1.46	0.97	4.41	57.4	0.081	0	0.081	0.409	65.2	300
2.0	1.40	44.8	1 in 32	2.32	0.32	2.32	72.5	0.113	0.081	0.194	0.194	39.1	225
2.1	0.01	49.1	1 in 80	1.77	0.46	2.78	67.9	0.045	0.105	0.150	0.344	64.9	300
3.0	0.98	48.5	1 in 49	1.43	0.56	2.56	70.5	0	0.129	0.129	0.129	25.2	150
2.2	1.65	54.3	1 in 33	2.74	0.33	3.11	65.5	0.101	0.073	0.174	0.647	117.7	300
1.3	1.22	27.7	1 in 23	3.29	0.14	4.55	56.9	0.121	0.235	0.356	1.412	223.2	300
4.0	0.88	54.9	1 in 62	1.66	0.55	2.55	70.5	0.093	0.093	0.186	0.186	36.4	225
4.1	0.58	45.7	1 in 79	1.48	0.52	3.07	66.3	0.069	0.040	0.109	0.295	54.2	225
1.4	0.52	22.9	1 in 43	2.77	0.14	4.69	55.8	0.069	0	0.069	1.776	275.3	375

Notes: Storm return period = one in one year. Roughness coefficient = 0.6 mm. Time of entry = 2 mm

Step 3   Determine the areas of the directly connected surfaces that contribute run-off to the pipe section being considered. Apply appropriate impermeability values (from Table 7.2) to each area and convert them into 'equivalent' impermeable areas (columns 9–12).

Step 4   Assume a suitable pipe size for each section (column 14) and using accepted tabular data (see reference 14) determine the full bore velocity (column 5) and the flow time in each section (column 6), calculated using columns 3 and 5.

Step 5   Calculate the time of concentration (column 7) for the raindrop to flow from the most remote part of the catchment that is contributing to the pipe length under consideration. This is the sum of the entry time (2 min in this example) plus the total time of pipe flow up to and including the piped section being considered. When two drains join, the time of concentration is assumed to be the greater time to the manhole concerned.

Step 6   Select a suitable storm-return period, and determine from appropriate data (Table 7.1 in this example) the intensity of rainfall (column 8) for a storm length equal to the time of concentration. As the concentration time varies from one pipe section to another, the design intensity will change accordingly.

Step 7   Calculate the expected peak rate of flow, Q, in each pipe length using the Rational formula, with I being obtained from column 8 and Ap from column 12.

Step 8   Compare the full-bore capacity of each pipe length section with the expected peak flow in that section. If any pipe length is found to be inadequate, assume a greater pipe diameter and/gradient, and repeat steps 4–8.

The Rational method of run-off estimation is very useful in explaining the complexities of the factors which need to be taken into account when carrying out run-off calculations for a drainage design. In practice, however, the formula may overestimate the maximum run-off, and the extent to which this will occur increases with the size of the catchment area. Hence, use of the formula is generally confined to urban areas with catchments that have relatively small proportions of pervious areas, and to smaller (ideally <80 ha) rural areas.

The following are some of the main assumptions underlying the Rational formula, which indicate why the outcomes of its use may be qualified by engineering judgement upon completion of the calculations.

1. The rainfall intensity I is selected for a storm of duration equal to the time of concentration t_c, and the entry time component of this time, t_e, is normally selected using measurements which assume that the raindrop moves perpendicularly to the ground contour lines. If this causes the t_c-value finally selected to be too low, it will result in the choice of too high a rainfall intensity with consequent overestimation of the run-off. Also, the frequency of the peak run-off is assumed to be the same as the frequency of the rainfall causing it.

2. The average I-value used is assumed to be steady throughout the catchment area throughout the storm. The catchments are usually small enough for this assumption to be reasonable for designs concerned with run-off from roads and cuttings only. However, if the area is very large and/or the topography is very variable, this assumption can lead to erroneous results.

3. The impermeability factors are essentially average antecedent wetness figures that are assumed to be constant, and independent of rainfall intensity, duration

and frequency. Also, they only try to take into direct account the ground-surface
type and slope, and assume that other influencing factors (e.g. surface storage
and other delay-producing characteristics) are taken into account by the use of
'average' figures.
4. The velocity of flow in the pipe drain is assumed to equal the full-bore velocity,
and to be constant throughout the time of concentration.

7.3.2 Advice Note HA 37/88 method

Until relatively recently, two main methods of dealing with surface run-off from
rural trunk roads were used in the UK, filter drains for roads in cuttings, and kerbs
and gullies connecting to subsurface pipes for roads on embankments. An alternative, formally recommended[15] in 1988, is to provide longitudinal surface channels
at the edge of the carriageway, beside the hardstrip or hard shoulder, to collect the
run-off and carry it to appropriate outlet points. The Advice Note HA 37/88
approach describes a method for determining the cross-section size or, alternatively, the outlet spacing for an edge-of-the-carriageway surface channel.

In this design approach it is assumed that (a) the inflow into the channel is uniformly distributed along a particular drainage length, (b) there is total run-off from
concrete and bituminous surfacings, (c) the rate of inflow per unit length of channel is equal to the width of the surface drained multiplied by the instantaneous
rainfall intensity, (d) the entry time for water flowing across the carriageway is
small in comparison with the time of flow along the channel, and (e) generally, the
geometric properties of the channel, and the road, do not vary with distance or
depth of flow between any two adjacent discharge outlets.

Kinematic wave theory, that is applicable to time-varying flows in shallow channels, enables the peak depth of flow at the downstream end of a section of drain to
be determined for a given intensity and duration of rainfall. For design purposes,
the storm duration that will result in the maximum possible depth of flow in the
channel for a given return period must be identified; this is done using the original
Bilham formula (Equation 7.5) but modifying the average intensity by two factors,
f_n and f_r, which correct the calculated I-value so as to allow for more recent[16]
regional rainfall information, i.e.

$$I_o = I f_n f_r \qquad (7.6)$$

where I_o = modified rainfall intensity, I = Bilham-determined rainfall intensity, f_n is
a factor that corrects for the effect of the return period, and f_r is a factor that
corrects for regional variations. For impervious road surfaces in the UK the greatest run-off flows tend to result from heavy short-period summer storms, so the
design method uses the 50 per cent summer profile, which recognizes that the peak
intensity at the mid-point of a storm is about 3.9 times the average intensity. Equation 7.6 was developed so as to be applicable to relatively long surface channels,
and it tends to over-design those with closely-spaced outlets. Hence, it is recommended that the design method should not be used for storm durations of less than
8 min.

The practical design method involves the selection of an appropriate design
curve from amongst 16 that cover the following combination of design variables:

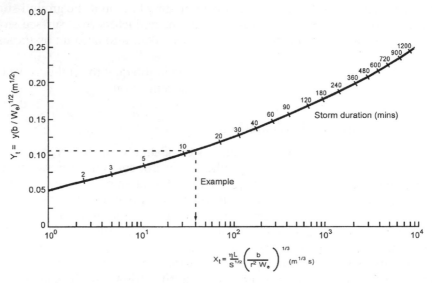

Fig. 7.4 Design curve for triangular surface channels subject to storms with a return period of 1 year and located in regional category A, used in the Advice Note HA 37/88 method[15]

(a) channel shape, i.e. triangular or rectangular, (b) storm return period, i.e. 1, 5, 10, 20 or 50 years for triangular channels, and 1, 5 or 10 years for rectangular channels, and (c) regional location, i.e. regional category A which refers to roads located in Scotland, Wales, and the Department of Transport's administrative regions of Eastern, Greater London, Northern, North West, South West, or West Midlands, and regional category B which refers to roads in Northern Ireland and the DoT regions of the East Midlands, South East, and Yorkshire and Humberside. Figure 7.4 shows one such design curve as an example; this is to be used in the design of a triangular channel for a road located in regional category A and subject to storms with a return period of 1 year. The critical durations of storms are also shown along the design curve in Fig. 7.4. The design curve in the figure is not dimensionless, i.e. the vertical ordinate Y_t, has units of $m^{1/2}$, whilst the units of X_t, the horizontal ordinate, are $m^{1/3}$ s. The ordinate values Y_t and X_t for triangular channels are obtained from the following equations:

$$Y_t = y\,(b/W_e)^{1/2} \tag{7.7}$$

and
$$X_t = (nL/S^{1/2})(b/r^2\,W_e)^{1/3} \tag{7.8}$$

where y=design depth of flow (m); b=effective cross-fall, i.e. rate of increase of surface width per unit depth=$b_1 + b_2$, where b_1 is the slope of the side of the triangular channel remote from the carriageway, and b_2 is the slope of the side adjacent to the carriageway; W_e=effective width of the whole catchment (m)=width of carriageway being drained plus the overall width of the channel and any other impermeable surface being drained plus an allowance (if appropriate) for run-off from a cutting; n=Manning roughness coefficient for the channel $(s/m^{1/3})$; L=drainage length of channel between two adjacent outlets on a continuous slope, or the distance between a point of zero slope and the downstream outlet (m); S=longitudinal gradient of channel (vertical fall per unit length of channel) (m/m); and r=hydraulic radius factor determined from

$$r=[(b_1+b_2)]/[(1+b_1^2)^{1/2}+(1+b_2^2)^{1/2}] \qquad (7.9)$$

The effective catchment width, W_e, in Equations 7.7 and 7.8 is calculated from

$$W_e=W+aC \qquad (7.10)$$

where W=width of the impermeable part of the catchment (m), C=average plan width of the cutting drained by the channel section being designed (m), and a=run-off coefficient for the cutting (from Table 7.5). The antecedent wetness categories given in Table 7.5 are assumed to depend upon the average annual rainfall at the site – as located in relation to Northern Ireland, Scotland, Wales, and the Department of Transport administrative regions in England – as follows: *low* – Eastern, Greater London regions; *medium* – East Midlands, Northern, South East, West Midlands, Yorkshire and Humberside regions; *high*–Northern Ireland, North West region, Scotland, Southwest region, Wales. The coefficients also allow for the relative steepness of road cuttings, which are likely to result in more run-off than from equivalent natural catchments.

Whilst a rectangular channel example is not shown here it should be noted that the design curves for these channels (see reference 15) are based on Y- and X-relationships that are different from the triangular channel ones given in Equations 7.7 and 7.8.

For design purposes, the appropriate curve is chosen from the 16 given, on the basis of the road's regional location and the return frequency of the storm of specified duration. If the locations of two adjacent outfall points have already been identified, the selected curve is then used to find the width or depth of channel that is required between them. Alternatively, if the dimensions of the channel have been selected, the curve is used to find the maximum spacing between any two adjacent outlets on a continuous slope or between a point of zero slope and the next downstream outlet.

Table 7.5 Run-off coefficients for cuttings, used in the Advice Note HA/37 method[15]

Soil type	Antecedent wetness	a-value
High permeability	low	0.07
	medium	0.11
	high	0.13
Medium permeability	low	0.11
	medium	0.16
	high	0.20
Low permeability	low	0.14
	medium	0.21
	high	0.26

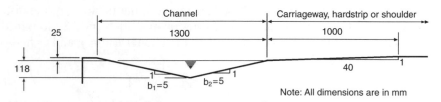

Fig. 7.5 Typical cross-sectional shape of a roadside triangular surface channel

All of the design curves were developed so as to be applicable to channels that are designed to cater for the run-off without water overflowing from the channel. Larger flows can be carried with some limited encroachment of water onto the hard verge and carriageway by setting the outside lip of the channel at a higher level than the inside edge. For example, for the channel shown in Fig. 7.5 the extra 25 mm surcharge depth will increase the flow capacity to around 1.6 times the capacity obtained with a design flow depth of 118 mm. If the 'normal' channel is designed to cater for storms with a frequency of occurrence of 1 year with flow contained within the channel cross-section (typical British practice[5]), the surcharged channel will be able to provide for storms with return periods varying from about 3.4 years if the storm duration is 5 min, to about 4.6 years if the storm duration is 2 h.

It might be noted here that it is common British design practice to allow a width of flooding from surface channel overflow of up to 1.5 m and 1.0 m on near-side shoulders and hardstrips, respectively, for a 5 year storm. Such flooding on longitudinally-sloping roads is normally most likely to occur where the flow depths in channels are normally greatest, e.g. adjacent to outfall points. Assuming that crests most often occur in cutting, and sags on embankments, the sags may also be critical locations for overflow flooding; in this case the level of the back of the channel should be set so that any more flooding is avoided by allowing the water to flow onto the verge and down the embankment slope.

In central reserves, surcharged surface channel levels should be set below carriageway levels so that any flooding occurs within the non-pavement 'dished' width of reserve as, for traffic safety reasons, it is particularly important that flood waters do not encroach onto the offside (high-speed) lanes.

The required outlet capacity at the downstream end of a channel section can be calculated from Manning's resistance equation[17]:

$$Q = (AR^{2/3}S^{1/2})/n \tag{7.11}$$

where Q = discharge (m³/s), A = cross-sectional area of flow (m²), R = mean hydraulic radius (m) = A/P, in which P = wetted perimeter of flow (m), S = longitudinal gradient of the channel (m/m), and n = Manning's roughness coefficient (see Table 7.6) whose value depends upon the nature of the surface over which the water is flowing, i.e. its surface texture, standard of construction, and the presence of deposited silt and grit.

For a depth of flow y in a triangular channel with side-slopes of 1:b_1 and 1:b_2 (vertical:horizontal), Equation 7.11 becomes

Table 7.6 Values of Manning's *n* used in the Advice Note HA 37/88 method[15]

Channel type	Condition	*n*-value
Concrete	average	0.013
	poor	0.016
Bituminous	average	0.017
	poor	0.021

$$Q = (bS^{1/2}/n)(r^2 y^8/32)^{1/3} \qquad (7.12)$$

where Q, b, S, n, r, y are as defined previously.

V, the flow velocity at the downstream end of a channel section, can be calculated by dividing the discharge, Q, by the cross-sectional area of flow, A.

The following illustrates the use of the Advice Note HA 37/88 method in the design of a triangular surface channel for a new road[15].

Problem Find the length of a 9.3 m wide carriageway located in a sandy cutting in the West Midlands, and subject to a storm return frequency of 1 year, that can be drained by a single triangular concrete channel that has the cross-sectional shape shown in Fig. 7.5, an overall width of 1300 mm, and the following design characteristics:

Side-slopes	$b_1 = 5$, and $b_2 = 5$
Effective cross-fall	$b = b_1 + b_2 = 10$
Design flow depth	$y = 0.118$ m
Design flow width	$B = By = 1.18$ m
Longitudinal channel gradient	$S = 1/100 = 0.01$
Manning roughness (ave condition)	$n = 0.013$

Solution The hydraulic radius factor is calculated from Equation 7.9:

$$r = 10/2(26)^{1/2} = 0.981$$

The width of the impervious part of the catchment is

$$W = 9.3 + 1.3 = 10.6 \text{ m}$$

The soil has a high permeability so, from Table 7.5, the run-off coefficient for the cutting is $a = 0.11$. As the drainage length is not yet known it is necessary to estimate the plan width of the cutting, i.e. the average width along the whole drainage length. The first estimate is $C = 15$ m but, if necessary, the calculation can be repeated later with a more accurate C-value. The effective catchment area is then found from Equation 7.10 as follows:

$$W_e = 10.6 + 0.11 \times 15 = 12.25 \text{ m}$$

As the road is located in regional category A, and the channel is to be designed for storms with a return period of 1 year, use the design curve at Fig. 7.4. The Y_t-value of the vertical ordinate in this figure is calculated from

$$Y_t = (b/W_e)^{1/2} = 0.118(10/12.25)^{1/2} = 0.107 \, \text{m}^{1/2}$$

The X_t-value then obtained from the design curve is $X_t = 38.9 \, \text{m}^{1/3}\text{s}$.

Re-arranging Equation 7.8 enables the drainage length L to be determined from

$$L = (S^{1/2}/n)(r^2 W_e/b)^{1/3} X_t$$

$$= (0.01^{1/2}/0.013)(0.981^2 \times 12.25/10)^{1/3} \times 38.9$$

$$= 316 \, \text{m}$$

The critical storm duration corresponding to the Y_t and X_t design coordinates is estimated from Fig. 7.4 as $D = 12 \, \text{min}$.

7.3.3 The TRRL hydrograph method

The Transport and Road Research Laboratory hydrograph method was developed in the 1950s and early 1960s as a simple sewer design tool for use in large urban areas. Whilst the data required by this method are no more than are required for the Rational method, it differs in that it involves the construction of a hydrograph, based on the well-established time–area diagram for a catchment, that takes into account both the variation of rainfall intensity with time during the storm and the storage capability of the drainage pipe system. Also, as the calculations were very difficult to do by hand for a drainage system of any significant size, from the beginning they were normally carried out with the aid of a digital computer – it was the first such design method of significance that was developed for the electronic computer.

The TRRL method, as originally developed, assumed that only impermeable areas contributed significantly to the peak flow in a stormwater pipe system. Using these areas a time–area diagram was constructed with the routeing velocity in each pipe assumed to be equal to the full-bore velocity. The flow was then calculated for a given rainstorm, after which the hydrographs derived from the time–area calculations were re-routed through a reservoir storage equation and a storage correction was applied to the system as a whole, assuming that in a correctly designed system the proportional depth of water is the same throughout the system.

Subsequently, however, the routeing procedure was applied to individual pipe sections, using the uniform proportional depth assumption, so that the hydrograph contributing to the pipe section under consideration is now added to the outflow hydrograph from the upstream pipe and the combined hydrograph is routed through the reservoir created by the pipe length being designed. If the peak flow is greater than the capacity of the pipe, the size of the pipe is increased by an appropriate amount, and the calculations repeated until a pipe that is sufficiently large is obtained.

7.3.4 Culverts

When a new road is being designed, it is usually found that its line crosses many small natural watercourses, and that culverts are needed to safely convey the stream waters under the pavement. A culvert is simply an enclosed channel that is open at both ends and used to convey the water through an embankment. Pipe and arch culverts are formed from both rigid and flexible materials, whereas box culverts – these have rectangular cross sections – are always made rigid. Flexible culverts are either thin-walled steel pipes or galvanised corrugated metal pipes that are designed to deflect under load; when deflection takes place the horizontal diameter of the culvert increases and compresses the soil laterally so that the passive resistance of the soil is mobilised to support the applied loads. Rigid culverts are composed of reinforced concrete, cast iron or vitrified clay and their load-carrying ability is primarily a function of the stiffness of the culvert wall(s).

To operate most effectively a culvert is placed in the natural stream-bed so that its alignment conforms with the original situation, even though this may result in the culvert being skewed rather than at a right angle to the road's centreline. If the stream meanders or if using the natural channel requires an inordinately long culvert, the stream is often diverted into a new channel prior to the culvert; however, the flow is then returned to its natural channel as soon as possible after passing through the culvert. The slope of a culvert should also conform closely to that of the natural streambed. Since the silt-carrying capacity of a stream varies as the square of its velocity, sedimentation will take place within the culvert if the speed of the water is reduced by flattening its slope. If the culvert slope is greater than that of the natural watercourse, the higher water velocity may cause scouring of the sides and bottom of the stream channel at the culvert outlet.

Once the need for, and locations of, culverts have been established, it is necessary to estimate the peak run-offs for the crossing streams without, normally, the benefit of any stream records, so that the culvert sizes required to safely pass the floodwaters can be determined. To resolve this problem, the Transport and Road Research Laboratory (now the Transport Research Laboratory) developed an empirical run-off formula which can be used to estimate the run-off from natural rural catchments in the UK[18]. The derived formula, which resembles the Rational formula but is different from it, is as follows:

$$Q = F_A A R_B / 3.6T \qquad (7.13)$$

where W = peak discharge (m^3/s); F_A = annual rainfall factor (dimensionless), given by $0.00127R_A$–0.321 where R_A is the average annual rainfall (mm), valid for R_A-values up to 2440 mm; A = catchment area (km^2); R_B = expected total rainfall (mm) for a storm duration equal to the time of concentration, valid for up to 48 h, and the selected storm frequency; and T = time of concentration (h) when all the catchment is contributing to the run-off, obtained from

$$T = 2.48(LN)^{0.39} \qquad (7.14)$$

where L = catchment length (km), i.e. the plan distance from the outfall to the upstream divide, measured approximately along the centre of the catchment, and N = a dimensionless slope number equal to the ratio L/Z, where Z is the rise in height from the outfall to the average height of the upstream divide (km).

Equations 7.13 and 7.14 were derived from data from five rural catchments with areas ranging from 2.77 to 21.30 km^2, differing slopes and shapes, and dissimilar average annual rainfalls. A feature common to all of the catchments is that they were all relatively impermeable and contained an underlying soil stratum of clay or boulder clay; thus if the catchment being designed for contains an area of pervious soil, only the impervious area should be used in Equation 7.13. Also, it was found that the *L*-distance sometimes required a 'cranked' line but, generally, the location of this line was not critical.

It might be noted that culverts across major roads in the UK are often designed on the basis of a 50- or 100-year storm frequency, whilst a 25-year return period may be used for culverts under less important roads.

7.4 Draining the carriageway

As noted previously, it is generally considered that driving becomes quite dangerous, due to loss of tyre–carriageway friction, when the water depth in the wheelpath of a car is in excess of around 2.5 mm. The aim therefore in road drainage design is to remove the surface water from the carriageway as quickly as possible, without causing other safety problems in the process.

7.4.1 Depth of surface water

Research[19] at the Transport Research Laboratory has resulted in the establishment of a formula relating water depth to rainfall intensity, and drainage length and slope for motorway-type surfaces, i.e. the flow depth above the tops of the surface texture asperities (d cm) is related to the length of flow path (l_f m), rainfall intensity (I cm/h), and flow path slope (1:N) by

$$d = 0.015(l_f I)^{0.5} N^{0.2} \qquad (7.15)$$

As the rainfall intensity is dependent upon hydrology, Equation 7.15 shows that the major geometric parameter affecting the flow depth is the flow path length. The slope of the flow path is of lesser importance as it is only the fifth root of N that is involved. In practical terms the equation indicates that the l_f-length should be kept as short as possible to reduce the water depth on the carriageway.

7.4.2 Carriageway and verge crossfalls

The draining of carriageways is usually achieved with the aid of *pavement crossfalls*. When the longitudinal gradient is nearly flat in comparison with the crossfall, the width of carriageway contributing to the run-off is more important than the steepness of the crossfall in terms of its influence upon the length and depth of flow; the main advantage of a steep crossfall in this case is its ability to reduce the amount of water that can pond in deformations in the surface of a flexible pavement, particularly in wheelpaths where rutting is most likely to occur. When the longitudinal gradient is steep, the result is a longer and deeper flow (see Table 7.7); however, in this instance, steepening the crossfall will result in a reduction in the flow path length, albeit the width of the contributing carriageway is still important.

Crossfall recommendations vary from country to country, and are mostly based on experience with the road-type under consideration. Unsealed gravel roads with, typically, A-crowns have crossfalls of 1: 30 to 1:25 (3.5 to 4 per cent) to minimize water penetration of the pavement; the surfacings of these roads must be graded

Table 7.7 Effect of longitudinal gradient on drainage length and water depth[19]

Crossfall	Longitudinal gradient	Resultant drainage slope	Increase in drainage length (%)	Increase in water depth (%)
1:40	level	1:40	0	0
1:40	1:100	1:37	11	4
1:40	1:40	1:28	44	10
1:40	1:20	1:18	122	27
1:40	1:10	1:9.8	311	64

regularly to maintain the crossfalls and remove potholes and depressions that would otherwise fill with water. Historical practice for sealed roads in the UK was to use minimum crossfalls of 1 to 50 (2 per cent) with relatively smooth surfacings, e.g. rolled asphalt and concrete, 1:50 to 1:40 (2 to 2.5 per cent) for coated macadams, and 1:35 (3 per cent) for block pavings. The current Department of Transport crossfall recommendation (see Fig. 8.4) for all new road construction is 1:40 (2.5 per cent).

Surfaced two-lane roads have either parabolic or curved cross-sections to ensure that the swaying of commercial vehicles is minimized as they cross and recross the crown of the road during an overtaking manoeuvre; the minimum crossfall is therefore the average from the crown to the carriageway edge. A common practice in the case of single carriageways with three or four lanes is to use a curved crown section for the central lane(s) and to use tangent sections for the outer lanes; this avoids the use of excessively steep outer-lane crossfalls. The crown is normally at the middle of the carriageway when the number of lanes is even, and at the 'centre-line' when the number of lanes is odd.

As the most effective way of reducing the water depth on a carriageway is to decrease the length of the flow path, each carriageway of a dual carriageway road should, ideally, be cambered. However, this solution is capital expensive as it requires drainage facilities on both sides of each carriageway.

Provided that no more than two lanes slope in the one direction, a more economical drainage solution for divided roads is to slope the whole of each carriageway from the central reserve toward the nearside shoulder; however, this means that the greatest water accumulation on each carriageway occurs on the lane with the heaviest traffic flow. If each two-lane carriageway is sloped towards the central reserve the drainage facilities can be concentrated in the central reserve, with a consequent reduction in capital costs; however, this may have traffic safety implications as the greatest depth of water on the carriageway is then on the high-speed offside lanes.

Verges, which are normally either paved (bituminous or concrete), stabilized aggregate, and/or turf (grass), should be taken into account when designing carriageway drainage as they also contribute to the run-off. The coarser the verge surface the slower the run-off velocity and the more likely it is that deep water may be caused to extend back onto the carriageway. A verge that stands proud of the carriageway because of careless design or construction is a potential hazard as it will cause run-off water to pond on the road surface.

Crossfalls on most verges are usually steeper than on carriageways, as the resultant increased velocity of the escaping water facilitates the passage of run-off from the adjacent traffic lanes. In the case of verges on the high side of one-way crossfall pavements and/or superelevated sections, the additional run-off effect can be quite significant; whilst, theoretically, this could be ameliorated by breaking back the verge crossfall this is usually not acceptable for road safety reasons and the verge crossfall is normally the same as on the adjacent lanes.

7.5 References

1. NAASRA, *Guide to the Design of Road Surface Drainage*. Sydney: National Association of Australian State Road Authorities, 1986.
2. The *Effects of Highway Construction on Flood Plains*, HA 71/95. London: HMSO, 1995.
3. Department of Transport, *Highway Construction Details*. London: HMSO, 1987.
4. Russam, K., *The Hydraulic Efficiency and Spacing of BS Road Gulleys*, RRL Report LR 236. Crowthorne, Berkshire: The Road Research Laboratory, 1969.
5. *Edge of Pavement Details*, Advice Note HA 39/89. London: Department of Transport, 1989.
6. Colyer, P.J. and Pethick, R.W., *Storm Drainage Design Methods: A Literature Review*, Report No. INT 154. Wallingford, Oxon: Institute of Hydrology, 1976.
7. Watkins, L.H., *The Design of Urban Sewer Systems*, Road Research Technical Paper No. 55. London: DSIR, 1962.
8. Transport and Road Research Laboratory, *A Guide for Engineers to the Design of Storm Sewer Systems*, 2nd edn, Road Note 35. London: HMSO, 1976.
9. Mulvaney, T.J., On the use of self-registering rain and flood gauges in making observations on the relation of rainfall and of flood discharges in a given catchment, *Transactions of the Institution of Civil Engineers of Ireland*, 1850, **4**, No. 2.
10. Kuiching, E., The relationship between the rainfall and the discharge of sewers in populous areas, *Transactions of the American Society of Civil Engineers*, 1889, **20**, No. 1.
11. Lloyd-Davies, D.E., The elimination of storm water from sewerage systems, *Proceedings of the Institution of Civil Engineers*, 1906, **164**, pp. 41–67.
12. Bilham, E.G., The classification of heavy falls of rain in short periods, *British Rainfall 1935*. London: HMSO, 1936, pp. 262–80.
13. Bartlett, R.E., *Surface Water Sewerage*, 2nd edn. London: Applied Science, 1980.
14. Ackers, P., *Tables for the Hydraulic Design of Storm Drains, Sewers and Pipelines*, Hydraulics Research Paper No. 4, 2nd edn. London: HMSO, 1969.
15. *Hydraulic Design of Road-edge Surface Water Channels*, Advice Note HA 37/88 (plus Amendment No. 1). London: Department of Transport, 1988.
16. *Flood Studies Report*, London: National Environment Research Council, 1975.
17. Manning, R., On the flow of water in open channels and pipes, *Transactions of the Institution of Civil Engineers of Ireland*, 1891, **20**, pp. 161–207.
18. Young, C.P. and Prudhoe, J., *The Estimation of Floodflows from Natural Catchments*, TRRL Report LR 565. Crowthorne, Berkshire: The Transport and Road Research Laboratory, 1973.

19. Ross, N.F. and Russam, K., *The Depth of Rain Water on Road Surfaces*, RRL Report LR 236. Crowthorne, Berkshire: The Road Research Laboratory, 1968.
20. Ackers, P., *Charts for the Hydraulic Design of Channels and Pipes*, HRSP DE2, 3rd edn. London: HMSO, 1969.

CHAPTER 8

Subsurface moisture control for road pavements

C.A. O'Flaherty

8.1 Why subsurface moisture control is so important

Water is normally required in the construction of a road to allow for the proper compaction of the foundation and of pavement courses of soil, to ensure the attainment of compaction levels upon which the design is based. As noted previously (Chapter 7), however, excess water is the enemy of road foundations and pavements, and good design ensures that extra water is prevented from entering the road construction. If it is not, the normal overall effect will be higher maintenance costs and a reduction in the intended design life of the pavement.

All pavement design procedures rely on subgrade soil tests, and the design determined as a consequence of the testing is only valid if (a) the foundation and pavement conditions assumed on the basis of these measurements are similar to those that actually pertain after construction and (b) they remain substantially unchanged during the design life of the road. Thus, for example, if the subgrade moisture content becomes much higher than that used in the design following, say, continuous rainfall over a poorly-drained road section, the result may well be a softening of the subgrade – which, in turn, will be reflected in pavement distress, e.g. surface cracking and the development of rutting and pot holes in the outer parts of the pavement and, especially, in the wheelpaths of heavy commercial vehicles.

If water enters the component layers of a soil – aggregate pavement, the pavement's strength may be so reduced that permanent local deformation occurs in the wheelpaths of vehicles. If the roadbase becomes saturated, pore pressures may develop under wheel loadings that result in decompaction and the development of potholes.

If the subgrade soil is an expansive clay and moisture is allowed to enter the foundation, the deterioration of the pavement may be exacerbated by damaging subgrade volume changes and differential heaving. If the moisture content of the expansive subgrade is too high at the start of construction, and it dries back to the equilibrium moisture content over time, the end result could be a loss of pavement shape and cracking of the road surfacing.

Edge rutting of weak verges caused by the outer wheels of commercial vehicles leaving the carriageway in, for example, heavy rain can result in water becoming ponded in the ruts, the infitration of water into the pavement through openings at the verge–surfacing interface, and a consequent loss of support for the road pavement edge.

The pumping of water and fines through cracks in rigid pavements, and distress associated with frost heave in both flexible and rigid pavements, are clear signs of inadequate road drainage.

Overall, if unwelcome water cannot be prevented from obtaining entry to a pavement, good design ensures that drainage mechanisms are provided for its swift removal so that its detrimental impacts on the subgrade and/or pavement are minimized.

An alternative to the prevention and remedial approaches to moisture entry is to design the pavement so that it is able to withstand the anticipated design traffic loads in the presence of a much higher moisture content in the foundation. The weakness in this approach, however, is that it is associated with very expensive pavement designs, e.g. if the pavement is designed for, say, saturated subgrade conditions when saturation is not normally likely to occur.

8.2 Protecting the road pavement and foundation

Whilst the cost of subsurface drainage systems is high in terms of material, construction and maintenance, the extended pavement life that can be anticipated as a result of their installation generally makes these systems economically worthwhile[1]. The evidence is clear: it is better to design the road so as to ensure that the excess moisture is kept out in the first instance – and, if it does get in, that subsurface drainage is provided to ensure that it is removed as quickly as possible.

Protection of the foundation is not always easy, however, because there are many ways whereby moisture can enter/leave the subgrade and pavement of a roadway (see Fig. 8.1).

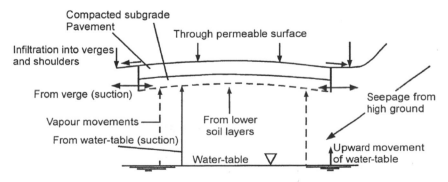

Fig. 8.1 Ways whereby water can enter and leave road pavements and subgrades

8.2.1 Preventing seepage from high ground

Seepage from high ground is common in hilly topography. It is especially noticeable in road cuttings that extend below the natural water table, so that spring inflows and seepage from water-bearing permeable strata have easy access to the pavement and subgrade. Seepage through the face of a cutting may also erode its slope, as well as being a potential cause of damage to the pavement, if it is left untreated.

In the case of the sidelong cutting shown in Fig. 8.2(a) the basic solution is to use a subsurface drain to intercept the seepage water flowing through the permeable layer on the uphill side of the road. The longitudinal cut-off drain shown in this figure is (i) located adjacent to the verge and parallel to the centreline of the road, (ii) founded in the underlying impermeable layer, and (iii) sealed at the top to prevent the entry of surface water (which, preferably, should be handled separately). Note also that the drain has a perforated pipe at its base that is laid on a slope to enable the more effective removal of the intercepted drainage water that trickles down through the free-draining backfill. A pipe's capacity is much greater than that of the intercepting granular material and, provided that the pipe is properly laid with enough slope, water will flow more swiftly through it, with less silt deposition, for disposal at a suitable outlet.

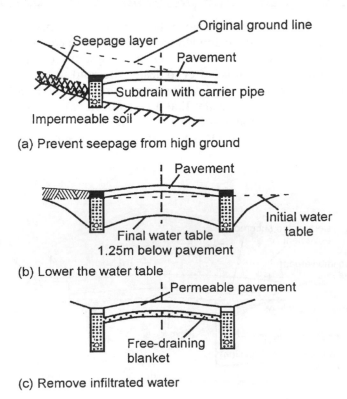

(a) Prevent seepage from high ground

(b) Lower the water table

(c) Remove infiltrated water

Fig. 8.2 Diagrammatic illustrations of the use of subdrains to prevent water intruding into a road's pavement or foundation

If the seepage layer in Fig. 8.2(a) is very thick, or if no underlying low-permeable layer can be economically found, the cut-off drain will need to be taken sufficiently deep to ensure that the lowered water table is enough below the formation that the upward movement of water through capillary action does not become a subsequent problem. If high lateral flows are expected from the seepage layer it may be appropriate to attach a vertical barrier membrane, e.g. plastic sheeting, to the pavement side of the drainage trench to ensure that the subsoil water does not penetrate beyond the trench in periods of high flow.

If the seepage layer is close to the surface, and it is not too thick, it may be possible to move the subsurface drain further uphill. Alternatively, it may be possible to use a roadside surface drain to intercept the water; however, as the lowered water table must be (for most British soils) at least 1 to 1.5 m below the formation, the use of a surface drain may require an open ditch that is so deep that, for traffic safety reasons, it will either have to be protected by a safety fence or formed so that it can be safely traversed by vehicles that might accidentally leave the carriageway.

Embankments across steep slopes are often built with benches cut into the natural ground. If this is not done properly and impervious embankment material on a bench (especially on a lower-down one) cuts off the flow of water through an in-ground seepage layer, high pore pressures may be developed that cause instability and the eventual failure of part of the embankment. Preventive drainage measures used (see Fig. 8.3) at such locations include (a) diverting surface run-off, (b) deep subsoil drains, (c) horizontal drains drilled into the hill, (d) benches with drainage layers and subsoil drains, and/or (e) providing a drainage blanket within the embankment or its foundation.

When a road cutting penetrates below the water table in permeable soil, and excessive seepage causes erosion of the excavated face, another drainage solution

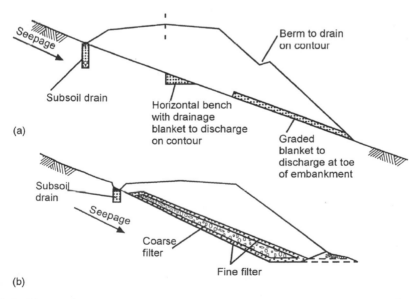

Fig. 8.3 Alternative drainage systems for embankments on hillsides[2]

may be to cover the slope with a graded granular filter, or a porous fabric under a free-draining blanket, in conjunction with a subsurface drain at the toe of the slope[3]. Cuttings through alternate permeable and impermeable layers, within groundwater not under artesian head, may be drained by sinking vertical stone-filled bores in advance of, or above, the slope of the cutting to allow gravity drain-age from upper aquifers to a lower one that can be drained. Where seepage in a cutting is confined to isolated permeable layers running horizontally, or at a fairly shallow gradient, it may be appropriate to achieve drainage by drilling into them from the face of the excavation, along the bedding, to allow the insertion of a per-forated drainage pipe protected by a prefabricated filter[4].

A not uncommon drainage problem can occur at the end of a cutting, where it may be found that longitudinal seepage from the cut region is moving downgrade and accumulating at the cut–fill transition, causing a weakening of the foundation. In such instances, it is advisable to install transverse cut-off drains in the subgrade in the cut region close to the transition.

Ideally, if complete soil survey data are available, decisions regarding the installation of cut-off drains are made in the design office prior to the construction. If, however, only incomplete information regarding subsurface seepage from high ground is available at the design stage, it is often better to wait until the actual conditions are exposed during construction and then to decide what to do, i.e. whilst it may be more expensive, a better engineering decision can usually be made in the field if it is based on known conditions.

8.2.2 Lowering the water table

In many instances a road pavement in flat terrain may be constructed on the nat-ural subgrade or on a low embankment so that the formation is relatively close to the natural water table. If the subgrade soil is fine-grained, the water may be drawn up into the subgrade through capillary (suction) action with consequent detri-mental effects upon pavement stability and design life.

The most common solution to this problem is to lower the water table by installing longitudinal subdrains in the verges on either side of the carriageway (see Fig. 8.2(b)), parallel to the centreline, to a depth that is greater than the desired minimum level of the lowered water table below the middle of the pave-ment. Alternatively, if the water table is already relatively deep and/or the sub-grade soil is coarse-grained, a satisfactory solution may be obtained by installing a single, deeper, longitudinal drain to an appropriate depth below the com-pacted subgrade under the centre of the pavement; the greater drain depth is usually necessary to ensure that the (steeper) surface of the lowered water table is below the minimum acceptable level at the outer edges of the pavement. In either case, the water that is below the water table is at greater than atmospheric pressure and will flow towards a drain, as the drain provides an opening in the saturated soil to atmospheric pressure – thereby enabling the water table to be lowered.

Recommendations as to the minimum acceptable depth to the free water table vary from country to country and, as might be expected, from soil type to soil type. Typical of these is the recommendation[2] that the formation level should be at least

1 m above a water table in non-plastic sandy soils, or up to 3 m in sandy or silty clays, to ensure that the water table does not have a dominating effect on the moisture conditions in the subgrade and pavement. The preferred practice in the UK is to keep the water table at least 1 to 1.5 m below the formation level.

Various ways of determining the drawdown effects of subsurface drains are available in the literature[5]. However, as the depth to which a drain should be laid, and the effectiveness of any particular type of installation, are mainly dependent on the pavement width and on the soil type, the most practical way of determining the suitability of a drainage proposal is to carry out a simple field trial. For example, for the installation shown in Fig. 8.2(b), two parallel 15 m long trenches would typically be dug to a depth of at least 0.6 m below the level to which it is desired to lower the water table, along the lines of the proposed subsurface drains. A transverse line of boreholes spaced at intervals of 1.5 to 3 m would then be sunk between the two ditches and extended on either side for a further distance of, say, 4.5 to 6 m. Observations would then be made of the water table levels in the boreholes before and after pumping the water from the trenches, after allowing sufficient time for equilibrium conditions to be established. Once the observations are graphed an estimate could then be made of the significance of the drawdown effects of the ditches, and decisions taken regarding the depth and spacing of the drains. Also, the pipe size would be estimated on the basis of the rate of pumping necessary to keep the trenches free of water.

Fine-grained soils with high water tables cannot normally be drained by installing gravity sub-drains. Instead of attempting to lower the water table in these soils the usual practice is to construct a higher embankment so that the formation is raised the desired height above the water level

8.2.3 Minimizing edge-of-the-carriageway moisture movements

Moisture conditions beneath a road pavement can be expected to vary as the seasons change, with the greatest variations occurring during the wet and dry seasons. Assuming that a road surfacing is impermeable, that the water table is well below the formation, and that the verge is not sealed, it can be expected that most of the moisture variations will be experienced in the upper 1 to 2 m of the subgrade soil within, say, 1 m of the edge of the pavement; the initial moisture regime under the remainder of the pavement width will remain relatively stable, irrespective of whether it was too wet, too dry, or about right in the first instance. Given that the outer wheelpath of vehicles is normally within this edge region, it is clear that considerable effort should be made to prevent the lateral movement of moisture within this failure-prone zone.

The more obvious moisture problem associated with providing grassed verges up to the edges of the carriageway occurs following continuous wet weather, when the verges are much wetter than the subgrade and pavement. This sets up a suction potential that causes moisture from the verge to move to the subgrade, and detrimental softening may then take place beneath the most heavily stressed portion of the pavement, i.e. the outer wheelpath. If the subgrade soil is an expansive clay, the additional moisture may be enough to cause the carriageway edge to lift and longitudinal cracking to be initiated.

A second effect occurs in very dry weather when the subgrade soil is much wetter than the verge soil, so that soil shrinkage occurs beneath the sides of the carriageway; this, in turn, may also lead to cracking. Keeping in mind that the lateral spread of a tree's root system is roughly equal to the diameter of its canopy or 'drip line', the drying-out of the subgrade will be accentuated if fast-growing tall trees and shrubs are allowed to grow in or close to the verge.

The movement of moisture between the verge and the carriageway edge is best minimized by interposing a full hard shoulder. The shoulder should have a good crossfall for run-off purposes, and be composed of construction materials that are relatively impermeable, well compacted, and resistant to erosion; ideally, it should comprise an extension of the pavement, and be sealed with a bituminous surfacing that is of a different colour or composition to that used in the carriageway. Also, the shoulder's joint with the carriageway should constantly be checked to ensure that it does not open, as gaps permit the infiltration of surface water. The shoulder should be well maintained so that depressions and potholes that might collect water are not allowed to develop on its surface.

If, for reasons of economy, a full hard shoulder cannot be constructed, a hardstrip should be provided next to the edge of the carriageway. There is a rough rule-of-thumb to the effect that the hardstrip should be as wide as economically possible but never less than twice the thickness of the roadbase.

Figure 8.4 shows the main geometric features of the current Department of Transport recommendations for the cross-sections of new rural motorways and single carriageways. Note that the motorway has a hardshoulder between the outer edge of each carriageway and the verge, and a lesser hardstrip width at each inner edge within the central reserve; the single carriageway has only a hardstrip included in the verge. Many existing single and dual carriageway roads that were constructed in the 1950s and 1960s, or earlier, were built without hard shoulders or hardstrips, and were provided with grassed verges only.

Hard shoulders were initially introduced for road safety reasons, i.e. to provide (i) a refuge area for vehicles forced to make emergency stops, (ii) recovery space for vehicles that inadvertently leave the carriageway, or deliberately do so during emergency evasive manoeuvres, (iii) help in the achievement of design horizontal sight distances, (iv) greater lateral clearance between side-by-side (or opposing) vehicles by encouraging those in the outer lane to drive closer to the edge of the carriageway, and (v) temporary extra lanes to cope with traffic blockages associated with road maintenance or carriageway reconstruction. It was then noted that roads with sealed shoulders achieved greater life expectancies than those without, and that this was partially due to the extra lateral support that the shoulders gave to the pavement, and partially to the improved protection that they provided in relation to lateral moisture movement.

8.2.4 Minimizing the effects of water entering through the pavement surface

Whilst some water can be expected to enter pavements from backups in ditches and from groundwater sources, most free water will enter through joints, cracks, and pores in the road surface[1]. Bituminous surfacings, which are generally

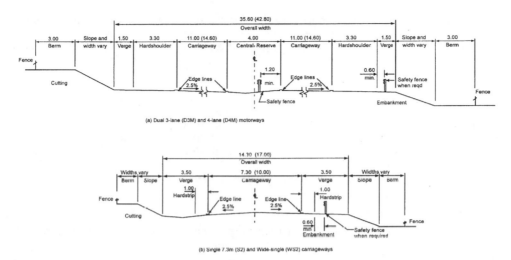

Fig. 8.4 Current Department of Transport cross-section recommendations for some new rural roads[7] (*Note*: All dimensions are in metres. D4M and WS2 dimensions are in brackets.)

expected to act as impermeable barriers to infiltrating surface water, are often relatively permeable because of mix design or construction techniques; many also develop cracks over the course of time, and these allow significant quantities of ponded water to enter and weaken the pavement and subgrade. In concrete roads, water can be expected to enter through joints and cracks which are not adequately sealed. For both bituminous and concrete roads poor-quality edge joints are an obvious source of water entry.

If water from the carriageway is likely to enter a pavement that is resting on a vulnerable subgrade the pavement design should ensure that it is removed immediately, before the water trickles down to the formation. This requires interposing a free-draining subbase across the full width of the pavement, between the roadbase and subgrade, and arranging for the collected water to be taken away, usually via carrier pipes (e.g. see Fig. 8.2(c)). (Also, the subgrade should be chemically stabilized if it is likely to be detrimentally affected by water.)

Whilst a subbase drainage layer is primarily intended to remove entering surface water, it will also stop capillary rise water from going into the pavement.

The basic requirements for this drainage course have been specified[2] as: (1) it should be highly permeable, with a permeability of at least 100 times that of the upper pavement and shoulder layers; (2) its constituent materials should have a low capillary rise, be resistant to breakdown under moist conditions over long periods, and not be susceptible to loss of, or blockage by, fine components carried in the surface water; (3) its thickness should be about twice the assessed capillary rise (expressed in mm of water); and (4) it should be laid with a crossfall to facilitate drainage to the outer edge of the pavement. Typical values of capillary rise for different classes of drainage materials are given in Table 8.1, together with an indication of the corresponding minimum layer thickness for each material.

Table 8.1 Typical values of capillary rise for materials used in drainage courses[2]

Material	Particle-size range (mm)	Capillary rise (mm)	Minimum thickness of drainage layer (mm)
Fine sand	0.05–0.25	300–1000	2000
Medium sand	0.25–0.5	150–300	600
Coarse sand	0.5–2	100–150	300
Well-graded sand	0.25–2	150–1000	2000
Fine gravel	2–6	20–100	200
Coarse gravel	6–20	5–20	100 (nominal)
One-sized aggregate	>5	<5	100 (nominal)

Drainage courses do not protect formations against rainwater during construction; hence, drainage must be ensured not only after, but also before and during the placement of a layer. As the materials used in drainage layers are deficient in fines, difficulty can be experienced in placing and compacting them. As the materials are also relatively open-graded drainage courses may need to be provided with filters to protect them, during construction, from intrusion by soft material from adjacent layers above and, especially, below.

8.2.5 Protecting against vapour movement

Movement of moisture in vapour form is associated with differences in vapour pressure arising from temperature and/or moisture content differences in various vertical positions in the subgrade.

Vapour movements due to temperature differences can be very important in climatic areas where there are substantial fluctuations in the daily temperature; however, because of the prevailing temperature, climate it is not considered to be a significant form of vapour moisture movement in the UK. Vapour movements due to moisture differences is normally not a problem either, i.e. it requires the soil to be relatively dry and this is also a moisture condition that is normally not very common in a temperate climate.

Moisture movement in the vapour phase is usually associated with coarse-grained rather than fine-grained soils – and, thus, the introduction of a coarse-grained drainage layer between the roadbase and subgrade will do little to protect the pavement against vapour movement. In practice, the most effective way of preventing vapour movement is to interpose a horizontal impermeable membrane between the top 0.5 m of the subgrade soil and the underlying uncompacted soil.

8.3 Designing and laying conventional longitudinal subdrains

A conventional longitudinal subsurface drain comprises a carrier pipe laid in the bottom of a trench and surrounded with a protective backfill material (Table 8.2)

Table 8.2 Two recommended ranges of grading of filter materials used in sub-drains in the UK[8]

BS sieve size (mm)	Percentage by mass passing a given sieve size for:	
	Type A	Type B
63	–	100
37.5	100	85–100
20	85–100	0–25
10	50–100	0–5
5	35–90	–
1.18	15–50	–
0.600	5–35	–
0.150	0–5	–

that acts as a filter. The trenches are relatively shallow, with vertical walls, and nowadays they are mostly dug with a ditching machine. The widths of the trenches are typically the external diameter of the pipe plus either 300 mm (for drains with <1.5 m cover below the finished level) or 450 mm (for drains with >1.5 m cover). The pipes are provided with outlets at regular intervals so that the collected water can escape into convenient watercourses or into a stormwater drain; typically these outlet locations are less than 90 m apart. The slopes of the pipes are normally as steep as possible, and never less than 1:200 (0.5 per cent).

To allow the water to enter, the pipes are normally either of (a) perforated concrete, plastic, unplasticized polyvinylchloride, or vitrified clay, (b) porous concrete, or (c) unperforated concrete or vitrified clay laid with open joints. Piping with either circular or slotted perforations is less liable to silting and is therefore to be preferred. Except when laid on a concrete bed the piping is better placed with the perforations downward so as to minimize clogging by fine particles. If the underlying soil is coarse-grained not much may need to be done to the trench bottom other than nominal shaping. If the drain is an intercepting one, and the pipe is being laid in an underlying soil of low permeability, it will usually be necessary to bed the pipe on at least 75 mm of compacted granular material to avoid plugging of the perforations and to ensure uniform support.

The size of a subsurface drainage pipe should be chosen so that it will not run full near its outlet and flood the surrounding filter material. Also, the pipe should be able to handle all of the water intercepted by the drain without causing a high head in the filter material; this, for example, would reduce the depth to which a water table could be lowered and the rate at which lowering would take place.

Ideally, the carrier pipes should be laid in fine weather on the same day that the trenches are dug. The backfill material is usually capped with a 150 mm layer of impermeable material (or pavement material, if the trench is to be covered by the pavement) to prevent surface water from entering the subsurface drains and, also, to minimize the risk of subsequent clogging of the filter by fines washed down from the surface.

The main purpose of the backfill in a conventional subdrain is to provide an effective water-intercepting/collecting space adjacent to the pipe. To fulfil this function satisfactorily the backfill is composed of imported filter material (see Table 8.2) that is coarse enough to allow the water to have easy access to the pipe, and fine enough to stop the infiltration of the surrounding soil; however, it must not be so fine-grained that it will substantially enter or block the pipe perforations, e.g. for the Type A material shown in Table 8.2 at least 15 per cent must be larger than the hole diameter, or greater than 1.2 times the slot width in the perforated pipe. Proper backfilling of the trench, with uniform layers not exceeding 225 mm loose depth, is essential to a filter's successful operation with the minimum of maintenance.

Excellent reviews of the literature[2,6] recommend that conventional filter materials protecting a pipe in any type of *granular soil* should meet the following requirements regarding (i) permeability, (ii) filtration, (iii) segregation, (iv) fines, and (v) pipe perforation. In relation to the following criteria it should be noted that the symbols D_{10}, D_{15}, D_{50}, D_{60}, and D_{85} refer to the 10, 15, 50, 60 and 85 per cent particle sizes, by mass, of the material concerned.

(i) D_{15} (filter)/D_{15} (soil) should not be less than 5.
(ii) D_{15} (filter)/D_{85} (soil) should not be more than 5,
       D_{15} (filter)/D_{15} (soil) should not be more than 20, and
       D_{50} (filter)/D_{50} (soil) should not be more than 25.

If, however, the surrounding soil is either uniform or well graded, then these filtration criteria will need to be varied. If the soil is uniform, i.e. if D_{60} (soil)/D_{10} (soil) is less than or equal to 1.5, then D_{15} (filter)/D_{85} (soil) should not be more than 6. If the soil is well graded, i.e. if D_{60} (soil)/D_{10} (soil) is greater than or equal to 4, then D_{15} (filter)/D_{15} (soil) should not be more than 40. If the surrounding soil is gap-graded, then the filter design (and, hence, the D-values) should be based on the grading of the soil portion finer than the gap in the grading. If the soil has layers of fine material the filter should be designed to protect against the intrusion of the fine particles from the finest layer.

(iii) To avoid segregation during placement, a filter material should not be gap graded, should not contain stones larger than 75 microns, and its uniformity coefficient (D_{60}/D_{10}) should not be greater than 20.
(iv)   No more than 5 per cent by mass of the filter material should be able to pass the 75 micron BS sieve, as otherwise the filter fines will migrate into the pipe.
(v)   For pipes with perforations, the diameters of circular holes should not be greater than D_{85} (filter), and the widths of slotted holes should not be greater than $0.83D_{85}$ (filter), so that the filter material will not be washed into the pipe and clog it. The minimum hole dimension is 0.8 mm (except in areas where bacterial slime is likely to develop in the pipe, in which case the minimum dimension should not be less than 3 mm).

If the surrounding soil contains appreciable amounts of *silt* or *clay*, application of the above criteria may indicate that the filter should be composed of a clean sand. A clean sand backfill will normally be permeable enough to allow the easy transmission of all the low velocity water flow from the soil whilst protecting the pipe against intrusive aggregations of clay particles.

8.3.1 Example of conventional filter design

Application of the above criteria is most easily illustrated by considering an example which requires determining (a) the gradation of the filter material needed between a perforated pipe and a coarse-grained surrounding soil with the gradation shown in Fig. 8.5, and (b) the maximum size of pipe perforations that may be used with the designed filter.

The first step is to check the extent to which the surrounding soil is uniform or well graded:

$$D_{60}/D_{10} = 0.15/0.075 = 2$$

This demonstrates that the soil is neither uniform nor well graded, so the exceptions noted in (ii) above do not apply, and the normal criteria should be used:

$$D_{15} \text{ (filter)}/D_{85} \text{ (soil)} = D_{15} \text{ (filter)}/0.21$$

Therefore D_{15} (filter) should not be greater than 1.05 mm. Alternatively

$$D_{15} \text{ (filter)}/D_{15} \text{ (soil)} = D_{15} \text{ (filter)}/0.085$$

Therefore D_{15} (filter) should not be greater than 1.7 mm.
 Use the lower of the two D_{15} values, i.e. 1.05 mm.
 The remaining criterion in (ii) above is

$$D_{50} \text{ (filter)}/D_{50} \text{ (soil)} = D_{50} \text{ (filter)}/0.14$$

Therefore D_{50} (filter) should not be greater than 3.5 mm.
 A filter material should now be selected that meets the above D_{15} and D_{50} requirements. In this instance a suitable backfill filter material could well have a

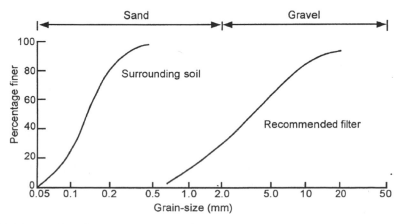

Fig. 8.5 Example of a conventional filter design to protect a perforated pipe laid in a trench in a coarse-grained soil

D_{85} size of about 9.5 mm. The proposed filter material (also shown in Fig. 8.5) should now be checked to ensure that it meets the permeability requirements in (i) above, i.e.:

$$D_{15}\,(\text{filter})/D_{15}\,(\text{soil}) = 1.05/0.085 = 12.35$$

This value exceeds 5, so the filter material will be sufficiently permeable.

Examination of the proposed filter shows that the requirements in (iii) and (iv) are also met.

The proposed filter material in Fig. 8.5 therefore meets all the requirements specified in (i) to (iv) above.

From Fig. 8.5 it can be seen that D_{85} (filter)=9.5 mm. Therefore the maximum size of circular holes in the pipe=9.5 mm and the maximum width of slot=0.83×9.5=7.9 mm.

If, by chance, the only pipes available have perforations that are larger than the above, then a more coarse filter material will need to be placed between the pipe and the designed filter, to create a two-stage filter. New gradation calculations for the coarser intervening filter will then need to be carried out, with the designed filter in Fig. 8.5 considered as the 'surrounding soil'.

8.4 French drains

The term *French drain* tends to be used loosely to describe diverse drains ranging from those containing only random rubble to those with well-designed filters (as in conventional drains) but without carrier pipes. Some French drains are open at the surface and are used to cope with both surface run-off and subsurface flow; most others, much more preferably, are used to collect either surface or subsurface water but not both.

A pipeless French drain with a well-designed permeable filter can be a very effective subsurface drain provided that it is sealed at the top to stop surface water from entering the filter. However, as water flows much more slowly through coarse granular material than through a pipe, the quantity of water to be handled must be small, and any detrimental impacts arising from poor drainage performance must be relatively unimportant, before a decision is taken to omit the carrier pipe.

8.5 Geotextiles in drains

Considerable use is now being made of commercially-available filter fabrics in subsurface drainage systems, particularly in locales where ideal granular filter materials have become more difficult and expensive to obtain.

In conventional drains, for example, a geotextile fabric may be used to line the sides of the trench prior to backfilling with a coarse aggregate; the fabric then acts as a primary filter between the surrounding soil and the secondary coarse aggregate filter material covering the pipe. Alternatively, a geotextile 'sock' is often fitted about a perforated carrier pipe to prevent fine material from entering the pipe. A French drain formed from coarse aggregate wrapped in a geotextile fabric may not need a pipe if the water flow is low.

The successful use of geotextile filters in conjunction with granular filters led to the development and use of commercial pre-fabricated filters that can be used

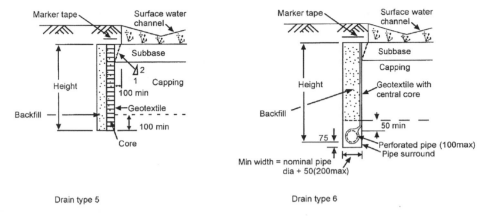

Fig. 8.6 Two examples of fin drains laid in narrow trenches (based on reference 7)

without granular material. Cross-sections of two of these filters are shown in Fig. 8.6.

The Type 5 fin drain is a thin, high-strength, box-shaped, pipeless, 'sandwich' system, the main features of which are two vertical geotextile fabric walls separated by horizontal polyethylene spacers. The fin drain is placed against the trench side-wall that is closest to the carriageway, and the as-dug soil is used to backfill the cavity on the other side; thus, the amount of waste material that has to be removed is much reduced compared with a conventional drain, no filter material has to be imported, and the placement of the drain is carried out more quickly. The sub-surface water enters the drain through the geotextile sidewalls, which act as filters, and flows along the enclosed bottom, which is laid at a conventional slope, to outlets located at intervals along its length.

The Type 6 drain shown in Fig. 8.6 comprises a highly permeable core faced on either side with a geotextile fabric that also envelops a perforated plastic pipe. Again, the geotextile material acts as a filter for the subsurface water; in this case, however, the water runs down the core and enters the carrier pipe for subsequent removal via appropriately spaced outlets.

8.6 References

1. Christopher, B.R. and McGuffey, V.C., *Pavement Subsurface Drainage Systems*, NCHRP Synthesis of Highway Practice 239. Washington, DC: Transportation Research Board, 1997.
2. NAASRA, *Guide to the Control of Moisture in Roads*. Sydney: National Association of Australian State Road Authorities, 1983.
3. Goodman, R.C. and Jeremiah, K.B.C., Groundwater investigation and control in highway construction, *Highways and Road Construction International*, 1976, **44**, No. 1798, pp. 4–7 and No. 1799, pp. 4–8.
4. Royster, D.L., Horizontal drains and horizontal drillings: An overview, *Transportation Research Record 783*, 1980, pp. 16–20.

5. Russam, K., *Sub-soil Drainage and the Structural Design of Roads*, RRL Report LR110. Crowthorne Berkshire: The Road Research Laboratory, 1967.
6. Spaulding, R., *Selection of Materials for Sub-surface Drains*, RRL Report LR346. Crowthorne, Berkshire: The Road Research Laboratory, 1967.
7. Department of Transport, *Highway Construction Details*. London: HMSO, 1987.
8. *Specification for Highway Works*. London: HMSO, 1998.

CHAPTER 9

Introduction to pavement design

C.A. O'Flaherty

9.1 Evolution of the road pavement

The pavements of the early Roman roads (see Introduction: a historical overview of the development of the road) can be divided into three main types, according to the quality of their construction: (1) *viae terrenae*, the lowest type, which were made of levelled earth; (2) *viae glareatae*, which had gravelled surfaces; and (3) *viae munitae*, the highest type, which were paved with rectangular or polygonal stone blocks. The pavements were normally constructed on firm ground, albeit the type of construction was varied to meet the needs of a particular location, e.g. when soft soils were met it was not uncommon to drive wooden piles and to construct the pavements on those piles. In the case of the *viae munitae*, the roads were normally flanked by longitudinal drains, the excavation from which was often used to form the embankment to raise the pavement. Also, in the case of the *viae munitae*, lime (and, in parts of mainland Europe, lime-pozzolan) mortars were used to fill the gaps between the surface slabs. It is reported[1] that the Romans achieved desired compaction levels by rolling their pavements with large and heavy cylindrical stones drawn by oxen or slaves and hand-guided by extended axle shafts.

Figure 9.1 shows sections through Roman road pavements built in England. In relation to Fig. 9.1(a) it might be noted that the term *pavement* probably originated with the word 'pavimentum' which means a rammed floor.

From Roman times until the 18th century, little new happened in the way of pavement construction until Pierre Tresaguet was appointed Inspector General of Roads in France in 1775. Prior to his appointment the 'science' of road construction, as generally practised, simply involved heaping more soil on top of the existing mud and hoping that traffic would ensure it eventually compacted into a hard surface. Tresaguet appreciated the effect of the moisture content of the subsoil upon foundation stability; thus, whilst he built pavements in trenches, he sought to minimize the effect of moisture intrusion by cambering the tops of foundations, making pavement surfaces as dense as possible, and digging deep side ditches alongside the road. He also tried to distribute the stresses transmitted to the foundation soil by spreading the stones over the entire foundation and ensuring that each foundation stone had at least one flat surface (Fig. 9.2(a)).

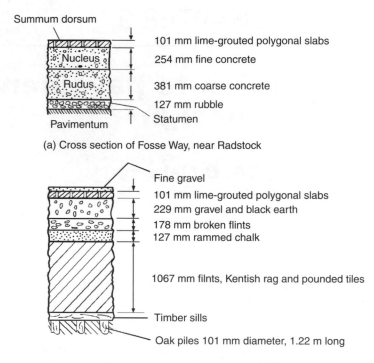

(a) Cross section of Fosse Way, near Radstock

(b) Cross section of pavement through Medway Valley, near Rochester

Fig. 9.1 Pavement sections through Roman roads in England

Thomas Telford was a Scotsman who started his career as a stonemason and eventually went on to found the Institution of Civil Engineers. In the course of his illustrious career Telford perfected a method of broken stone construction (Fig. 9.2(b)) which built upon but refined Tresaguet's, e.g. he used level foundations but raised the pavement above ground level to reduce drainage problems, shaped the foundation stones so that they fitted more closely together, and made the pavement as dense as possible to minimize moisture penetration.

One of Telford's most renowned achievements as a road-maker was the reconstruction of the London–Holyhead road, which was started in 1815, and his thoroughness is reflected in the technology used on that project. He first laid a foundation layer of hand-packed stones with each stone being placed with its broadest end downward; these stones varied in depth from 229 to 179 mm at the centre of the pavement and reduced to between 127 and 51 mm at the haunches. The specification required that the top face of each foundation block be not more than 101 mm wide, and that the interstices between adjacent stones be filled with fine chippings. The central 5.5 m of the pavement width was then covered with two layers of stones – these were 101 mm and 51 mm thick, respectively – with individual stones being sized so that each passed through a 63.5 mm diameter ring. This central width formed the 'working' portion of the pavement and its layers were left to be compacted by traffic. The 1.8 m wide side portions were composed of broken stone or clean gravel, and levelled to give a crossfall of not more than

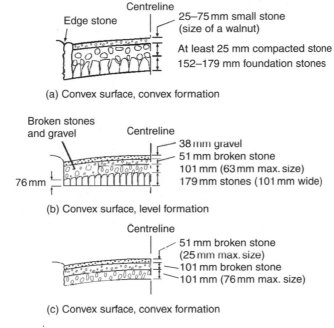

Fig. 9.2 Pavement cross-sections of historical importance: (a) Tresaguet's; (b) Telford's; and (c) McAdam's (not to scale)

1 in 60. A 38 mm thick binding layer of gravel was then placed over the entire width of pavement and watered in, with consolidation being again left to traffic and the weather. As the foundation layer's interstices were big enough to admit water percolating down from the surface, cross-drains were often provided beneath this layer, at intervals of about 91 m, so that any such water could be discharged into side ditches.

Whilst Thomas Telford was first and foremost a civil engineer who had a remarkable aptitude for road-making, his contemporary, John Louden McAdam, was the first true road specialist.

Roads fascinated McAdam and, when made a Trustee of his local turnpike, he took a great interest in all matters relating to roads and their administration; his pastime was to travel about the southwest of England examining roads and devising means for their betterment. He is best remembered for the economical method of road construction (Fig. 9.2(c)) that he advocated and which, in modified form, bears his name (*macadam*) today. The two road-making principles that he emphasized were (a) *'that it is the native soil which really supports the weight of the traffic; that while it is preserved in a dry state it will carry any weight without sinking'* and (b) *'put broken stone upon a road, which shall unite by its own angles so as to form a solid hard surface'*.

Under McAdam's method of construction the road foundation was shaped to the intended surface camber, thereby giving good side drainage to both the foundation and carriageway surfaces, as well as ensuring a uniform pavement thickness across the entire width of the road. The amount of crossfall that he called for was about half that required by Telford. Also, McAdam had little respect for costly stone-paved foundations and, instead, considered that two 101 mm layers of 76 mm broken stone could satisfactorily provide a pavement of similar strength; on top of these he placed a finishing layer of angular fragments not greater than 25.4 mm in size, which were consolidated by ramming them into the interstices and then by the traffic.

Whilst McAdam is renowned for his construction method, his roads were often inferior to Telford's. Nonetheless, the greatness of his construction was that, for efficiency and cheapness, it was a considerable improvement over the methods used by his contemporaries.

McAdam's real claim to fame should, perhaps, be based more legitimately upon his eventual defeat of the then hostility to increasing the traffic upon roads, and his advocacy of the importance of effective road maintenance and administration. He treated with scorn the controversy about whether vehicle sizes and loads should be regulated, emphasized that the road construction methods were at fault, and that pavements could be constructed to carry any type of traffic. He also urged the creation of an (advisory) central road authority, and stressed that the essential feature of any effective road service was the trained professional official who would be paid a salary that enabled him to be above corruption, and who would give his entire time to his duties and be held responsible for any undertaking.

It is of interest to note that both Telford and McAdam regarded compaction by traffic as being sufficient for their stone pavements. They recognized that rolling produced a smoother and more permeable surface, but were concerned that it would also break down the individual stones, destroy the mechanical interlock and, hence, reduce the pavement life. In the 1830s, however, the French added rolling as an integral part of McAdam's process, which they had adopted; however, their horses had difficulty in pulling rollers heavy enough to provide the desired high contact pressure and it was not until the late 1850s that this problem was resolved with the development and use, initially in Bordeaux and then in Paris, of the self-propelled steam-roller.

Internationally, the demand for surfaced roads that arose before and after World War 1 was fuelled by a ready supply of cheap bitumen which, for a long time, was a waste product of the burgeoning oil-refining industry that was supplying petrol to ever-growing numbers of motor vehicles, especially in the USA. Nonetheless, prior to World War II little concentrated attention was directed towards the problems of pavement design, mainly because the roads then in use were seen as being structurally satisfactory for the traffic volumes and loads of the era. Greater emphasis was placed on improving the quality of materials used in pavements, and on developing more economic construction methods using the equipment then available. In many instances 'design' methodology simply relied upon the use of standard thicknesses of pavement for particular classifications of road. Given that in Great Britain, for example, normal subgrade strengths are now known to vary over a range of at least 25 to 1, this approach meant that, if a road's classification designated it, the same thickness of construction was as likely to be placed on a weak foundation as on a strong one, so that too thin a pavement might be placed on a weak foundation and an unnecessarily thick one on a strong foundation.

With the onset of World War II scientific investigation began to be focused on pavement design because of the urgent need to quickly construct great lengths of road and airport runways for the heavy traffic loads demanded by the war effort.

This need gave rise to the development of empirical design methods which related subgrade quality with flexible pavement thickness and commercial vehicle volumes, e.g. methods based on the California Bearing Ratio (CBR) test[2,3] and on the soil classification test[4]. An implicit assumption underlying the use of these methods was that the same design thickness of pavement spread the applied wheel loads to a similar extent, irrespective of the quality of the materials used in the pavement.

As post-war standards of living improved, the 1950s and 1960s saw the acceleration of world-wide growth in the numbers and uses of private and commercial motor vehicles. Greater use of the roads by a better-educated general public and by efficiency-conscious business enterprises meant that more was demanded from road pavements – and from the governments and road engineers responsible for their construction. This created great pressures for better road systems, intensive road research, and more economic (and scientific) pavement design methodologies. As a consequence of experience and research, it then became generally accepted that a well-designed, well-constructed, modern trunk road pavement should at least meet the following basic performance criteria: (i) the finished carriageway should have good skid resistance and provide the motorist with a comfortable and safe ride, (ii) the pavement should be able to carry its design traffic without excessive deformation, and its component layers should not crack as a result of the stresses and strains imposed on them by heavy commercial vehicles and/or climatic conditions, and (iii) a pavement's foundation (including its subbase and any capping layer that might be required to protect the subgrade) should have enough load-spreading capability for it to provide a satisfactory platform for construction vehicles whilst the road is being built.

Before discussing some basic design developments related to the above criteria, it is necessary to clarify the terminology used in relation to a road pavement, as well as the roles and functions of the pavement's individual component layers.

9.2 The component layers of a road pavement: types and functions

A road *pavement* is a structure of superimposed layers of selected and processed materials that is placed on the basement soil or subgrade. The main structural function of a pavement is to support the wheel loads applied to the carriageway and distribute them to the underlying subgrade. The term *subgrade* is normally applied to both the in-situ soil exposed by excavation and to added soil that is placed to form the upper reaches of an embankment.

Modern pavement design is concerned with developing the most economical combination of pavement layers that will ensure that the stresses and strains transmitted from the carriageway do not exceed the supportive capacity of each layer, or of the subgrade, during the design life of the road. Major variables affecting the design of a given pavement are therefore the volume and composition of traffic, the subgrade environment and strength, the materials economically available for use within the pavement layers, and the thickness of each layer.

For discussion purposes, pavements can be divided into two broad types; flexible pavements and rigid pavements (see Fig. 9.3).

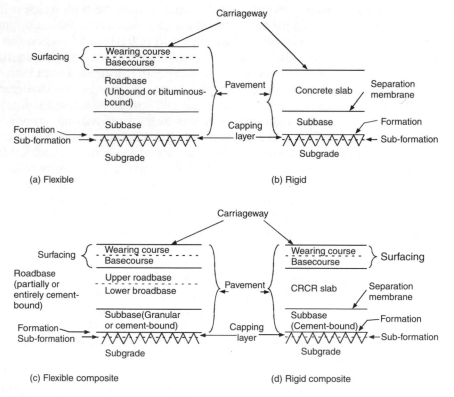

Fig. 9.3 Basic elements of new flexible and rigid types of pavement (not to scale)

9.2.1 Components layers of flexible pavements

Internationally, it is generally accepted that a *flexible pavement* is any pavement other than a Portland cement concrete one. It should be clearly understood, however, that this definition is essentially one of convenience and does not truly reflect the characteristics of the many types of construction masquerading as 'flexible' pavements. In practice, the classical flexible pavement has a number of unbound and/or bituminous-bound granular layers that are topped by a surface layer which, most commonly (but not exclusively), is bitumen-bound. In the UK the term *flexible construction* is a general term that is used to describe[5] a pavement construction consisting of surfacing and roadbase materials that are bound with a bituminous binder.

The uppermost layer of a flexible pavement is called the *surfacing*. The primary function of this layer is to provide a safe, smooth, stable riding surface, i.e. a carriageway, for traffic; its secondary functions are to contribute to the structural stability of the pavement and protect it from the natural elements. The majority of modern-day surfacings in the western world are bitumen-bound; only minor road surfacings are now composed of soil–aggregate materials. When a surfacing is composed of bituminous materials it may comprise a single homogeneous layer or *course*; more usually, however, with heavily-trafficked roads, two distinct sublayers known as a wearing course and a basecourse are laid in separate operations.

The *wearing course* forms the uniform carriageway surface upon which vehicles run. Ideally[6], it should (i) offer good skid resistance, (ii) allow for the rapid drainage of surface water, (iii) minimize traffic noise, (iv) resist cracking and rutting, (v) withstand traffic turning and braking forces, (vi) protect the underlying road structure, (vii) require minimal maintenance, (viii) be capable of being recycled or overlaid, and (ix) be durable and give value for money. No one material meets all of these requirements so, in practice, the selection of a wearing course material depends on the design needs at each site. The *basecourse* (in mainland Europe this is called a *binder course*) is a structural platform which regulates (i.e. makes even) the top of the underlying roadbase, thereby ensuring that the wearing course has a good riding quality when built; it also helps to distribute the applied traffic loads. If the wearing course is impervious, the basecourse can be composed of a more permeable material.

The *roadbase*, which provides the platform for the surfacing, is the main structural layer in a flexible pavement. As the stresses induced in a flexible pavement by the applied wheel loads decrease with depth, the main function of the roadbase is to distribute the loads transmitted to it so that the strength capacities of the weaker subbase and subgrade are not exceeded. Roadbases in flexible pavements are normally designed to be very dense and highly stable, and to resist fatigue cracking and structural deformation.

If a pavement is formed with layers of bitumen- and cement-bound materials, the structure is often referred to as a 'composite' pavement. The rationale for building a composite pavement is to combine the better qualities of both flexible and rigid pavements. In the UK, a *flexible composite* pavement is defined as having its surfacing and upper roadbase (if used) constructed from bituminous materials, supported on a roadbase or lower roadbase of cement-bound material.

A *subbase* is very often present in a flexible pavement as a separate layer beneath the roadbase. Whether it is present, or how it is used, depends upon its intended function(s). As a structural layer within the pavement the subbase further distributes the applied wheel loads to the weaker subgrade below. Whilst the subbase material is of a lesser quality (and, thus, is normally cheaper) than the roadbase material, it must be able to resist the stresses transmitted to it via the roadbase and it must always be stronger than the subgrade soil.

Another major function of the subbase is to act as a working platform for, and protect the subgrade from, site and construction vehicles as a pavement is being built. This is especially important when the subgrade is of poor quality, e.g. clayey or silty, as the critical load-carrying period is when the heavy wheel loads used in the laying and compaction of the roadbase are applied to the subbase during construction. Whilst they are few in number the magnitude of these loads may be great. If the subgrade is strong, e.g. granular, a subbase may not be needed.

A well-graded dense subbase may be used (with or without a geotextile filter) to prevent the upward infiltration of fine-grained subgrade soil into a roadbase. This 'subgrade intrusion' function is especially important during construction, when site traffic and compaction loadings are high. It may also be used to prevent moisture from migrating upward from the subgrade into, say, a soil–aggregate roadbase, or to protect a vulnerable subgrade from downward frost action.

An open-graded subbase may be used as a drainage layer to pass moisture that falls during construction or which enters the pavement after construction. Removal of the water is best ensured by extending the subbase through the shoulders into longitudinal drains located at their edges; these drains should not be allowed to clog-up and should have periodic outlets that are well maintained[7]. In the case of a pavement with a dense bituminous surfacing and a bituminous-bound roadbase, the amount of water infiltrating after construction may be small, and the need for a drainage layer is lessened. If, however, both the surfacing and roadbase are very pervious, an open-graded subbase may be needed to protect the subgrade.

When the subgrade soil is weak a *capping layer* may be created to provide a working platform for equipment used to lay the subbase. This is most commonly done by improving the top of the subgrade, e.g. by adding a layer of imported material that is stronger than the subgrade soil or by stabilizing the upper reaches of the subgrade with, say, lime or cement.

The interface between the subbase and the subgrade (or between subbase and the capping) is termed the *formation*. Its cross-section is normally shaped to reflect the cross slope(s) of the carriageway, to assist in the lateral drainage of water that might accumulate within the pavement. The basement soil surface in contact with the underside of the capping is termed the *sub-formation*.

In the UK the *pavement foundation* is a term that is used to describe the subbase and any material beneath it.

One furher point which should be noted here is that, as a result of the harmonization of European standards that is currently under way in countries associated with the European Economic Community (EEC), the Comité Européen de Normalisation (CEN) has determined that the above terminology will be varied so that, in future, what is now known in the UK as a 'wearing course' will be called a 'surface course', today's 'basecourse' will in future be termed a 'binder course', and a 'roadbase' will become a 'base course'. However, in practice, it will inevitably be some time before current usages and Standards will be fully superseded and, hence, to minimize confusion the traditional UK terms are still used in this text.

9.2.2 Component layers of a rigid pavement

The cross-section of a rigid roadway comprises a pavement superimposed upon the subgrade, and most usually this pavement is composed of a cement concrete slab on top of a subbase. The pavement is described as being a *rigid pavement* because the slab is composed of pavement-quality concrete with considerable flexural strength which enables it to act as a beam and bridge over any minor irregularities in the surface of the layer beneath. In this context the *concrete slab* substitutes for the combined surfacing and roadbase in a new flexible pavement. As the traffic runs directly on the top of the slab the concrete surface must provide a smooth comfortable ride and have good skid resistance under all weather conditions.

In the UK, only a new continuously reinforced concrete roadbase (CRCR) with a bituminous surfacing is termed[5] a *rigid composite* pavement. If the concrete slab in an existing rigid pavement is covered with a bituminous surfacing, e.g. as may happen following the structural maintenance of a pavement, the rigid pavement is simply described as 'with' a bituminous surfacing or a bituminous overlay.

Concrete slabs in rigid pavements are either jointed unreinforced (i.e. plain), or reinforced. Reinforced concrete slabs are usually described as being jointed reinforced or continuously reinforced.

Jointed unreinforced concrete (URC) and *jointed reinforced concrete (JRC)* slabs are provided with transverse and longitudinal joints. *Continuously reinforced concrete* slabs , i.e. *CRCPs*, are normally provided with longitudinal joints but have no transverse joints (other than construction joints). As there are no transverse joints CRCP slabs develop transverse shrinkage cracks at (typically) 1 m to 4 m spacings.

As indicated in Fig. 9.3(b) and (d) a *separation membrane* is placed between the bottom of the concrete slab and the top of the subbase in a rigid pavement. With unreinforced and jointed reinforced slabs the membrane is normally an impermeable polythene sheeting whose functions are to (a) reduce friction between the slab and subbase and allow free movement of the slab to occur as a result of temperature and/or moisture changes in the concrete, (b) prevent underlying material from being mixed into the bottom of the freshly-poured concrete, and/or (c) prevent the loss of moisture and fine material from the concrete mix into a porous subbase (or the subgrade). With continuously reinforced slabs (CRCP and CRCR) a bituminous spray is substituted for the polythene and applied to the surface of the subbase prior to concreting, to prevent water loss from the mix. As a consequence the long central lengths of these slabs are fully restrained and only the ends have to be anchored to cater for seasonal movements.

The function of a *subbase in a concrete pavement* is to provide a uniform, stable and permanent support for the slab when subgrade damage is anticipated from one or more of the following: frost action, poor drainage, mud-pumping, swell and shrinkage, and construction traffic.

Practices regarding how to prevent frost action under concrete slabs vary from country to country. Generally, however, they involve ensuring that the water table is well below the formation and that a sufficient subbase thickness of non-frost-susceptible granular or stabilized material is provided between the concrete slab and the frost-susceptible subgrade within the frost-penetration zone

Mud-pumping can occur at slab joints, edges and cracks when there is free water in the pavement, a subgrade soil with a high clay content that is able to go into suspension, and traffic flows involving the frequent passage of heavy wheel loads. If one of these three ingredients is missing pumping is unlikely to occur. Hence, and depending upon the circumstances prevailing, the likelihood of pumping is minimized if the pavement has an open-graded granular subbase which will act as a drainage layer when there is a danger of water entering the pavement and accumulating beneath the concrete slab. The drainage layer needs to be extended through the shoulder to allow the water to escape into, normally, a subsurface drain. Alternatively, mud-pumping can be prevented by providing a subbase composed of a well-graded compacted material, or a cemented layer, that is essentially impervious to water.

In many regions of the world (but not in the UK) large areas of very expansive soils are quite common. The excessive shrinking and swelling associated with these clayey soils during dry and wet seasons often result in non-uniform subgrade support, with consequent distortion of the concrete pavement and a severe loss of carriageway smoothness and riding quality. Obviating this problem often requires the chemical stabilization of the subgrade (i.e. the creation of a capping layer) and/ or the construction of a deep granular subbase to form a pavement of sufficient thickness and mass to weigh down the subgrade and minimize its upward expansion.

Whatever its other function(s) a subbase in a rigid pavement must be able to act as a working platform for the construction equipment that will lay the concrete, with minimum interruption from wet weather.

The *subgrade* beneath a concrete pavement must be sufficiently stable to withstand the stresses caused by construction and compaction traffic whilst the subbase and concrete slab are being laid, and to provide the uniform support required by the pavement throughout its life. If the subgrade is strong then, technically, a subbase can be omitted from the pavement; in practice, however, this rarely occurs in the UK.

9.3 Some basic considerations affecting pavement design

The number of considerations relating to pavement design that might be addressed under a heading such as this are literally legion. What must be stressed, however, is that pavement design is basically concerned with protecting the subgrade, and the various courses within the pavement structure, from excessive stresses and strains imposed by commercial vehicles and that it is the wheel loads of these heavy vehicles (and not of cars) that are the primary contributors to pavement distress. However, it must also be appreciated that all heavy vehicles do not cause equal distress (because of variations in wheel loads, number and location of axles, types of suspension, etc.) and, also, that the damage that they cause is specific to the pavement's properties, and to operating and environmental conditions.

Of the various forms of pavement distress, fatigue (which leads to cracking) and rutting (i.e. permanent deformation) are of great importance.

The following paragraphs, which are not in any order of priority, are intended to introduce the young engineer to some basic considerations with which he/she should be familiar in respect of the design of road pavements.

9.3.1 Influences of vehicle tyre pressure and speed

The tyre–carriageway contact area beneath a wheel is approximately elliptical in shape. *Tyre inflation pressure*, being easy to measure, is often used as a proxy for contact pressure when applying elastic theory to flexible pavement design. However, research has shown that under normal combinations of wheel load and inflation pressure (a) the average contact pressure is less than the inflation pressure, (b) at constant inflation pressure the contact pressure increases with load, and (c) at constant load an increase in inflation pressure causes a contact pressure increase.

High tyre inflation pressures, especially over-inflated wide-base single tyres, greatly increase the fatigue damage caused to flexible pavements[8]. Tyre pressure has only a moderate influence upon rigid pavement fatigue.

The effects of high contact pressures are most pronounced in the upper layers of a flexible pavement and have relatively small differential effects at greater depths, i.e for a given wheel load the tyre inflation pressure has little effect upon the total depth of pavement required above a subgrade, but it influences the quality of material used in, for example, the surface course.

In practice, contact pressures of around 500 kPa are normally used in mechanistic methods of flexible pavement design.

Generally, there is a decrease in pavement deflections with increasing *vehicle speed*; this is most obvious in pavements carrying vehicles at very low speeds. High speeds result in a reduction in the time that a moving wheel load 'rests' on the road pavement, and this reduced exposure can reduce fatigue and rutting of the visco-elastic materials in a flexible pavement. The speed effect – which is more noticeable when the roadbase comprises bituminous-bound materials vis-à-vis cement-bound ones – suggests that, in concept, greater thicknesses and/or higher qualities of pavement materials should be considered for roads in urban areas and on uphill gradients. However, an analysis of research[8] has also indicated that, overall, the effects are relatively minor and that flexible pavement fatigue remains fairly constant with speed in most cases, albeit rutting decreases as speed increases.

The resilient moduli of some pavement materials and subgrades depend upon the rate at which they are loaded, with the measured modulus increasing with increased rates of loading. Thus, the loading rate that is chosen for testing purposes should reflect typical speed values of the commercial vehicles for which the pavement is being designed.

9.3.2 Importance of flexible pavement thickness and material

Figure 9.4 shows how the stresses at the top and bottom of a flexible pavement change when the tyre-contact pressure is kept constant and the load applied to the smooth-treaded tyre of a test vehicle is progressively increased. Note especially, in the right-hand diagram, that as the wheel load increases the vertical stress at the formation also increases. Thus, if the stress transmitted to the subgrade is not to be increased, the thickness of the pavement needs to be made greater. However, it is not only the total thickness of a flexible pavement but also the quality and thickness

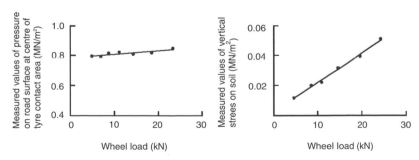

Fig. 9.4 Effect of changing the applied wheel load[9]

of the material in each pavement course that determines the stress distribution and resultant deformation.

Figure 9.5 presents some dynamic stress measurements taken at pavement sections over a uniform clay subgrade before the road was opened to general traffic. The only differences between the sections related to the roadbase materials. Note that the stresses transmitted through the rolled asphalt section are the lowest, lean concrete is next, whilst the soil–cement and crushed stone wet-mix sections are the least successful in spreading wheel loads. When the road was opened to heavy traffic it was found that the stress under the rolled asphalt roadbase did not increase with time whereas that under the lean concrete doubled during the first year of traffic and increased again during the second year. The deterioration of the load-spreading properties of the lean concrete roadbase was considered to be due to excessive cracking of the layer arising from tensile stresses.

The data in Fig. 9.6 show how the deformation history of a pavement of fixed material composition is influenced by the thickness of each layer as well as by its total thickness. Note that all pavement sections exhibit an initial phase of relatively rapid deformation, followed by a second phase during which the rate of deformation is lower and relatively constant. The total deformation accumulated after the passage of a given number of standard axles is very dependent upon the behaviour during the first phase.

Other studies[12] have also shown that a slight increase in course thickness (and, hence, in initial capital outlay) results, for any given traffic load, in an appreciably lower probability of pavement distress and lower subsequent maintenance costs. Conversely, a slight decrease in course thickness results, for the same traffic load, in a higher probability of distress and higher subsequent maintenance costs.

9.3.3 Flexible pavement deterioration and failure mechanisms

An inadequate road pavement does not fail suddenly. As indicated in Fig. 9.7, it is generally considered to begin to deteriorate after entering service and then, gradually, to get worse as time progresses until a failure condition (defined by unacceptable levels of rutting, pot holes, general unevenness, cracking, etc.) is reached.

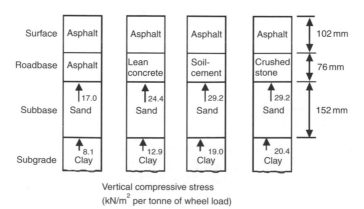

Fig. 9.5 Effect of roadbase material upon subgrade stress[10]

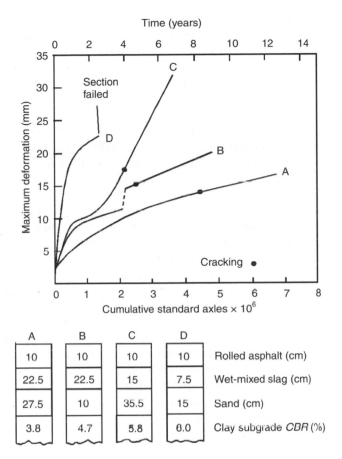

A	B	C	D	
10	10	10	10	Rolled asphalt (cm)
22.5	22.5	15	7.5	Wet-mixed slag (cm)
27.5	10	35.5	15	Sand (cm)
3.8	4.7	5.8	8.0	Clay subgrade *CBR* (%)

Fig. 9.6 Deformation history of a flexible pavement constructed with bases of varying thickness at Alconbury Hill[11]

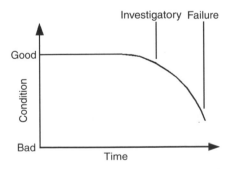

Fig. 9.7 Pavement deterioration cycle

The structural deterioration of a flexible pavement can be divided into four main phases which have been described[13] as follows:

1. a phase when a new or strengthened pavement is stabilizing, when its strength is variable but generally increasing;

2. a stability phase when strength may remain stable or slowly increase or decrease, and the rate of structural deterioration may be predicted with some confidence;
3. an investigatory phase when a pavement's structural deterioration becomes less predictable, when strength may either continue as before or gradually or rapidly decrease. (Previously termed a 'critical' phase, this is now called an 'investigatory' phase as it indicates a period during which monitoring is required to ensure that the next phase is not reached); and
4. a failure phase when the pavement deteriorates to a condition from which it can be strengthened only by total reconstruction. Whilst this failure phase can be quite long, and a pavement can have several years of life left in it before major maintenance is required, it can also be quite short.

Whole life costing studies have shown that intervention during the investigatory phase is more economic than reconstruction after failure at the end of a full design life. Thus, if a pavement's performance is closely monitored so that the point in time is identified at which the deterioration begins to accelerate, i.e. with the onset of the investigatory phase (see Fig. 9.7), and if maintenance in the form of replacing the bituminous surfacing or providing an overlay is then carried out to utilize the remaining existing strength of the pavement and preserve its structural integrity, the time to 'failure' can be very much extended. In the UK the investigatory phase is usually expected after about 20 years in the case of fully flexible and flexible composite trunk road pavements that are conventionally-designed using the determinate-life design procedures discussed in Chapter 14; on average, up to 15 per cent of such pavements can be expected to achieve the investigatory condition by the time their 20 year design load has been carried. In practice, about 15 per cent of a suitable treatment length must have entered the investigatory phase before strengthening is deemed to be justified.

Rutting is caused by deformation in one or more pavement courses. Deformation which occurs within the upper courses of the pavement is termed *surface rutting*; this begins to affect the structural integrity of a pavement when it becomes large. Deformation which arises in the subgrade, due to the inadequate load-spreading abilities of the bituminous-bound and granular courses in place in the pavement, is termed *structural deformation*; this is commonly the main deformation component and, if allowed to continue, will usually result in the break-up of the pavement. In the UK[13], a rut of 20 mm or severe (usually longitudinal) cracking in the wheelpath of a trunk road are regarded as indicators of a pavement section that has failed and probably requires reconstruction to fix it (see also Table 20.1).

Major determinants of the rate at which pavement deterioration takes place include[14] the topography and subgrade soil, pavement materials and thicknesses, drainage (surface and subsurface), quality of construction and maintenance, environment (rain, frost, solar radiation), traffic (volume, axle loads and configuration), and road condition.

Under traffic loading the various courses in a bituminous-bound pavement are subject to repeated stressing and the possibility of damage by fatigue cracking is usually considered to continually exist. When a wheel load passes over a flexible pavement, each course in the pavement responds in the same general way: an applied stress pulse is caused by the wheel mass whilst the resultant horizontal strain consists of resilient and permanent components. The permanent strain component, although tiny for a single-load application, is cumulative and becomes substantial after a great number of load applications. An excessive accumulation of these permanent strains from all layers can lead to fatigue cracking and to pavement failure.

Fatigue damage to flexible (and rigid) pavements is most directly associated with large wheel loads and inadequate pavement thickness. Fatigue damage varies over a range of 20:1 with typical variations in axle loads, and over the same range with typical variations in pavement thickness[8]. Other vehicle properties have a smaller but still significant influence on fatigue.

Fatigue cracking in an inadequate bituminous-bound layer is generally considered to originate at the bottom of the layer, with its onset controlled by the maximum tensile strain, ρ_t, that is repeatedly generated by the passage of commercial vehicles (Fig. 9.8). As the cracks propagate upward through the bituminous-bound courses to the carriageway there is a progressive weakening of these structural layers which, in turn, increases the level of stress transmitted to the lower layers and contributes to the development of structural deformation in the subgrade. As time progresses and the number of applied axle loads increase, the transmitted stresses increase, the development of permanent structural deformation in the subgrade is accelerated, and this is reflected eventually in surface rutting. If the compressive strain in the subgrade, ρ_z, is limited, experience has shown that excessive rutting will not occur unless poor bituminous mix design or inadequate compaction is involved[13]. Experience has also shown that if the depth of rutting at the top of a bituminous-surfaced pavement is more than about 15 mm below the original carriageway level, cracking is likely to occur and water can then enter the pavement and accelerate its deterioration.

A recently-published research review[6] of the design of flexible pavements in the UK has queried the applicability to all flexible pavements of damage mechanisms that are either a fatigue phenomenon that causes a gradual weakening and eventual cracking at the underside of the roadbase or structural deformation originating in the subgrade. This work concluded that the 'critical condition' concept, based on pavement deflection, is not necessarily applicable to thick well-built bituminous-bound pavements that are constructed above a defined threshold strength, and

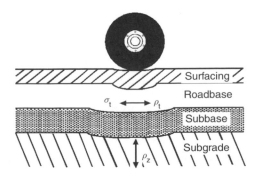

Fig. 9.8 Conceptual response of pavement layers to load

that long structural service lives (of at least 40 years) can be expected from these pavements if distress, which is manifested in the form of cracks and ruts in the surface course(s), can be detected and remedied before it starts to affect the structural integrity of the pavement.

An examination of 'long-life' pavements with bituminous roadbases found a discontinuous relationship between the measured rate of rutting and pavement thickness. Note that the data in Fig. 9.9, which describes rutting in roads with DBM roadbases, form two clusters; the first cluster relates to pavements with less than about 180 mm of bituminous-bound material which deformed under traffic at a high rate, whilst the second set relates to thicker pavements that deformed at a rate that is about two orders of magnitude less than the first cluster. The sharp difference between the two sets of data were interpreted as reflecting a threshold effect. No correlation was found between the rutting rate and pavement thickness, above the threshold thickness, in the range 180 mm to 360 mm. It was concluded that, for the thicker pavements, nearly all the rutting was due to non-structural deformation within the upper layers and that the traffic-induced strains in the subgrade were too low to cause structural damage.

The effects of subgrade strength upon rutting were also examined in the review and it was found that most of the rutting in pavements with a strong foundation occurred in the upper bituminous layers, and that this rutting also was not an indicator of structural deficiency. Table 9.1 shows that the rate of surface rutting is greater for pavements on subgrades with CBRs less than 5 per cent than on those with CBRs greater than 5 per cent, irrespective of the thickness of bituminous-bound cover. (This table also shows the predicted lives to 10 mm rut depth, assuming a linear relationship with traffic.)

The data in Fig. 9.10 are crucial to understanding why, in well-built thick bituminous-bound flexible pavements that are sufficiently strong to resist structural

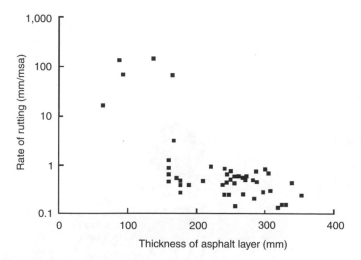

Fig. 9.9 Measured rates of rutting of trunk roads with dense bitumen macadam (DBM) roadbases[6]. (Note: Similar results were obtained for pavements with hot rolled asphalt (HRA) roadbases)

Table 9.1 Comparison of rates of surface rutting in flexible pavements on weak and strong subgrades[6]

Roadbase type	Rate of rutting (mm/msa):			Life to 10 mm rut (msa)
	Mean	Std. dev.	Sample size	
HRA or DBM (Subgrade CBR <5%)	0.58	0.18	20	17
HRA or DBM (Subgrade CBR >5%)	0.36	0.12	21	28
Lean concrete	0.38	0.14	28	28

MSA – million standard axles

damage in their early life, the conventionally-accepted mechanisms of pavement deterioration (roadbase fatigue and structural deformation) are far less prevalent than surface-initiated deterioration that is associated with excessive ageing of the surface course(s).

As is illustrated in Fig. 9.10(a) the penetration of the 100 pen binder used in the roadbase reduced from about 70 pen, shortly after laying, to between 20 and 50 pen after some 15 to 20 years 'curing', whilst Fig. 9.10(b) shows the corresponding changes in the elastic stiffness modulus with time. These data indicate that time curing allowed the roadbase bitumen to gradually harden, with the result that there were in-step increases in the the roadbase stiffness modulus; as the pavements became stiffer their load-spreading abilities also improved steadily with time, so that there were consequent reductions in the traffic-induced stresses and strains in the roadbase and subgrade and, therefore, fatigue and structural deformation did not become prevalent in the structures. Thus, as these bituminous-bound pavements are most vulnerable to structural damage during their early life, it can be expected that they will have a very long life if they are built initially to an above-threshold strength that is sufficient to withstand the traffic-imposed stresses and strains during this early period.

The review also concluded that foundations built to the then current British design specifications were generally adequate for long-life pavements (albeit the use of stronger foundation materials, e.g. using CBM1 or CBM2 or the strengthening of Type 1 subbase material with cement, would reduce the risk of damage from construction traffic, especially under adverse wet weather conditions). In the case of the structural layers the review concluded that pavements did not need to be built thicker than that required by the then current design for 80 msa in order to achieve a very long structural life; however, a thickness of less than 200 mm bituminous-bound material was not recommended for even lightly trafficked pavements if they had to last for 40 years, as thin bituminous-bound pavements were at risk of structural deformation and the rapid propagation of surface-initiated cracks through the full thickness of bituminous material.

As a consequence of the review, new British design curves derived for standard roadbase macadams were proposed for long-life fully-flexible pavements.

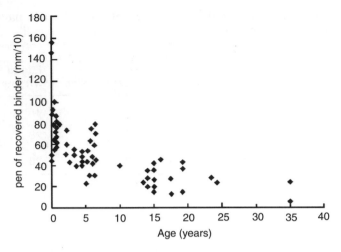

(a) Change in pen of recovered binder with time

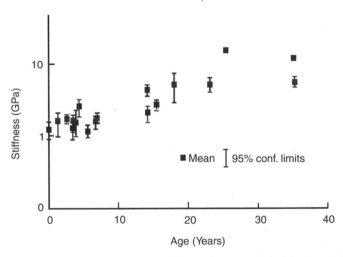

(b) Change in roadbase stiffness of in-service road over time

Fig. 9.10 Measured changes with time of recovered bitumen penetration and elastic stiffness modulus for long-life pavements with dense bitumen macadam roadbases manufactured with a nominal 100 pen bitumen[6]

9.3.4 Rigid pavement deterioration and failure

The major factor influencing fatigue cracking in concrete slabs and in lean concrete bases is σ_t the maximum tensile stress at the bottom of the cemented layer (see Fig. 9.8). Also, cracking of concrete slabs is encouraged by the enhanced stresses generated at the corners and edges of the slab as well as by temperature stresses.

In practice, four types of cracking can be described in relation to concrete pavements: (1) hair cracks, a usual slab feature, which are most clearly seen when concrete is drying out after rain; (2) fine cracks, which are up to 0.5 mm wide at the slab surface; (3) medium cracks, which are >0.5 mm wide and usually accompanied by a loss of aggregate interlock across the fracture; and (4) wide cracks which exceed 1.5 mm and are associated with a complete loss of aggregate interlock. Cracks allow water to enter the subbase and subgrade and lead to deformation and loss of structural support and, in cold climates, to an accelerated rate of crack formation associated with ice formation. Debris particles that enter cracks cause stress concentrations when the pavement flexes; this encourages spalling in the vicinity of the cracks which affects riding quality. If a crack opens to the extent that the ability to transfer load is lost then faulting will occur, i.e. the slab on one side of the crack will be displaced relative to the other and a 'step' is caused in the carriageway.

Whilst, internationally, there has been a tendency to associate the failure of a concrete road with a fixed amount of cracking per length of carriageway, this failure definition should be treated with caution as its practical application can vary considerably.

In the UK[13] individual bays of an unreinforced concrete (URC) pavement are assumed to have failed if any one of the following defects is present: (a) a medium or wide crack crossing the bay longitudinally or transversely; (b) a medium longitudinal and medium transverse crack intersecting, both starting from an edge and longer than 200 mm; or (c) wide corner cracking, more than 200 mm radius centred on the corner. A medium crack is defined as more than 0.5 mm wide and accompanied by a partial loss of aggregate interlock across the fracture, whilst a wide crack is >1.5 mm wide and is associated with complete loss of aggregate interlock. On average up to 30 per cent of URC bays can be expected to have failed by the end of a pavement's design life.

In the UK also, the failure condition of a jointed reinforced concrete (JRC) pavement, which represents the end of its serviceable life, occurs when the rate of cracking begins to increase rapidly. Individual bays of JRC pavements are judged to have failed when the length of wide cracking per bay exceeds one lane width. On average, up to 50 per cent of the JRC bays can be expected to have failed by the time that the design traffic load has been carried, and individual slabs will need to be replaced in the interim (in addition to the resealing of joints, and any arris and thin bonded repairs that are necessitated).

The number of 'old' continuously reinforced concrete(CRCP) and rigid composite (CRCR) pavements built in the UK is limited, so it is difficult to define the condition anticipated at the end of design life; nonetheless, these pavements are not normally expected to need structural strengthening before the end of design life (assuming that they are designed in accordance with current standard practice[13]). Structural strengthening of both pavement types is required when: (i) most cracks are wide, reinforcement is showing signs of corrosion, and the subbase and subgrade is affected by water penetration; and (ii) settlement has resulted in a profile which seriously affects surface water drainage.

Roughness in the road surface excites the axle loads of commercial vehicles, and increases the fatigue damage caused to rigid (and flexible) pavements[8]. (However, roughness does not appear to systematically affect the rutting damage of flexible pavements.)

9.3.5 Pavement deterioration and the fourth power law

The heaviest axles in a stream of commercial vehicles cause a disproportionate amount of damage to a flexible pavement, especially if they are badly loaded (see, for example, reference 12). By comparison, the structural damage caused by lighter vehicles (e. g. cars and commercial vehicles less than about 15 kN unladen weight) is negligible.

A major objective of the AASHO Road Test[15] was to determine the relative damaging effects of different commercial vehicle axle loads. In this major American study an axle with a load of 18000 lb (8160 kg) was defined as a standard axle and given a damaging factor of unity, and the number of repetitions of this axle load which caused the same amount of structural damage or wear to different flexible pavement sections as were caused by various other axle loads were determined using statistical analyses. The conclusion from the study was that the relative damaging effect of an axle load was approximately proportional to the nth power of the load, and that $n = 4$ irrespective of the type or thickness of pavement. Subsequent re-analysis of the data[16] confirmed the thrust of the conclusion but suggested that n lay in the range 3.2 to 5.6. A more recent analysis has reported[6] that n varies between 2 and 9, depending upon the degree and mode of pavement deterioration and its condition at the time that the comparison is made.

Whatever the exact value that should be used in a given circumstance, the concept of a *standard axle load* is very important as it allows the variety of commercial vehicle loads encountered in traffic to be replaced by equivalent numbers of a single axle load, which is easier to handle in pavement design. Thus, if n is taken as equal to 4, the 'fourth power law' can be used to very clearly illustrate the considerable influences upon pavement distress of heavy axle loads, e.g. it demonstrates that a 6 t axle load (which is about 75 per cent of the standard axle load) only causes damage equivalent to less than 0.3 of a standard axle, whereas a 10 t load (which is about 25 per cent more than a standard load) causes damage equivalent to 2.3 standard axles.

The load applied through any wheel assembly of a properly-loaded commercial vehicle should be half the load on the axle to which the assembly is attached. However, if a commercial vehicle has a heavy payload that is not correctly placed in the vehicle, the effect will be to significantly increase the axle and/or wheel load and, consequently, the pavement damage factor.

Figure 9.11 is an actual transverse deformation profile for one carriageway of a busy dual carriageway road in Great Britain. It shows that the greatest deformation is along the wheelpaths of the nearside (left) lane, which carries the greatest proportion of heavy commercial vehicles. The farside traffic lane, which mainly serves overtaking vehicles – mostly cars and light commercial vehicles – is little stressed. These measurements were taken over time, i.e. after 1, 2, 3 and 4 million standard axles (msa) had passed the test site.

Figure 9.11 also suggests the importance of carrying the roadbase through the verge and of providing a shoulder with an impervious covering. Shear failure and lateral displacement of the pavement and/or subgrade occurs more easily if the roadbase is not extended beyond the carriageway edge, whilst the lack of an impervious shoulder means that rainwater can find immediate ingress to the subgrade from the verge.

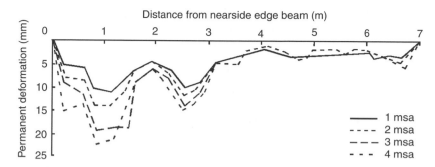

Fig. 9.11 Typical transverse deformation profiles on the same pavement section (composed of 100 mm rolled asphalt surfacing, 250 mm wet-mix slag roadbase, and 170 mm sand subbase) at four different levels of traffic flow[17]

(The approximate national average road wear factors derived in the UK for various commercial vehicle classes using the fourth power law, and based on a standard axle applying a force of 80 kN, are shown in Fig. 9.13 in Subsection 9.3.12.)

9.3.6 Importance of ambient temperature

Temperature conditions should be taken into account when designing a pavement, particularly when high axle loads and temperatures are involved.

In general, the performance of a bituminous-bound flexible pavement deteriorates with rising temperature, and the resistance to permanent deformation/rutting drops rapidly. Generally, therefore, harder bituminous binders and lower binder contents are used in pavement surfacings in hot climates; conversely, softer binders and higher contents are used in cold climates.

Complicating matters is the fact that high ambient temperatures accelerate the rate of in-service hardening or 'ageing' of the bitumen in a bituminous mix, and this results in an increase in mix modulus that is most evident at or close to the surface of a pavement; this modulus is detrimental to, and probably accounts for, the high incidence of surface cracking in geographical regions with high ambient temperatures[18].

The characterization of pavement layers, and subgrade, for mechanistic analysis purposes is complicated by the fact that the capability of most flexible pavement materials that are used to sustain and distribute load is greatly influenced by temperature conditions, e.g. Table 9.2 shows how the modulus of a bituminous material is highly dependent on temperature and duration of loading. Temperature-induced modulus variations also affect the ability of a bituminous material to resist cracking when subjected to repetitions of tensile stress or strain. The structural capacities of underlying unbound granular layers and of the subgrade are also affected by temperature-induced variations in the load-spreading capability of the bituminous layer.

The thermal gradient in a concrete slab is most important because of the thermal stress created in the slab. The need to insulate a cement-bound layer from thermal

Table 9.2 Modulus values determined at different temperatures and loading frequencies for the bituminous surfacing used in the AASHTO Road Test[19]

Temperature °F (°C)	Dynamic modulus (psi × 10^5) for loading frequencies of:		
	1 Hz	4 Hz	16 Hz
40 (4.4)	9.98	12.3	14.0
70 (21.1)	2.32	3.53	5.30
100 (37.8)	0.62	0.89	1.29

effects, and prevent reflection cracking, frequently dictates the thickness of an overlying bituminous surfacing in a composite pavement.

If a concrete slab in a rigid pavement is uniformly maintained at a constant temperature it will rest flat on its supporting layer. If, however, the slab is in place during a hot day and then the temperature drops significantly in the evening, the initial reaction of the top of the slab will be to try to contract. As the thermal conductivity of the concrete is fairly low the bottom will stay at its initial temperature and the slab corners and edges will tend to curl upward. If the temperature conditions are reversed the slab's tendency will be to warp downward. In either case temperature stresses will be induced which can result in the slab cracking, usually near its centre. Hence, warping stresses need to be taken into account in the slab design, especially if large diurnal temperature variations are anticipated.

As a rigid pavement's temperature increases (or decreases) over the course of, say, a year, each end of the slab will try to move away from (or toward) its centre. If the slab is restrained the friction generated between its underside and its supporting foundation will generate compressive or tensile stresses in the slab, depending upon whether it is increasing or decreasing in temperature; hence (and excepting continuously reinforced slabs), a plastic separation layer is normally put between the slab and its supporting layer to reduce this friction. The greater the 'long-term' temperature difference between when the slab is laid and that which it experiences during the various seasons of a given year, the more important it is that this temperature-friction stress be taken into account in design.

9.3.7 Roles of joints in concrete pavements

Joints are deliberate planes of weakness used to control the stresses resulting from (a) expansion and/or contraction volume changes associated with variations in temperature and, to a lesser extent, moisture content, and (b) warping induced by temperature and moisture content differentials between the top and bottom of a concrete slab.

Figure 9.12 illustrates the main types of joints used in concrete pavements: expansion, contraction, warping and construction joints.

Expansion joints (Fig. 9.12(a)) are transverse joints that provide space into which thermal expansion of adjacent slabs can take place when the concrete temperature

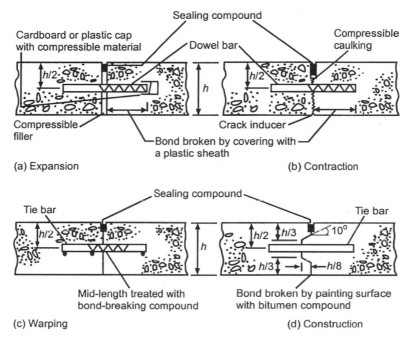

Fig. 9.12 Some typical joints used in concrete pavements (not to scale)

rises above that at which the slabs were laid. If construction takes place in hot weather expansion joints may not be necessary; however, their use is assumed in the UK when specifying the modulus for unreinforced and jointed reinforced slabs that are built in winter. Each expansion joint is vertical, is kept open with a firm compressible filler, and is sealed at the top.

Smooth steel bars known as dowels (a) provide for load-transfer across expansion joints, and (b) keep adjacent surfaces at the same level during slab movement. The dowels are located at mid-slab depth and parallel to the centreline of the pavement; they are placed across the joint openings so that one-half of the length of each is fixed in one slab, whilst the other half is provided with a protective coat (usually a plastic sleeve) to break the bond with the concrete so it can slide within the adjacent slab. Each dowel's sliding end normally terminates in a tight-fitting waterproof cap containing an expansion space equal to the width of the joint gap.

There is ample evidence to indicate that distress at expansion joints is mainly associated with dowels that do not function properly.

Contraction joints (Fig. 9.12(b)) are the most commonly-used transverse joint in jointed concrete pavements. As concrete is weak in tension the function of a contraction joint is to enable a slab to shorten under controlled conditions when its temperature falls below that at which it was laid, and to expand again by up to the same amount when the temperature rises. As with expansion joints, contraction joints are only provided at right-angles to the axis of the pavement as it is not normally necessary to allow for transverse contraction and expansion of a concrete slab in a road pavement.

A common type of contraction joint involves placing a crack inducer at the bottom of a slab and a surface groove at the top, to narrow the slab and deliberately create a vertical plane of weakness. As the slab thickness is thus reduced by some 25–35 per cent at that point, controlled cracking results from the tensile stresses focused there. If the contraction joints are closely spaced (say, <4.5 m apart) the crack opening may be small enough for the interlocking aggregate particles at the faces to provide for load transference without the need for load-transfer bars. More commonly, however, contraction joints are provided with dowel bars similar to those used with expansion joints, except that a receiving cap is not provided, i.e. each bar is covered with the plastic sheath for two-thirds of its length so that, when a slab contracts, the free end of the dowel leaves a gap into which it can return when the slab again expands to its original length.

Warping joints (Fig. 9.12(c)) – these are also known as *hinge joints* – are simply breaks in the continuity of the concrete in which any widening is restricted by tie bars, but which allow a small amount of angular movement to occur between adjacent slabs.

Transverse warping joints are used only in unreinforced concrete pavements, to control cracks which result from the development of excessively high longitudinal warping stresses, e.g. in long narrow slabs. In reinforced slabs the warping stresses – which are associated with the development of temperature gradients in the slabs – are controlled by the reinforcing, and load transfer is maintained by aggregate interlock. As the longitudinal movements in plain concrete slabs are usually fairly small, it is not uncommon to allow up to three consecutive contraction joints to be replaced by warping joints with tie bars.

Longitudinal warping joints control the irregular cracking that occurs as a result of thermal warping and loading stresses within wide concrete pavements. This cracking, which should never be allowed in the wheel tracks of vehicles, is often induced so as to coincide with the intended lanelines on the carriageway.

The tie bars used in warping joints are intended to keep the adjacent slab surfaces at the same level, and to hold the joints tightly closed so that load transference is obtained via face-to-face aggregate interlock. Unlike the smooth dowels used with contraction joints, tie bars are firmly anchored at either end. They are normally located at the mid-depth of the slab; however, because they do not have to allow for movement great accuracy in their placement is not essential. Tie bar spacings are given in the literature[21].

Construction joints are those other than expansion, contraction or warping joints that are formed when construction work is unexpectedly interrupted, e.g. by mechanical breakdown or bad weather, at points where joints are not normally required by the design. Good construction planning ensures that end-of-day joints coincide with predetermined contraction or expansion joint positions; however, the pavement's structural integrity is better maintained if they are at contraction joints. When transverse construction joints coincide with contraction joints in undowelled pavements, the joints should be keyed, whereas they should be dowelled in dowelled pavements. Transverse construction joints that are located between contraction joints should be keyed and tie bars provided (Fig. 9.12(d)).

When the full width of a pavement is not laid in one concreting operation, a formal longitudinal construction joint has to be established between the two abutting concrete slabs. These slabs are tied with tie bars.

Irrespective of the type of joint, the following should apply to all joints to enable them to properly fulfil their functions: (1) long-life sealing materials should be applied at the time of construction to make the joints both waterproof and able to withstand repeated contraction/expansion of the concrete so that foreign materials, e.g. grit, cannot enter and hinder the free movement of adjacent slabs in hot weather; (2) joints should not cause uncomfortable riding conditions that are structurally undesirable and/or detract from the riding quality of the road, e.g. due to excessive relative deflections of adjacent slabs or to repeated impacts by vehicles driven over numerous transverse ridges of sealant that project above the carriageway; (3) joints should not be the cause of unexpected, undesigned, structural weaknesses in a pavement, e.g. transverse joints on either side of a longitudinal joint should not be staggered from each other as transverse cracking will be induced in line with the staggered joints, and no joint should unintentionally be constructed at an angle of less than 90° to an adjacent joint or edge of a slab; and (4) joints should interfere as little as possible with the continuous placing of concrete during construction.

Filler boards provide the gaps for expansion joints at the time of construction, and support for the joint sealing compound over the design life of the pavement. The filler board material should be capable of being compressed without extrusion, be sufficiently elastic to recover its original thickness when the compressive force is released, and not affect/be affected by the covering sealant. Holes have to be accurately bored or punched out of the filler boards in order to provide a sliding fit for sheathed dowels.

Joint seals (also termed *sealants*) should ideally[20] (a) be impermeable, (b) be able to deform to accommodate the rate and amount of movement occurring at the joint, (c) be able to recover their properties and shape after cyclical deformations, (d) be able to bond to faces of joints and neither fail in adhesion nor peel at areas of stress concentration, (e) not rupture internally, i.e. not fail in cohesion, (f) be able to resist flow or unacceptable softening at high service temperatures, (g) not harden or become unacceptably brittle at low service temperatures, (h) and not be adversely affected by ageing, weathering, or other service factors, for a reasonable service life under the range of temperatures and other environmental conditions that can be expected. No one sealant material meets all these requirements; hence, successful joints are those that get regular sealant maintenance.

The *spacings of transverse joints* are a reflection of the capacity of a slab to distribute strain, rather than allow damaging strain concentrations. Joint spacings in plain (unreinforced) slabs normally depend upon concrete strength, the coefficient of thermal expansion of the aggregate, the climatic conditions during construction, and the in-service environmental regime. Limestone aggregate, which has a lower coefficient of thermal expansion than other types of aggregate, causes less expansion/contraction of a slab; consequently, greater joint spacings can be used with concrete slabs containing this aggregate. The effectiveness of reinforcement as a distributor of strain increases with the amount used; hence, greater joint spacings can also be used with larger amounts of reinforcing steel.

9.3.8   Reinforcement in concrete pavements

The primary function of the reinforcing steel in a concrete slab in a rigid pavement is not to contribute to the flexural strength of the concrete, but to control the amount and scale of cracking in the slab.

Whilst concrete is strong in compression it is weak in tension; hence, repeated stressing from commercial vehicle and temperature loading will eventually lead to crack initiation. By resisting the forces which pull the cracks apart the steel reinforcement ensures that the interlocking faces of the cracks remain in close contact, thereby maintaining the structural integrity of the slab. The tight contact also makes more difficult for water on the carriageway to find its way through the cracks into, and soften, the subbase and subgrade. The entry of foreign material and (in cold climates) freezing water, both of which promote crack widening, is also minimized.

Typically, a reinforced concrete slab in a road pavement is rarely wider than 4.65 m, and its length exceeds its width. With this configuration reinforcement is normally only required to control transverse cracking in the slab, as longitudinal cracking seldom occurs. Reinforcement is therefore placed in the longitudinal direction and transverse steel is normally only provided for ease of construction, e.g. to give rigidity to mesh fabrics or to support and space deformed bars, except when there is a risk of differential settlement or where it is required to act as tie bars at joints in jointed reinforced concrete pavements. Also, if the 4.65 m width is exceeded, e.g. when constructing a three-lane carriageway in two equal widths, extra transverse steel may be needed to control longitudinal cracking.

As the reinforcing steel is not intended to resist the induced flexural stresses, its exact location within a jointed reinforced slab is not critical provided it is reasonably close to the upper surface of the slab (because of its crack-control function), and it is well-bonded and protected from corrosion induced by salt penetrating through cracks. As it is convenient for construction, practice in the UK is to place the steel 60–70 mm below the surface in single layer slabs; with two-layer construction, e.g. when using two different aggregates and/or an air-entrained upper layer, the steel is laid on top of the lower concrete slab. With continuously reinforced slabs the steel is normally placed at the mid-depth of the slab to minimize the risk of corrosion from water percolating down through the cracks[22].

9.3.9   Bituminous vs concrete pavements

In 1969 the Road Research Laboratory (now the Transport Research Laboratory) released a report[23] of a study involving estimates of whole-life costs, including initial costs and subsequent maintenance and traffic delay costs discounted to present-day values, for various bituminous and concrete pavements. Some conclusions reached in that study, which related to the construction costs of that era, were: (i) initial construction costs tended generally to be higher for concrete than for flexible pavements, especially for rural secondary and housing estate roads, but there was much overlap in costs for all road types, and (ii) for rural motorway and arterial roads for which equal construction tenders were received, flexible pavements were cheaper for the first 20 years when maintenance and associated traffic costs were also taken into account, but concrete roads were cheaper after 20 years, whereas for rural secondary and housing estate roads there was no significant difference.

Some advantages claimed for bituminous pavements generally include the following, when compared with concrete pavements:

1. In new major roads bituminous surfacings generally provide a better riding quality when opened to traffic, especially if the transverse joints in the concrete slab are closely spaced and not well formed.
2. Bituminous surfacings are traditionally considered to be quieter and are preferred for use in locales where noise is deemed a problem. (However, the reader should be aware that a low noise concrete paving has now been developed that gives a much quieter ride[24].)
3. Bituminous pavements can be opened to traffic as soon as compaction is completed and the surfacings have cooled to the ambient temperature, whereas concrete ones formed from conventional mixes cannot be opened until they have gained sufficient strength. (However, it should be noted that a 'fast-track' process[25] has now been developed which uses rapid-hardening cement in association with high-temperature curing to allow concrete pavements to be opened to traffic within 12 hours of their construction.)

Some advantages now claimed for concrete roads include the following:

1. Concrete pavements are generally better able to cope with unexpected loads and fuel spillages in industrial estates and service areas. If a new housing or industrial estate is being built and it is a requirement that the roads be constructed first so that they can be used by both construction traffic and subsequent general traffic, concrete is considered by many to have a clear advantage.
2. Concrete roads are generally able to maintain an adequate skid resistance under heavy traffic for longer than bituminous surfacings.

9.3.10 Importance of controlling moisture

As is discussed on various occasions elsewhere in this text, a certain amount of water must be present in subgrades, and in pavement courses containing soil, to lubricate the soil particles and ensure that the design densities are achieved when compaction is applied. Additional, unplanned, water should always be prevented from entering a pavement because of its deleterious impacts upon, in particular, unbound courses and cohesive subgrades, and the consequent reduction in the structural conditions upon which the pavement's design is based. The equilibrium moisture content of the soil, which is commonly the moisture amount used in the UK when determining the strength of a subgrade, should therefore not be exceeded whilst the formation is being established and a subbase and capping is being constructed, or during the service life of the pavement.

During construction of the pavement, any aggregate to be utilized should ideally be put in place before rain can enter and soften the foundation. If excess moisture is already in the foundation a subsurface escape route should be provided for this water; where possible this drainage should be kept separate from the surface run-off drainage.

When the water table is high and the subgrade is moisture sensitive (e.g. if the *PI* >25) the installation of a granular, preferably aggregate, drainage blanket is beneficial. Geosynthetic separators may also be used to stop pores in the blanket from being clogged by fines from other adjacent fine-grained layers.

It is essential that appropriate drainage measures always be put in place to minimize the likelihood of surface water causing surface damage to a road or verge, or finding its way into the pavement foundation.

9.3.11 Design life, service life, and design strategy

In practice, a new pavement is designed to last for a selected number of years, i.e. for its *design life*. At the end of this period it is assumed that the pavement will have deteriorated to its terminal or failure condition and, in concept, it will then have to be rehabilitated/reconstructed to restore its structural integrity and serviceability. Current British practice is to use a standard design life of 40 years for new bituminous and concrete pavements as, with appropriate maintenance, this is seen to result in a minimum whole-life cost.

In the UK rigid and rigid composite pavements, and fully flexible pavements (i) with design traffic loads that are heavy in relation to the capacity of the load layout or (ii) where whole-life costing is to be taken into account in the analysis of the pavement design options, are always designed initially to cope with the anticipated traffic volumes over the full 40 years. However, in the case of other flexible pavements, and flexible composite pavements, the practice since 1984 has generally been to stage the pavement construction so that the first stage is designed to carry the traffic loading predicted over the first 20 years, when the 'investigatory' performance condition (Fig. 9.7) is expected to be reached. At the investigatory condition the roadbase and foundation courses should still be substantially intact from a structural aspect and if major maintenance, in the form of overlaying and/or partial reconstruction, is then carried out the re-designed pavement should be able to cope with the traffic loading predicted at that time for the next 20 years of its design life. In addition to the major maintenance it can also be expected that surface dressing or resurfacing will also be needed at about 10 and 30 years (depending, of course, upon a given site's need for skidding resistance).

It is argued by many road engineers that no new rigid and flexible types of trunk road pavement should be stage-constructed, and that the initial design life should always be 40 years as (a) road users are increasingly unwilling to tolerate the frequent delays and accidents associated with the lane closures on busy bituminous roads that result from the use of a 20-year Stage 1 life, (b) the capital cost of the extra initial thickness needed to double the flexible pavement's initial life is relatively small in the context of the overall construction cost of a new road, and (c) the extra initial construction costs associated with a long-life design would be more than compensated for by reductions in whole-life costs. The governmental view is that appropriate staged construction can minimize the whole-life cost and provide the required serviceable life.

A pavement's *service life* is of indefinite duration. It may equal the initial design life or, more likely, it may comprise a succession of design lives for each of which the pavement is provided with the structural capacity needed to satisfactorily carry a specified traffic loading. For example, at the end of the second 20 years of a two-stage flexible pavement's design life a new 'investigatory' condition may be reached, at which time a further major maintenance operation may be carried out to extend its design life to, say, 60 years, and so on.

Design strategy is primarily concerned with the timing of capital investment in a pavement. For example, one design strategy for a particular road might be to consider a long design life and build a high capital cost (e.g. concrete) pavement that will not require further capital investment for, say, 40 years; in this case the pavement's design life and its service life might be assumed the same for economic analysis purposes. In response to pressure for scarce funds to be used elsewhere, however, a contrasting strategy might be to build a low-cost (e.g. soil aggregate) pavement with a much shorter initial design life, and schedule further major rehabilitation investments at staged intervals during the road's service life. The choice of pavement type and quality, and which strategy to follow, would then be made following an economic comparison on the basis of whole life costs involving a consideration of works costs and road user costs. Included in the works costs component are the costs of new construction and of maintenance, and the pavement's residual value (i.e. salvage value). Included in the user costs component are the costs of traffic delay, skidding accidents, accidents at roadworks, fuel consumption and tyre wear, and a residual allowance. In this comparison, the alternatives would be compared on the basis of their present worth, and future costs and salvage value would be discounted to their present day values (see reference 26 for useful discussions on economic analyses).

9.3.12 Determining the design traffic loading

A successful outcome to the pavement design process in any given instance is dependent upon the accuracy with which the total number of equivalent standard axle loads, and their cumulative damage or *wear* effects, can be predicted for the design lane(s) over the period of the design life that is selected. The longer the design life the more difficult it is to be absolutely sure of the exactness of the outcome of this exercise.

In the UK a standard axle load (see Subsection 9.3.5) is now assumed to be that which applies a force of 80 kN to the pavement. In practice, only commercial vehicles in excess of 15 kN unladen weight are taken into account in the equivalency calculations, and wear from private cars is deemed negligible. The heavy vehicles are divided in terms of their total damage effects over the pavement's design life into either the seven individual categories or, more commonly, the three grouped categories shown in Fig. 9.13. The national average wear factor[38] for each vehicle category approximates a fourth power law.

In concept, the total design traffic load is the sum of the cumulative equivalent (80 kN) standard axles in each of the classes or categories shown in Fig. 9.13. The cumulative design load for each class or category is calculated from[27]:

$$T = 365 \, F \, Y \, G \, W \times 10^{-6} \qquad (9.1)$$

where T = design equivalent standard axle load (msa); F = present traffic flow, i.e. the annual average daily traffic flow (AADF) in one direction in the design lane for

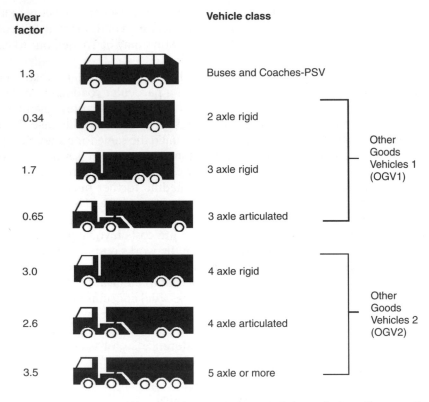

Fig. 9.13 The commercial vehicle categories, and the relative damage factors, used in pavement design in the UK (based on reference 27)

the commercial vehicle class or category on the day of opening; Y=design life of the pavement, from when it is opened (years); G=growth factor for each class or category; and W=wear or damage factor (from Fig. 9.13).

For multi-lane single carriageway roads, and for dual carriageway roads, the proportion of commercial vehicles in the most heavily-trafficked lane (i.e. the left lane in the UK) is required when determining F for use in Equation 9.1. The thickness design eventually determined for the most heavily-trafficked lane of a single or dual carriageway is normally also applied across the full width of the carriageway and shoulder.

In general, the greater the commercial vehicle traffic flow on a given carriageway of a dual carriageway road the smaller will be the proportion using the left-hand lane (see Fig. 9.14); for flows greater than 30 000 cv/d, a proportion of 50 per cent is normally taken. However, it must be emphasized that different proportions may apply in other countries.

The growth factor, G, which represents the proportional difference between the average vehicle flow over the entire design period and the present flow, or the flow at the time of opening in the case of a new road, is dependent on the growth rate for the commercial vehicles under consideration during the pavement's design life. The growth factor derived for future traffic in Great Britain is determined from

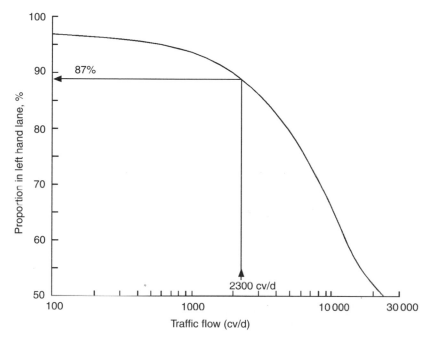

Fig. 9.14 Estimating the proportion of commercial vehicles using the design left-hand lane for dual carriageways in Great Britain[27]

Fig. 9.15. (Another diagram is used if past traffic is being calculated.) The National Road Traffic Forecast (NTRF) growth lines shown for the grouped PSV+OGV1 and OGV2 vehicles in Fig. 9.15 are normally used unless actual growth rate data are available for specific groups of vehicle – in which case these are used instead of the grouped growth lines.

Again, it might be emphasized that the NRTF data represent national estimates and may not necessarily apply to a given road. Also, the NRTF data were derived for Great Britain and it can be expected that different growth rate factors will be applicable to other countries.

9.4 Flexible pavement design methods

Flexible pavement design methods can conceptually be divided into three main groups: (1) methods based on precedent; (2) methods based on empirical strength tests of the subgrade, which are combined with measured results from experimental test roads whose performance was monitored to failure under either controlled or normal traffic flows; and (3) mechanistic methods based on structural analysis and calibrated using empirical data from test roads.

9.4.1 Methods based on precedent

Basically, methods based on precedent rely on the use of standard thicknesses of pavement with particular classifications of road. Many lightly-trafficked pavements

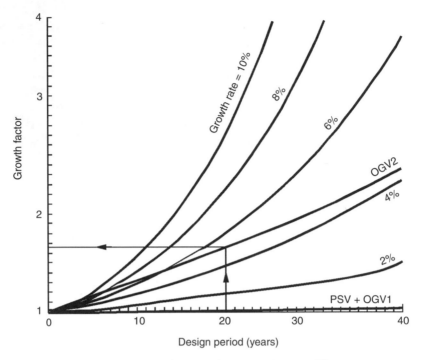

Fig. 9.15 Estimating the growth factor for future traffic[27]

have been successfully built on the basis of previous experience, and there is no reason why this approach should not continue to be used in locales where local knowledge of soils and traffic conditions is very strong. However, the road engineer should be very wary of applying the precedent approach to wide geographical areas and/or varied traffic volumes, as the inevitable result would be instances where inadequate pavements are built on weak subgrades and excessively thick ones on strong subgrades.

9.4.2 Methods based on a soil strength test

These methods normally use some form of penetration or bearing test to determine the subgrade strength, and then rely on empirically-derived test road results (supplemented by structural analysis to rationalize and extend the data) to determine the thickness design for different pavement layer materials. Two methods that are of particular interest are the UK[13] and AASHTO[28] design procedures.

The UK method, which relies on California Bearing Ratio (CBR) tests of the subgrade in combination with analyses of British road test data, is discussed in Chapter 14. The American Association of State Highway Officials (AASHTO) method, which is primarily based on results from the AASHTO Road Test[15], is briefly described here as it has introduced new concepts to, and influenced pavement design in, many countries outside the United States.

The objective of the controlled test road investigation by the AASHTO was to determine relationships between the three main elements of pavement design;

loading, structural capacity, and performance. Major outputs included design equations for flexible (and concrete) pavements as well as load and structural equivalence factors. Prior to the actual road tests an important concept, that of a *Present Serviceability Index (PSI)*[29], was developed as a means of measuring, in numerical terms, a pavement's performance and condition at a given time. Panels of experienced drivers were used to subjectively assess the performance of sections of flexible (and concrete) pavements on the public road network that were in various states of serviceability, with each assessed section being rated on a scale of 0 to 5, with 0 and 5 representing Very Poor and Very Good ratings, respectively. Using multiple linear regression analysis, the subjective performance ratings were then combined with objective measurements of pavement condition on the same sections to develop serviceability equations for both types of pavement. The equation developed for flexible pavements was

$$PSI - 5.03 - 1.91 \log(1 + SV) - 0.01(C + P)^{0.5} - 1.38RD^2 \qquad (9.2)$$

where SV = average longitudinal slope variance, i.e. longitudinal roughness, in the two wheeltracks; C = cracked area (ft^2/1000 ft^2 of pavement); P = patched area (ft^2/1000 ft^2 of pavement); and RD = average rut depth (in) for both wheeltracks measured under a 4 ft straightedge.

When the pavement serviceability concept was subsequently applied to the Road Test's flexible pavements the average initial *PSI* was 4.2; then, the test track pavements were subjected to continued traffic flows and sections were considered 'terminal' and taken out of service when the *PSI* reduced to 1.5; this represented a *PSI* loss of 2.7 under traffic. *PSI*-values of 2.5 and 2.0 are now commonly specified as terminal conditions for high-volume and low-volume roads, respectively, with the AASHTO design method.

Thickness design equations were developed from the Road Test data which related a pavement's *Structural Number* (i.e. this is an abstract thickness index reflecting the relative strength contribution of all layers in the pavement) to the number of axle load applications, the *PSI* loss, the resilient modulus of the subgrade, and the drainage characteristics of unbound roadbase and subbase layers. The initial design equations were subsequently modified and the following is now used[28] in the design of flexible pavements.

$$\log W_{18} = Z_R S_0 + 9.36 \log(SN + 1) - 0.20$$
$$+ \log[\#PSI/(4.2 - 1.5)]/[0.4 + (1094/SN + 1)^{5.19}]$$
$$+ 2.32 \log M_R - 8.07 \qquad (9.3)$$

where W_{18} = design equivalent standard axles loads; Z_R = overall standard deviation: S_0 = combined standard errors of the traffic and performance predictions; $\#PSI$ = design serviceability loss, i.e. the difference between the initial and terminal PSIs; M_R = effective subgrade resilient modulus (psi); and SN = a structural number = $(a_1 D_1 + a_2 D_2 m_2 + a_3 D_3 m_3)$, where D_1, D_2 and D_3 = thickness (in) of the surfacing, roadbase and subbase layers, respectively, a_1, a_2 and a_3 = corresponding layer structural coefficients, and m_2 and m_3 = drainage coefficients for unbound roadbase and subbase layers.

A range of structural coefficients have since been developed for use with differing materials in the various pavement layers. Drainage coefficients recommended for the unbound layers are based on the quality of the drainage (rated from 'excellent' to 'very poor') and the percentage of time the pavement is exposed to moisture contents approaching saturation (ranging from <1% to >25%). Values assigned to Z_R and S_0 vary with the desired level of reliability, e.g. they range from 0.40 to 0.50 for Z_R (with a low value being selected when traffic predictions are reliable), whilst S_0-values vary according to the functional classification of the road (typically, 80 to 99.9 and 85 to 99.9 for rural and urban motorways, respectively, and 50 to 80 for both rural and urban local roads). Soil support is characterized in terms of the resilient modulus, M_R, and relationships are provided which allow the layer structural coefficients, a_1, a_2 and a_3, to be determined from modulus test data. The M_R-value is representative of the monthly variations in modulus over a 1 year period and the effect these have on damage accumulation; hence, the term 'effective' subgrade modulus.

Various combinations of layer thickness are then used to provide the overall structural number, SN, that is determined from Equation 9.3, and a procedure is described in the *AASHTO Guide*[28] for selecting the layer thicknesses that will provide a structurally-balanced pavement. The steps involved in the procedure are: (1) determine the subgrade's M_R-value; (2) select the design serviceability loss *#PSI*; (3) estimate the equivalent standard axle loads for the design life; (4) select appropriate Z_R- and S_0-values; (5) determine the structural number, SN, required for the pavement; (6) select the pavement material type and vary the thickness of individual layers until an SN-value is calculated that is equal to or greater than the required SN.

The basic principle inherent in the AASHTO design process is that the quality of the material in each course, and each course's thickness, should be sufficient to prevent overstressing (and early failure) of the underlying weaker layers.

9.4.3 Mechanistic methods

Whilst the mechanistic approach has been a component of concrete pavement design for many years, its application to flexible pavement design is much more recent. The philosophy underlying the mechanistic approach is that the design of a pavement structure should be handled in the same way as other civil engineering structures: (i) specify the loading; (ii) estimate the size of components; (iii) consider the materials available; (iv) carry out structural analysis using theoretical principles; (v) compare critical stresses, strains or deflections with allowable values; (vi) make adjustments to materials or geometry until a satisfactory design is achieved; and (vii) evaluate the economic feasibility of the result. If this approach is adopted the design task in relation to a flexible pavement then becomes one of proportioning the structure so that the critical levels of stress or strain are not exceeded during the pavement's design life. This, in turn, requires knowledge of the engineering properties of the pavement materials, including (a) effective stiffness modulus, which governs load-spreading behaviour, (b) deformation resistance, which covers rutting behaviour, and (c) fatigue resistance, which relates to cracking behaviour.

Interest in the mechanistic approach in the UK was accelerated in the 1990s by the changing international environment in relation to pavement construction and maintenance. The main influencing elements were the development of European Standards for materials – which opened the UK to contractual competition (notably from France) – and implementation of the government's privatization policy involving DBFO (design, build, finance and operate), as well as significant findings from research and a more open approach from governmental agencies to the implementation of new ideas and procedures[30].

It is generally accepted that Burmister[31,32,33] was the first to develop, in the early 1940s, a theoretical framework for the structural analysis of layered elastic systems that was reasonably close to actual conditions in a flexible pavement. In the mechanistic design procedures in use today the stresses and strains induced in the subgrade and pavement layers are generally based on a generalization of Burmister's theoretical approach.

Flexible pavement design using various mechanistic methods is discussed in detail in Chapter 15, and the reader is referred to Fig. 15.1 which outlines the underlying design approach. In relation to these methods, the following may be noted: (1) the pavement and subgrade are modelled as a multi-layer linear elastic system, with each layer characterized by its modulus of elasticity and Poisson's ratio; (2) the traffic loading predicted for the design period is quantified in terms of the cumulative number of equivalent standard (80 kN) axles; (3) a pavement's response to loading is evaluated for the case where a 40 kN dual wheel load acts normal to the pavement surface; (4) the stresses and strains considered to be the main determinants of specific types of pavement deterioration are calculated for a particular combination of layer materials and thicknesses; (5) the modes of deterioration normally considered are fatigue cracking and wheeltrack deformation (i.e. rutting); (6) a performance model is used to predict the number of repetitions of the 40 kN wheel load that the pavement can withstand before reaching a defined terminal condition; and (7) the pavement geometry is adjusted as necessary until the performance prediction matches the design loading.

9.5 Rigid pavement design considerations

Most countries now have standard approaches to rigid pavement design which, world-wide, are based on accumulated engineering experience gained by study and appraisal of existing concrete pavements, controlled trial roads and experimental road tests under normal traffic, accelerated tests on existing or specially built pavement sections, laboratory experiments, and theoretical and rational analyses. The standard approach used in new trunk road design in the UK is discussed in Chapter 14, whilst basic theoretical/semi-theoretical methods that can be used elsewhere, e.g. in industrial estate and off-road pavements, are covered in Chapter 16.

The following is an introductory discussion of some factors that are taken into account when designing a rigid pavement.

9.5.1 Wheel load stresses

A concrete slab will fail under load when the applied load or bending moment is so great that the developed flexural stress exceeds the modulus of rupture. When a vehicle travels over a concrete pavement the stresses caused vary with the position of the wheels at a given time and hence, in concept, the stresses in each slab should be analysed for when the wheel loads are at all points on the slab, so that the most severe stresses can be evaluated and used for design purposes. Commonly-used methods of analysis are based on that derived by Dr H. M. Westergaard in the 1920s.

In his original analysis[34] Westergaard assumed that: (1) a concrete slab acts as a homogeneous, isotropic, elastic solid in equilibrium; (2) subgrade reactions are vertical only, and proportional to the deflections of the slab; (3) the subgrade reaction per unit of area at any given point is equal to the deflection at that point multiplied by a constant k, termed the 'modulus of subgrade reaction' and assumed to be constant at each point, independent of the deflection, and the same at all points within the area of consideration; (4) slab thickness is uniform; (5) a single wheel load at the interior or at the corner of the slab is distributed uniformly over a circular area of contact and, for the corner loading, that the circumference of this circular area is tangential to the edge of the slab; and (6) a single wheel load at the slab edge is distributed uniformly over a semi-circular contact area, with the diameter of the semi-circle being at the edge of the slab. He initially examined three critical conditions of single wheel loading; these were at the corners, edges and interior of the slab (see Fig. 9.16), and developed the following three equations:

$$\sigma_c = (3P/h^2)[1 - (2^{1/2}a/l)^{3/5}] \tag{9.4}$$

$$\sigma_i = (0.31625P/h^2)[4\log_{10}(l/b) + 1.0693] \tag{9.5}$$

and

$$\sigma_e = (0.57185P/h^2)[4\log_{10}(l/b) + 0.3593] \tag{9.6}$$

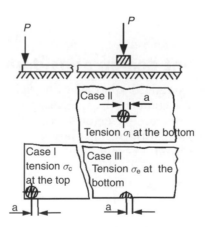

Fig. 9.16 Three cases of loading considered by Westergaard in his original analysis

where σ_c=maximum tensile stress (lbf/in^2) at the top of the slab, in a direction parallel to the bisector of the corner angle, arising from a load applied at the unprotected corner; σ_i=maximum tensile stress (lbf/in^2) at the bottom of the slab directly under the load, when the load is applied at a point in the interior of the slab at a considerable distance from the edges; σ_e=maximum tensile stress (lbf/in^2) at the bottom of the slab directly under the load at the edge, and in a direction parallel to the edge; P=point load (lbf/in^2); h=thickness of the slab (in); μ=Poisson's ratio for concrete (=0.15); E=modulus of elasticity of the concrete (lbf/in^2); k=modulus of subgrade reaction (lbf/in^2/in); a=radius of the load contact area (in), noting that the area is circular for corner and interior loads and semi-circular for edge loads; b=radius of equivalent pressure distribution at the bottom of the slab (in)= $(1.2a^2+h^2)^{1/2}-0.675h$; and l=radius of relative stiffness (in)=$[Eh^3/12k(1-\mu^2)]^{1/4}$, which relates the modulus of subgrade reaction to the flexural stiffness of the slab.

The most critical situation illustrated by the above equations (which were subsequently modified by Westergaard and by others) relates to the corner loading where, due to local depressions of the underlying material e.g. from mud pumping or warping of the slab, the slab corners can become unsupported and, in extreme circumstances, behave as cantilevers and break. Edge loading produces stresses that are slightly less than those caused by corner loading, whilst a load at the interior of the slab, away from edges and corners, generates the least stress.

In the case of a wheel load applied at the slab edge, at a considerable distance from any corner, the edge deflects downwards immediately under the load and upwards at a distance away. The critical tensile stress is on the underside of the slab directly below the centre of the (semi-) circle, and the tensile stresses at the upper surface of the edge at a distance away are much smaller than the tensile stress at the bottom of the slab below the centre.

The least critical situation considered by Westergaard relates to when the load is applied at the interior of the slab. Here, the critical stress is the tensile stress at the bottom of the slab under the centre of the circle, except when the circle radius is so small that some of the vertical stresses near the top become more important; this exception is not a problem, however, in the case of a wheel load applied through a rubber tyre.

The modulus of subgrade reaction, k, is a measure of the stiffness of the supporting material. Whilst the values of k are important, they vary widely, depending upon the soil density, moisture condition and type (e.g. see Table 9.3); however, it should also be noted that a fairly large change has but a relatively small influence upon the calculated stress in a concrete slab.

Table 9.3 Modulus of subgrade reaction values for various soil types[35]

Soil type	k (MPa/m)
Plastic clays	14–27
Silts and silty clays	27–54
Sands and clayey gravels	54–81
Gravels	81+

Table 9.4 Equivalent foundation moduli for a range of typical foundations for concrete pavements[36]

Subbase:						Subgrade:		
Upper layer			Lower layer					Equivalent modulus of foundation
Type	Depth (mm)	Modulus (MPa)	Type	Depth (mm)	Modulus (MPa)	CBR (%)	Modulus (MPa)	(MPa)
Granular Type 1	150	150	Capping layer	600	70	1.5	23	68
				350	70	2.0	27	65
	225	150	None	–	–	5.0	50	89
Lean concrete (C10)	150	28 000	Capping layer	600	70	1.5	23	261
				350	70	2.0	27	268
				150	70	5.0	50	358
			None	–	–	15.0	100	683
Lean concrete (C15)	150	35 000	Capping layer	600	70	1.5	23	277
				350	70	2.0	27	285
				150	70	5.0	50	383
			None	–	–	15.0	100	732

Table 9.4 summarizes equivalent foundation moduli for various subbase and subgrade combinations that, research has shown, can provide all-weather working platforms for pavement construction. In this table equivalence is expressed in terms of a uniform elastic foundation of infinite depth that provides the same surface deflection under a standard wheel load as that of the actual combination foundation.

It might also be noted that Westergaard's radius of relative stiffness is now regarded as a useful guide to determining the desirable length of an unreinforced concrete slab. Research indicates that intermediate transverse cracking in a plain slab can be expected when the ratio of the slab length to the radius of relative stiffness is grater than 5. With these slabs, therefore, the length should be chosen so that these intermediate cracks are eliminated.

9.5.2 Temperature-warping stresses

The temperature-warping effect was also considered by Westergaard[37]. Assuming that the temperature gradient from the top to bottom of a slab was a straight line, he developed equations for three different cases, the simplest of which (for an infinitely large slab) was

$$\sigma_0 = E\varepsilon\, t/2(1-\mu) \tag{9.7}$$

where σ_0 = developed tensile stress (lb/in^2); E = modulus of elasticity of concrete (lb/in^2); ε = coefficient of linear thermal expansion of concrete per °F; μ = Poisson's ratio for concrete; and t = temperature difference between the top and bottom of the slab (°F). Since then, of course, it has been shown that the temperature gradient is closer to a curved line, and this results in calculated stress values that are much lower than those derived by Westergaard.

Even though it has been shown that a temperature gradient of 5°C can occur between the top and bottom of a 150 mm slab in Great Britain, the stresses induced by temperature-warping are not as detrimental as might be expected. The reasons for this are as follows: (i) at slab corners, where the load stresses are actually the greatest, the warping stresses are negligible as the tendency of a slab to curl at these locations is resisted by only a very small amount of concrete; (ii) whilst significant warping stresses can be developed at the interior of the slab and along its edges which, under certain circumstances, are additive to load stresses, slabs normally have a uniform thickness which copes with the corner loading needs, so that the margins of strength present in the interior and edges are usually enough to offset the warping stresses at these locations; and (iii) long-term studies indicate that in the UK the temperature at the bottom of a slab exceeds that at the top more often than the reverse, so that the warping stresses in the interior and at the edges are more frequently subtractive than additive.

9.5.3 Temperature-friction stresses

Stresses are also generated in concrete pavements due to long-term changes in the temperature of the slab (Subsection 9.3.6), which cause each end of the slab to try to move away from or toward its centre as the pavement temperature increases or decreases. If cooling takes place uniformly, a crack may occur about the centre of the slab. If the slab expansion is excessive and adequate expansion widths are not provided between adjacent slabs, *blow-ups* can result in adjacent slabs being jack-knifed into the air.

Assuming that adequate widths of joint are provided, the stresses due to such long-term temperature changes could be considered negligible if there was no friction between the slab and its supporting layer. In fact, however, much friction can be developed between them, i.e. as the slab tries to expand it is restrained by friction and compressive stresses are produced at its underside, and as it tries to contract the friction restraint causes tensile stresses in the bottom of the slab.

The stresses resulting from friction restraint are only important when the slabs are quite long, say >30 m or if they are laid (unusually in the UK) at very high temperatures, say >32 °C. They are only critical when conditions allow them to be applied at the time that the combined loading and warping stresses from other sources are at their maximum. As the maximum tensile stress due to frictional restraint only occurs when a slab is contracting, and as the warping stresses from temperature gradients are not at their maximum at this time, the net result is that, in practice, these restraint stresses are often neglected when calculating the maximum tensile stresses in a conventional concrete slab for thickness design purposes in the UK.

9.5.4 Moisture-induced stresses

Differences in moisture content between the top and bottom of a slab also cause warping stresses. This is due to concrete's ability to shrink when its moisture content is decreased, and to swell when it is increased.

Generally, it can be assumed that the effects of moisture will oppose those of temperature, e.g. in summer a slab will normally shorten rather than lengthen and, on drying out from the top, will warp upward rather than downward. Overall, it is likely that the effects of moisture change on slab stresses are more important in hot regions with pronounced wet and dry seasons compared with the UK's more equitable climate.

9.6 References

1. Lay, M.G., *Ways of the World*. Sydney: Primavere Press, 1993.
2. Porter, O.J., The preparation of subgrades, *Proceedings of the Highway Research Board*, 1938, **18**, Part 2, pp. 324–392.
3. Symposium on Development of the CBR Flexible Pavement Design Method for Airfields, *Transactions of the American Society of Civil Engineers*, 1950, **115**, pp. 453–589.
4. Steel, D.J., Discussion to classification of highway subgrade materials, *Proceedings of the Highway Research Board*, 1945, **25**, pp. 388–392.
5. *Structural Design of New Road Pavements*: Departmental Standard HD 14/87. London: Department of Transport, 1987.
6. Nunn, M.E., Brown, A., Weston, D. and Nicholls, J.C., *Design of Long-life Flexible Pavements for Heavy Traffic*, TRL Report 250. Crowthorne, Berkshire: The Transport Research Laboratory, 1997.
7. Dawson, A.R., Brown, S.F., and Barksdale, R.D., Pavement foundation developments, *Highways and Transportation*, 1989, **3**, No. 36, pp. 27–34.
8. Gillespie, T.D., Karamihas, S.M., Sayers, M.W., Nasim, M.A., Hansen, W., Ehsan, N. and Cebon, C., *Effects of Heavy-vehicle Characteristics on Pavement Response and Performance*, NCHRP Report 353. Washington, DC: National Academy Press, 1993.
9. Whiffin, A.C. and Lister, N.W., The application of elastic theory to flexible pavements, *Proceedings of the First International Conference on Asphalt Pavements*, 1962, pp. 499–521.
10. Thompson, P.D., Croney, D. and Currer, E.W.H., The Alconbury Hill experiment and its relation to flexible pavement design, *Proceedings of the Third International Conference on the Structural Design of Asphalt Pavements*, 1972, **1**, pp. 920–937.
11. Cooper, K.E. and Pell, P.S., *The Effect of Mix Variables on the Fatigue Strength of Bituminous Materials*, TRRL Report LR633. Crowthorne, Berkshire: The Transport and Road Research Laboratory, 1974.
12. *Heavy Trucks, Climate and Pavement Damage*. Paris: OECD, 1989.
13. *Pavement Design*, HD 26/94. London: The Department of Transport, 1994, plus Amendment No. 1, March 1995, Amendment No. 2, February 1996, and Amendment No. 3, February 1998.
14. *Road Deterioration in Developing Countries – Causes and Remedies*. Washington, DC: The World Bank, 1988.
15. *The AASHO Road Test*, Highway Research Board Special Reports No. 61A-E. Washington, DC: The Highway Research Board, 1962.
16. Addis, R.R. and Whitmash, R.A., *Relative Damaging Power of Wheel Loads in Mixed Traffic*, TRRL Report LR979. Crowthorne, Berkshire: The Transport and Road Research Laboratory, 1981.

17. Lister, N.W. and Addis, R.R., Field observations of rutting and their practical implications, *Transportation Research Record 640*, pp. 28–34, Washington, DC: The Transportation Research Board, 1977.
18. Mc Elvaney, J., Cracking of bituminous surfacings in regions of high ambient temperature, *Highways and Transportation*, 1991, **38**, No. 7.
19. Finn, F., Saraf, C., Kulkarni, R., Nair, K., Smith, W. and Abdullah, A., The use of distress prediction subsystems for the design of pavement structures, *Proceedings of the Fourth International Conference on the Structural Design of Asphalt Pavements*. Ann Arbor, Michigan: The University of Michigan, 1977.
20. Hodgkinson, J.R., *Joint Sealants for Concrete Road Pavements*, Technical Note 48. Sydney: The Cement and Concrete Association of Australia, 1982.
21. *Specification for Highway Works*. London: HMSO, 1998.
22. Garnham, M.A., The development of CRCP design curves. *Highways and Construction*, 1989, **12**, No. 36, pp. 14–18.
23. The *Cost of Constructing and Maintaining Flexible and Concrete Pavements Over Fifty Years*. RRL Report LR256. Crowthorne, Berkshire: The Road Research Laboratory, 1969.
24. Swan R., Quiet concrete revisited, *World Highways*. 1994, **3**, No. 1, pp. 39–42.
25. Franklin, R.E., Walker, B.J., and Hollands, P.M., *Fast Track Paving: Laboratory and Full-scale Trials in UK*, Research Report 355. Crowthorne, Berkshire: The Transport Research Laboratory, 1992.
26. Peterson, D.E., *Life-cycle Cost Analysis of Pavements*, NCHRP Report 122. Washington, DC: The Transportation Research Board, 1985.
27. Traffic assessment, HD 24/96, *Design Manual for Roads and Bridges*, Vol. 7, Section 2. London: The Department of Transport, February 1996.
28. AASHTO *Guide for Design of Pavement Structures*. Washington, DC: The American Association of State Highway and Transportation Officials, 1993.
29. Carey, W.N. and Irick, P.E., Performance of flexible pavements in the AASHO Road Test, *Proceedings of the International Conference on the Structural Design of Asphalt Pavements held at the University of Michigan*, Ann Arbor, Michigan, 1962.
30. Brown, S.F., Developments in pavement structural design and maintenance, *Proceedings of the Institution of Civil Engineers Transport*, 1998, **129**, pp. 201–206.
31. Burminster, D.M., The theory of stresses and displacements in layered systems and applications to the design of airport runways, *Proceedings of the Highway Research Board*, 1943, **23**, pp. 126–144.
32. Burmister, D.M., The general theory of stresses and displacement in layered systems, *Journal of Applied Physics*, 1945, **16**.
33. Burmister, D.M., Applications of layered system concepts and principles to interpretations and evaluations of asphalt pavement performances and to design and construction, *Proceedings of the International Conference on the Structural Design of Asphalt Pavements*. Ann Arbor, Michigan: The University of Michigan, 1962.
34. Westergaard, H.M., Stress in concrete pavements computed by theoretical analysis, *Public Roads*, 1926, **7**, No. 2, pp. 25–35.
35. Yoder, E.J., Design principles and practices – Concrete pavements, *Proceedings of the Australian Road Research Board*, 1978, **9**, Part 1, pp. 149–171.

36. Mayhew, H.C., and Harding, H.M., *Thickness Design of Concrete Roads*, TRRL Research Report 87. Crowthorne, Berkshire: The Transport and Road Research Laboratory, 1986.
37. Westergaard, H.M., Analysis of stresses in concrete pavements caused by variations in temperature, *Public Roads*, 1928, **7**, No. 3, pp. 54–60.
38. Robinson, R.G., *Trends in Axle Loading and their Effect on Design of Road Pavements*, TRRL Research Report 138. Crowthorne, Berkshire: The Transport and Road Research Laboratory, 1988.

CHAPTER 10

Earthworks and unbound bases for pavements

C.A. O'Flaherty

10.1 Establishing the horizontal and vertical alignment

A basic set of plans for a road section between any two fixed points will generally include the following:

1. Details of the proposed section, such as the chainages and locations of all pegs; bearings of tangent lines, and radii and other geometrical data affecting the lay-out of all horizontal curves; constraints such as streams, railways, buildings, property fences, public utility lines, and other structures contained within the right-of-way; existing and new drainage structures; benchmark locations; and other details to be considered in the course of construction. Contours showing the topographical nature of the terrain may also be included.
2. A longitudinal profile showing the required ground surface or formation, and the natural ground or existing road line; existing and required elevations at marked pegs; required vertical curve data; grade percentages for gradients; required culvert, bridge and stream-bed elevations, and other data needed to construct the vertical alignment.

The selection of the optimum horizontal and vertical alignment, to enable the preparation of the above documentation, requires good engineering judgement following a detailed examination of a number of alternative alignments which allow for the site's constraints and the application of design standards appropriate to the road type. The end objective is to have a 'flowing' alignment that is attractive to drivers whilst also being economical to construct and protective of the environment in which it is located. Much of the work involved in comparing alternative alignments involves repetitive manipulations of detailed levelling and earthworks data and this is now done with the aid of the electronic computer using, in the UK, sophisticated computer-aided design (CAD) packages[1] such as the *British Integrated Program System* for Highway Design (*BIPS*) and the *MOSS* (an abbreviation of *MO*delling *Sy*Stems) packages.

In practice, selecting the optimum alignment involves considering a number of horizontal alignments in the first instance – each consisting of a series of tangent, i.e. straight, sections joined by transition and circular curves – and the most appropriate vertical alignment is then fitted to each one (see reference 2 for details of alignment features considered in design). The vertical alignment is composed of tangents joined by parabolic summit or valley, i.e. crest or sag, curves. The optimum alignment is controlled by a series of constraints, e.g. the minimum radii and lengths of the horizontal and vertical curves, properties of transition curves, sight distance requirements at horizontal and vertical curves, maximum (for vehicles) and minimum (for drainage) gradients, and level-control points.

In many locations the vertical and/or horizontal alignment may also be controlled by adjacent topographical features or man-made culture. If the terrain is low-lying and swampy the line must be fixed well above natural ground level. Existing bridges and intersections with other roads require the new road to be graded to meet them. In built-up areas the line may be controlled by street intersection and footpath requirements, drainage needs, and sensitivities associated with adjacent property values. In rolling topography, especially in rural locales, the selection of the alignment may be strongly influenced by the earthmoving that has to be carried out, and by the amounts and types of imported fill material that have to be hauled to the site; ideally, the final line selected should result in a balancing of cut with fill, and minimize the amounts of earthworks that have to be wasted from, or imported to, the site.

10.2 Earthworks quantities

The determination of earthworks quantities is based on cross-section data gathered in the field or interpolated from digital ground control models. These cross-section data normally indicate the extent of the excavation (i.e. cut) from cuttings, and filling (i.e. fill) for embankments, at regular intervals – say every 15–30 m – and where major ground irregularities and changes occur along the selected alignment. As noted above, current practice makes maximum use of computing technology in the determination of earthwork quantities so that what used to be very tedious and repetitive calculations (often delegated to the most junior engineers in the design office) are now handled much more quickly and effectively. Nonetheless, the following discussion assumes that earthwork quantities are still determined 'manually' as it is on the basis of fundamentals such as these that the computer programs were developed.

10.2.1 Determining cross-section areas

When the ground surface is level or regular, the area of a cross-section is easily determined by dividing the enclosed space into triangles and trapesiums and using standard formulae in the calculations.

If the ground surface is irregular, the area of a cross-section can be determined using the coordinate method. By geometry, it can be shown that the enclosed area in Fig. 10.1(a) is given by

$$0.5[y_1(x_4-x_2)+y_2(x_1-x_3)+y_3(x_2-x_4)+y_4(x_3-x_1)] \qquad (10.1)$$

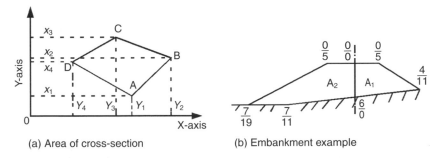

(a) Area of cross-section (b) Embankment example

Fig. 10.1 Determining the area of an embankment's cross-section using the coordinate method

and from this the following simple rule can be postulated:

> Moving about the enclosed figure in an anticlockwise direction, multiply each ordinate (y-value) by the algebraic difference between the prior and following abscissae (x-values), find the algebraic sum of the products and divide the result by 2. If the enclosed figure is to the left of the y-axis, movement should take place in a clockwise direction.

To illustrate the use of the above rule, consider the embankment cross-section shown in Fig. 10.1(b). The embankment has a top width of 10 m. For convenience the cross-section is divided into two parts, A_1 and A_2. Note that the coordinates of each corner are written in fractional form and referenced to the top of the embankment and to the centreline; thus the 'fraction' 7/11 means that the reference point is 7 m below the top of the embankment and 11 m from the centreline. For area A_1, begin at any reference point and proceed anticlockwise about the enclosed area; thus

$$A_1 = 0.5[4(0-5) + 0(11-0) + 0(5-0) + 6(0-11)] = -43\,\text{m}^2$$

For area A_2 begin at any point and proceed clockwise about the area. Thus

$$A_2 = 0.5[7(0-19) + 7(11-5) + 0(19-0) + 0(5-0) + 6(0-11)] = -157\,\text{m}^2$$

Then, $A_1 + A_2 = -200\,\text{m}^2$

The minus sign indicates that an embankment is being considered; if the calculations were carried out for a cutting a positive answer would be obtained. In the case of a side-hill section involving both cut and fill, the vertical reference axis should be taken through the point of intersection of the formation or carriageway and the natural ground-level so that the cut and fill areas can be determined separately.

10.2.2 Determining earthwork volumes

The simplest way of measuring volume is by means of the *trapezoidal* or *average-end-area method*. Thus, if the areas delineated by points A,B,C and D, and I,J,K and L, in Fig. 10.2(a) are denoted by A_1 and A_2, respectively, then

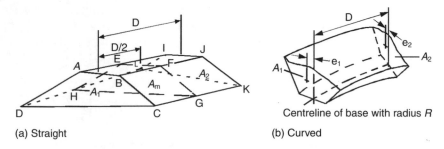

(a) Straight   (b) Curved

Fig. 10.2 Determining volumes on (a) straight and (b) curved sections of embankment

$$V=0.5D(A_1+A_2) \tag{10.2}$$

where V=volume (m³), D=distance between the end areas (m), and A_1 and A_2=end areas (m²) of the embankment.

For a series of successive cross-sections spaced at uniform distances D apart, with areas $A_1, A_2, A_3, A_4, \ldots, A_n$, the volume enclosed between the first and last sections is given by

$$V=D[0.5(A_1+A_n)+A_2+A_3+A_4+\ldots+A_{n-1}] \tag{10.3}$$

Equation 10.3 is exact only when, unusually, the end areas are equal; when they are not equal the results obtained are greater than the true values – albeit the total error on a long line may not be more than a few per cent.

When more precise results are needed, and the field data are sufficiently exact to warrant them, volumes may be determined with the *prismoidal formula*. Referring again to Fig. 10.2(a), if A_1 and A_2 are as before and the mid-section cross-sectional area EFGH is denoted by A_m, then

$$V=D(A_1+4A_m+A_2)/6 \tag{10.4}$$

For a series of successive and equally spaced cross-sections with areas $A_1, A_2, A_3, \ldots, A_n$, and when n is an odd number, the volume enclosed between the first and last sections is given by

$$V=D(A_1+4A_2+2A_3+4A_4+2A_5+\ldots+2A_{n-2}+4A_{n-1}+A_n)/6 \tag{10.5}$$

Where the need for precise results justifies the use of the prismoidal formula, it will also warrant correcting the volumes on curves for errors caused by assuming the centreline to be a straight line. On curves the cross-sections are taken as near radially as possible, so that the volume being examined is a curved solid between two non-parallel plane ends (Fig. 10.2(b)). This volume can be determined using Pappus' second theorem:

> If a plane area rotates about an axis in its own plane which does not divide it into two parts, the volume of the solid thereby formed is equal to the area multiplied by the length of the path of the centre of mass of the area.

Thus, in Fig. 10.2(b), let A_1 and A_2 be two adjacent end areas that are at a distance D from each other, and let the centroids of A_1 and A_2 be at distances of e_1 and e_2, respectively, from the centreline which has a radius of curvature R. If the average area is $0.5(A_1 + A_2)$, then by Pappus

$$V = 0.5(A_1 + A_2) \times (\text{arc distance between the centroids}) \tag{10.6}$$

But,

$$D = R\theta \tag{10.7}$$

where D = length of curve, R = radius of curve, and θ = angle subtended at the centre of the curve. Therefore

$$\theta = D/R \text{ radians} \tag{10.8}$$

The average eccentricity is $0.5(e_1 + e_2)$; hence
arc distance between centroids $- \theta[R + 0.5(e_1 + e_2)]$

$$= D[1 + (e_1 + e_2)/2R] \tag{10.9}$$

Thus the volume is given approximately by

$$V = 0.5D(A_1 + A_2)[1 + (e_1 + e_2)/2R] \tag{10.10}$$

Thus it can be seen that if D is small and R is large, the correction factor $D(e_1 + e_2)/4R$ is relatively unimportant.

If the eccentricity of the centroid of the average section is within the inside of the curve, the correction factor is subtracted from, instead of being added to, the volume as determined by ignoring the curve.

10.3 Balancing earthworks quantities

Ideally, the selection of the optimum horizontal and vertical alignment should result in the volume of material excavated within the limits of the road scheme being equal to the amount of fill required in embankment, so that there is no need to waste good on-site excavation soil or to import expensive borrow material from elsewhere. In engineering practice, however, the ideal does not always happen, and good excavation material may have to be wasted because it is uneconomical to do otherwise, whilst material that is unsuitable for use in embankments may have to be discarded. In addition, issues associated with the swelling, bulking, and shrinkage of soil materials used in earthworks (see Table 4.3) also have to be addressed.

Before discussing the mass-haul diagram, which is fundamental to the earthworks balancing process, it is useful to explain the terms swell, bulking, shrinkage, and haul as they apply to earthworks.

10.3.1 Swelling

When a given volume of soil or rock is excavated, it will bulk-up so that $1\,m^3$ of material before excavation becomes more than $1\,m^3$ when carried 'loose' in a hauling vehicle (see Fig. 10.3). The swelling must be taken into account, when assessing the amount of transport required, for costing and construction purposes. In this

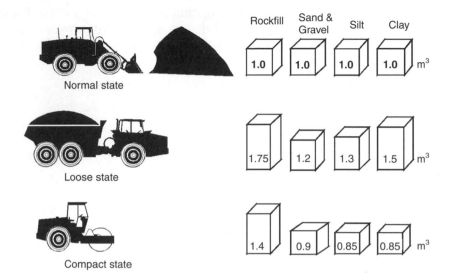

Fig. 10.3 Typical bulking coefficients for common materials encountered in roadworks[3]

context the ratio of the loose volume to the in-situ or non-excavated volume is termed the *swelling factor*.

Figure 10.3 shows that when hauled material is laid in an embankment and made dense by compaction, the net result for most soils is a shrinkage whereas, in the case of rocks, this will not be so.

10.3.2 Shrinkage

The term 'shrinkage' is almost universally used to explain the situation whereby a unit volume of excavation soil will occupy less space when placed in a compacted embankment, whilst *shrinkage factor* is the term used to describe the relationship between the two volumes.

If the shrinkage factor (SF) is defined as the ratio of a volume of embankment to a volume of excavation with the same mass of soil, then

$$SF = V_f/V_e \tag{10.11}$$

where V_f = volume of embankment, and V_e = volume of excavation. But

$$M = D_f V_f = D_e V_e \tag{10.12}$$

where M = mass of dry soil in a given volume of either excavation or embankment, and D_f and D_e = dry densities of embankment and excavation materials, respectively. Therefore

$$SF = (M/D_f)/(M/D_e) = D_e/D_f \tag{10.13}$$

This means that if the density of the undisturbed excavation soil is 1520 kg/m³, and the embankment density is designed to be 1680 kg/m³, then 1 m³ of excavated soil will occupy 0.905 m^3 in the embankment, and $SF = 0.905$. The *% shrinkage*, expressed on the basis of a unit excavation volume, is then 9.5.

However, payment for earthworks is usually based on excavation quantities; hence, for design purposes, earthworks calculations are often determined on the basis of the volume of excavated soil needed to occupy a given volume of embankment containing the same mass of material. Using the same method of analysis as before, it can be shown that in this case

$$SF = D_f/D_e \qquad (10.14)$$

Then, using the same example as above, $SF = 1680/1520 = 1.111$. In other words, 1.111 m^3 of loose excavation is needed to obtain 1 m^3 of embankment. The *% shrinkage* expressed on this basis is 11.1.

Whatever the basis of comparison, shrinkage for most soils is usually quite small and, in some instances (e.g. as noted previously with rocks), there may be a negative shrinkage, i.e. a net bulk up.

10.3.3 Haul

In earthworks calculations the term 'haul' can have a dual meaning. Whilst it is commonly used to refer to the distance over which material is transported, it is also used to describe the volume-distance of material moved.

In earthworks contracts the contractor may be paid a specified price for excavating, hauling and dumping material provided that the haulage does not exceed a certain distance; this distance is called the *free-haul*. The free-haul will vary according to the project but, in the UK, it typically can be as short as 150 m on small road schemes and more than 350 m on motorway projects. Within the free-haul distance the contractor is paid a fixed amount per cubic metre, irrespective of the actual distance through which the material is moved. When the haulage distance is greater than the free-haul, the contractor may be paid a higher rate for the *overhaul*, i.e. the unit overhaul price may be based on the cost per station-metre of moving the material beyond the free-haul distance.

When haul distances are great, it may be more economical to waste good excavation material and import an alternative fill material from a more convenient source, rather than pay for overhauling. On any given scheme, the *economic-haul* distance will vary considerably as it depends on the availability of both suitable borrow sites and of nearby dump sites at which the excavated material can be wasted.

The economic-overhaul distance can be determined from

$$a + bL = c + a \qquad (10.15)$$

and

$$L = c/b \qquad (10.16)$$

where a = cost of on-site haulage per m^3 (including excavating, hauling, and wasting material) within the free-haul distance, b = cost per m^3 per station of overhaul and dumping in embankment, c = cost of borrow material per m^3 (this includes the material cost plus the cost of excavating, hauling and tipping in embankment), and L = economic-overhaul distance in stations.

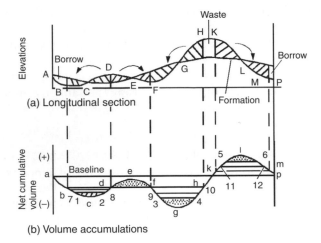

Fig. 10.4 An example of a mass-haul diagram: (a) a longitudinal section showing the ground contour, the proposed formation, and areas of cut and fill, and (b) the volume accumulations for mass haul

If the free-haul distance is denoted by *F* stations, then the economic-haul distance is given by $(F+c/b)$ stations.

10.3.4 The mass-haul diagram

To enable earthworks contractors to submit proper bids for a scheme it is normal practice to include the following information in the design plans:

1. earthworks cross-sections at identified, necessary, locations which indicate accurate levels and the line of the existing ground and of the proposed formation, as well as the areas of cut and fill at each location; and
2. a mass-haul diagram that indicates where excavated material can be economically moved to form embankments, and the amounts of cut and fill at particular locations along the line of the road.

The mass-haul diagram (Fig. 10.4) is a simple graphical representation of the earthworks involved in a road scheme, and of the way in which they can be most economically handled. It shows the algebraic-accumulation of earthworks volume (excavation positive, fill negative) at any point along the proposed centreline and from this the economical directions of haul and the positioning of borrow pits and spoil heaps can be worked out. Some characteristics of the mass-haul diagram at Fig. 10.4(b) are:

1. All volumes are expressed in terms of excavated volumes and embankment data are adjusted to take this into account.
2. The ordinate at any station represents the earthworks accumulation to that point. The maximum ordinate (+) indicates a change from cut to fill, whilst the minimum ordinate (−) represents a change from fill to cut, proceeding along the

centreline from the arbitrarily-assumed origin. These maximum and minimum points may not necessarily coincide with the apparent points of transition indicated in the longitudinal section; this depends on whether or not there are side-hill transitions at these points.

3. A rising curve at any point indicates an excess of excavation over embankment material, whilst a falling curve indicates the reverse. Steeply rising (or falling) curves indicate major cuts (or fills), whereas flat curves show that the earthworks quantities are small.

4. The shapes of the mass-haul loops indicate the directions of haul. Thus, a convex loop shows that the haul from cut to fill is from left to right, whilst a concave loop indicates that the haul is from right to left.

5. The curve starts with zero-accumulated earthworks and the baseline is the zero balance line, i.e. when the curve intersects this line again the total cut and fill will balance. A line that is drawn parallel to the baseline so as to cut a loop is called a 'balancing line', and the two intersection points on the curve are called 'balancing points' as the volumes of cut and fill are balanced between them.

6. The area between a balance line and the mass-haul curve is a measure of the haul (in station-m^3) between the balance points. If this area is divided by the maximum ordinate between the balance line and the curve, the value obtained is the average distance that the cut material needs to be hauled in order to make the fill. (This distance can be estimated by drawing a horizontal line through the mid-point of the maximum ordinate until it intersects the loop at two points; if the loop is 'smooth' the length of this line will be close to the average haul.)

7. Balance lines need not be continuous, i.e. a vertical break between two balance lines merely indicates unbalanced earthworks between the adjacent termination points of the two lines. Adjacent balance lines should never overlap as it means using the same part of the mass-haul diagram twice.

In Fig. 10.4(b) the economic-haul limits are drawn as the balance lines bd, fh, and km; this indicates that, generally, the earthworks volumes are not only balanced in volume but economically as well. The direction of haulage is shown by the arrows in Fig. 10.4(a); note that the haulage is downhill to the embankments so that the emptied vehicles can travel uphill to the excavation sites.

The limits of free-haul are indicated by the balance lines 1–2, 3–4, and 5–6. The free-haul station-meterage is indicated by the dotted areas 1c2, 3g4, and 5l6. In this case, by chance, the balance line d–f is equal to the free-haul distance and, hence, the area def is also free-haul station-meterage.

The overhaul volume for section BCD is given by the difference between the ordinates from c to b–d and from c to 1–2. The average length of overhaul is estimated by drawing the balance line 7–8 through the median of the overhaul ordinate; as the curve is smooth, the points 7 and 8 lie directly below the centres of mass of the overhaul volumes, and the average distance that this excavated material is moved is given by 7–8. Since the free-haul is given by 1–2, the average overhaul is the distance 7–8 minus the distance 1–2.

In practice, selection of the optimum horizontal and vertical alignment rarely results in the earthworks being exactly balanced from beginning to end of the project. In Fig. 10.4 the earthworks are not fully balanced, and it can be seen that borrow material will have to be imported for the embankments between A and B and between M and P, and that the required quantities are given by the ordinates at b and m, respectively. Between H and K the excavated material will have to be wasted to a spoil tip as it is uneconomical to overhaul it for use in the embankment.

10.4 Excavation and earthmoving equipment

Nowadays earthwork operations are highly mechanized and carried out most economically with the aid of very efficient machines, some of which have been developed for specific tasks whilst others have a multiplicity of uses. When one also considers the variety of machines produced by different manufacturers – some of which have distinctive technical advantages over others, whilst others again are little different – it is understandable why the choice of equipment to use on a construction project can appear very daunting to the young engineer.

The space available here does not permit more than a brief overview of the main types of equipment used in earthworks and the reader is encouraged to read about the detailed characteristics of different machines in manufacturers' literature, construction magazines, and professional publications, e.g. reference 4 provides excellent descriptions of generic excavation and earthmoving equipment and the principles underlying their usage in various circumstances.

Table 10.1 lists some common types of equipment and typical uses to which they are put.

A *backacter*, also known as a *backhoe*, has a heavy bucket that is connected to the machine's arm by means of a knuckle joint which simulates the action of a human wrist and allows excavation to take place to depths of 5 m or more, digging towards the machine, below ground level. Backacters can be hydraulically- or rope-operated. The hydraulically-operated machines can dig deeper ditches and work in harder soils.

A *bulldozer* is essentially a track-mounted (crawler) or rubber-tyred tractor that is normally fitted at its front with a straight or angled dozer blade to push earth. Crawler machines are slower than rubber-tyred units, but can handle much rougher terrain. The cross-sections of most straight blades are actually curved so that, as the machine advances, the hardened steel edge digs into the soil and the loose material is pushed upward and forward. Whilst the straight blade is used for general purpose earthmoving, the angled blade is used to push material sideways. A 'universal' blade has large wings for moving large loads over longer distances. The bulldozer is also used to push scrapers, using a reinforced blade that is fitted with a stiffened push plate and shock absorbers. When the ground is very hard a hardened steel ripping tool (usually with up to three teeth) is often attached to the rear of a bulldozer and used to 'plough' the ground to depths up to 1 m deep; the most effective results are obtained if ripping is carried out downhill in the direction of the slope and, in the case of rocks, in the direction of the dip of bedding planes.

The most common *dragline crane* is essentially a crawler-mounted crane that has a revolving superstructure and is fitted with a rope-operated excavation bucket that can be cast out from a long boom. The bucket, which may be perforated for underwater use, has hardened cutting teeth and is filled by dragging it along/in the

Table 10.1 Typical uses of some excavation and earthmoving equipment

Equipment	Typical uses
Backacters (Backhoes)	Excavating below ground level in confined spaces in firm soil, e.g. when digging drainage trenches; as small 'handling cranes', e.g. when laying drainage pipes
Bulldozers	Opening up access roads; grubbing and clearing vegetation; stripping top soil; ripping stony and/or non-cohesive soils, and soft and/or stratified rocks with planes of weakness, prior to excavation; shallow excavating; moving earth over short distances (e.g. <100 m); pushing scrapers; spreading and rough-grading earth previously moved by scrapers, trucks or wagons
Dragline crane	Excavating below ground level in non-confined spaces in soft or loose soils, e.g. digging large trenches or large foundations; dredging underwater deposits of gravel
Face shovels	Excavating firm material above ground level, e.g. in cuttings, and previously-blasted rock in quarries
Front-end loaders	Miscellaneous localized earthworks operations, e.g. digging and/or filling shallow trenches, excavating for manholes, loading loose materials onto trucks and wagons, stockpiling
Graders	Accurate finishing work, e.g. trimming the subgrade and establishing the formation, and shaping shoulders, ditches and backslopes; maintaining haul roads. Also, for blending materials, including water
Scrapers	Earthmoving operations which involve self-loading, hauling over various distances (e.g. <3 km), dumping and spreading of materials
Trucks and wagons	General haulage operations over long distances (typically 1–10 km)

ground towards the machine. As a rule-of-thumb the dragline is able to dig to a depth below its tracks of roughly from one-third to one-half the length of the boom whilst the throw of the bucket beyond the radius of the boom may be of a similar length, depending on the skill of the operator[4]. The dragline bucket may be replaced with a grab bucket, which has two half-shells hinged at the centre and fitted with interlocking teeth, or a clamshell bucket which is similar to the grab but has no teeth. The *grab crane* is used to 'spot' excavate, for example, deep shafts, whilst the *crane with clamshell* is used to stockpile loose materials such as sand.

The *face shovel* also has a rotating superstructure mounted on a crawler- or wheeled-tractor, and has a bucket attached to the boom that is either rope-operated or hydraulically-operated. When in use the bucket is positioned at the bottom of the face being excavated, pressed into it, and pulled upward so that it fills with soil as it rises; the bucket is fitted with hardened teeth to assist in the penetration of the face.

The *front-end loader*, also commonly known as the *loader shovel*, can do work normally associated with both the face shovel and a smaller bulldozer. In fact, the tracked loader looks very much like a bulldozer and its pushed bucket is a very effective excavator; however, the wheeled loader does not have good traction and is therefore more effectively used to dig in loose soils and/or for stockpiling work.

The *motor grader* is composed of a 2- or 3-axle frame with a centrally-mounted 3–5 m blade, that is carried by four or six wheels. When the bottom of the blade is operated in the horizontal position the grader can, very effectively, trim surfaces and/or push spread excavated earth over short distances: to slice off a layer of cohesive soil (e.g. to trim a subgrade) the blade is set with its top tilting backwards; to trim a compacted granular material the blade top is tilted forward; to grade loose soils the blade is set vertically or with its top slightly forward. If the blade is tilted to the side, or rotated at the side through variable positions to the near-vertical, it can be used to form ditches and to shape the sides of embankments and cuttings.

The *scraper* is essentially a metal box on wheels that has a bottom that can be lowered so that one end digs into the ground when excavating. As the scraper moves forward, a layer of earth (typically 150–300 mm deep, depending upon the setting of the cutting edge) 'boils' upward over the edge into the bowl of the box; when the bowl is full the bottom is raised and the scraper then transports the soil to wherever it is to be spread. Loading is most efficiently carried out downhill, in the direction of travel. At the spreading site, e.g. an embankment, the edge of the bowl is again lowered to an above-ground height that will ensure that the desired layer thickness of soil is deposited as an ejector plate pushes from the back of the bowl as the scraper continues to move forward. Scrapers used on large projects are usually motorized and capable of achieving high speeds on haul roads; these may/may not be assisted by pusher-dozers during the digging process. Scrapers used on smaller projects, with short haul-distances, are often non-motorized and not self-loading, i.e. each is towed by a separate tractor, usually a bulldozer, during the digging, transporting and spreading process.

10.5 Compaction specifications

A soil spread in the course of earthworks by machines such as scrapers and bull-dozers is relatively loose and must be compacted into a closely-packed dense mass with a minimum of air voids before the layers of a pavement are placed upon it. Good earthworks compaction increases bearing capacity and slope stability, and reduces settlements and undesirable volume changes, whilst uniform compaction ensures the uniform behaviour of the pavement and prevents differential settlements. In all instances good control of the compaction process is essential to ensuring good outcomes for the roadworks.

Two types of earthworks specification are used in the UK[6] to ensure that compaction for roadworks is properly carried out. These are termed method speci-fications and performance specifications.

With *method specifications*, which are extensively used in the UK, detailed instructions are given to contractors in relation to the layer thickness and number of passes of the compaction equipment to be used with each type of material and equipment (see, for example, Table 6.5), as well as any other type of control that has to be observed. In this case the onus is on the supervising engineer to ensure that the contractor has carried out the work in the manner required by the speci-fication; if any deficiencies are subsequently determined and the engineer has already agreed that the work was carried out according to the specification, then these deficiencies may be deemed the responsibility of the engineer.

With *performance specifications* contractors may select their own methods of compaction to achieve a specified outcome; this outcome is typically specified in terms of dry density, air void content, and/or some form of strength criterion. In this case the onus is on the contractor to choose the compaction equipment for the material to be compacted, the number of equipment passes, and the layer thicknesses, and it is the responsibility of the supervising engineer to carry out regular tests (normally using on-site laboratory facilities) to ensure that the specified outcome is met by the contractor.

10.6 Compaction equipment for earthworks

Many factors influence the dry densities achievable in the course of earthworks compaction, including the soil type and moisture content (see Chapter 4), the type of compaction equipment used and the number of passes made by each machine, and the layer depth of soil being compacted. In concept the compaction equipment should normally be chosen on the basis of the soil to be densified; in practice, however, that decision is often influenced by the equipment available at the time. Whatever the equipment used the overall compaction objective is to obtain a uniform densely-compacted platform upon which the pavement layers can be placed.

Soils contaminated with hazardous chemicals, peats, highly organic soils, and (especially in wet climates) clays with high liquid limits, should never be used in embankments.

In essence, the following are the four main ways in which earthworks materials are compacted by various types of equipment:

1. Using heavy static weights to press the particles together, e.g. smooth-wheeled rollers. Grid rollers act similarly to smooth wheeled rollers when compacting coarse materials; however, with fine-grained soils they compact partly by direct contact pressure and partly by a kneading action. Rubber-tyred rollers have an action that is a cross between the static pressure of a smooth-wheeled roller and the kneading action of a tamping roller.
2. Kneading the particles whilst at the same time applying pressure, e.g. tamping rollers.
3. Vibrating the materials so that the particles are shaken together into a compact mass, e.g. vibrating rollers.
4. Pounding the soil so that the particles are forced together, e.g. impact rollers or tampers.

Table 10.2 provides overall guidance as to which type of equipment, number of passes, and layer thickness are commonly used with various earthworks' materials in the UK. (Detailed specifications are available in the 600 Series, Table 6/1 of reference 6.)

Whatever the types of compaction equipment used it must be emphasized that, when building an embankment, the earthworks contractor should be encouraged to route hauling equipment as evenly as possible over the whole surface of the deposited earthworks, so as to minimize the development of local rutting and related damage caused by concentrated repetitive tracking of heavy haulage equipment. A motor grader or bulldozer should be used as needed to smooth the surface and allow higher vehicle speeds.

Table 10.2 Typical compaction characteristics for natural soils, rocks and artificial materials used in earthworks construction[5]

Material	Major divisions	Subgroups	Suitable type of compaction plant	Minimum number of passes for satisfactory compaction	Maximum thickness of compacted layer (mm)	Remarks
Rock-like materials	Natural rocks	All rock fill (except chalk)	Heavy vibratory roller not less than 180 kg per 100 mm of roll Grid roller not less than 800 kg per 100 mm of roll Self-propelled tamping rollers	4–12	500–1500*	If well graded or easily broken down, then this can be classified as a coarse-grained soil for the purpose of compaction. The maximum diameter of the rock fragments should not exceed two-thirds of the layer thickness
		Chalk	See remarks	3	500	This material can be very sensitive to weight and operation of compacting and spreading plant. Less compactive effort is needed than with other rocks
	Artificial	Waste material	Burnt and unburnt colliery shale	Vibratory roller. Smooth-wheeled roller. Self-propelled tamping roller	4–12*	300	–
		Pulverized fuel ash	Vibratory roller. Self-propelled tamping roller. Smooth-wheeled roller. Pneumatic-tyred roller			Includes lagoon and furnace bottom ash
		Broken concrete, bricks, steelworks slag, etc.	Heavy vibratory roller. Self-propelled tamping roller. Smooth-wheeled roller			Non-processed sulphide brick slag should be used with caution

	Type of soil	Compacting equipment			Notes
Coarse soils	Gravels and gravelly soils	Well-graded gravel and gravel–sand mixtures; little or no fines. Well-graded gravel–sand mixtures with excellent clay binder. Uniform gravel; little or no fines. Poorly-graded gravel and gravel–sand mixtures; little or no fines. Gravel with excess fines, silty gravel, clayey gravel, poorly-graded gravel–sand–clay mixtures	Grid roller over 540 kg per 100 mm of roll. Pneumatic-tyred roller over 2000 kg per wheel. Vibratory plate compactor over 1100 kg/m² of baseplate. Smooth-wheeled roller. Vibratory roller. Vibro-rammer. Self-propelled tamping roller	3–12*	75–275*	—
	Sands and sandy soils	Well-graded sands and gravelly sands; little or no fines. Well-graded sands with excellent clay binder	As above	As above	As above	—
	Uniform sands and gravels	Uniform gravels; little or no fines. Uniform sands; little or no fines. Poorly-graded sands; little or no fines. Sands with fines, silty sands, clayey sands, poorly-graded sand–clay mixtures	Smooth-wheeled roller below 500 kg per 100 mm of roll. Grid roller below 540 kg per 100 mm of roll. Pneumatic-tyred roller below 1500 kg per wheel. Vibratory roller. Vibrating plate compactor. Vibro-tamper.	3–16*	75–300*	—
Fine soils	Soils having low plasticity	Silts (inorganic) and very fine sands, rock flow silty or clayey fine sands with slight plasticity. Clayey silts (inorganic). Organic silts of low plasticity	Tamping (sheepsfoot) roller. Smooth-wheeled roller. Pneumatic-tyred roller. Vibratory roller over 70 kg per 100 mm of roll. Vibratory plate compactor over 1400 kg/m² of baseplate. Vibro-tamper. Power rammer	4–8*	100–450*	If moisture content is low it may be preferable to use a vibratory roller. Tamping (sheepsfoot) rollers are best suited to soils at a moisture content below their plastic limit

(Continued on p. 282)

Table 10.2 (*continued*)

Material	Major divisions	Subgroups	Suitable type of compaction plant	Minimum number of passes for satisfactory compaction	Maximum thickness of compacted layer (mm)	Remarks
Fine soils	Soils having medium plasticity	Silty and sandy clays (inorganic) of medium plasticity. Clays (inorganic) of medium plasticity.	As above	As above	As above	–
		Organic clays of medium plasticity				Organic clays are generally unsuitable for earthworks
	Soils having high plasticity	Micaceous or diatomaceous fine sandy and silty soils, plastic silts Clay (inorganic) of high plasticity, 'fat' clays	As above	As above	As above	Should only be used when circumstances are favourable
		Organic clays of high plasticity				Should not be used for earthworks

Notes: The information in this table should be taken only as a general guide; when the material performance cannot be predicted, it may be established by earthworks trials. This table is applicable only to fill placed and compacted in layers; it is not applicable to deep compaction of materials in-situ. Compaction of mixed soils should be based on that subgroup requiring most compactive effort
*Depending upon type of plant

10.6.1 Smooth-wheeled rollers

Current smooth-wheeled rollers are either single-axle rollers that are towed by tractors, or tandem or three-wheeled rollers that are self-propelled. Typically these rollers range in size from 1.7 to 17 t deadweight. These rollers apply static compaction, and their mass can be increased by adding damp sand or water ballast to the hollow steel drum roll, or by placing ballast weights on the roller frame. With articulated-steering types of three-wheeled roller the three drums have about the same static linear load, and the rear rolls overlap the front one; this enables a uniform compaction to be obtained across the entire roller width and ridges are not formed during the compaction process.

The main factors affecting the compacting performance of smooth-wheeled rollers are the mass per unit width under the compaction rolls and the width and diameter of each roll. The mass per unit width and the roll diameter control the pressure near the surface of the material being compacted whilst the gross mass of each roll affects the rate at which this pressure decreases with depth. Also, the greater the roller mass and the smaller its diameter the greater the wave of material pushed ahead of the roller as it moves forward, and the less even the compacted surface that is left behind.

Smooth-wheeled rollers are most effectively used to compact gravels, sands and materials that need a crushing action; however, they are of no value on rock fill. They have difficulty in maintaining traction on moist plastic soils, and they often cause a 'crust' to be formed on the surface of these soils which can prevent proper compaction of the material beneath.

10.6.2 Grid rollers

This roller typically consists of a box-girder framework within which two rolls are mounted end-to-end. The surface of each roll is composed of a heavy square-patterned steel mesh, whilst the core of the roll is a conical container with its minimum diameter to the outside of the roller. The mass of a typical grid roller varies from 5.5 t net to 15 t when ballasted; the roll cores can hold sand or water ballast, whilst the chassis is also able to carry specially-designed ballast weights. The core shapes enable soil that passes through the grid during compaction to be ejected outside the path of the roller.

When used to compact coarse materials such as marls or soft sandstones, grid rollers have a crushing action that can reduce relatively large boulders (often obtained during ripping) to macadam-sized fragments. They are particularly effective when it is desired to break and force larger stones below the subgrade surface, leaving the top few centimetres of the layer to be graded and compacted (by other equipment) to a uniform and smooth surface[7].

When used with fine-grained soils the grid roller's compaction action partly involves direct contact pressure and partly a kneading action.

10.6.3 Tamping rollers

It is reported[19] that the concept for a tamping roller came to a contractor in 1900 when he saw the compaction caused by a flock of 10 000 sheep crossing a scarified oiled road in California and this led him to develop a tamping roller for compacting cohesive soils – which is why a tamping roller is now also known as a *sheepsfoot roller*. A deadweight tamping roller is a machine with at least one cylindrical steel roll with steel projections or 'feet' extending in a radial direction outward from the surface of the drum; the drums have diameters ranging from 1.0–1.8 m and masses, when loaded, of 2.7–27 t. The feet can have different lengths and tapers, and can be circular, square or rectangular in shape to suit different soil types; however, experience has shown that the area of each tamping foot should be at least 0.01 m^2 and the sum of the areas of the feet should exceed 15 per cent of the end of the roll swept by the ends of the feet. The rollers can be self-propelled but, more commonly, they are towed by track or rubber-tyred tractors with up to three rollers being towed in tandem.

The tamping roller is designed to compact small areas at high load concentrations, and to knead the soil as it compacts from the bottom up. During its first pass over a lift of loose soil, the feet penetrate to near the base of the layer so that the bottom material is well compacted. Then, as more passes are made, the feet penetrate to lesser depths as the soil's density and bearing capacity are increased. Eventually the soil becomes so well compacted that the feet of the roller 'walk out' of the layer.

The contact pressure applied by each tamping foot is that given by dividing the mass of the roller by the total area of one row of tamping feet. This pressure should be as large as possible but should not exceed the bearing capacity of the soil as, otherwise, the roller will sink into the ground until the drum comes into contact with the earth.

In general, the tamping roller is most effectively used to compact fine-grained soils. However, it produces a greater amount of air voids in such soils than either smooth-wheeled or rubber-tyred rollers; this can be detrimental to soils when a large air void content increases the soil's liability to attract moisture. A factor that limits the tamping roller's use in the wet climes of the British Isles is that, when high dry densities are sought, the optimum moisture contents required by these rollers are often lower than the natural moisture contents of fine-grained soils in the field.

10.6.4 Rubber-tyred rollers

Rubber-tyred rollers can be divided into three classes: medium, heavy, and super-heavy weight. The medium class includes both tow-type and self-propelled units up to 12 t total rolling capacity. The heavy class extends up to 50 t loads whilst the super-weights (which are mainly used to compact airport runways and taxiways, and to proof-roll road subgrades) include rollers up to 200 t. With all classes, ballast boxes containing water, sand or pig iron are carried over or between the axles to ensure that the desired loads are achieved.

Generally, rubber-tyred rollers in the medium or heavy class have two axles, with the rear axle having one less wheel than the front one, e.g. five rear and four front wheels or four rear and three front wheels; they have a wheel load of up to 3000 kg, which is adequate for most soil compaction. A large tyre with a big contact area has a better compaction effect than a smaller tyre with the same ground contact pressure; the wide-faced tyres are arranged so that full ground coverage is achieved, with no ridges, by always having an overlap between the front and rear wheels. Individual wheels are capable of independent vertical movement so that the rollers are able to negotiate uneven ground whilst maintaining a constant load on each tyre. Some rollers are fitted with wobbly wheels which provide an additional kneading action to the soil.

The main factors affecting the performance of a rubber-tyred roller are the wheel load, and each tyre's size, ply and, especially, inflation pressure. The inflation pressure (which is correlated with the ground contact pressure) mainly affects the amount of compaction achieved near the surface whilst the wheel load primarily influences the compaction depth.

Controlled studies[8] show that there is little point in using tyre-inflation pressures in excess of 275–345 kN/m2 with wheel loads of 5 t, when compacting clayey soils. With granular soils there is some advantage to be gained by using the highest tyre-inflation pressure and the heaviest wheel load practicable, consistent with not over-stressing the soil.

10.6.5 Vibratory rollers

Vibratory rollers typically range in size from 0.5 to 15 t, and can be pedestrian-guided, towed or self-propelled. Many vibratory roller designs (including both smooth-wheeled and tamping rollers) have been developed by different manufacturers and are now on the market; most commonly these use a rotating eccentric weight installed inside the roller drum to bring about the vibrations. All types achieve densification by producing vibrations in the material being compacted at a frequency in the range of the material's natural frequency. These vibrations momentarily reduce the internal friction between particles, following which the impact of the compaction machine – which is all the greater because of the accelerations of the vibrator unit – forces the particles closer together.

Frequency is the number of roll impacts per unit of time, measured in hertz (vibrations/s). *Amplitude* is the maximum movement of the drum from the axis, usually expressed in millimetres. Laboratory and field tests indicate that frequencies in the 25–40 Hz range have maximum compaction effect on soil, and that a change within this range will not significantly affect compaction effort[17]. High amplitudes are desirable on materials that require a high compaction effort, e.g. coarse rock fill and dry clay soils. A vibratory roller compacting large volumes of soil and rock in thick layers in an embankment would typically have an amplitude in the range 1.6–2.0 mm whereas 0.8–2.0 mm would be used to compact granular pavement bases.

When the rolls of a smooth-wheeled or sheepsfoot vibrating roller are not vibrating the results obtained are as those from a static-weight roller of the same mass. When the rolls are vibrating higher densities and a better depth effect can be achieved than with static compaction; consequently, specification requirements can be obtained with fewer roller passes. Repeated passes of a vibrating roller at high speed are less effective than fewer passes at low speed, for a given thickness of lift.

Initially, vibratory compaction was considered suitable for compacting only rockfill, sand and gravel; however, vibratory techniques have been developed so that they can now also be effectively used with fine-grained soils.

Reference 9 provides a useful introduction to the theory underlying the use of vibrating rollers in earthworks.

10.6.6 Impact compactors

Impact compaction is normally only used on soils and where rollers cannot be used. For example, when trenches have to be backfilled or soil compacted near abutments, walls or columns, it is often not possible to use roller compactors and recourse may be made to a smaller portable impact compactor such as the *power rammer* or *impact stamper*. Generally of about 100 kg mass, the power rammer is actuated by an explosion in its internal-combustion engine (which is controlled by the operator) and this causes the machine to be driven upward in the first instance; the subsequent downward impact of the baseplate causes the densification and constitutes one 'pass' of the compactor. *Dropping-weight compactors* are machines which use a hoist mechanism to drop a dead weight from a controlled height; there are self-propelled machines with mechanical traversing mechanisms that are able to compact soil in trenches and close to structures.

An *impact roller* with a 'square' concrete-filled drum, has been developed overseas for, primarily, proof rolling and, in large sites, for thick lift compaction of in-situ sandy-type soils and subgrades. One such machine[18] has rounded sides and corners on its 'roll' so that, when towed by a tractor – typically at a speed of 12 km/h – the drum rises on one corner and then falls down onto its side, giving a considerable impact to the soil. The impact roller should not be used with thin lifts or pavement courses as it actually loosens the top 100 mm or so of a soil as it compacts. A grader is normally used to finish a surface compacted by an impact roller and a vibratory roller is then used to compact the loose upper layer.

10.7 Constructing embankments on soft soils

One of the more challenging problems that can confront a road engineer is that of constructing an embankment and pavement on a compressible soil, e.g. over peats, marls, and organic and inorganic silts and clays, so that there is little or no subsequent settlement. The soft soil deposits may be localized, as in a peat bog or stream crossing, or they may involve large areas such as tidal marshes or glacial lake beds; they may occur at ground surface or be covered by a desiccated crust or a more recent soil deposit. Table 4.3 (Column 9) shows that these soils usually provide poor road foundations.

Some settlement can be tolerated in almost all embankments built on soft soils; it is detrimental only if it results in pavement fracture or excessive roughness. The total amount of settlement that can be tolerated after construction of the pavement is not well defined, albeit experience would suggest[10] that 0.15–0.3 m can be tolerated on long embankments if the settlement variations are uniformly distributed along the embankment length. Differential settlements in local areas (>25 mm over a 10 m length) can be a cause of considerable concern.

Current methods of embankment construction on soft soils can be grouped as follows: (1) increasing the shear strength and reducing the compressibility of the problem soil by preloading, using full or stage construction and/or accelerated construction techniques; (2) removing the problem soil and replacing it with a more stable material; and (3) using synthetic fabrics at the interface of the embankment and the soft ground. Drainage measures, which are also often considered, may have only limited potential.

10.7.1 Preloading by full or stage construction

Preloading by direct construction of an embankment is a common way of prestressing a soft soil (and of improving the load-carrying properties of uncompacted fill materials such as domestic, industrial and mining wastes, and opencast mining backfill). The purpose of preloading is to consolidate the soft material, not to displace it. If an embankment is properly constructed on a soft subsoil, the subsoil will compress considerably, its moisture content will decrease, and its dry density and shear strength will increase – and the whole construction will be quite stable.

The consolidation of a soft foundation soil beneath an embankment has four components:

1. *Settlement arising from the elastic compression of the subsoil* By elasticity is meant the property of a material which allows it to compress under load, and then to rebound to its original shape when the load is removed. Whilst no soil can be said to be truly elastic, some possess elasticity to the extent that they can cause fatigue damage to overlying pavements, e.g. silts and clays with sizeable amounts of flat flaky particles such as mica, certain diatomaceous earths, and those (including the more fibrous peats) containing large quantities of organic colloids.
2. *Settlement arising from the primary consolidation of the subsoil* When the embankment load is applied there is a reduction in the volume of the soil voids as the subsoil attempts to consolidate. When the soil is saturated the particles pressing about the pores cause the load to be transferred to the incompressible water for a length of time which depends upon the type and compaction state of the subsoil. If the subsoil is a sand and has a high coefficient of permeability, the pore water will be expelled quickly and settlement will rapidly occur. Settlement of a clay subsoil takes place over a longer time; the rate is relatively rapid at the start of primary consolidation but decreases with time. With peat the time needed for primary consolidation can be very much longer, i.e. whilst the settlement time for a clayey soil can be assumed to vary with the square of the length of the drainage path, with the coefficient of permeability remaining constant, this is not so with peat. Because of the greater change in the voids content of peat, its permeability decreases significantly with time and this, coupled with the continuing decrease in pore-water pressure, means that the settlement of peat can take many years, depending upon the depth and type of peat being stressed. (It might be noted here that peat can have moisture contents of 100 to 1800 per cent, plastic indices of 150 to 400, and shear strengths <4 kPa. By contrast, London clay's moisture content, *PI*, and shear strength ranges are 20 to 40 per cent, 35 to 65, and 50 to 500 kPa, respectively.)
3. *Settlement arising from secondary consolidation* After the excess pore-water pressures are largely dissipated settlement will still continue at a rate that produces a linear relationship with the logarithm of time. In practice, secondary

consolidation is assumed to begin when 90 per cent of primary consolidation, as predicted by Terzaghi's theory of consolidation, is complete.

4. *Settlement arising from plastic deformation*   Regions in the subsoil approaching a state of failure cause some further consolidation of a soft soil.

Whilst the subsoil will gain strength with the progress of consolidation, a serious stability failure may occur if the rate of increase in strength is less than the shearing stresses generated during the construction of the embankment. Such a failure is usually manifested by a lateral displacement of the foundation soil, a bulging upward of the adjacent soil, and excessive settlement of the embankment.

One way of reducing the risk of subsoil failure during construction of the embankment is to use a *lightweight fill material* (see reference 11). Whilst pulverized fuel ash (pfa) has been effectively used in Britain to reduce stresses by over 50 per cent compared with normal embankment material, its use can be expensive, however, especially if the pfa has to be hauled a long way to the site. With or without the use of lightweight fill, and provided that extra land width is available, *flatter sideslopes* and the building of *berms* will help prevent lateral displacement of the foundation soil.

Probably the most usual way of avoiding subsoil instability is to *stage-construct* the embankment. This is a long-term process whereby the embankment is first built to a height which is limited by the shearing strength of the foundation, and further construction is then delayed until pore-water pressure measurements show that the extra stress induced in the subsoil has dissipated. When the extra stress is reduced to a safe value, an additional height of fill is then put into place without inducing foundation failure and the process is repeated until the full embankment is constructed.

10.7.2 Preloading by accelerated construction

If the construction programme is such that it requires the embankment to be proceeded with, but it is anticipated that settlement will still be ongoing after completion of construction, it may be decided to accelerate the settlement using either, or a combination of, a surcharge, vertical sand or wick drains, horizontal drains, or dynamic consolidation.

A commonly-used, economical, way of accelerating the rate of settlement is to add a *surcharge* of an extra height of fill, so as to attain, during construction, the full settlement predicted under the embankment and pavement. Then, when the desired settlement has been achieved, the surcharge is removed and the pavement constructed in the normal way.

A major danger associated with the use of a surcharged embankment is subsoil instability. Also, care needs to be taken with some subsoils to ensure that they are not over-consolidated by the surcharge load; the result could be that swelling will subsequently occur and that this will affect the long-term riding quality of the pavement surface.

The consolidation process can also be accelerated by installing vertical drains, the theory being that by providing shorter drainage paths, the time taken for the pore-water to escape is speeded up and settlement time is reduced.

The classic *sand-drain method* involves boring vertical drainage holes through the first embankment layer that is put in place and down through the subsoil until firm soil is reached. The holes, which commonly are 500–750 mm diameter and 9–18 m deep, are spaced about 3–6 m apart and filled with a clean uniform coarse sand; the tops of the vertical drains and the initial construction layer are then covered with a 1–1.5 m thick blanket of the same sand. The normal embankment is then constructed on top of the horizontal sand blanket and a surcharge added as necessary.

A measure of the beneficial effect of sand drains upon the rate of consolidation of an alluvium soil can be gained from Fig. 10.5.

A 100 mm by 3 mm *vertical cardboard-wick drain*, with ten drainage channels preformed internally, may be used instead of a 50 mm diameter sand column. During installation the drain is enclosed within a mandrel (i.e. a hollow tube with a closed flap at the bottom) which is driven into the subsoil; when the mandrel is extracted the flap opens and the wick remains behind. Advantages of these drains over conventional sand drains include lower cost, better maintenance of the vertical continuity of each drain, and an acceleration of consolidation due to closer spacing of the drains. The limitations of the installation equipment mean, however, that the maximum depth to which these drains can be installed is about 20 m.

The *sand-wick drain* is composed of a preformed sand-filled plastic sleeve, about 65 mm in diameter, that is made of water-permeable material. The advantages of

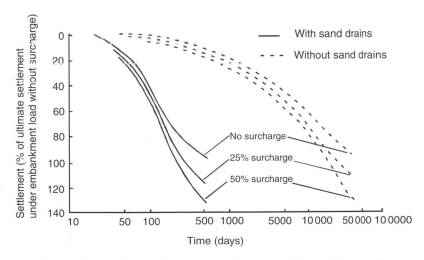

Fig. 10.5 Comparison of the effects of various accelerated consolidation measures upon the settlement of an embankment over a hypothetical alluvium soil[12]
Notes: 1. Compressible soil depth = 10 m with one-way drainage; 2. coefficients of both vertical and horizontal consolidation = 1 m²/year; 3. spacing of the 225 mm diameter water-jetted sand drains without surcharge = 1.48 m to a square pattern; 4. embankment height = 5 m, with 2:1 sideslopes; 5. loading applied at a uniform rate over a period of 60 days without surcharge; 6. surcharge = 25 or 50% of the embankment height.

this system are that only relatively small-diameter boreholes are required and, as the sand is kept within a sleeve, the continuity of the drain is maintained even if large lateral displacements occur within the soft subsoil. The use of sand-wick drains installed by vibratory displacement driving is most economical when the area to be treated is greater than $2000\,m^2$ and the subsoil depth does not exceed 3 m.

10.7.3 Removal and replacement

The most reliable way of building a road across soft ground is to completely remove the soft material and then construct a stable embankment in its place, on the firm strata beneath. The depth at which removal and replacement ceases to be the most economical treatment for a given site depends on the soil type and the groundwater conditions encountered as well as, of course, on the anticipated construction and maintenance costs. As a general guide, however, this approach is most economical when the depth of soft soil does not extend beyond 5–6 m.

Whilst peat bogs and marshes are easily excavated, the removal and replacement approach can become expensively impracticable in the case of extensive deposits of compressible clays or of soft deposits buried beneath a mantle of good soil. Also, the excavated soil must be replaced by good fill, so that the availability of suitable borrow material is an important criterion affecting the decision to use this method. Another influencing factor is whether the subsoil material is to be taken from below the water table, as special procedures may be required to dewater the excavation or to place the fill under water.

The three most common ways of removing a soft soil and installing an embankment in its place are by mechanical excavation and replacement, displacement by gravity, and displacement by blasting.

When the depth of unstable material is less than 5 m it may be economical to use *mechanical excavators* to remove the soil, and the embankment is then built in the empty space on firm stratum as the excavation proceeds. Depending upon the site conditions, the excavators may be able to work from the underlying firm stratum or from the embankment as it is built forward.

When the unstable soil is very soft and is between, say, 5 and 15 m deep, the *side-tipping displacement-by-gravity method* may be used. In this case a shallow ditch is first cut along the proposed centreline of the road and through the surface mat of the soft soil to the depth of the water table, to form a plane of weakness within the unstable soil. Fill material is then added symmetrically on either side of this line, until its mass displaces the underlying soil to the side. Trenches may be excavated on either or both sides of the embankment to assist in the lateral displacement of the soft soil.

If the underlying material is too stiff to be displaced by the embankment alone, it may have to be softened using water jets. The usual way of doing this is to sink 25 mm diameter pipes through the embankment and down to near the bottom of the soft material; water is then forced into the soil under high pressure as the pipes are slowly withdrawn.

The displacement method is straightforward but slow. If the process is not carefully controlled pockets of unstable soil may be trapped under the embankment and this may give rise to further settlement over time.

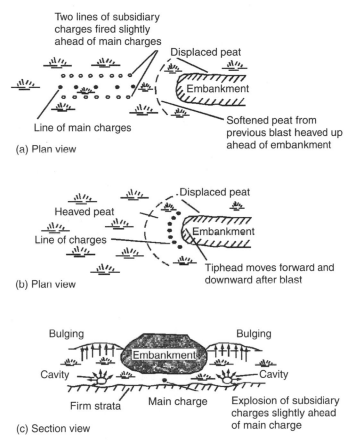

Fig. 10.6 Methods of bog blasting: (a) trench-shooting; (b) toe-shooting; and (c) underfill

Displacement-by-blasting methods have been mostly used when constructing embankments over peat. The main methods are trench-shooting, toe-shooting, and underfill blasting. With the *trench-shooting method* explosive charges are placed near the bottom of the soft soil, usually by jetting, in longitudinal rows in front of the embankment being built (Fig. 10.6(a)). When the charges, which are spaced at distances of one-half to two-thirds the thickness of the soft soil, are fired an open trench is created, further fill is then placed in this, and the embankment is advanced. This blasting method is best used for peat depths of less than 6 m, and when the peat is fairly stiff so that it is not likely to slip back into the excavated trench.

With the *toe-shooting method*, the end of the embankment is advanced, and its head-height raised, until the peat is partially displaced from below and heaved up ahead of it. As shown in Fig. 10.6(b) charges are then placed at the bottom of the peat immediately ahead of the embankment; when these are fired simultaneously the peat is displaced and the fill drops into the void left by the explosion. This blasting method is particularly suitable for use with peats that are so soft that they can nearly flow by gravity displacement alone and the depth is less than about 6 m. For greater depths (up to 15 m) the peat can be displaced using *torpedo-blasting*; this is the same as toe-shooting except that, instead of all the charges being set off at the bottom of the peat, they are attached at regular intervals to long posts that are placed vertically in the peat.

With the *underfill method* of bog blasting the full embankment is placed over the peat, and charges are placed beneath it near the bottom of the peat, so that a row of main charges is below the centre of the embankment and a row of subsidiary charges is beneath each edge. The subsidiary charges are detonated just before the main charges; this causes the peat to be blasted outward and the fill material then drops neatly into the continuous series of cavities left by the explosions. The under-fill method is especially effective when a sound soil overlies an unstable one. If the peat is over 9 m deep, wide embankments are usually constructed by first building a narrow embankment and then widening it, either on one or both sides, by further underfill blasting.

10.7.4 Using synthetic fabrics

From the 1960s, and particularly since the 1980s, considerable use has been made of geotextile and geogrid fabrics in embankment construction[13] to perform one or more of four major functions, i.e. separation, filtration, reinforcement, and drainage. In this context separation can be defined as the introduction of a flexible barrier between two dissimilar materials (e.g. fine soil and aggregate) so as to prevent intermixing, and maintain or improve the integrity of both materials. The filtration function refers to the ability to allow movement of water through the fabric face (normal to the plane) whilst retaining the soil on the upstream face. By reinforcement is meant the ability to increase the strength or stability of the soil (or rock) with which the fabric is being used. The drainage function is the ability to transmit water in the plane of the fabric.

The great majority of *geotextiles* are synthetic thermoplastics. Natural fibre geotextiles have a tendency to biograde and are only used where this property is desired (e.g. for temporary erosion control before vegetation growth). All geo-textiles are porous to water flow across their manufactured plane. Large numbers of geotextiles are available commercially for use in road construction, and most of these are described as woven or non-woven. The weaving process for woven multifilament (i.e. regularly-arranged multi-thread) fabrics produces a geotextile that has high tensile strength, a high modulus, low creep, and relatively low water permeability. When arranged in horizontal planes in an embankment, this type is most commonly used where reinforcement is required to resist outward movement of the soil mass; it transfers stress to the soil primarily through friction. Non-woven geotextiles are formed by bonding a loose mat of randomly-arranged filaments, with the type of bonding differentiating the various types within this category. As the pore sizes and tensile strengths of non-woven fabrics are generally lower than woven ones, they are most often used for drainage and separation functions especially if soil filtration is required.

Made from similar polymers as geotextile filaments, *geogrids* are plastics that are formed into a very open net-like configuration. Thus, for example, a series of holes are punched in a thick extruded sheet of high density polyethylene to form 'geosheets', whilst thick webs of multiple filaments are cross-laid and bonded at the crossover points to form 'geowebs'. High-strength geogrids arranged in horizontal planes in a backfill may be used when reinforcement is required to resist the out-ward movement of the soil mass; they transfer stress to the soil through passive soil resistance on transverse members of the grid and friction between the soil and horizontal surfaces of the geogrid.

Figure 10.7 illustrates how geotextiles can be used in the construction of embankments over soft ground. Note that the role of the reinforcing high-strength woven geotextile (Fig. 10.7(a)) is to reduce the risk of a slip circle failure of the embankment and foundation soils, and to reduce lateral spreading and cracking of the embankment; however, it will not significantly reduce embankment settlements associated with time-dependent consolidation of uniformly-weak foundation soils. If the foundation has a non-uniform weakness, e.g. if it contains lenses of clay or peat, the geotextile's reinforcing function is to bridge the weak spots and reduce the risk of localized failure and/or to reduce differential shrinkage. In the separa-tion and filter layer usage (Fig. 10.7(b)) the non-woven geotextile will deform with-out developing high tensile stresses, whilst preventing the soil from passing into the granular drainage layer but allowing the water to do so, thereby preventing a build-up of excess pore-water pressures and a corresponding loss of foundation strength.

10.7.5 Drainage measures

General drainage, other than shallow drains for the removal of surface water, should not be carried out close to a road on an embankment on a soft soil *after* the pavement is constructed. The reason for this is that the construction of new ditches will remove some of the lateral support for the road and the underlying material may be pushed into them, causing deformation of the pavement and blocking of the ditches so that they require frequent maintenance and regrading. In addition, the ditches may have to be extended over long distances to obtain a suitable out-fall, as soft soils are often found in low ground.

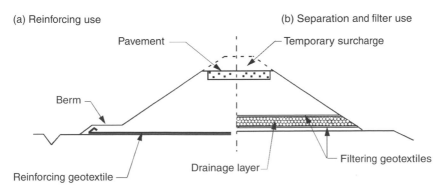

Fig. 10.7 Use of geotextiles in embankment construction over soft ground[13]

Deep land drains have also been used to lower the groundwater, accelerate consolidation, and improve the shear strength of soft soils prior to construction. They are an economical way of treating subsoils with an area of less than 5000 m² and a depth not exceeding 6 m.

10.8 Materials used in embankments

As noted previously in relation to Table 10.2 comprehensive advice regarding materials that may be used in embankment construction in the UK is given in Series 600, Table 6/1 of reference 6. A summary listing of potential applications of selected fill is given in Table 10.3, together with characteristic qualities for satisfactory performance and possible criteria that might be used to assess the quality of each material. The value of local knowledge should not be underestimated in relation to the use of the materials in this table. Engineers in regional areas generally have considerable knowledge regarding the aggregates, rock, and granular materials that are available to them and, from experience, are aware of the roadwork

Table 10.3 Applications and requirements of selected fill in earthworks[20]

Application	Qualities for satisfactory performance	Length of required performance	Possible criteria of quality	Frequently used materials
Replacement for unsuitable material below ground or formation	General suitability as fill	Long-term	Common fill can be used in this context but selected fill is often preferred	Soils suitable as embankment fill. Granular soils as selected fill
Filling below standing water	Underwater strength and permeability	Long-term	Hardness, grading, durability	Rock fill; gravel; hard chalk
Drainage layer	Permeability	Short- or long-term	Hardness, durability (long-term only)	Rock fill; gravel
High strength fill, e.g. to form steeper side slopes	High angle of shearing resistance	Long-term	Hardness, durability	Rock fill; granular soil; hard chalk
Lightweight fill	Low maximum density	Long-term	Bulk density	Pulverized fuel ash
Granular fill to structures	Medium to high strength	Long-term	Grading, plasticity, sulphate and pyrite content (inertness)	Rock fill; granular soil
Capping layer	Strength and bearing capacity	Long-term	Hardness, grading, durability	Well-graded granular soil; rock
Working platform	Strength and permeability	Short-term	Plasticity, grading	Free-draining granular soil; rock

applications in which each can be successfully used or, alternatively, the type of test needed to assess the quality of local materials for a particular application.

The reinforcement of earth, defined[14] as the inclusion of resistant elements in a soil mass to improve its mechanical properties, is a proven, technically attractive, and cost-effective technique for extending the use of soil as a construction material. The process is most commonly used in the construction of retaining walls and abutment structures, to repair slope failures and to retain excavations, and for the stabilization of slopes. The use of reinforced soil enables steeper slopes to be utilized, with a consequent reduction in the width of embankment right-of-way that is required; this is especially useful when an existing carriageway is being widened within a constricted right-of-way. Reinforced soil construction generally causes less traffic disruption than conventional construction techniques, as site preparation requirements are normally minimal and construction can be carried out easily and quickly.

An earth reinforcement system has three main components: metallic or non-metallic (e.g. geotextile or plastic) reinforcements, imported backfill or in-situ soil, and facing elements. Strip reinforcing systems are formed by placing the reinforcing strips in horizontal planes between successive lifts of backfill, whilst grid reinforcing systems comprise metallic bar mats, wire mesh, or polymeric tensile resisting elements arranged in rectangular grids that are also placed in horizontal planes in the backfill. Facing elements are commonly needed to retain fill material and to prevent slumping and erosion of steep faces. Facing elements include precast concrete panels, prefabricated metal sheets and plates, gabions, welded wire mesh, shotcrete, seeded soil, masonry blocks, and looped geotextiles.

The transfer of stress between soil and reinforcements takes place via two basic mechanisms: (1) friction between plane surface areas of the reinforcement and soil, and (2) passive soil bearing resistance (or lateral bearing capacity) on reinforcement surfaces oriented normal to the direction of relative movement between the soil and its reinforcement. Both are active in many earth reinforcement systems, but the relative contribution of each is indeterminate (on the basis of current knowledge). Strip, rod and sheet reinforcements transfer stress to the soil mainly by friction, whilst deformed rod and grid reinforcements transfer stress mainly through passive resistance or through both passive resistance and friction.

In-situ reinforcement systems, i.e. soil *nailing*, normally use steel bars, metal tubes, or other metal rods that resist not only tensile stresses but also shear stresses and bending moments.

10.9 Preparing the subgrade

The first step in preparing the subgrade prior to constructing either a flexible or rigid pavement is to remove all extraneous man-made debris or natural vegetation from the site. This will minimize subsequent construction stoppages that might otherwise arise from the need to remove known obstacles from below and above ground, as well as unwanted soil and vegetative growths.

In rural areas the removal of unwanted surface materials (known as *clearing*) is often a relatively straightforward exercise. The initial approach to clearing the site in such locales may involve removing from the right-of-way trees, hedges, underbrush, other vegetation, man-made construction, and rubbish. Generally, heavy chains drawn between two crawler tractors can be used to remove much of the vegetative matter at ground surface. The debris is removed from the site and disposed of by burying in approved waste-disposal pits or, when appropriate, by burning. Trees, hedges, and appropriate shrubbery that are outside the road construction width and do not interfere with construction traffic or the safe operation of vehicles on the finished road, may be saved within the right-of-way; however, they often need to be protected from construction vehicles, e.g. using temporary fences whilst the road is being built.

Grubbing is the term used to describe the removal of the stumps and roots of trees, bushes and hedges. Blasting may be required in some instances to remove tree roots that cannot be dug out by backacters. Holes left after the grubbing process should be backfilled with acceptable material and compacted prior to compacting the subgrade as a whole, so as to avoid interference with subsequent compaction and/or possible settlement. The objective is to ensure that the subgrade is sufficiently dense and uniformly stable to withstand the stresses transmitted to it via the pavement.

Topsoil is the general term used to describe surface soils that contain enough humus to support plant growth without resort to artificial fertilization. Using appropriate equipment, the topsoil should be stripped to a depth that will ensure the removal of all vegetative matter within the roadway width; if left, it will eventually decay and leave voids that may be the cause of subgrade or low-embankment settlement. The topsoil is normally stored in soil heaps adjacent to the site and reserved for subsequent re-use, e.g. in verges, on embankment berms, and on the slopes of cuttings. The soil heaps should be shaped to shed rainwater, and located where they will not obstruct the natural drainage channels.

If the site contains man-made construction, as is likely in built-up areas, it will need to be demolished and removed. Backacters fitted with appropriate attachments can be used to demolish smaller structures, superficial obstructions, and flexible pavements, and to load lorries, etc. Bulldozers can be used to break up masonry and concrete and push it into stockpiles. Special equipment and techniques may be required to demolish, to the depth prescribed, existing concrete pavements and significant structures and obstructions (e.g. tall buildings, deep foundations, underground storage facilities, cellars, etc.) that are typically found in right-of-ways in urban areas. Holes remaining after the removal of the debris should be properly cleaned out and filled. To permit free drainage, holes should be made in at least 10 per cent of the area of slabs and basements that the specification says need not be removed, but which are liable to hold water. Disused soil and surface water drains, sewers, cables and ducts together with any bed or haunch within 1 m of the formation should be removed.

A major (and expensive) concern in road construction in both urban and rural areas, but most particularly in built-up locales, relates to the installation, relocation and/or maintenance of *public utility services*. Typically these services may include above- and/or below-ground electrical and telecommunication mains, and underground pipes carrying sewage, stormwater, water, gas and oil. The problems of relocation may be compounded by the fact that some of the underground services are the heritage of bygone days, so that plans indicating their exact locations are inaccurate or no longer in existence. Nonetheless, good site planning ensures that existing services are kept operating during the road construction period; if they are not, there will normally be significant financial and political reverbrations.

It is appropriate to note here that new services which lie within the right-of-way should be installed prior to construction rather than after. Whilst the use of common trenching for new and/or relocated underground utilities is commonly proposed, it is difficult to implement. Whilst it seems logical, its adoption is compromised by differing professional views relating to access requirements for maintenance and emergencies, and to the need for minimum clearances between the lines of particular services. Ideally, as many services as possible should be placed under footways and verges rather than under carriageways. In practice, however, stormwater drains and foul sewers, i.e. the lowest placed of the utilities (in case of leakage), are often put under the carriageway, as they rarely need to be uncovered for repairs. Water mains should always be beneath the depth of frost penetration for the site locale; they tend to be put on the high side of an urban street to minimize flood damage to adjacent properties, and on the lower side of a rural road to minimize scouring, in the event of a burst water main. All electric power lines should be above the water mains; they should also be horizontally away from them so as to avoid corrosion resulting from the water mains being within the electric field of the power lines. Communication mains, which may contain ducts for telephone, telegraph, and emergency and police lines, need also to be away from electric power mains and above the water mains.

In heavy clay soils the entire plastic B-horizon should be removed so that the C-horizon can be used as the foundation soil.

If the subgrade soil comprises fill material the top 300 mm of the embankment, and preferably the top 750 mm, should be compacted to ensure that its final state is that used in the design of the pavement.

If the subgrade soil in a cut section is relatively coarse and loose it may be quite easy to compact it to a substantial depth. However, if the soil is a heavy clay a light roller may only be used to smooth the subgrade surface as such undisturbed clays cannot usually be compacted beyond their natural state, i.e. the exposed clays will tend to gain moisture content and rolling with a heavy roller will only result in the exudation of moisture with a resultant remoulding and weakening of the soil.

If the road is being constructed through a rock cutting a 100–150 mm thick layer of suitable material should be installed between the pavement and the rock (a) to level the irregularities that occur in the rock surface during blasting, and (b) to provide a cushion for the transition between the rock and the adjacent soil. However, before the cushion is laid grooves should be cut in the foundation rock to allow for the lateral drainage of any water accumulations then or in the future. When the subgrade passes from cut to fill, especially if the change is from rock to soil, there should be no abrupt change in the degree and uniformity of compaction; rather, a soil layer should be installed to allow any differential effects to occur gradually, e.g. at the 'point' of change the transition layer depth should be at least 1 m, and feathered back to about 150 mm over a distance of around 14 m on the rock.

After the subgrade has been compacted its surface will still be fairly rough, and it must be graded to the desired formation shape. The *formation* is normally reflective of the final carriageway shape (a) to help ensure that each subsequent pavement layer is constructed to its design thickness and (b) to allow water that gets into the subgrade to drain away. This final shaping, which is usually done by a motor grader, should be delayed until the subbase (or capping layer) is about to be laid. When the grading is finished any loose material should be taken away and any surface irregularities removed by light rolling with smooth-wheeled or rubber-tyred rollers. Preferably, no traffic should be allowed onto the formation prior to the placement of the next pavement layer; this should be carried out as soon as possible, to minimize the danger of rain softening the subgrade.

10.10 Unbound capping and subbase layers

The recommended[15] British practice in relation to *subbases* in all flexible and flexible composite pavements is to use 225 mm and 150 mm of Type 1 granular material (see Table 6.2) on subgrades with California Bearing Ratios of 5–15 per cent and >15 per cent, respectively, for pavements with design traffic loadings (one direction) of 100 to 10 000 cv/d. Type 1 material, which is composed of crushed rock or concrete, or well-burnt colliery shale, and may contain up to 12.5 per cent by mass of natural sand that passes the 5 mm BS sieve, must not be gap-graded; also, the material passing the 425 micron sieve must be non-plastic and the 10 per cent fines value must be more than 50 kN. In the case of less heavily-trafficked roads (<400 cv/d), Type 2 subbase material (Table 6.2), which also includes gravel and natural sand, may be used instead of Type 1, provided that its plasticity index does not exceed 6, its 10 per cent fines value is also greater than 50 kN, and its CBR is 30 per cent or more.

Both types of subbase material are often brought on-site by truck, spread by a motor grader, and compacted with vibrating rollers. Type 1 material is assumed to be free-draining and no moisture content is specified for its compaction. Type 2 is compacted to 5 per cent air voids at its optimum moisture content as determined from the vibrating hammer method of test[15].

With the 'method'-type specification (rather than the 'recipe'-tyre specification) the number of passes required to compact Types 1 and 2 materials in both 110 mm and 150 mm layers are the same as those specified in Table 6.5. However, the maximum compacted thickness permitted for Types 1 and 2 materials is 225 mm (which is different from that in Table 6.5), and the numbers of passes specified with the various compaction machines for this thickness is given in reference 6.

On subgrades weaker than 5 per cent it is not uncommon in the UK to also use an additional *capping layer* of cheaper granular material to a lower-quality specification than that used in the subbase. Common practice in this respect is to use 150 mm of either Type 1 or Type 2 subbase material (depending upon the design traffic) on top of 350 mm of capping if the subgrade CBR is more than 2 and less than or equal to 5 per cent, or on top of 600 mm of capping if the CBR is less than or equal to 2 per cent. The capping layer typically comprises material with a minimum CBR of 15 per cent, e.g. rock or as-dug gravel found on or adjacent to the site, or waste or recycled materials obtained at low cost. Whilst capping material stabilities are less than those required for subbase usage, their gradation limit requirements are not dissimilar.

When the subgrade CBR is less than about 2 per cent the contractor may have difficulty in properly placing and compacting the capping material, especially if the subgrade has a high moisture content, due to excessive rutting of the subgrade and the intermixing of the softer soil with the imported material. In such instances consideration should be given to placing a geotextile or geogrid fabric on top of the subgrade soil to separate it from the capping layer; this will also allow construction equipment to gain easier access to the site.

10.11 Unbound roadbases

Roadbases in major roads in Britain are now normally bound with bitumen or cement. In car parks and on lightly-trafficked rural roads, however, it is not uncommon for unbound macadam roadbases (and subbases) to be used in pavements, e.g. crusher-run, dry-bound, or pre-mixed water-bound (wet-mix) macadams. Aggregates used in these macadam roadbases are most usually non-flakey crushed rock or gravel or crushed slag. All primarily rely for their strength and resistance to deformation upon the interlocking of individual crushed particles and upon the friction between the rough surfaces in contact.

Because of the dependence placed upon particle interlock and friction, the edges of an unbound pavement must be able to resist the lateral thrust of the macadam aggregate as it is being compacted. This can be facilitated by laying steel or timber sideforms alongside the edges or by constructing and compacting temporary (small) earth embankments and using the blade of a grader to form vertical side supports. Compaction is normally begun at the edges and carried out parallel to the centreline, so as to lock the outer stones, and then extended progressively towards the centre. In the case of superelevated curves, however, compaction is initiated at the lower edge and continued progressively toward the higher edge. In either case rolling is continued until there is no obvious aggregate creep ahead of the roller.

These macadam materials should be protected from the weather before compaction as wetting or drying obviously changes the moisture content, and may cause particle segregation. The materials should be spread evenly at the site, preferably by a bituminous paving machine or by a box spreader. If smooth-wheeled or rubber-tyred rollers are used for compaction, then the compacted layers should not be more than about 150 mm deep; a good rule-of-thumb says that the compacted layer thickness should not be more than 1.5 times the largest aggregate in the layer if good particle interlock is to be achieved. If vibratory rollers are used then single layers of up to 225 mm compacted thickness can be satisfactorily laid.

10.11.1 Crusher-run macadam

This macadam consists of natural stone aggregate that is generally passed through two stages of crushing to give a graded material of 50 mm or 75 mm maximum size. The material is then transported to the site, moistened (to <1 per cent) and compacted in layers up to 225 mm thick.

Whilst crusher-run macadams are relatively cheap, they have the disadvantage that particle segregation can occur en route from the crushing plant to the site. If the travel distance is long, the result of the segregation may be that the material has excessive fines in some pavement locations when spread, and severe shortages in others. Also, it is difficult to get high dry densities with this material, and subsequent traffic-induced deformations are not unlikely in the early life of the pavement. Consequently, crusher-run macadam bases are now used only in pavements where it is known that the traffic volumes and loads will be low and there is good local knowledge of how to work with this material.

10.11.2 Dry-bound macadam

With dry-bound macadam the constituent crushed stone or slag aggregates are transported to the site in two separate sizes; a coarse aggregate that is normally either 37.5 mm or 50 mm nominal size, and fine screenings graded from 5 mm to less than 10 per cent passing the 0.075 mm sieve. In this way aggregate segregation during stockpiling and transporting is minimized, and a more uniform construction is obtained at a relatively moderate cost.

At the site the dry coarse material is spread to a uniform thickness of 75 to 100 mm and preliminary rolling (two passes) and shaping is then carried out with an 8–10 t smooth-wheeled roller. After correcting for any depressions or projections, the surface is progressively 'blinded' with dry fine aggregate using either a vibrating roller or plate compactor to ensure that the voids between the coarse aggregate particles are filled ('choked'), thereby increasing the layer's dry density, increasing its internal friction, and maintaining the interlock. The blinding process is continued until no more fines are accepted, after which the surface is broomed to remove excess fines, leaving the coarse aggregate standing 3–6 mm proud. After the side supports have been repaired and built-up as necessary, additional macadam layers are laid as before until the desired thickness is achieved.

Whilst strong pavements can be obtained with properly constructed dry-bound macadam bases, their main drawback relates to the fact that construction *must* be carried out in dry weather. Whilst moisture has little effect on the compaction of the coarse aggregate, it is practically impossible to fill the voids by vibration, and achieve high dry densities, if the fines are damp or wet. Because of this moisture problem, as well as the fact that the construction method is relatively slow and labour-intensive, use of this type of base is now limited to small projects, and a less weather-susceptible unbound macadam, e.g. wet-mix, is usually preferred on larger projects.

10.11.3 Premixed water-bound (wet-mix) macadam

Premixed water-bound macadam, known as *wet-mix*, is prepared by mechanically mixing a measured amount of water with the graded aggregate to ensure that it is at the optimum moisture content for maximum dry density (determined by the vibrating hammer test[16]) at the start of field compaction. Its grading is similar to that for Type 1 subbase material (compare Table 10.4 with Table 6.2), and it has the same 10 per cent fines value; however, it has a smaller maximum aggregate size. The graded aggregate is generally derived from a combination of individual sizes, but suitable crusher-run material can also be used.

Neither rain nor sunshine nor drying winds should be allowed to affect the moisture content of the wet-mix between the times of mixing and compaction, e.g. it

Table 10.4 Grading limits for wet-mix materials

BS sieve size (mm)	Percentage by mass passing
50	100
37.5	95–100
20	60–80
10	40–60
5	25–40
2.36	15–30
0.600	8–22
0.075	0–8

should be sheeted if it has to be transported significant distances by truck. Relatively little segregation should take place whilst in transport or being spread because of the cohesive films of moisture about the particles. The wet-mix is normally spread by a paving machine and compacted immediately with a vibrating roller. High dry densities are more easily obtained as the moisture films allow the particles to slide over each other into their interlocking positions during compaction.

Wet-mix requirements in relation to use in roads in the UK are described in reference 6.

10.12 References

1. Leake, G.R., Introduction to computer-aided design of junctions and highways, Chapter 21 in O'Flaherty, C.A. (ed.), *Transport Planning and Traffic Engineering*. London: Arnold, 1997.
2. O'Flaherty, C.A., Geometric design of streets and highways, Chapter 19 in O'Flaherty, C.A., *Transport Planning and Traffic Engineering*. London: Arnold, 1997.
3. Kindberg, J., Compaction: The bottom line, World Highways, Jan/Feb, 1992, pp. 59–61.
4. Harris, F., *Modern Construction and Ground Engineering Equipment and Methods*, 2nd edn. London: Longman, 1994.
5. BS 6031:1981, *Code of Practice for Earthworks*. London: British Standards Institution, 1981.
6. Highways Agency, *Specification for Highway Works*. London, HMSO, 1998.
7. Morris, P.O., *Compaction – A Review*, ARR Report 35. Vermont South, Victoria: Australian Road Research Board, 1975.
8. Lewis, W.A., *Investigation of the Performance of Pneumatic-tyred Rollers in the Compaction of Soils*, Road Technical Paper No. 45. London: HMSO, 1959.
9. Jeffries, T.G., Fill compaction by vibrating rollers: Theory and practice, *Highways and Road Construction*, 1973, **41**, No. 1767, pp. 33–6.
10. *Construction of Embankments*, NCHRP Synthesis of Highway Practice No. 8. Washington DC: Highway Research Board, 1971.
11. PIARC Technical Committee on Earthworks, Drainage and Subgrade (C12), *Lightweight Filling Materials*. Paris: PIARC – World Road Association, 1997.

12. Lewis, W.A., Murray, R.T. and Symons, R.F., Settlement and stability of embankments constructed on soft alluvial soil, Part 2, *Proceedings of the Institution of Civil Engineers*, 1975, **59**, pp. 571–93.
13. Hudson, K. and East, G.R.W., *Geotextiles*, Transit New Zealand Research Report No. 6. Wellington, NZ: Transit New Zealand, 1991.
14. Mitchell, J.K. and Villet, W.C.B., *Reinforcement of Earth Slopes and Embankments*, NCHRP Report 290. Washington, DC: Transportation Research Board, 1987.
15. *Structural Design of New Road Pavements*, Departmental Standard HD 14/87. London: Department of Transport, 1987.
16. BS 5385: Part 1, *Testing of Aggregates: Compactibility Test for Graded Aggregates*. London: British Standards Institution, 1980.
17. Kindberg, J., Persson, I. and Lakey, B., Compaction; making the choice, *World Highways*, 1992, Mar/April, pp. 75–78.
18. Anon., Compaction: Squaring the circle, *World Highways*, 1992, July/Aug, p. 62.
19. Lay, M., A history of compaction, *Routes*, 1994, No. 284, pp. 59–64.
20. Perry, J. and Parsons, A.W., *Assessing the Quality of Rock Fill: A Review of Current Practice for Highways*, Research Report 60. Crowthorne, Berkshire: Transport and Road Research Laboratory, 1986.

CHAPTER 11

Premixed bituminous-bound courses: standard materials

M.J. Brennan and C.A. O'Flaherty

Since the 1960s a substantial body of research knowledge has been built up in relation to the design and use of bituminous-bound materials in road pavements. This has provided the impetus for the current trend towards the use of mechanical test methods in bituminous mix design to measure particular properties of mixes and ensure that they meet desired performance-related specifications. Nonetheless, current practice in many countries is still biased towards the use of standard 'recipe-type' specifications that are based on historical technical and commercial experiences. Indeed, the recipe-type specifications may well continue to be relied on for minor roads where 'hi-tech' performance-based specifications are not required.

11.1 Advantages and disadvantages of standard 'recipe-type' specifications

Standard recipe specifications prescribe mixes in terms of their aggregate size and grading, bitumen type and content, and methods of mixing, laying and compacting, and do not normally require any mechanical testing of the finished product. Their main advantages are as follows:

(i)   they are widely accepted and they are easy to use;
(ii) they are based on mixes of established composition that have been proven in practice for particular purposes;
(iii) experienced road engineers have no difficulty in selecting suitable mix specifications to meet their design needs;
(iv) the supplier at the mixing plant is normally well-organized to meet recipe needs; and
(v)   the supervising engineer can easily test for compliance with the specification at various stages in the construction process, thereby ensuring good quality control.

Recipe specifications are criticized for the following reasons:

(i)   the performance of a mix under traffic can depend as much upon workmanship as upon composition, and it is hard to prescribe workmanship in a recipe;
(ii) if the road environment within which the recipe is being applied is different from whence it was derived, modifications may be required and there is no guidance as to what these should be or their likely impact;
(iii) if the quality control checks show that the material supplied does not meet the recipe, the seriousness of the variation upon performance cannot be evaluated even if the transgression is a minor one; and
(iv) recipes usually prevent the use of cheap locally-available pavement materials, thereby resulting in increased construction costs.

Overall, it is probably true to say that it is now generally considered that, at a time of rapid change, the disadvantages of standard recipe specifications outweigh the advantages. Consequently, there is now a tendency to move away from their use in, especially, major roadworks.

11.2 Harmonization of European standards

As noted elsewhere, major changes are currently occurring in specifications as a result of legislation associated with greater European union.

The Comité Européen de Normalisation (CEN) is the association of national standards organizations of the European Union (EU) and of the European Free Trade Association (EFTA). The CEN has responsibility for developing common European Standards (EN), so as to facilitate trade and communication between member states; its national members are therefore committed to conferring the status of a national standard on approved ENs and to withdraw existing standards that conflict with them.

The standard recipe-type specifications used in most EU countries will be affected by this process. Indeed, it is likely that all recipe specifications will eventually be superseded by performance-based ones, as it is intended to include performance parameters in the specifications, which will put limits on such mix properties as dynamic stiffness, deformation, and fatigue[1]. However, pending agreement on fundamental methods of testing for these properties, the use of recipe specifications will continue. In the interim, it can be expected that there will be a continued emphasis on a rationalization and harmonization of standard specifications used in different countries.

It should also be repeated here that CEN has also agreed that the generic name 'asphalt' will be used to describe all bituminous mixes used in pavements and that local terminology, e.g. 'coated macadam' and 'bituminous materials', will be phased out. Also, the terminology currently used in the UK to describe the various pavement layers (see Fig. 9.3) will be harmonized so that what is now known as the 'wearing course' will, in future, be called the 'surface course', today's 'basecourse' will be termed the 'binder course', and the 'roadbase' will become the 'base course'. However, in practice, it will be some time before the current terminology and British Standards will be superseded and, hence, to avoid confusion and to facilitate interpretation of current practices, the traditional UK terms will be used here.

11.3 Mechanisms by which asphalts and coated macadams distribute traffic stresses

The apparent complexity of the number and variety of recipe compositions used in the UK and in other countries can seem daunting to a person introduced to them for the first time. Consequently, the emphasis in this chapter will be upon explaining the main features of particular types of mixes, and why and where they are used. Before doing so, however, it is most important to understand the main differences in the stress distribution mechanisms of asphalts and coated macadams, which arise from their aggregate gradings.

Asphalt mixes tend to have an intermediate range of particle sizes missing from their total aggregate gradings. A typical asphalt grading might consist of a single-sized coarse aggregate blended with fine aggregate and filler to produce a gap-graded material, or a graded coarse aggregate blended with fine aggregate and filler to give a semi-gap-graded mix. Because they are not well graded, and therefore lack good aggregate interlock, asphalt mixes mainly derive their strength and stability from, and distribute the applied wheel stresses through, the fines–filler–binder mortar.

Both the filler and binder contents of asphalt mixes are relatively high. A large amount of a high-viscosity binder must be available to coat the extensive surface areas of filler and fine aggregate particles, and a high filler content is needed to stiffen the binder, so that the fines–filler–binder combination, when hardened, produces a mortar with a high stiffness modulus which is able to resist wheel-track deformation. Generally, the role of the coarse aggregate in an asphalt mix is to bulk the material and to provide additional stability to the hardened mortar; in this context the coarse particles can be seen as 'floating' within the mortar.

As the binder used in the mortar must be relatively hard, asphalts must be hot-mixed, hot-laid and, where compaction is required, hot-compacted. Compacted asphalt layers have low air void contents, are nearly impervious to the entry of water, and are very durable.

Carriageways need to have a coarse surface texture if they are to have good skid resistance. Because the high mortar content of an asphalt results in a fine-textured surfacing, it is very often necessary for a surface dressing of pre-coated chippings to be applied to a wearing course composed of an asphalt material to improve its skid resistance.

Whilst *coated macadam* materials tend to have higher coarse aggregate contents and to be more continuously-graded than asphalt materials, the major difference between the two is that with the coated macadam mixes it is aggregate friction and interlock that generally controls the rate of material deformation whereas with asphalt materials it is the stiffness of the mortar. In other words, the strength and stability of a coated macadam material is primarily derived from, and stresses are distributed through, particle-to-particle contact, inter-particle friction and aggregate interlock. With coated macadams, therefore, it is important that the aggregate particles be sufficiently tough to ensure that they do not break down under rollers when they are being compacted, or under the subsequent repeated impact and crushing actions of traffic.

The role of the binder in coated macadam mixes is mainly to lubricate the aggregate particles during compaction, whilst also acting as a bonding and water-proofing agent when the pavement is in service. However, with dense coated macadam materials, the filler also causes an increase in the binder viscosity; this reduces the risk of the binder flowing from the aggregate during transport, as well as helping the binder to fill small voids.

As the aggregate surface area to be coated by the binder in a coated macadam mix is relatively small (by comparison with an asphalt material), the binder content can also be fairly small. As the binder's primary role is that of a lubricant, a low-viscosity binder can normally be used (except with heavy-duty dense macadams), and macadams can therefore be laid at temperatures that are lower than for asphalts and they are easily compacted. Some macadams can be warm- or cold-laid, depending upon the binder used.

The voids content of a coated macadam material should be within the specified range after compaction. This is especially important in the case of a wearing course material which is likely to be susceptible to fretting if the voids content is too high and to deformation if it is too low. A high voids content makes the binder more susceptible to accelerated ageing (i.e. hardening due to, mainly, oxidation) and, due to the greater permeability of the material, more vulnerable to loss of adhesion with the stone. Ultimately, too high a voids content can lead to brittle fracture of a wearing course at low pavement temperatures.

Compared with asphalts, coated macadam materials used in road pavements generally: (1) have gradations that tend more towards those given by Fuller's curve (Equation 5.14); (2) utilize lower amounts of filler and fine aggregate and smaller quantities of softer binder; (3) have higher void contents; (4) are more permeable and less durable; and (5) cost less to produce and lay.

It is also important to appreciate that many materials have stress-distribution characteristics that overlap those described above in relation to the classical asphalt and coated macadam materials. With some bituminous materials the mortar stress-distribution mechanism predominates, albeit a considerable component of the resistance to deformation is also provided via the stone-to-stone contact mechanism; with others the reverse is true. For example, *asphaltic concrete*, which originated in the USA and is now widely used throughout the world, and *stone mastic asphalt*, which was developed in Germany, and is now used throughout northern Europe and in the USA, are classical examples of mechanically-laid bituminous materials that rely mainly for their stability on aggregate-to-aggregate contact, but have relatively low void contents after compaction, and gain much durability and some strength from their mortars.

Figure 11.1 illustrates the differing compositions/stress distribution mechanisms of mastic asphalt, asphaltic concrete and pervious macadam.

11.4 Standard 'recipe-type' specifications currently used in the UK

Table 11.1 lists the main bituminous materials covered by British Standards and the uses to which they are put in road pavements, whilst Table 11.2 provides a

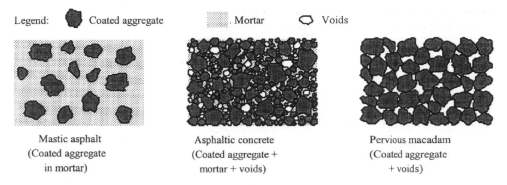

Fig. 11.1 Diagrammatic representations of wearing course materials

Table 11.1 The main categories and uses of recipe-type premixed bituminous materials currently covered by British Standards[2-5]

Type	Description	Usage
Asphalt	Mastic asphalt	Wearing course
	Hot rolled asphalt	
	-low stone (35% or less)	Wearing course
	-high stone (50% or more)	Wearing course, basecourse and roadbase
Coated macadam	Dense tar surfacing	Wearing course
	Dense bitumen macadam	Wearing course, basecourse and roadbase
	Close- and fine-graded macadam	Wearing course
	Medium-graded macadam	Wearing course, single course
	Open-graded macadam	Wearing course, basecourse
	Pervious macadam	Wearing course

summary assessment of their compositions. Hot rolled asphalt, dense bitumen macadam, and pervious macadam (now known as 'porous asphalt') are the main recipe-type premixed bituminous materials used in trunk roads in the UK.

11.4.1 Mastic asphalt

The use of mastic asphalt on roads in the UK dates back to the 19th century; the records show that a number of the major streets in London were paved with mastic asphalt in 1874. The mastic asphalt that is used today at specialized road pavement locations in the UK is most commonly a hot-mix mortar of fine limestone aggregate and a 50/50 mixture of refinery bitumen and fluxed Trinidad lake asphalt (15–25 pen grade), with added clean coarse aggregate. No more than 55 per cent

Table 11.2 General compositions of various standard recipe-type bituminous mixes for wearing courses (based on reference 6)

	Macadams:					Asphalts:	
	Open- and medium-graded	Pervious	Close-graded	Dense	Fine-graded	Rolled	Mastic
% coarse aggregate	High	High	Medium	Medium	–	Low	Low
% fine aggregate	Low	Low	Medium	High	High	High	High
% filler	Low	Low	Medium	Medium	Medium	High	High
% binder	Low	Low	Low	Medium	Medium	High	High
Binder grade	Soft	Soft/ medium	Medium	Medium /hard	Soft	Hard	Very hard
% voids	High (20%+)	High	Medium (10–20%)	Low (2–10%)	Medium	Low	Nil (<1%)

and no less than 45 per cent of the fine aggregate must pass the 75 micron sieve and no more than 3 per cent may be retained on the 2.36 mm sieve. The coarse aggregate (which must have an aggregate crushing value (ACV) no greater than 28) must pass the 14 mm BS sieve and be retained on the 10 mm sieve.

Features of mastic asphalt are that (i) it contains a relatively small amount of coarse aggregate and a relatively high content of a hard bitumen, (ii) the volume of the virtually-voidless mortar formed by the fines, filler and bitumen substantially exceeds the volume of the voids in the skeleton of the remainder of the aggregate (Fig. 11.1), and (iii) it can be poured whilst hot and requires no roller compaction after being laid.

The mastic mix is applied at temperatures between 175 and 230°C and, usually, hand-trowelled onto the pavement using wooden floats to form a very dense, virtually impermeable, durable wearing course with less than 1 per cent air voids. The layer is typically 40 to 50 mm thick in roads. After compaction from traffic, mastic asphalt develops a very fine-textured surface that has poor resistance to skidding. Thus, 20 mm or 14 mm pre-coated chippings, with a flakiness index of not more than 25, are normally applied to the running surface whilst the asphalt is still sufficiently plastic for the chippings to be partially, but firmly, embedded under rolling.

Because of the hand-laying, mastic asphalt is usually very costly; hence, its use in road pavements is now mostly confined to highly-stressed locations where the need for a maintenance-free long-life surfacing is overriding, e.g. in tunnels, at busy bus stops and loading bays, and on bridge decks. Also, as repair work can be easily carried out and surfacings restored to desired levels without compaction, 20 to 30 mm thick layers of mastic asphalt are sometimes used in urban areas in footpaths which have public utilities below the surface.

It might also be noted here that *Gussasphalt* (see Table 11.4), which is a machine-laid graded-aggregate mastic asphalt with a more substantial amount of coarse aggregate, is used as a wearing course in northern Europe, especially in autobahn and ring road pavements in Germany.

11.4.2 Hot rolled asphalt

The first British Standard for rolled asphalt was published in 1928 and since then the recipe has been revised on many occasions, on the basis of the considerable experience gained over the years regarding its use on different types of road under varying traffic and weather conditions. Today's hot rolled asphalt (HRA) surfacing is a dense, low air void content (typically 3–6 per cent), nearly impervious, durable, gap-graded material composed of fines, filler and bitumen (which form a mortar) in which coarse aggregate is dispersed. It contains little medium-sized (2.36 mm to 10 mm) aggregate and has a relatively high binder content. The primary function of the coarse aggregate in the mixture is to extend the bulk of the mortar and lower the asphalt's cost. Whilst the coarse aggregate contributes to the stiffness of the mix, it is the quality of the fines–filler–bitumen mortar that mainly controls the hot rolled asphalt's mechanical stability and stress distribution qualities.

Crushed rock, gravel or slag are permitted as coarse aggregate (>2.36 mm) in HRA wearing courses and basecourses. The maximum aggregate size in either layer is controlled by its thickness, i.e. the usual practice is to limit the maximum size of stone to one-third to one-half the thickness of the compacted layer as, if too large a size is used, the aggregate will not compact satisfactorily under the roller.

HRA wearing courses can have coarse aggregate contents of 0, 15, 30, 35 or 55 per cent. The 0 per cent coarse aggregate mixes, which are mainly used as footpath and playground surfacings, are often termed *sand carpets*.

The basecourse mixes have high (50 or 60 per cent) coarse aggregate contents. They are typically laid 50 mm or 55 mm thick beneath an HRA wearing course, and 60 mm thick beneath a porous asphalt wearing course, in a trunk road.

The need for strong roadbase materials that would not crack gave rise to hot rolled asphalt (and dense bitumen macadam) being used in roadbases in flexible pavements. HRA roadbase layers typically have 60 per cent coarse aggregate and use lesser amounts of bitumen due to having lower fine aggregate and filler contents; they have large maximum stone sizes (e.g. 28 mm or 40 mm) as they can be laid in layers that are up to 150 mm thick.

Whilst crushed rock and crushed slag fines are used in hot rolled asphalt, the fine aggregates that were most commonly used until relatively recently were natural sands, i.e. the grading envelopes used in various HRA recipes were historically developed to maximize the use of commonly available sands. Sand particles are normally fairly rounded and provide a more workable bituminous mix on site; crushed rock fines, which have been more used in recent years, are more angular and elongated and give a harsher and more stable mix, albeit at the expense of some workability. The gradings of the sand and rock fine aggregates must be such that no more than 5 and 10 per cent by mass, respectively, is retained on the 2.36 mm sieve, and no more than 8 and 17 per cent, respectively, can pass through the 75 micron sieve. Gap-graded hot rolled asphalt wearing courses containing sand fines are usually termed Type F (for Fine). A Type C (for Coarse) material, which is normally associated with the use of crushed rock or slag fines, is characterized by a coarser fine aggregate grading.

At least 85 per cent of the total filler material must pass the 75 micron sieve. Most of the filler is in the form of added limestone dust, hydrated lime or Portland cement, but some is also provided from the coarse and fine aggregate components. The filler stiffens the bitumen so that the fine aggregate particles are coated with, and the fine aggregate voids are filled with, a filler–bitumen mixture that is stronger and less susceptible to temperature variations than the bitumen alone.

Refinery bitumens (35, 50, 70 or 100 pen) are most commonly used in rolled asphalt, but a Trinidad lake asphalt–refinery bitumen mixture may also be used; the inclusion of lake asphalt has the effect of enhancing the skid-resistant properties of the wearing course. The grade of binder selected for use in hot rolled asphalt mixes is mainly decided by the traffic and weather conditions. Experience has shown that a 50 pen bitumen is generally suitable for most trunk road (including motorway) wearing courses, basecourses and roadbases. A hard (35 pen) bitumen is commonly used on roads with very heavy traffic and at severely stressed local sites such as bus stops, whilst softer (70 or 100 pen) bitumens are used on more lightly trafficked roads. Other things being equal, softer bitumens are used in the UK's cool northern climes vis-à-vis the warmer south.

HRA wearing courses with high filler and fines contents need high bitumen contents, because of the greater surface areas that have to be coated with binder. Conversely, the bitumen contents of HRA basecourses and roadbases are usually less than those of wearing courses, as they have lesser fines and filler contents. Lower binder content mixes are also associated with surface courses at sites under heavy traffic and with higher ambient air temperatures. Higher binder content mixes are used in surfacings at locations that do not dry out easily when wetted and/or that carry light traffic, i.e. where durability, not stability, is the main design criterion. If an excessive bitumen content is used with a given aggregate grading, the result, after compaction, is a low voids content material, with an associated 'over-filling' of the aggregate skeleton and a reduction in deformation resistance.

Table 11.3 shows some selected compositions of hot rolled asphalt mixes extracted from those specified in BS 594. In this table, a mix designated as 30/14 has 30 per cent coarse aggregate and its nominal coarse aggregate size is 14 mm.

HRA wearing courses containing 35 per cent or less of coarse aggregate have a relatively smooth surface texture upon completion of compaction. Hence, a prescribed surface dressing of precoated 20 mm or 14 mm chippings is usually added immediately after the initial compaction by the screed of the asphalt paver, to increase its resistance to skidding after final compaction. This process requires careful judgement, however, to achieve the correct combination of rate of spread and amount of rolling[7], i.e. an excessively high spread rate leads to a quick loss of chippings under traffic whilst too much rolling results in excessive embedment of the chippings in the asphalt and a consequent loss of texture depth. An HRA wearing course with a high (say, 55 per cent) coarse aggregate content has sufficient 'natural' surface texture for it to be used without a surface dressing in more lightly trafficked road pavements, provided it meets specified criteria relating to aggregate abrasion, polished stone, and 10 per cent fines values.

Next to mastic asphalt (which is now little used) hot-mixed hot-laid rolled asphalt is deemed by many engineers to be the most stable and durable surfacing mix used in the UK today. Experience has shown that HRA performs satisfactorily, providing adequate deformation resistance, skid resistance and durability for a wide range of traffic and climatic conditions[17]. A correctly chosen and properly

Table 11.3 Some hot rolled asphalt mix compositions (extracted from reference 3)

Usage	Basecourse/Roadbase		Wearing course*
Designation	50/20	60/20	30/14
Grading:			
% by mass passing BS sieve			
28 mm	100	100	
20	90–100	90–100	100
14	65–100	30–65	85–100
10	35–75	–	60–90
6.3	–	–	–
2.36	35–55	30–44	60–72
0.600	15–55	10–44	45–72
0.212	5–30	3–25	15–50
0.075	2–9	2–8	8–12
% aggregate passing 2.36 mm and retained on 0.6 mm			
Maximum	–	–	14
Bitumen content:			
% by mass of total mix ± 0.6%			
Crushed rock or steel slag	6.5	5.7	7.8
Gravel	6.3	5.5	7.5
Blastfurnace slag of bulk density:			
1440 kg/m^3	6.6	5.7	7.9
1280	6.8	6.0	8.1
1200	6.9	6.1	8.2
1120	7.1	6.3	8.3
Layer depth (mm):			
Thickness	45–80	45–80	40

*Type F mix

constructed HRA surfacing used in a trunk road can be expected to have a maintenance-free life of more than 10 years, provided that the pavement otherwise remains sound. Wearing courses and basecourses of hot rolled asphalt are sufficiently stiff and durable to be considered, structurally, as part of the combined bituminous layer thickness in flexible and flexible composite pavements in trunk roads[7].

The aggregate gradings used in HRA wearing courses, basecourses, and roadbases must be prepared according to the standard recipe specifications as prescribed in the British Standard. In the case of the basecourses and roadbases, the bitumen content must be as prescribed by the recipe; however, in the case of the wearing courses, the road engineer has the choice of either using a recipe bitumen content or of selecting a suitable binder content on the basis of laboratory test results (see Chapter 12).

The main usage of hot rolled asphalt is in the surfacings and roadbases of trunk roads (including motorways) and, to a lesser extent, city streets[8]; however, because of its superior qualities, HRA is also used in many other roads. Whilst hot rolled asphalt layers are dense, have good resistance to fatigue cracking, and are easier to

Table 11.4 Effectiveness of surfacing treatments in meeting desired properties[16]

(a) Materials and their ability to meet desired properties

Material#		Desired property:										
		Suitability for re-profiling	Deformation resistance	Resistance to cracking	Spray reducing	Noise reducing	Skid resistance	Texture depth	Initial cost	Durability	Speed of construction	Quality of ride
Thick wearing course	Rolled asphalt	✓✓✓	✓✓✓ (✓✓✓*)	✓✓✓✓	✓✓	✓✓	✓✓✓✓	✓✓✓✓	✓✓✓	✓✓✓✓✓	✓✓	✓✓✓✓
	Porous asphalt	✓✓✓	✓✓✓✓✓	✓✓✓✓	✓✓✓✓✓	✓✓✓✓✓	✓✓✓✓	✓✓✓✓✓	✓✓	✓✓✓	✓✓✓	✓✓✓✓✓
	Asphalt concrete/ Dense bitumen macadam	✓✓✓	✓✓✓✓	✓✓✓	✓✓	✓✓✓	✓✓✓	✓✓	✓✓✓✓	✓✓✓	✓✓✓	✓✓✓
	Mastic asphalt/ Gussasphalt	✓✓✓	✓✓	✓✓✓✓✓	✓	✓✓✓	✓	✓	✓	✓✓✓✓✓✓	✓✓	✓✓✓
	Stone mastic asphalt	✓✓✓✓✓	✓✓✓✓✓	✓✓✓✓	✓✓✓✓	✓✓✓✓	✓✓✓✓	✓✓✓✓	✓✓✓	✓✓✓✓✓	✓✓✓	✓✓✓✓
Thin wearing course	26–39 mm thick	✓✓✓	✓✓✓✓	✓✓✓	✓✓✓✓	✓✓✓✓	✓✓✓✓	✓✓✓✓	✓✓✓	✓✓✓	✓✓✓	✓✓✓✓
	18–25 mm thick	✓✓	✓✓✓✓	✓✓✓	✓✓✓	✓✓✓	✓✓✓✓	✓✓✓	✓✓✓✓	✓✓✓	✓✓✓✓	✓✓✓✓
	<18 mm thick	✓	✓✓✓	✓✓	✓✓	✓✓✓	✓✓✓✓	✓✓✓	✓✓✓	✓✓	✓✓✓✓	✓✓✓✓
Veneer treatment	Surface dressing	n/a	n/a	✓✓✓	✓✓✓	✓	✓✓✓✓	✓✓✓✓✓	✓✓✓✓✓	✓✓	✓✓✓✓✓	†

High-friction systems	n/a	n/a	✓✓	✓✓✓✓	✓✓	✓✓✓✓✓	✓	✓✓✓	n/a
Slurry surfacing	✓✓‡	n/a	✓✓	✓✓✓	✓✓	✓✓✓✓✓	✓✓	✓✓✓	n/a

✓ = least advantageous   to   ✓✓✓✓ = most advantageous

\# Some of these materials will have a limited laying season.

* The deformation resistance of hot rolled asphalt can be enhanced by designing to conform to Clause 943 of reference 9.

† The quality of ride for surface dressing will depend on the design of the surface dressing, the aggregate size(s) employed and the evenness of the substrate.

‡ Slurry surfacing can give a useful improvement to the profile of the type of surface to which it is applied, for which this rating is appropriate – for other types of surfacing, it may not be appropriate.

(b) Explanations of 'desired properties' in Table 11.4(a)

Suitability for re-profiling: *Deformation resistance:*	The suitability of the material for regulating or re-profiling an existing surfacing The ability of the material to resist the effects of heavy traffic to create ruts in the wheel paths during hot weather
Resistance to cracking:	The ability of the material not to crack or craze with age, particularly in cold weather and in areas of high stress
Spray reducing:	The ability of the material to form a surfacing that minimizes the amount of water thrown up by the wheels of passing traffic into a driver's line of sight in wet conditions
Noise reducing:	The ability of the material to form a surfacing that reduces the noise generation at the tyre/surfacing interfaces and/or increases the noise absorbed
Skid resistance:	The ability of the material to form a surfacing that can achieve a high mean-summer SCRIM coefficient
Texture depth:	The ability of the material to form a surfacing that can achieve a high texture depth, with particular reference to the requirement for high-speed trunk roads of greater than 1.5 mm using the Sand-patch method
Initial cost:	The initial cost to supply, lay and compact an area with the material
Durability:	The ability of the material to remain in place and retain its other properies under the prevailing traffic and climatic conditions
Speed of construction:	The time required between closure and re-opening of the road when surfacing it with the material
Quality of ride:	The ability of the material to form a surfacing which gives the driver a comfortable ride

lay and more compact than bitumen macadams, they can be susceptible to wheel-track deformation under channelled heavy traffic, especially when it is moving slowly, e.g. uphill and at locations subject to acceleration and braking forces. (This has led to consideration of a rutting criterion for the mix design approach instead of the usual Marshall stability requirement, and the use of a British Standard wheel tracking test[17] as a compliance test for some contracts.)

The extent to which hot rolled asphalt wearing courses can effectively meet desired trunk-road properties, by comparison with other surfacing materials, is summarized in Table 11.4(a). Expanded descriptions of the headings used in Table 11.4(a) are given in Table 11.4(b).

Requirements for transporting, laying, and compacting of various bituminous materials (including hot rolled asphalt) are given in the *Specification for Highway Works*[9]. Generally, on-site hot rolled asphalt is too stiff to be workable at bitumen viscosities in excess of 30 Pa.s, and too fluid to compact at less than 5 Pa.s.

11.4.3 Dense tar surfacing

Prior to the 1960s road tar was the main bituminous binder used in pavement surfacings in the UK, and dense tar surfacing (DTS) was the premier tar product used in the wearing courses of roads. As a consequence of the discovery of North Sea gas and the subsequent closure of essentially all town gasworks, the competitiveness and availability of road tar was much reduced and its use in road pavements in Great Britain is now considerably less. Also limiting tar's use are (i) its temperature sensitivity compared with bitumen, e.g. tar has a narrower range of working temperatures and it hardens more rapidly than bitumen during mixing, as well as in the road pavement; and (ii) there is evidence to suggest that tar is a health hazard.

Overall, dense tar surfacings are now only used as wearing courses at sites where they have a special advantage over bitumen-bound materials, e.g. in the driveways of, and where vehicles stand or park at, motorway service areas because of tar's greater resistance to softening from petrol, diesel and/or oil spillages.

Although generally classified as a coated macadam, DTS corresponds closely to hot rolled asphalt wearing course material in that (a) when rolling is complete, a dense and impermeable surfacing is obtained that is able to carry heavy traffic flows over long periods of time, and (b) it primarily relies for its strength and stability upon the stiffness of the mortar in the mix. It is a hot-mix hot-laid wearing course material composed of coarse aggregate (35 or 55 per cent crushed rock, slag or gravel >3.35 mm), fines (sand, crushed rock or slag), filler (fine mineral dust from limestone, crushed rock or slag) and high-viscosity road tar (usually 54°C evt). For good compaction, the maximum stone size (either 14 mm or 10 mm) is normally about one-third of the layer thickness. Finer aggregate gradings and lower tar contents are required for (the more workable) mixes containing sand fines vis-à-vis crushed rock or slag fines, because of the differing characteristics of the two types of fines. DTS tars have viscosities that are as high as can practically be used without fuming during mixing and without hardening too quickly; this limits the tars used to those with viscosities at or below 54°C evt.

When a dense tar surfacing with 35 per cent stone is used in a road pavement, precoated 20 mm or 14 mm chippings are normally added to the running surface to

Table 11.5 Comparison of properties of some of the (now) lesser-used standard bitumen macadam wearing course materials (based on reference 6)

	Open- and medium-graded macadam	Close-graded macadam	Fine-graded macadam	Mastic asphalt
Structural contribution	Low	Medium	Low	High
Deformation resistance	Low	Medium	Low	Medium/high
Weather resistance	Low	Medium	Low	Very high
Texture depth	Good	Poor	Poor	Good*
Spray reduction	Good	Poor	Poor	Poor*
Noise reduction	Good	Poor	Poor	Poor*
Workability	High	Low	High	Low

* With surface dressing chippings

provide additional skid resistance. The surface dressing is not needed with 50 per cent stone DTS, which is used at hardstanding areas.

A DTS wearing course is normally laid on top of a basecourse of rolled asphalt or densely-coated macadam with a high-viscosity binder. However, the underlying material must be cold before the DTS material is placed upon it as the DTS laying temperature can cause the basecourse binder to soften, with consequent movement under rolling.

11.4.4 Coated macadams

A variety of standard coated macadams are used in the UK and these are generally described in terms of their nominal maximum aggregate size and grading, and their intended usage. Broadly, dense, close-graded, medium-graded, and open-graded coated macadam materials are all continuously-graded and rely mainly on aggregate-to-aggregate contact for their strengths; also, they tend to have progressively decreasing proportions of fines, densities and stabilities as they become more open-graded. Fine-graded macadam and porous asphalt (previously 'pervious macadam') are special types of coated macadam that have particular uses. (It might also be noted here that stone mastic asphalt is a 'new' type of coated macadam that is currently gaining favour in the UK.)

Nowadays, bitumen is by far the main type of binder used in coated macadam mixtures, albeit tar was heavily used prior to the 1960s. The general compositions of the various types of coated macadams are summarized in Table 11.2. The effectiveness of porous asphalt and dense bitumen macadam in meeting desired wearing courses properties is summarized in Table 11.4; the effectiveness of the remaining (less-used) macadams is given in Table 11.5.

Dense bitumen macadam (DBM) – some compositions of which are not dissimilar to the asphaltic concrete referred to above in relation to Fig. 11.1 – is the predominant type of bitumen-bound material used in the basecourses and road-bases of trunk roads (including motorways) in the UK. Conventional DBMs are well graded (see Table 11.6) with, typically, 3–8 per cent air voids when compacted; hence, they are practically impervious to moisture movement downward (from the surface) and upward (from the subgrade and subbase). The high fines content and

Table 11.6 Compositions of dense bitumen macadam (DBM) and heavy duty macadam (HDM) recipe mixes comprising crushed rock, fines and filler

Nominal size (mm)	DBM		HDM	
	28	40	28	40
Gradation: *% by mass of aggregate passing BS sieve*				
37.5 mm	100	95–100	100	95–100
28	90–100	70–94	90–100	70–94
20	71–95	–	71–95	–
14	58–82	56–76	58–82	56–76
6.3	44–60	44–60	44–60	44–60
3.35	32–46	32–46	32–46	32–46
0.300	7–21	7–21	7–21	7–21
0.075	2–9	2–9	7–11	7–11
Bitumen content:				
% by mass of total mix	3.4–4.6	2.9–4.1	3.4–4.6	2.9–4.1
Grade of binder:				
Penetration (dmm) @ 25°C	100	100	50	50

the use of a relatively low amount of soft (most commonly 100 pen) bitumen results in this material having very good load-spreading properties and a high resistance to deformation. However, the ability of conventional DBM to resist fatigue cracking from repeated loading is suspect, compared with other standard materials (e.g. hot rolled asphalt) that are used in busy trunk roads. To help overcome this problem a superior dense bitumen macadam, DBM50, was developed for use in basecourses and roadbases; it has the same gradations and binder contents as the conventional DBMs (Table 11.6) but uses a harder (50 pen) bitumen.

Dense bitumen macadam wearing courses (6 mm size) are used only on more lightly-trafficked roads, as they are not able to provide the deep texture needed for high-speed skid resistance under heavy traffic[6].

Heavy-duty macadam (HDM) is a strong macadam material that has been developed for use in roadbases and basecourses of very heavily trafficked roads. HDM is a dense bitumen macadam (Table 11.6) that uses a higher filler content and a harder bitumen grade (50 pen) to give a roadbase material with a stiffness that is two or three times that of conventional DBMs and has good resistance to both fatigue cracking and deformation; this allows a 10–15 per cent reduction in roadbase thickness for an equivalent performance or an extended life for the same thickness. The reduced thickness option that results from use of heavy-duty macadam can facilitate reconstruction work at road locales where 'headroom' space is constrained, e.g. under bridges.

It might also be noted here that research is currently being carried out to evaluate the load-spreading abilities of stiffer dense bitumen macadams incorporating lower penetration binders, e.g. 35, 25 and 15 pen bitumens, in roadbases; these long-life materials are termed high modulus base (HMB) materials. It is reported[16] that savings in material cost of over 25 per cent result from the use of HBM15 instead of conventional DBM in the construction of a pavement with the same design life.

The maximum compacted thickness of a dense bitumen macadam layer is 200 mm; if thicknesses greater than 200 mm are required, they must be laid in two or more layers. The general rule-of-thumb is that the compacted layer thickness should be 2.5 times the size of the nominal maximum aggregate. (If a thickness of less than 1.5 times the nominal maximum aggregate size is laid, it can be expected that some crushing of the larger stones will take place under the roller, thereby changing the mix gradation and reducing its stability.)

Dense tar macadam (DTM) is similar to conventional DBM but utilizes a grade C50 or C54 tar binder instead of bitumen when used in the basecourses or road-bases of trunk roads. DTM, which is more temperature susceptible than dense bitumen macadam, is now limited in its usage in the UK, for reasons given above in relation to dense tar surfacing (DTS).

Close-graded bitumen macadam wearing course (14 mm and 10 mm sizes) mixes have void contents that are fairly high, and they are relatively pervious. They are used only in lightly-trafficked roads (carrying less than 2.5 msa in 20 years) because they do not have the high skid-resistance and long-term durability required for more heavily-trafficked pavements.

Medium-graded macadams (6 mm size) and *open-graded macadams* (14 and 10 mm sizes) have low fines contents and, after compaction, these wearing courses have high air voids (typically, 15–25 per cent), which facilitates their laying and curing. The mix binder is normally either a 200 or 300 pen grade bitumen, or a cutback bitumen, depending upon the traffic intensity and the time of the year that the material is being laid. These macadams are easy to mix, lay and compact, i.e. they have good workability. However, they have low strength and thus are used mainly in off-road situations, e.g. car parks, driveways, footpaths, etc. Because they are pervious, the pavement must be waterproofed by a surface dressing above or below the wearing course.

Rainwater on carriageways is a hazard to motorists as it reduces skidding resistance. Many wet road accidents are also associated with water spray that is thrown up by fast-moving vehicles, especially commercial vehicles, which obscures the vision of drivers of adjacent or closely-following vehicles. Both of these factors are of concern in a wet island such as Great Britain where, it is reported[8] that, in any given year, the roads are wet for about a number of hours equal to 100 times the rainfall in inches – which means that many roads are wet for up to 50 per cent of the time. To reduce the number of accidents, a particular open-graded coated macadam material has been developed in the UK as a rapid draining, anti-splash, wearing course for use on heavily-trafficked roads; this material is termed porous asphalt.

Porous asphalt (PA), which was previously known as *pervious macadam*, is an open-graded wearing course material that facilitates the rapid drainage of rainwater from a carriageway; it is composed of gap-graded aggregates that are held together by bitumen to form a matrix through which water can drain laterally under the influence of the pavement's crossfall and longfall. The void content of this open-graded material is about 18 to 25 per cent when compacted; thus, the rain water falling on the carriageway can be absorbed in the open voids, and it then flows through interconnected voids in the wearing course (and shoulders) material for collection in roadside edge-drains. As it is highly pervious, the porous asphalt

Table 11.7 Aggregate grading limits[9] and binder contents[11] for 20 mm porous asphalt

(a) Grading limits

BS sieve (mm)	Percentage by mass of total aggregate passing*
28	100
20	100–95
14	75–55
6.3	30–20
3.35	13–7
0.075	5.5–3.5

*The mid-points of the grading limits should be targeted

(b) Target bitumen contents

| Traffic flow (commercial vehicles/lane/day) | Bitumen: | |
	Grade (pen)	Percentage by mass of total mix
<1500	200	3.4 (min)
	200 + modifier*	4.5
1500–3000	100	3.4 (min)
	100 + modifier*	4.5
>3000	100 + modifier*	4.5

* With pre-blended modified binders the base bitumen grade should be 100–200 pen

layer must be laid on top of an impermeable basecourse (usually a 60 mm thick layer of HRA, DBM, DBM50, or HDM) so that the surface water cannot penetrate any further downward into the pavement.

Whilst PA was primarily developed to enable water to be removed quickly from the road surface, it has a number of ancillary benefits: (i) it reduces the glare reflected from a wet carriageway at night, (ii) it enables carriageway markings, which otherwise would be covered by water, to be more visible in wet conditions at night, (iii) through acoustic absorption, it reduces the level of noise that is generated by traffic in dry weather (typically, to half that generated by an HRA wearing course), and (iv) it reduces the incidence of noise that is additionally caused by the generation of spray by tyres in wet weather.

The grading specification for the most commonly used porous asphalt mix, and the targeted bitumen contents to be used with it, are given in Table 11.7. The coarse aggregate is crushed rock or steel slag, and the fine aggregate may be crushed rock or steel slag or (under certain conditions) natural sand. At least 2 per cent by mass of the total aggregate must be hydrated lime filler, to stiffen the bitumen and reduce the risk of it stripping from the aggregate. Porous asphalt's durability is helped by using soft binders and as high a bitumen content as possible. Too much binder and/or an excessive mixing temperature causes binder drainage and mix segregation whilst the PA is being transported from the mixing plant to the site, leading to patches in the finished wearing course that are either too rich or too lean in bitumen. Too little bitumen leads to thinner binder films that are prone to premature hardening, and result in a shorter life for the PA layer.

20 mm PA is normally laid to a thickness of 50 mm: this thickness optimizes its compaction time during laying, and its rainwater storage capacity and spray-reducing life whilst in service. It might also be noted that only steel-wheel tandem drum rollers are permitted for compaction; vibrating rollers cannot be used because of the danger of aggregate crushing, rubber-tyred rollers knead and close the surface and affect the material's drainage performance, and three-wheeled rollers leave marks that are difficult to remove.

Porous asphalt can be more expensive initially than hot rolled asphalt. Even though the amount of bitumen required is less, and precoated surface dressing chippings are not needed, a greater quantity of high quality aggregates is required to form the wearing course and maintain the same level of skid resistance. Also, PA is less stiff and durable than other competing bituminous layers and contributes less to the structural strength of the pavement, e.g. a 50 mm layer of porous asphalt is considered to be structurally equivalent to only 20 mm HRA or 25 mm DBM; hence, the overall thickness of a new flexible pavement has to be increased by about 25 mm when porous asphalt is used instead of, say, HRA in a wearing course[7] – and this can have consequential effects on some earthworks and drainage designs. (It is reported[16] that more recent research has indicated that the structural equivalence of PA relative to HRA may be closer to 1.5:1 rather than 2:1.)

Whilst porous asphalt is less durable than hot rolled asphalt, due to long-term hardening of the bitumen and consequent fretting of the surface aggregate, its durability can be improved by the use of higher binder contents in conjunction with an appropriate fibre additive (which reduces binder drainage whilst the material is being transported). Porous asphalt is much more effective than HRA in reducing surface spray, particularly in its early life; however, this advantage is gradually reduced as the material compacts and voids become clogged with rubber from tyres, dirt and detritus. As noted above, a beneficial by-product of the use of PA is that some tyre noise is absorbed in the surface voids so that there is a significant reduction in the traffic noise associated with high-speed roads, e.g. the average reduction in dry road surface noise levels, compared with conventional surfaces, is around 4 dB(A) and 3 dB(A) for 'light' and 'heavy' vehicles, respectively[11]; however, the compaction and clogging effect can cause half of the initial sound reduction to be lost over the design life[10].

In general, porous asphalt wearing courses are best used on new, well-designed and constructed, high-speed roads without kerbs and with few junctions, e.g. motorways and rural dual carriageways[11]. Their use is also worthwhile at accident sites on existing roads where skidding accidents attributed to wet weather spray have occurred[7], and on new road bypasses where it is anticipated that noise will be a problem to adjacent householders. PA wearing courses should not be used at road locations where (a) there is much acceleration, braking or turning, as the material is weak in shear, (b) excessive droppings of detritus and oil from vehicles are expected, (c) free drainage cannot be accommodated at the low side(s) of the carriageway, (d) the pavement is likely to be subsequently dug up by statutory authorities, e.g. in urban streets, and (e) a 30 mph (48 km/h) speed limit is enforced, as the spray and noise level reductions achieved at low vehicle speeds are negligible.

A porous asphalt wearing course constructed with 100 pen bitumen has an expected life of 7 to 10 years, at traffic levels up to 4000 commercial vehicles per lane per day, compared with about 10 years for similar wearing courses built with hot rolled asphalt[11]. PA is difficult to 'patch' as this often impedes the essential drainage through the material; consequently, the best maintenance treatment is replacement.

Fine-graded macadam (previously known as *fine cold asphalt*) is a wearing course material that is not strong, has a high void content, and is relatively pervious when first laid; consequently, it is best laid on a sound impervious layer. Eventually, however, the fine-graded macadam, whose nominal maximum aggregate size is 3 mm, seals itself under traffic. The mix is normally made with 200 or 300 pen bitumen or 100 s cutback bitumen, which means that it must be laid and compacted warm (at 80–100°C); however, cutback materials are also produced for cold laying and depot storage. This macadam is laid as a 15–25 mm thick carpet and the layer on which it is placed must have good regularity, to avoid subsequent problems associated with it being laid to varying thicknesses. If used in a road pavement, fine-graded macadam needs a surface dressing of pre-coated chippings to give skid resistance under traffic; most commonly, however, it is used in driveways and footpaths where skid resistance is normally not a problem.

Stone mastic asphalt (SMA) – also known as *Splittmastixasphalt* – is a wearing course material that is hot-mixed (at 150–190°C), hot-laid, and hot-compacted. Originally devised in Germany in the 1960s as a proprietary wearing course that would withstand the abrading action of the studded tyres then used by vehicles during the snow season – the use of studs was subsequently banned in 1975 – SMA was found to be much more durable, and more resistant to deformation under heavy traffic, than asphaltic concrete. Consequently, it was further developed as a surfacing for autobahns and other major roads and was eventually included in the German specifications. It is the material that is now most widely used in wearing courses on heavily trafficked roads in Germany, and has also been used in modified forms in trunk roads in many other countries in northern Europe. Even though it was only introduced into Great Britain in 1993, its use has grown significantly because of its excellent reputation; experience has shown that, when correctly laid, it is an extremely durable and rut-resistant material[12]. A draft specification has now been prepared to govern its usage in the UK (see Appendix A of reference 13).

Stone mastic asphalt has been described[14] as a gap-graded aggregate bound with bitumen that is stabilized with additives during manufacture, transport, laying and compaction. Its main features are a self-supporting interlocking stone skeleton that is achieved with a high, relatively single-sized, coarse aggregate (>2 mm) content, an aggregate gradation that has a pronounced gap in the 3.0 mm to 0.5 mm range, and a rich mortar composed of fine aggregate, filler and bitumen. The relative quantities of aggregate and mortar are selected so that the spaces in the stone skeleton are almost filled with mortar, to give a voids content within the range 2–4 per cent. In this way, the coarse aggregate framework is effectively locked into place by the mortar, with continued good contact between the particles being ensured. This mechanism endows stone mastic asphalt with good durability and excellent resistance to plastic deformation under heavy traffic.

Table 11.8 Compositions of stone mastic asphalt (SMA) mixes[13]

BS sieve size (mm)	Nominal size of aggregate (mm):	
	14	10
Mass of aggregate passing sieve (%):		
20	100	
14	90–100	100
10	35–60	90–100
6.3	23–35	30–50
2.36	18–30	22–32
0.075	8–13	8–13
Binder content:		
Mass of total mix (%)	6.5–7.5	6.5–7.0

It is important that the point-to-point contact be maintained between the coarse aggregate particles during and after compaction, as it is through the interlocking aggregate framework that the wheel load stresses are primarily distributed. If the volume of the mortar is smaller than the volume of the voids in the compacted coarse aggregate, the skeleton will not be spread (and the mortar will not be compacted) as the brunt of the mass of the compacting roller will be borne by the stone. If, however, the mortar content is too high, i.e. if the mortar volume is greater than the void space available after compaction, the stone skeleton will be forced apart with consequent instability and flushing of the bitumen. In practice, relatively little compaction is needed to ensure that the wearing course's coarse aggregate framework is properly established. Saturation of the fines and filler with a relatively high bitumen content results in the mortar-coated aggregate framework being locked into place under the roller whilst the voidless mortar nearly fills the voids between the aggregate particles without the need for compaction. It is the mortar that gives the SMA mix its durability against ageing and weathering[15].

Draft BS specification gradations and bitumen contents for stone mastic asphalt are shown in Table 11.8. The binder is normally either 50 or 100 pen bitumen. It is essential that stabilizing agents be added to the bitumen to prevent the high binder content, and the mortar, from draining from the stone between mixing and compaction, i.e. whilst the SMA is being transported and handled. Cellulose fibres, which are normally incorporated at a minimum rate of 0.3 per cent by mass of the total mix, are predominantly used for this purpose; modified binders, which also improve performance as well as prevent binder drainage, are mainly used in the remaining mixes. Considerable care must be taken in the production of stone mastic asphalt as the wearing courses are very sensitive to variations from the recipes.

Compared with the other recipe specifications described above, stone mastic asphalt is a relatively new material in pavements in the UK. It has been demonstrated that SMA can be laid successfully, meet the 1.5 mm minimum texture depth (anti-skidding) requirements, and produce less tyre noise than hot rolled asphalt[16].

The indications are that its toughness and durability, its ability to resist cracking and fatigue, and the good surface drainage and noise characteristics associated with its coarse and open surface texture, will see stone mastic asphalt being more heavily used in the foreseeable future.

11.5 Possible future standard wearing course materials

In recent years, following the realization that the design thicknesses of bituminous layers in pavements should primarily be related to the early life properties of the main structural courses, the trend has been towards using stiffer bituminous-bound roadbases in heavily-trafficked trunk roads. The long-life successes that have been achieved with well-built thick pavements constructed above a threshold strength[16] has given rise to new concepts regarding the thickness of, and materials used in, a wearing course.

Bituminous wearing courses in trunk roads are most commonly laid at thicknesses of 40 mm, 45 mm or, in adverse weather conditions, 50 mm, with the thicker courses being used to attain better compaction and improve durability. The primary function of these wearing courses is to provide safe durable smooth riding surfaces for traffic (see Subsection 9.2.1) and, even though they are relatively thick, only secondarily are they expected to act as load-spreading layers, i.e. the stress distribution capability is seen to be primarily the role of the lower layers in the pavement. The successful results obtained with deformation-resistant high-modulus roadbase materials has, however, given rise to the view that these might be laid to greater thicknesses, e.g. to within 20–30 mm of the carriageway, and thin high-quality *renewable* wearing courses then put in place above them. In other words, the stiff underlying materials would be regarded as forming the permanent courses whereas the thin wearing course would be seen as expendable, to be replaced as demanded by rehabilitation needs, e.g. to restore skid resistance or riding quality, or to remedy surface cracking. Table 11.9 shows interrelationships between surfacing and roadbase thicknesses for different high-quality roadbase materials,

Table 11.9 Effect, upon necessary thickness of different roadbase materials, of using thin surfacings of varying thickness[16]

Thickness of surfacing (mm)*	Increase in roadbase thickness (mm):			
	DBM	DBM50	HDM	HMB15
50	− 10	− 9	− 8	− 6
40	0	0	0	0
30	+ 10	+ 9	+ 8	+ 6
20	+ 20	+ 18	+ 16	+ 13
10	+ 30	+ 27	+ 24	+ 20
0	+ 40	+ 36	+ 32	+ 26

* Surfacing includes HRA, SMA and thin surfacings; porous asphalt is not included.

whilst Table 11.4 describes the effectiveness of different depths of thin wearing courses upon desired surfacing properties. (The 'veneer' treatments shown in Table 11.4 are shown here for comparative reasons; they are unlikely to be used in long-life trunk roads except, possibly, for maintenance purposes.)

Various proprietary thin surfacing materials that are laid at nominal thicknesses of less than 40 mm have been developed and widely used in France to restore surface characteristics and to surface pavements that have received a strengthening overlay. It is reported[16] that a number of these proprietary materials have been introduced into the UK and that their long-term performance capabilities are currently being evaluated as potential surfacing materials at the initial construction stage.

It might be noted, however, that if the use of thin wearing courses in the pavements of heavily-trafficked roads is to become the norm, then closer tolerances will need to be achieved in relation to the profile of the surface of the underlying structural material. Also, it is not unlikely that standard specifications might be developed for them.

11.6 References

1. Farrington, J., The production of Europe harmonised specifications for bituminous mixtures. In *The Asphalt Year Book 1996*. Stanwell, Middlesex: Institute of Asphalt Technology, 1996, pp. 31–2.
2. BS 1447:1988, *Specification for Mastic Asphalt (Limestone Fine Aggregate) for Roads, Footways and Pavings in Buildings*. London: British Standards Institution, 1988.
3. BS 594:1992, *Hot Rolled Asphalt for Roads and Other Paved Areas*: Part 1 – *Specification for Constituent Materials and Asphalt Mixtures*; and Part 2 – *Specification for the Transport, Laying and Compaction of Rolled Asphalt*. London: British Standards Institution, 1992.
4. BS 5273: 1975, *Dense Tar Surfacings for Roads and Other Paved Areas*. London: British Standards Institution, 1975.
5. BS 4987:1993, *Coated Macadam for Roads and Other Paved Areas*: Part 1 – *Specification for Constituent Materials and for Mixtures*; and Part 2 – *Specification for Transport, Laying and Compaction*. London: British Standards Institution, 1993.
6. *Bituminous Mixes and Flexible Pavements*. London: British Aggregate Construction Materials Industries, 1992.
7. *Pavement Design*, HD 26/94, In *Design Manual for Roads and Bridges*. London: HMSO, 1994 (amended March 1995, Feb 1996 and Feb 1998).
8. Whiteoak, D., *The Shell Bitumen Handbook*. Chertsey, Surrey: Shell Bitumen, 1990.
9. *Specification for Highway Works*. London: HMSO, 1998.
10. Wright, M., Thin and quiet?: An update on quiet road surface products, *H & T*, 2000, **47**, No. 1/2, pp. 14–8.
11. *Pavement Construction Methods*, HD 27/94, In *Design Manual for Roads and Bridges*. London: HMSO, 1994.
12. Asphalt – making the best use of our roads, *H & T*, 2000, **47**, No. 1/2, pp. 19–20.

13. Nunn, M.E., *Evaluation of Stone Mastic Asphalt (SMA): A High Stability Wearing Course Material*, PR 65. Crowthorne, Berkshire: The Transport Research Laboratory, 1994.

14. Huning, P., Stone mastic asphalt justified, *European Asphalt Magazine*, 1995, No. 3/4, pp. 9–13.

15. Liljedahl, B., Mix design for heavy duty asphalt pavements. In *The Asphalt Yearbook 1992*. Stanwell, Middlesex, Institute of Asphalt Technology, 1992, pp. 18–24.

16. Nunn, M.E., Brown, A., Weston, D. and Nicholls, J.C., *Design of Long-life Pavements for Heavy Traffic*. TRL Report 250. Crowthorne, Berkshire: The Transport Research Laboratory, 1997.

17. BS 598: Part 110, *Method of Test for the Determination of the Wheel-tracking Rate*. London: British Standards Institution, 1996.

CHAPTER 12

Design and construction of hot-mix bituminous surfacings and roadbases

S.E. Zoorob

12.1 Why design bituminous paving mixtures?

The overall objective for the design of bituminous paving mixtures is to determine, within the limits of the project specifications, a cost-effective blend and gradation of aggregates and bitumen that yields a mix having the following properties: (a) sufficient bitumen to ensure a durable pavement; (b) sufficient mixture stability to satisfy the demands of traffic without distortion or displacement; (c) sufficient voids in the total compacted mixture to allow for a slight amount of bitumen expansion due to temperature increases, without flushing, bleeding and loss of stability; (d) a maximum void content to limit the permeability of harmful air and moisture into the mixture and the layers below; and (e) sufficient workability to permit efficient placement of the mixture without segregation and without sacrificing stability and performance. Also, for wearing course materials, the mix should (i) have sufficient aggregate texture and hardness to provide good skid resistance in bad weather conditions, (ii) produce an acceptable level of tyre/road noise, and (iii) provide a surface of acceptable ride quality.

Ultimate pavement performance is related to durability, impermeability, strength, stability, stiffness, flexibility, fatigue resistance, and workability. The goal of mix design is to select a unique design bitumen content that will achieve an appropriate balance among all of the desired properties.

Selection of the aggregate gradation and the grade and amount of bitumen can be accomplished by two general methods: a standard 'recipe' approach, or an engineering design approach.

12.2 Standard 'recipe' approach

As implied by its name, the 'recipe' approach for selecting the types and proportions of the materials comprising a bituminous mixture is a 'cookbook' procedure. It is based wholly on experience, not on engineering principles. Although it is technically not a design method, it remains a popular method for proportioning the materials in a mixture, particularly where it has been tried and tested over many years.

In this method the type and gradation of aggregate, the grade of bitumen, the proportions of bitumen and aggregate, and the method of construction (i.e., mixing, placement and compaction) are all described in standard specifications. Thus, the procedure involves consulting a set of specifications to produce a mixture which, for the most part, has been shown historically to provide acceptable performance.

However, such a method is subject to several limitations. The conditions to which the in-service mixture is subjected (e.g. traffic, climate, etc.) may not be the same as those that existed when the specifications were developed. Additionally, with the trend in the highway industry towards inclusion of innovative materials (e.g. polymer-modified bitumens, reclaimed asphalt, fibre-modified mixes, etc.) in bituminous paving mixtures, specification by recipe may not be suitable for such materials. Furthermore, the performance of the mixture is not only dependent on its composition but also on workmanship during production, which is difficult to specify. Checking for compliance is relatively simple but assessing the seriousness or practical implications for minor non-compliance requires much experience and judgement.

Current recipe specifications do not fully take into account the required engineering properties of the mixture which are strongly dependent on volumetric composition. Recipe specifications are based on the aggregate grading curve and since such curves cannot take into account aggregate properties and packing characteristics, except in the most general way, they are not very reliable, particularly when different aggregate types are used in the mixture. For example, significant differences in bitumen film thickness and distribution of voids can occur in mixtures using the same gradation (by mass) but different aggregate types having relatively small differences in specific gravity. Workability and viscous stiffness of the mixture are also influenced by aggregate gradation, texture, and particle shape, factors which cannot be accounted for by grading curves alone.

12.3 Engineering design approach

The objectives of a mixture design can be achieved by ensuring the mixture contains an appropriate quantity of bitumen to adequately coat all of the aggregate particles and provide good workability so that, when the mixture is compacted, it possesses adequate stiffness, deformation resistance, and air voids.

Many empirical and semi-empirical design procedures have been devised which first attempt to evaluate various properties of bituminous mixtures and then base the bitumen content determination on these evaluations. Some of the more widely known of these design procedures are the Marshall[1], Hveem[1], SHRP Superpave[2], LCPC[3], CROW[4], Texas[5], EXXON[6], and University of Nottingham methods[7,8], as well as many others[12,29].

Regardless of the method of mix design and testing procedure adopted, there is consensus that it is important to analyse the volumetric properties of the bituminous mix in its compacted state. The objective of volumetric analysis (in any mix design procedure) is to aid the designer in determining amongst other factors, the efficiency of bitumen utilisation in the mix, the degree and efficiency of compaction, and the quantity of air voids present in the mix. These and other volumetric parameters have been shown to have a direct influence on the mix stiffness, stability and durability.

Sections 12.4 and 12.5 contain detailed explanations of the procedure and terminologies used in volumetric analysis which is common to, and forms an integral part of, all mix design procedures.

12.4 Outline of procedure for analysing a compacted paving mixture

The steps involved in analysing a compacted bituminous mixture, and the BS (British Standard), AASHTO (American Association of State Highway and Transportation Officials), or ASTM (American Society for Testing and Materials) test methods used at each step, are as follows.

1. Measure the bulk specific gravities of the coarse aggregate (AASHTO T 85 or ASTM C 127) and of the fine aggregate (AASHTO T 84 or ASTM C 128); also BS 812, Pt. 2: 1995.
2. Measure the specific gravity of the bitumen (AASHTO T 228 or ASTM D 70[23]) and of the mineral filler (AASHTO T 100 or ASTM D 854 or BS 812, Pt. 2: 1995).
3. Calculate the bulk specific gravity of the aggregate combination in the paving mixture.
4. Measure the maximum specific gravity of the loose paving mixture (ASTM D 2041, BS DD 228: 1996).
5. Measure the bulk specific gravity of the compacted paving mixture (ASTM D 1188 or ASTM D 2726 or BS 598, Pt. 104: 1989).
6. Calculate the effective specific gravity of the aggregate.
7. Calculate the maximum specific gravity of the mix at other bitumen contents.
8. Calculate the bitumen absorption of the aggregate.
9. Calculate the effective bitumen content of the paving mixture.
10. Calculate the per cent voids in the mineral aggregate in the compacted paving mixture.
11. Calculate the per cent air voids in the compacted paving mixture.
12. Calculate the per cent air voids filled with bitumen.

12.5 Terms used in bituminous mix design

It is very important that the engineer engaged in the design of a bituminous material be very clear on the terminology that is used and its precise meaning. The following are the main terms and their meaning.

12.5.1 Specific gravities of aggregates

Mineral aggregate is porous and can absorb water and bitumen to a variable degree. Furthermore, the ratio of water to bitumen absorption varies with each aggregate. The three methods of measuring aggregate specific gravity (ASTM bulk, ASTM apparent, and effective) take these variations into consideration. The differences between the specific gravities arise from different definitions of aggregate volume.

The *Bulk Specific Gravity* (G_{sb}) is the ratio of the mass in air of a unit volume of a permeable material (including both permeable and impermeable voids normal to the material) at a stated temperature to the mass in air of equal density of an equal volume of gas-free distilled water at a stated temperature (see Fig. 12.1).

The *Apparent Specific Gravity* (G_{sa}) is the ratio of the mass in air of a unit volume of an impermeable material at a stated temperature to the mass in air of equal density of an equal volume of gas-free distilled water at a stated temperature.

The *Effective Specific Gravity* (G_{se}) is the ratio of the mass in air of a unit volume of a permeable material (excluding voids permeable to bitumen) at a stated temperature to the mass in air of equal density of an equal volume of gas-free distilled water at a stated temperature (see Fig. 12.1).

When the total aggregate consists of separate fractions of coarse aggregate, fine aggregate, and mineral filler, all having different specific gravities, the bulk specific gravity for the total aggregate is calculated from:

$$G_{sb} = \frac{P_1 + P_2 + \cdots + P_n}{\dfrac{P_1}{G_1} + \dfrac{P_2}{G_2} + \cdots + \dfrac{P_n}{G_n}} \tag{12.1}$$

where G_{sb}=bulk specific gravity of the total aggregate; P_1, P_2,...P_n=individual percentage by mass of aggregate; and G_1, G_2,...G_n=individual bulk specific gravities

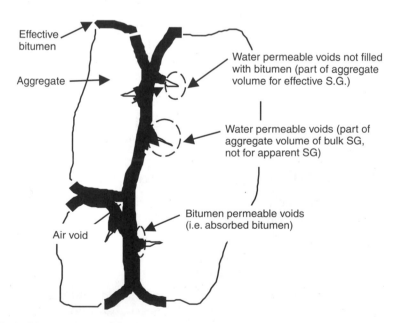

Fig. 12.1 Illustration of bulk, effective, and apparent specific gravities, air voids, and effective bitumen content, in compacted bituminous paving mixture

of aggregate. The bulk specific gravity of mineral filler is difficult to determine accurately; however, if the apparent specific gravity of the filler is substituted, the error is usually negligible.

When based on the maximum specific gravity of a paving mixture, G_{mm} (as measured using ASTM D 2041), the effective specific gravity of the aggregate, G_{se}, includes all void spaces in the aggregate particles except those that absorb bitumen. G_{se} is determined from

$$G_{se} = \frac{P_{mm} - P_b}{\dfrac{P_{mm}}{G_{mm}} - \dfrac{P_b}{G_b}} \qquad (12.2)$$

where G_{se} = effective specific gravity of aggregate; G_{mm} = maximum specific gravity (ASTM D 2041) of paving mixture (no air voids); P_{mm} = percentage by mass of total loose mixture = 100; P_b = bitumen content at which ASTM D 2041 was performed, percentage by total mass of mixture; and G_b = specific gravity of bitumen.

Note: The volume of bitumen absorbed by an aggregate is almost invariably less than the volume of water absorbed. Hence, the value of G_{se} should normally lie between its bulk and apparent specific gravities.

12.5.2 Maximum specific gravity of mixtures with different bitumen contents

When designing a paving mixture with a given aggregate, the maximum specific gravity, G_{mm}, at each bitumen content is needed to calculate the per cent of air voids for each bitumen content. While the maximum specific gravity can be determined for each bitumen content by ASTM D 2041, the precision for the test is best when the mixture is close to the design bitumen content. Also, it is preferable to measure the maximum specific gravity in duplicate or triplicate.

After calculating the effective specific gravity of the aggregate from each measured maximum specific gravity and averaging the G_{se} results, the maximum specific gravity for any other bitumen content can be obtained as shown below. For all practical purposes, G_{se} is constant because the bitumen absorption does not vary appreciably with variations in bitumen content, thus:

$$G_{mm} = P_{mm} \bigg/ \left(\frac{P_s}{G_{se}} + \frac{P_b}{G_b} \right) \qquad (12.3)$$

where G_{mm} = maximum specific gravity of paving mixture (no air voids); P_{mm} = percentage by mass of total loose mixture = 100, P_s = aggregate content, percentage by total mass of mixture, P_b = bitumen content, percentage by total mass of mixture, G_{se} = effective specific gravity of aggregate, and G_b = specific gravity of bitumen.

12.5.3 Bitumen absorption

Absorption is expressed as a percentage by mass of aggregate rather than as a percentage by total mass of mixture. Bitumen absorption, P_{ba} is determined using:

$$P_{ba} = 100 \frac{G_{se} - G_{sb}}{G_{sb} G_{se}} G_b \qquad (12.4)$$

where P_{ba} = absorbed bitumen, percentage by mass of aggregate; G_{se} = effective specific gravity of aggregate; G_{sb} = bulk specific gravity of aggregate; and G_b = specific gravity of bitumen.

12.5.4 Effective bitumen content of a paving mixture

The effective bitumen content, P_{be}, of a paving mixture is the total bitumen content minus the quantity of bitumen lost by absorption into the aggregate particles. As it is the portion of the total bitumen content that remains as a coating on the outside of the aggregate particles, it is this bitumen content which governs the performance of a bitumen paving mixture. The formula is:

$$P_{be} = P_b - \frac{P_{ba}}{100} P_s \qquad (12.5)$$

where P_{be} = effective bitumen content, percentage by total mass of mixture; P_b = bitumen content, percentage by total mass of mixture; P_{ba} = absorbed bitumen, percentage by mass of aggregate; and P_s = aggregate content, percentage by total mass of mixture.

12.5.5 Per cent *VMA* in compacted paving mixture

The voids in the mineral aggregate, *VMA*, are defined as the intergranular void space between the aggregate particles in a compacted paving mixture that includes the air voids and the effective bitumen content, expressed as a per cent of the total volume (Fig. 12.2). The *VMA* is calculated on the basis of the bulk specific gravity of the aggregate G_{sb} and is expressed as a percentage of the bulk volume of the compacted paving mixture. Therefore, the *VMA* can be calculated by subtracting the volume of the aggregate determined by its bulk specific gravity from the bulk volume of the compacted paving mixture. If the mix composition is determined as per cent by mass of total mixture:

$$VMA = 100 - \frac{G_{mb} P_s}{G_{sb}} \qquad (12.6)$$

where *VMA* = voids in mineral aggregate, percentage of bulk volume; G_{sb} = bulk specific gravity of total aggregate; G_{mb} = bulk specific gravity of compacted mixture (AASHTO T 166, ASTM D 1188 or D 2726); and P_s = aggregate content, percentage by total mass of mixture.

Or, if the mix composition is determined as a per cent by mass of aggregate:

$$VMA = 100 - \frac{G_{mb}}{G_{sb}} \times \frac{100}{100 + P_b} 100 \qquad (12.7)$$

where P_b = bitumen content, percentage by mass of aggregate.

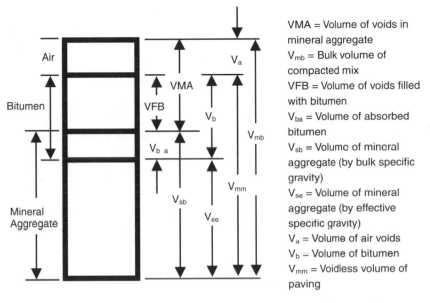

Fig. 12.2 Representation of volumes in a compacted bituminous mix

12.5.6 Per cent air voids in the compacted mixture

The air voids, V_a, is the total volume of the small pockets of air between the coated aggregate particles throughout a compacted paving mixture, expressed as a per cent of the bulk volume of the compacted paving mixture (Fig. 12.2). The volume percentage of air voids in a compacted mixture can be determined using:

$$V_a = 100 \times \frac{G_{mm} - G_{mb}}{G_{mm}} \tag{12.8}$$

where V_a = air voids in compacted mixture, percentage of total volume; G_{mm} = maximum specific gravity of paving mixture (as determined earlier or as measured directly for a paving mixture by ASTM D 2041); and G_{mb} = bulk specific gravity of compacted mixture. (*Note*: The Asphalt Institute mix design criteria do not apply unless *VMA* calculations are made using bulk specific gravity, and air void content calculations are made using effective specific gravity.)

12.5.7 Per cent voids filled with bitumen in the compacted mixture

The voids filled with bitumen, *VFB*, is the percentage of the intergranular void space between the aggregate particles (*VMA*) that are filled with bitumen (Fig. 12.2). The *VFB*, which does not include the absorbed bitumen, is determined using:

$$VFB = \frac{100(VMA - V_a)}{VMA} \tag{12.9}$$

where *VFB* = voids filled with bitumen, percentage of *VMA*; *VMA* = voids in mineral aggregate, percentage of bulk volume; and V_a = air voids in compacted mixture, percentage of total volume.

12.6 Marshall method of mix design

The Marshall method of mix design is intended both for laboratory design and field control of bituminous hot-mix dense-graded paving mixtures. Originally developed by Bruce Marshall of the Mississippi State Highway Department, the US Army Corps of Engineers refined and added certain features to Marshall's approach and it was then subsequently formalized as ASTM D 1559 and AASHTO T 245.

12.6.1 Outline of method

The Marshall method (which is now also used in the UK[28]) uses standard cylindrical test specimens that are 64 mm high by 102 mm diameter. These are prepared using a specified procedure for heating, mixing and compacting the bitumen–aggregate mixture. The two principal features of the Marshall method of mix design are a density voids analysis and a stability-flow test of the compacted test specimens.

The *stability* of the test specimens is the maximum load resistance, in newtons, that the standard test specimen will develop at 60°C when tested as outlined in the following paragraphs. The *flow value* is the total movement or displacement, in units of 0.25 mm, occurring in the specimen between no load and the point of maximum load during the stability test.

12.6.2 General

When determining the design bitumen content for a particular blend or gradation of aggregates by the Marshall method, a series of test specimens is prepared for a range of different bitumen contents so that the test data curves show well defined relationships. Tests are normally planned on the basis of 0.5 per cent increments of bitumen content, with at least two bitumen contents above the expected design value and at least two below.

The anticipated design bitumen content can be estimated in the first instance, using any or all of the following sources: experience, computational formula, or performing the centrifuge kerosene equivalency and oil soak tests in the Hveem procedure[1]. Another quick method to arrive at a starting point for testing is to use the filler-to-bitumen ratio guideline (normal range of 0.6 to 1.2); the anticipated design bitumen content (per cent by total weight of mix) is then estimated as approximately equivalent to the percentage of aggregate in the final gradation passing the 75 μm sieve. Equation 12.10 is one example of a computational formula:

$$P = 0.035a + 0.045b + K c + F \qquad (12.10)$$

where *P* = approximate mix bitumen content, percentage by weight of the total mix; *a* = percentage (expressed as a whole number) of mineral aggregate retained on the 2.36 mm sieve; *b* = percentage of mineral aggregate passing the 2.36 mm sieve and retained on the 75 micron sieve; *c* = percentage of mineral aggregate passing 75 μm sieve; $K=0.15$ for 11–15% passing the 75 μm sieve, 0.18 for 6–10% passing the 75 micron sieve, and 0.20 for 5% passing the 75 μm sieve; and $F=0$ to 2.0%, based on the absorption of light or heavy aggregates (in the absence of other data, $F=0.7$ is suggested).

12.6.3 Preparation of test specimens

The aggregates are first dried to a constant weight at 105 to 110°C and then separated, by dry sieving, into the desired size fractions. The amount of each size fraction required to produce a batch that will give a 63.5 ± 1.27 mm high compacted specimen is then weighed in a separate pan for each test specimen; this is normally about 1.2 kg of dry aggregates. At least three specimens have to be prepared for each combination of aggregates and bitumen. The pans are then placed in an oven until the dry aggregates achieve the required mixing temperature. When the batched aggregates have reached the mixing temperature, a mixing bowl is charged with the heated aggregates, the required quantity of bitumen is added, and mixing is carried out until all the aggregate particles are fully coated.

If the viscosity of the bitumen is too high during mixing, the aggregate will not be properly coated and if the viscosity is too low, the bitumen will coat the aggregate easily but may subsequently drain off the stone during storage or transportation; thus, for satisfactory coating the viscosity should be about 0.2 Pa.s. During compaction, if the viscosity is too low, the mix will be excessively mobile, resulting in pushing of the material in front of the roller; high viscosities will significantly reduce the workability of the mix and little compaction will be achieved. It is widely recognized that the optimal bitumen viscosity for compaction is between 2 Pa.s and 20 Pa.s. These temperatures can be estimated from a plot of the viscosity versus temperature relationship for the bitumen to be used.

12.6.4 Compaction of specimens

Depending upon the design traffic category (light, medium or heavy) that the compacted mix is expected to withstand, 35, 50 or 75 blows are applied, respectively, with the compaction hammer to each end of the specimen. The traffic classifications are: light = traffic conditions resulting in a design equivalent axle load (EAL) $<10^4$, medium = design EAL between 10^4 and 10^6, and heavy = design EAL $>10^6$. After compaction, the specimens are allowed to cool in air at room temperature until no deformation results when removing each from the mould.

12.6.5 Test procedure

In the Marshall method, each compacted test specimen is subjected to the following tests and analysis in the order listed: (a) bulk specific gravity test, (b) stability and flow test, and (c) density and voids analysis. The bulk specific gravity test may be performed as soon as the freshly compacted specimens have cooled to room temperature, after which the stability and flow tests are performed.

Prior to the stability and flow testing, the specimens are immersed in a water bath at 60 ± 1°C for 30 to 40 minutes. The Marshall testing machine, a compression

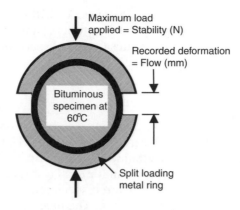

Fig. 12.3 The Marshall stability test

testing device, is designed to apply loads to test specimens through cylindrical segment testing heads at a constant rate of vertical strain of 51 mm per minute (Fig. 12.3). Loading is applied until specimen failure occurs. The Marshall testing machine is equipped with a calibrated proving ring for determining the applied testing load. The force in newtons required to produce failure of the test specimen is recorded as its Marshall stability value and the magnitude of deformation of the specimen at the point of failure is recorded as the flow value. The point of failure is defined by the maximum load reading obtained.

12.6.6 Preparation of test data

Measured stability values for specimens that depart from the standard 63.5 mm thickness have to be converted to an equivalent 63.5 mm value by means of a correlation ratio obtained from Table 12.1. Note the conversion factor may be applied on the

Table 12.1 Stability correlation ratios

Specimen volume (cm³)	Approx. specimen thickness (mm)	Correlation ratio	Specimen volume (cm³)	Approx. specimen thickness (mm)	Correlation ratio
277–289	34.9	3.33	457–470	57.2	1.19
290–301	36.5	3.03	471–482	58.7	1.14
302–316	38.1	2.78	483–495	60.3	1.09
317–328	39.7	2.50	496–508	61.9	1.04
329–340	41.3	2.27	509–522	63.5	1.00
341–353	42.9	2.08	523–535	65.1	0.96
354–367	44.4	1.92	536–546	66.7	0.93
368–379	46.0	1.79	547–559	68.3	0.89
380–392	47.6	1.67	560–573	69.8	0.86
393–405	49.2	1.56	574–585	71.4	0.83
406–420	50.8	1.47	586–598	73.0	0.81
421–431	52.4	1.39	599–610	74.6	0.78
432–443	54.0	1.32	611–625	76.2	0.76
444–456	55.6	1.25			

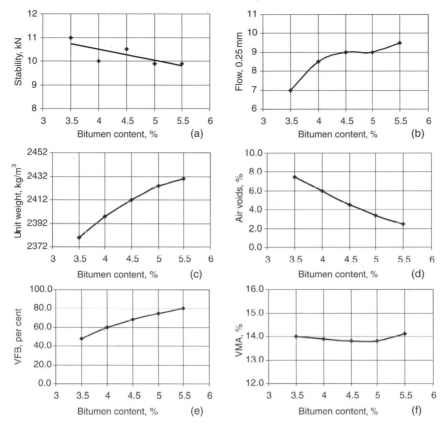

Fig. 12.4 Test property curves for hot mix design data by the Marshall method

basis of either measured thickness or measured volume. The flow values and the converted stability values for all specimens at a given bitumen content are then averaged.

As indicated in Fig. 12.4, a 'best fit' curve is plotted for each of the following relationships: stability vs bitumen content, flow vs bitumen content, unit weight of total mix vs bitumen content, percentage air voids (V_a) vs bitumen content, percentage voids filled with bitumen (*VFB*) vs bitumen content, and percentage voids in mineral aggregate (*VMA*) vs bitumen content. These graphs are used to determine the design bitumen content of the mix (Subsection 12.6.8).

12.6.7 Some comments on the test data

By examining the test property curves information can be gained about the sensitivity of the mixture to bitumen content.

The test property curves follow a reasonably consistent pattern for dense-graded bituminous paving mixes, but variations will and do occur. Trends generally noted are: (a) the stability value increases with increasing bitumen content up to a maximum after which the stability decreases; (b) the flow value consistently increases with increasing bitumen content; (c) the curve for unit weight of total mix follows the trend similar to the stability curve, except that the maximum unit weight normally (but not always) occurs at a slightly higher bitumen content than the maximum stability; (d) the per cent air voids, V_a, steadily decreases with increasing bitumen content, ultimately approaching a minimum void content; (e) the per cent voids in the mineral aggregate, VMA, generally decreases to a minimum value then increases with increasing bitumen content; and (f) the per cent voids filled with bitumen, VFB, steadily increases with increasing bitumen content, due to the VMA being filled with bitumen.

12.6.8 Determination of preliminary design bitumen content

The design bitumen content of the bituminous paving mix is selected by considering all of the data discussed previously. As an initial starting point, the Asphalt Institute recommends choosing the bitumen content at the median of the percentage air voids criteria in Table 12.2, i.e. 4 per cent; in the example given in Fig. 12.4(d), this would equate to a bitumen content of 4.7 per cent. All of the

Table 12.2 Marshall Mix design criteria

Marshall method Mix criteria	Light traffic: Surface and base		Medium traffic: Surface and base		Heavy traffic: Surface and base	
	Min	Max	Min	Max	Min	Max
Compaction: No. of blows to each end of specimen	35		50		75	
Stability, N	3336		5338		8006	
Flow, 0.25 mm	8	18	8	16	8	14
% air voids	3	5	3	5	3	5
% voids in mineral aggregate (VMA)			See Table 12.3			
% voids filled with bitumen (VFB)	70	80	65	78	65	75

Table 12.3 Minimum values for voids in mineral aggregate (VMA)

Nominal maximum particle size* (mm)	Design air voids, %		
	3.0	4.0	5.0
1.18	21.5	22.5	23.5
2.36	19.0	20.0	21.0
4.75	16.0	17.0	18.0
9.5	14.0	15.0	16.0
12.5	13.0	14.0	15.0
19.0	12.0	13.0	14.0
25.0	11.0	12.0	13.0
37.5	10.0	11.0	12.0

calculated and measured mix properties at this bitumen content (stability = 10.2 kN, flow = 9 × 0.25 mm, *VFB* = 70 per cent, and *VMA* = 14 per cent) are then compared with the relevant mix design criteria in Tables 12.2 and 12.3. If any design criteria are not met, then some adjustment or compromise is required, or the mix may need to be redesigned.

In relation to Table 12.3 it should be noted that the nominal maximum particle size is one size larger than the first sieve to retain more than 10 per cent of the aggregate. For example, if 11 per cent by weight of the aggregates is retained on the 12.5 mm sieve, the nominal maximum particle size for that mix is 19 mm.

In the UK, BSI 1990[28] specifies bituminous mix design, using the Marshall test, for hot rolled asphalt. In this case the bitumen contents corresponding to maximum mixture density, maximum compacted aggregate density, and stability, are determined and the mean value calculated. The engineer can determine the 'target bitumen content' of the mixture by adding extra bitumen, typically 0 to 1.0 per cent, to allow for such properties as workability and durability. Finally, depending upon the design traffic category and the bituminous mix type, limiting criteria exist for stability, flow and other volumetric properties in a manner similar to the Asphalt Institute's recommendations in Table 12.2.

12.6.9 Selection of final mix

The selected mix design is usually the most economical one that will satisfactorily meet all of the established design criteria. However, this mix should not be designed to optimize one particular property. For example, mixes with abnormally high values of stability are often less desirable because pavements with such materials tend to be less durable, and may crack prematurely, under heavy volumes of traffic. This situation is especially critical if the subbase and subgrade materials beneath the pavement are weak and permit moderate to relatively high deflections under the actual traffic.

The design bitumen content should be a compromise selected to balance all of the mix properties. Normally, the mix design criteria will produce a narrow range of acceptable bitumen contents that pass all of the guidelines (see Fig. 12.5), and the bitumen content selection is then adjusted within this narrow range to achieve a mix property that will satisfy the needs of a specific project. Different properties are more critical for different circumstances, depending on traffic loading and volume, pavement structure, climate, construction equipment, and other factors. Thus, the balancing process that is carried out prior to establishing the final design bitumen content is not the same for every pavement and for every mix design.

In many cases the most difficult mix design property to achieve is the minimum amount of voids in the mineral aggregate. The goal is to furnish enough space for the bitumen so it can provide adequate adhesion to bind the aggregate particles, but without bleeding when temperatures rise and the bitumen expands. Normally, the curve exhibits a flattened U-shape, decreasing to a minimum value and then increasing with increasing bitumen content, as shown in Fig. 12.6(a).

One might expect the *VMA* to remain constant with varying bitumen content, thinking that the air voids would simply be displaced by bitumen. In reality, the

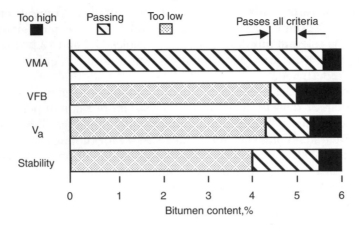

Fig. 12.5 An example of the narrow range of acceptable bitumen contents arising from the Marshall test

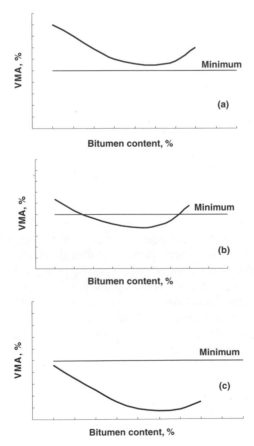

Fig. 12.6 Relationship between VMA and the specification limit

total volume changes across the range of bitumen contents. With the increase in bitumen, the mix actually becomes more workable and compacts more easily (meaning more weight is compressed into less volume) and, up to a point, the bulk density of the mix increases and the *VMA* decreases.

At some point as the bitumen content increases (the bottom of the U-shaped curve) the *VMA* begins to increase because relatively more dense material (aggregate) is displaced and pushed apart by the less dense material (bitumen). Bitumen contents on the 'wet' or right-hand increasing side of this *VMA* curve should be avoided, even if the minimum air void and *VMA* criteria are met. Design bitumen contents in this wet range have a tendency to bleed and/or exhibit plastic flow when placed in the field; i.e. any additional compaction from traffic leads to inadequate room for bitumen expansion, loss of aggregate to aggregate contact and, eventually, 'rutting' and 'shoving' in high traffic areas. Ideally, the design bitumen content should be selected slightly to the left of the low point of the *VMA* curve in Fig. 12.6(a), provided none of the other mixture criteria are violated.

In some mixes, the bottom of the U-shaped *VMA* curve is very flat, meaning that the compacted mixture is not very sensitive to bitumen content in this range. In the normal range of bitumen contents, compactability is more influenced by aggregate properties. However, at some point the quantity of bitumen will become critical to the behaviour of the mix, and the effect of bitumen will dominate as the *VMA* increases drastically.

If the bottom of the U-shaped *VMA* curve falls below the minimum criteria level required for the nominal maximum aggregate size of the mix (Fig. 12.6(b)), it is an indication that changes to the job mix formula are necessary and that the aggregate grading should be modified to provide additional *VMA*.

The design bitumen content should not be selected at the extremes of the acceptable range, even though the minimum criteria are met. On the left-hand side the mix would be too dry, prone to segregation and, probably, too high in air voids. On the right-hand side, the mix could be expected to rut.

If the minimum *VMA* criteria are completely violated over the entire bitumen content range, i.e. if curve is completely below minimum (Fig. 12.6(c)), a significant redesign and/or change in materials sources is normally warranted.

It should be emphasized that the desired range of air voids (3 to 5 per cent) is the level desired after several years of traffic and that this range does not vary with traffic (Table 12.2). This air void range will normally be achieved if the mixture is designed at the correct compactive effort (which is a function of the traffic load) and the per cent air voids immediately after construction is about 8 per cent, as some consolidation can be expected under traffic.

Research has shown that *asphaltic concrete* mixtures that ultimately consolidate to less that 3 per cent air voids are likely to rut and shove if placed in heavy traffic locations. Several factors may contribute to this occurrence, e.g. an arbitrary or accidental increase in bitumen content at the mixing facility or an increased amount of ultra-fine particles passing the 75 μm sieve (beyond that used in the laboratory) which acts as a bitumen extender (see also Subsection 12.7.12).

Problems can also occur if the final air void content is above 5 per cent or if the pavement is constructed with over 8 per cent air void initially. Brittleness, premature cracking, ravelling and stripping are all possible under these conditions.

Although the *VFB*, *VMA* and V_a are all interrelated, and only two of the values are necessary to calculate the third, including the *VFB* criteria helps prevent a mix design with marginally acceptable *VMA*. The main effect of the *VFB* criteria is to limit maximum levels of *VMA*, and, subsequently, maximum levels of bitumen content.

VFB also restricts the allowable air void content for mixes that are near the minimum *VMA* criteria. Mixes designed for lower traffic volumes will not pass the *VFB* criteria with a relatively high percentage of air voids (5%) even though the air void criteria range are met. This ensures that less durable mixes are avoided in light traffic situations.

Mixes designed for heavy traffic will not pass the *VFB* criteria with relatively low percentage air voids (less than 3.5 per cent), even though that amount of air voids is within the acceptable range. Because low air void content can be very critical in terms of permanent deformation, the *VFB* criteria helps to avoid those mixes that would be susceptible to rutting in heavy traffic situations.

12.7 Advanced mix design methods

To augment the standard Marshall method, many agencies have adopted laboratory design procedures, methods and/or systems that they have found suitable for their conditions. The advantage of these systems is that the agencies can develop clear criteria that are backed up by performance data from actual pavements. However, these agencies also have to conduct many experiments to achieve this experience, which is only applicable to the materials and environmental conditions tested. New products and materials require additional experimentation.

12.7.1 Determination of optimum bitumen content using advanced testing

The expectations of road users with respect to the smoothness and safety aspects of road surfacings are continually rising, and road engineers are expected to constantly improve bituminous mix performance and durability by adopting improved mix designs and types. Thus, in addition to the standard Marshall method requirements, very many advanced mix design methods employ sophisticated characterization tests to improve the accuracy and reliability of the optimum bitumen content determination for particular conditions. For example, the mixes may have to withstand severe traffic intensities and loading situations. The surfacings of pavements in semi-arctic regions and in extremely hot climates must be able to withstand low temperature cracking and excessive rutting, respectively. Extreme cases also occur where road surfaces are exposed to large temperature fluctuations between day and night, resulting in thermal fatigue.

The optimum binder content determination of a dense bituminous mixture may be improved by compacting all test specimens with a gyratory compactor and averaging the binder contents corresponding to the following parameters[12]: maximum stability, maximum density, minimum voids in the mineral aggregate (limits in Table 12.2 are adhered to), minimum permeability and maximum creep stiffness.

The optimum value thus obtained should lie within 3 to 5 per cent porosity, below 14 units of flow and within the *VFB* limits set in Table 12.2.

Depending on the level of performance and durability expected from the compacted bituminous mix, additional minimum mix property requirements may be set for indirect tensile stiffness, Marshall quotient, retained strength after moisture conditioning, rate of wheel tracking and fatigue life.

Subsections 12.7.2 to 12.7.12 provide brief descriptions of some of the more advanced bituminous mix testing methods. The list is not exhaustive and for more information, the reader is referred to the large body of literature that exists on the topic.

12.7.2 Bituminous mix compaction using gyratory compactors

As illustrated in Fig. 12.7, the main characteristic of a gyratory compactor is that it facilitates the application of an axial static pressure at the same time as it subjects a specimen to dynamic shear 'kneading motion'. The bituminous mixture is thus subjected to shearing forces similar to those encountered under the action of a roller during field compaction[9] (details of the test are given in CEN test method PrEN 12697–31).

The energy applied by a gyratory compactor should be close to the energy applied in the field and, consequently, compaction details (angle/speed of gyration and axial pressure) vary from one country to another. A good example is the US Superpave gyratory compactor which produces 600 kPa compaction pressure for 150 mm diameter specimens, and rotates at 30 rev/min with the mould positioned at

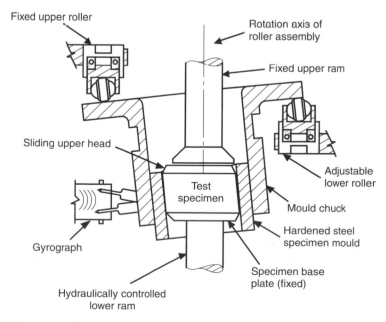

Fig. 12.7 Cross section of a gyratory compactor

a compaction angle of 1.25°. The required number of gyrations using the Superpave compactor is determined from the traffic level expected on the pavement and the design 7 day maximum air temperature for the site, as shown in Table 12.4.

In Table 12.4, N_d (N-design) is the number of gyrations required to produce a density in the mix that is equivalent to the expected density in the field after the indicated amount of traffic. In the mix design process, a bitumen content is selected that will provide 4 per cent air voids when the mix is compacted to N_d gyrations. N_i (N-initial) is a measure of mixture compactibility. Mixtures that compact too quickly are believed to be too tender during construction and may be unstable when subjected to traffic. A mix that has 4 per cent air voids at N_d should have at least 11 per cent air voids at N_i; mixtures that fail this requirement are often finer mixtures, and often tend to have a large amount of natural sand.

N_m (N-maximum) is the number of gyrations required to produce a density in the laboratory that should never be exceeded in the field, whereas N_d provides an estimate of the ultimate field density. N_m provides a compacted density with a safety factor which ensures that the mixture does not densify too much and have in-place voids that are so low that it leads to rutting. The air voids at N_m are required to be at least 2 per cent, as mixtures that have less air voids at N_m are believed to be more susceptible to rutting than those with more than 2 per cent air voids.

The number of gyrations for N_d was established on the basis of laboratory and field testing which compared in-place density with laboratory density for various numbers of gyrations. Once N_d was established for each traffic level and air temperature, the numbers of gyrations for N_i and N_m, were determined from Equations 12.11 and 12.12:

$$N_i = (N_d)^{0.45} \tag{12.11}$$

$$N_m = (N_d)^{1.10} \tag{12.12}$$

To summarize, for optimum aggregate gradations and bitumen contents, the Superpave gyratory compaction requirements are as follows: (i) at N_i the paving mix must attain 89 per cent of theoretical maximum specific gravity G_{mm} or less, (ii) at N_d the mixture must attain 96 per cent of G_{mm}, and (iii) at N_m the mixture must attain <98% of G_{mm} or an air voids content >2 per cent. By monitoring the mixture densification curve during gyratory compaction, a first attempt at optimum

Table 12.4 An example of the number of initial (N_i), design (N_d) and maximum (N_m) gyrations required for two selected traffic levels and high air temperature environments

| | Design 7 day max. air temperature (°C) | | | | | |
| | <39 | | | 39–41 | | |
Traffic ESALs	N_i	N_d	N_m	N_i	N_d	N_m
$<3 \times 10^6$	7	86	134.	8	95	150
$<1 \times 10^7$	8	96	152	8	106	169

bitumen content determination is achieved by selecting a bitumen content that will produce 4 per cent air voids content at N_d.

12.7.3 Mix ageing

To adequately characterize the ageing of a bituminous material, samples that are being evaluated in the laboratory must be aged to simulate the in-place properties of the mixture. Since the rate of ageing of the bitumen in a mixture is affected by such mixture properties as in-place air voids, the entire mixture must be aged.

The Superpave short-term ageing procedure for bituminous mixtures involves exposure of the loose mix, immediately after mixing and prior to compaction, to 4 h at 135°C in a forced draft oven. The process is intended to simulate the ageing of the mixture during production and placement in the field. It also allows the bitumen to be absorbed into the coarse aggregate, to simulate what happens during construction.

With the Superpave long-term ageing procedure, compacted specimens that have undergone short-term ageing (in the loose form) are subsequently exposed to 85°C temperature in a forced draft oven. The time for which the compacted specimens are kept in the oven is varied according to the duration of the working life of the pavement being simulated. The recommended time is 2 days, which is considered to correspond to roughly 10 years of pavement service. After ageing the samples in this way, they are tested in a condition similar to that in the roadway and, thus, should provide a better measure of expected performance.

12.7.4 Mix workability

The problem of measuring the workability of bituminous mixes is not new. For example, the measurement of the torque required to mix the mineral aggregates with bitumen has been proposed[10], as have parameters obtained from the triaxial test of bituminous mixtures[11]. Another method that has been used to assess the workability of gap-graded mixes[12] involves monitoring the specimen volume reduction during gyratory compaction so that the percentage air voids in the compacted mix can be calculated at any given number of revolutions. The rate of reduction of air voids with increasing compactive effort is then used as an indicator of the mix workability.

12.7.5 Measurement of permeability

Permeability measures the capacity of a porous medium to transmit a fluid, whereas the usual measure of air voids in a bituminous mixture provides no information about their availability to the external environmental forces that produce disintegration. It can therefore be argued that permeability is an advanced measure of durability.

'Permeability' can be defined as the volume of fluid of unit viscosity passing through, in unit time, a unit cross-section of a porous medium under the influence of a unit pressure gradient. Defined in this way, the permeability of a porous medium is independent of the absolute pressure or velocities within the flow system, or of the nature of the fluid, and is characteristic only of the structure of the medium. However, it has physical meaning only if the flow is of a viscous rather than a turbulent character. (*Note*: The term fluid includes both gases and liquids.)

In dimensions, permeability corresponds to an area, carrying the dimensions of (length)^2. The problems generally encountered in engineering deal with the flow of water in cases where the unit weight and viscosity of the water vary within fairly narrow limits. Thus, it has become customary to use a factor called the 'coefficient of permeability' when dealing with this problem; this is related to the intrinsic permeability by the expression:

$$k = K\frac{\gamma}{\eta} \qquad (12.13)$$

where K = intrinsic permeability (cm^2), k = coefficient of permeability (cm/s), γ = unit weight of permeating fluid (g/cm^3) and η = viscosity of fluid (g-s/cm^2).

Table 12.5 is an example of a bituminous material classification that is based on the coefficient of permeability.

If the pore system of a bituminous sample is fully saturated with water, and if sufficient pressure is applied, water may flow

$$V = \frac{K \times a \times \Delta p}{\eta \times L} \qquad (12.14)$$

where V = flow rate (m^3/s), K = intrinsic permeability (m^2), a = cross-sectional area of the specimen (m^2), Δp = fluid pressure head across the specimen (N/m^2), η = viscosity of the fluid (Ns/m^2) and L = length of specimen (m). (Note that the intrinsic permeability (K) has a unit of area, and is dependent on the properties of the porous media only, as the viscosity of the fluid is included in the equation; this is in contrast to the coefficient of permeability which has a unit of velocity, and is dependent on both the properties of the fluid and of the porous media.)

Air permeability measuring techniques of dense bituminous materials are mostly based on ideas that were originally developed to measure the permeability of mortars and concretes using differential pressure techniques. Air permeability testing is non-destructive, allows the determination of permeability in a very short period[13] and permits the specimens to be further tested for measurements of other parameters e.g. stability or creep stiffness. The test is also very effective in detecting variations in binder content and, therefore, can be used to control mix composition. Small changes in air voids content cause very large changes in permeability.

An air permeability apparatus used with bituminous mixes needs a specimen holder that will confine and seal the curved sides of the cylindrical compacted specimen,

Table 12.5 Classification of bituminous mixtures in terms of permeability[13]

k (cv/s)	Permeability
1×10^{-8}	Impervious
1×10^{-6}	Practically impervious
1×10^{-4}	Poor drainage
1×10^{-2}	Fair drainage
1×10^{-1}	Good drainage

whilst maintaining the flat ends free. Pressurized gas (usually air) is then introduced at one flat end of the specimen and, because of the side confinement, gas flow is then only possible through the interconnected air voids within the specimen and not around the sides. Other requirements are an accurate pressure gauge, a stable gas supply and an accurate flow meter at the downstream side. Knowing the viscosity of the gas at the test temperature, pressure and flow rate readings can be entered into Equation 12.14 to determine the specimen permeability.

In dense mixtures, air voids have to be kept to a minimum while leaving sufficient voids for the bitumen to expand and to accommodate void reduction due to over-compaction from traffic. On the other hand, a porous asphalt mix[14] must incorporate sufficient voids to maintain a permeable structure for adequate drainage, i.e. have adequate *water permeability*, but not so much as to impair its strength and stability. Using the apparatus in Fig. 12.8, the coefficient of vertical permeability k_v (m/s) can be calculated from

$$k_v = \frac{4 \times Q_v \times l}{\Delta h \times \pi \times D^2}(m/s) \qquad (12.15)$$

where; Q_v=vertical flow rate through the specimen (m^3/s), l=thickness of the specimen (m), Δh=height of water column (m), and D=diameter of the specimen (m). (*Note*: The water permeability values for porous asphalt gradations used in the UK are normally expected to be between 0.5×10^{-3} and 3.5×10^{-3} m/s.)

12.7.6 Indirect tensile stiffness modulus

Much work has gone into the development of economic and practical means of measuring the structural- and performance-related properties of bituminous materials. The stiffness modulus, which is considered to be a very important perfor-

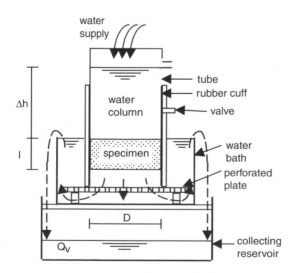

Fig. 12.8 Apparatus for measuring the vertical water permeability of bituminous specimens

mance property of the roadbase and basecourse, is a measure of the load-spreading ability of the bituminous layers; it controls the levels of the traffic-induced tensile strains at the underside of the lowest bituminous bound layer which are responsible for fatigue cracking, as well as the stresses and strains induced in the subgrade that can lead to plastic deformations. The non-destructive indirect tensile test[15] (see Fig. 12.9) has been identified as a potential means of measuring this property. With this test the stiffness modulus, S_m (MPa), is given by:

$$S_m = \frac{L(\nu + 0.27)}{Dt} \qquad (12.16)$$

where L =peak value of the applied vertical load (N), D =mean amplitude of the horizontal deformation obtained from two or more applications of the load pulse (mm), t =mean thickness of the test specimen (mm), and ν =Poisson's ratio (=0.35).

During testing (which is carried out at a standard 20°C), great attention is given to ensuring that the magnitude of the load pulses applied do not cause the specimen to deform outside the elastic recoverable range, as this may result in irrecoverable plastic deformations or microcracks. To accomplish this, the rise time – this is the time taken for the applied load to increase from zero (i.e. from when the load pulse commences) to its maximum value – is set at 124±4 ms. The load pulse, defined as the period from the start of an application of load until the start of the next application of load, is equated to 3.0±0.05 s. The target load factor area is 0.6; this is shown as the shaded area in Fig. 12.10. The peak load value is adjusted to achieve a peak transient horizontal deformation of 0.005 per cent of the specimen diameter.

12.7.7 Creep stiffness

Shear deformations resulting from high shear stresses in the upper portion of a bituminous layer appear to be the primary cause of rutting in flexible pavements.

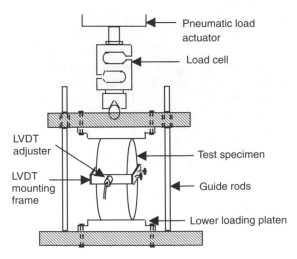

Fig. 12.9 Indirect tensile stiffness modulus test configuration

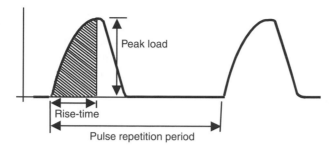

Fig. 12.10 Form of load pulse, showing the rise time and peak load

Repeated applications of these stresses under conditions of comparatively-low mix stiffness lead to the accumulation of permanent deformations at the pavement surface.

Under the centreline of a wheel load, the bituminous composite, according to elastic multilayer theory, is subjected to a compressive stress state at the top portion of the layer, and a predominantly tensile stress at the bottom. However, the application of a tensile stress to reproduce the theoretical stress condition in the bottom of the layer is quite difficult, and not suitable for use in a routine design method for evaluating the rutting characteristics of bituminous mixes.

Near the centre of the bituminous layer, at the neutral axis, the lateral stress is zero, and an axial compressive stress state exists that is similar to that in an unconfined compression test. The existence of a simple unconfined compressive stress state near the centre of the layer suggests that such a condition may be reasonably close to the average of the tensile and compressive stress states within the layer. Thus, while the unconfined creep test does not duplicate representative stress states in the upper portion of the bituminous layer it, nevertheless, is a useful test to be considered as part of a procedure for rutting evaluation for very slow moving or static loads.

The creep test is carried out either in the static or dynamic mode of loading. Each test typically lasts two hours (1 h loading plus 1 h recovery), and gives results which allow the characterization of the mixes in terms of their long-term deformation behaviour[16]. Tests are normally conducted at 40 or 60°C and a typical value of applied creep stress is 0.1 MPa. During the test, axial deformation is continuously monitored as a function of time. Thus, knowing the initial height of the specimen, the axial strain, ε, and, hence, the stiffness modulus S_{mix}, can be determined at any loading time from Equation 12.17:

$$S_{mix} = (\text{Applied stress} \div \text{Axial strain}) \qquad (12.17)$$

Better correspondence, in terms of mix ranking, with respect to creep deformation or strain rate has been shown to exist between the repeated load axial creep test[22] and the wheel tracking test than the ranking produced by the static creep test[7]. For the repeated load test, the applied stresses are as before, but the loading regime varies slightly, i.e. pulse width = 1 s, pulse period = 2 s, and the test is ended after 3600 pulses (giving an accumulated loading time of 1 h).

Unless testing is conducted purely for the sake of ranking the performance of samples of identical gradations, specimen confinement (in a similar manner to triaxial test apparatus in soils mechanics) during creep loading is recommended. This is especially important for porous asphalt mixes which have been shown to fail in the unconfined creep mode of testing and yet when well designed, can perform as well as dense graded mixes in the wheel tracking test. The actual amount of confinement required varies, but it is hypothesized to be dependent upon the angle of internal friction of the aggregates and mix cohesion, which are influenced by the aggregate gradation and angularity and the stiffness of the binder.

12.7.8 Resistance to disintegration

The Cantabro test of abrasion loss is used to quantitatively assess the resistance to disintegration of Marshall-sized porous asphalt specimens in the laboratory. By varying the bitumen content for a given porous asphalt gradation, the minimum amount of bitumen required to ensure resistance against particle losses resulting from traffic can be determined. The procedure consists of subjecting a specimen to impact and abrasion in the Los Angeles Abrasion drum (omitting the steel balls) at a controlled temperature. The aim of the test is to determine the Cantabro loss; this is the percentage weight loss after 300 drum revolutions in relation to the sample's initial weight. Typically, a 30 per cent Cantabro abrasion loss is the limit for mix acceptance at 18°C, or 25 per cent at 25°C. Currently, the test is being developed as European Standard draft test method PrEN 12697–17.

12.7.9 Indirect tensile strength test

The indirect tensile strength test (ITS) is performed by loading a cylindrical specimen with a single or repeated compressive load which acts parallel to and along the vertical diametral plane (see ASTM D4123 and PrEN 12697–23). The loading configuration shown in, for example, Fig. 12.11(e) develops a relatively uniform tensile stress perpendicular to the direction of the applied load and along the vertical diametral plane, which ultimately causes the specimen to fail by splitting along the vertical diameter. Based upon the maximum load carried by a specimen at failure, the indirect tensile strength is calculated from the following equation:

$$ITS = \frac{2P_{max}}{\pi \times t \times d} \tag{12.18}$$

where ITS = indirect tensile strength (N/mm^2), P_{max} = maximum applied load (N), t = average height (i.e. length) of the test specimen (mm), and d = diameter of the specimen (mm).

The indirect tensile test provides two mixture properties that are useful in characterizing bituminous mixes. The first of these is tensile strength, which is often used to evaluate the water susceptibility of mixtures, i.e. the tensile strength is measured before and after water conditioning of samples to determine the retained tensile strength as a percentage of the original tensile strength (see also Subsection 12.7.11). Tensile strength is also sometimes used to help evaluate the cracking potential of a bituminous mixture. The second property derived from this

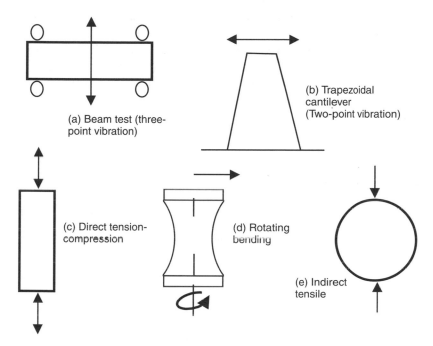

Fig. 12.11 Different configurations for measuring the fatigue life of bituminous specimens

test is tensile strain at failure, which is more useful for predicting cracking potential. Mixes that can tolerate high strains prior to failure are more likely to resist cracking.

12.7.10 Wheel tracking test

This is a very severe test and is the preferred method for ranking wearing course mixtures that are expected to perform under high temperatures and/or heavily stressed areas associated with channelled traffic and very heavy loads. Wheel tracking illustrates conclusively, as opposed to conventional creep testing, the advantages of using polymer modified bitumens for wearing courses[17].

Briefly, the wheel tracking test involves subjecting a 50 mm thick slab of material – this may be either manufactured in the laboratory or a 200 mm diameter core taken from a wearing course after laying – to a rolling standard wheel load, which traverses the specimen at a constant temperature of, normally, 45°C. The test measures rutting under the wheel over a period of time[18].

Equation 12.19, which was developed[20] from test site results, suggests the number of commercial vehicles travelling at 60 mph that are required to form a 10 mm rut at the end of a 20-year design life:

$$\text{Maximum } WTR = \frac{14\,000}{C_v + 100} \tag{12.19}$$

where WTR = wheel tracking rate (mm/h) and C_v = number of commercial vehicles per lane per day. As the effect of loading time on mix stiffness and, thus, deformation is significant, a correction factor needs to be applied to Equation 12.19 for speeds other than 60 mph.

The Marshall quotient (kN/mm) is defined as the ratio of Marshall stability (kN) to Marshall flow (mm). The relationship between the Marshall quotient (MQ) and the WTR has been shown to be statistically significant, e.g. one investigation[20] (involving a hot rolled asphalt with 30 per cent stone content) found that an equation of the form $WTR = 15 \times e^{-1.2\,MQ}$ had a correlation coefficient of 0.85.

12.7.11 Adhesion testing

Detachment of bitumen from aggregate (i.e. stripping) is associated with mixes that are permeable to water. If stripping occurs, it can result in a loss of internal cohesion and, possibly, disintegration of a surfacing. There is little risk of stripping in low void content bituminous mixes. However, materials that are permeable to water, even those that are relatively dense, are at risk of stripping. The degree of risk is, a function of the affinity between the aggregate and bitumen and its consequent ability to resist the displacing effect of water.

Immersion mechanical tests involve measurements of changes in a mechanical property of a compacted bituminous mix after its immersion in water. Thus, the ratio of the property after immersion to the initial property is an indirect measure of stripping. Possibly the most popular measurement is the 'retained Marshall stability' i.e. the ratio (expressed as a percentage) of the Marshall stability of a bituminous specimen after wet conditioning to that of an identical specimen that is not subjected to the conditioning process. The ratio of the indirect tensile strengths of conditioned and dry specimens is another measure that is also used.

Immersion trafficking tests recognize that traffic can play an important role in stripping. A test which simulates the effect of traffic is the immersion-wheel tracking test. The test configuration is as described in Subsection 12.7.10, except that the specimens in this test are kept immersed horizontally in a water bath at 40°C during testing.

12.7.12 Over-compaction

In service, the performance of bituminous mixes are adversely affected by their inability to resist over-compaction, especially during the initial stages of their lives. Voids closure due to the kneading action of traffic, which is a reflection of the resistance to over-compaction, can be simulated in the laboratory by compacting bituminous specimens to refusal density using either a gyratory compactor or a vibrating hammer. At refusal, the material is considered to be at the ultimate state of compaction which can be achieved in practice.

With this test, specimens prepared at binder contents that are at and slightly above the optimum bitumen value are over-compacted by a gyratory compactor using a compaction effort corresponding to the maximum expected traffic intensity (typically up to 350 revs). By monitoring the specimen height during over-compaction, the air voids content of the specimen can be calculated and plotted against the number of compactive gyrations up to 350 revolutions. The rate at which the air voids content reduces with compactive effort is an indication of the susceptibility of the mix to traffic densification, e.g. steep slopes indicate rut susceptible mixes. Limits are also set for the air voids content at refusal density, e.g. a minimum of 1 per cent. The effects of variations in aggregate gradation, coarse and fine aggregate types and angularity, filler content, temperature and bitumen type on the rate of densification, can thus be assessed.

12.7.13 Fatigue testing

Under traffic loading the layers of a flexible pavement structure are subjected to continuous flexing. The magnitude of the strains is dependent on the overall stiffness and nature of the pavement construction, but analysis confirmed by in-situ measurements has indicated that tensile strains of the order of 30×10^{-6} to 200×10^{-6} occur under a standard wheel load. Under these conditions the possibility of fatigue cracking – this is the phenomenon of fracture under repeated or fluctuating stress having a maximum value generally less than the tensile strength of the material – exists.

Fatigue tests are carried out by applying a load to a specimen in the form of an alternating stress or strain and determining the number of load applications required to induce failure of the specimen. As indicated in Fig. 12.11, a number of tests have been developed for the measurement of the dynamic stiffness and fatigue characteristics of bituminous mixes; these include bending tests using beams or cantilevers, compressive, compressive/tensile (push/pull), and indirect tensile tests[8,21]. With bending tests the maximum stress occurs at a point on the surface of the specimen and its calculation, using standard beam bending formulae, depends on the assumption of linear elasticity.

In *controlled stress testing*, which is more applicable to thick bituminous construction (>150 mm), the peak value of the cyclic load applied to the specimen is kept constant whilst monitoring the resultant strains. In *constant stress testing*, the measured lives do not usually contain a large amount of crack propagation time and the end point of a test is very definite, i.e. complete fracture of the specimen.

With controlled *strain testing*, which is more applicable to thin surfacings (≤50 mm), the damage accumulates during testing, and the stress required to maintain the initial strain gradually decreases after crack initiation as the stiffness of the mix is effectively decreased. When a crack initiates, a reduction in stress occurs around the crack which enables the same strain level to be maintained; consequently, crack propagation is relatively slow with controlled strain testing as compared with the controlled stress mode of failure. Failure is not where the sample literally fails but is at an arbitrary end point, i.e. the sample is usually deemed to have failed when the load required to maintain that level of strain has fallen to 50 per cent of its initial value. Therefore, in general, controlled strain testing gives a greater fatigue life than controlled stress testing.

Fatigue performance can be assessed either on the basis of applied stress or the basis of the resultant initial strain. Because of the scatter of test results associated with fatigue testing, it is normal to test several specimens at each stress or strain level and the results are plotted as stress or strain vs cycles to failure on a log–log graph. Constant stress fatigue life characteristics are shown in Fig. 12.12 for the same mix (a hot rolled asphalt wearing course) at different temperatures. Note that the lines are essentially parallel and show longer fatigue lives at lower temperatures (higher stiffness values).

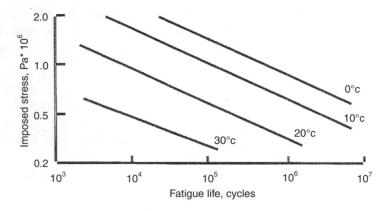

Fig. 12.12 Typical fatigue lines (stress criteria)

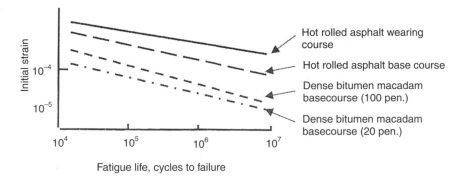

Fig. 12.13 Fatigue properties of different mixes (controlled stress)

If it is accepted that fatigue life is primarily controlled by the magnitude of strain then, under controlled stress testing conditions, stiffness will largely determine the position of the stress–life line, with higher stiffness values giving longer lives. This suggests that improved fatigue performance on the basis of stress–life may simply be a reflection of increased stiffness.

Consequently, if results are analysed on the basis of strain, the resulting strain–life relationships may be considered a better basis for the comparison of different mixes, with life at a given strain being taken as a measure of fatigue performance.

If the fatigue results are replotted in terms of resultant strain (Fig. 12.13) it is found that, for any one mix type, the results from different stiffnesses coincide. The general relationship defining the fatigue life is as follows:

$$N_{\mathrm{f}} = c \times \left(\frac{1}{\varepsilon_{\mathrm{t}}}\right)^{m} \tag{12.20}$$

where N_{f}=number of load applications to initiate a fatigue crack, ε_{t}=maximum value of applied tensile strain, c and m=factors depending on the compositions and properties of the mix and m is the slope of the strain–fatigue life line.

12.8 Construction methods for hot-mix hot-laid bituminous materials

Bituminous paving mixes made with bitumen are prepared at an asphalt mixing facility or 'central mixing plant'. Aggregates are blended, heated and dried, and mixed with bitumen to produce a hot asphalt paving mixture[24–27]. The basic steps in the construction of bituminous wearing courses, basecourses or roadbases may be summarized as follows: (1) prepare the mixture, (2) prepare the underlying or levelling course, (3) transport and place the course mixture, (4) construct joints, and (5) compact and finish.

12.8.1 Preparing the mix at a central mixing plant

A central mixing plant is the plant or 'factory' at which the bituminous paving mixture is produced, in a process involving the assembly, weighing, blending, and mixing of the aggregates and bitumen, and ending with the discharge of the mixture into hauling units for transport to the job site.

Central mixing plants for the production of bituminous paving mixtures are frequently described as being portable, semi-portable, or stationary. *Portable plants* can be either relatively small units that are self-contained and wheel-mounted or larger mixing plants whose component units are easily moved from one place to another. *Semi-portable plants* are those in which the separate units are more difficult to dismantle, but can be taken down, transported on trailers, trucks, or railway cars to a new location, and then reassembled, in a process that may require a few hours or several days, depending on the scale of plant involved. *Stationary plants* are those that are permanently constructed in one location, and are not designed to be moved from place to place. The capacities of these plants range up to about 400 t/h of material, and the majority of hot mixes are prepared at temperatures ranging from 110 to 185°C.

Two general types of central mixing plants are in common use: drum mix plants and batch plants.

A schematic diagram of the components in a typical *drum hot-mix plant* is shown in Fig. 12.14. The flow of materials is basically from left to right, after the aggregates have been moved from the stock-pile area to the metal cold feed bins via a front end loader. The number of bins depends on the number of different aggregates to be used in the bituminous mixture; however, most facilities have a minimum of four bins. Each bin has slanted sides, and an attached vibrator (if gravity feed is inadequate), to ensure a constant downward flow of aggregate. Typically, an

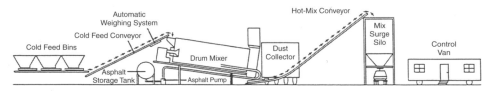

Fig. 12.14 A basic drum hot-mix plant[25] (*Courtesy of the Asphalt Institute*)

adjustable gate and a variable speed feeder are located at the bottom of each bin to proportion the materials to meet the job mix formula gradation. A gathering conveyor brings the job-mix material to a cold feed elevator which, in turn, moves it to the drum mixer. An automatic weighing system on this conveyor continuously weighs the amount of aggregate plus moisture going into the drum mixer. In the control room, a correction to the total weight is made for the moisture content of the aggregates so that the proper amount of bitumen can be pumped into the drum mixer.

A typical drum mixer has a 'parallel flow' design whereby the aggregates move in the same direction as the exhaust gases, i.e. the aggregate is fed into the drum at the burner end and then dried and heated as it moves down towards the discharge end.

The bitumen, which is pumped from a bitumen storage tank, enters the drum at a point about one-third the drum length from the discharge end. When the bitumen is added into the drum, it is pumped into the bottom of the drum at about the same location that the mineral filler and/or baghouse fines are reintroduced. Adding the bitumen and dust in close proximity allows the bitumen to trap and coat most of the fines before they are picked up by the high-velocity exhaust gas stream. The purpose of the dust collection system is to ensure that external dust emission requirements are met.

The bitumen (plus fines/dust) coats the aggregates as they move down the lower third of the drum, and the hot-mix material then exits the drum through a discharge chute into a conveyor system which carries it to a surge silo from which it is discharged into trucks, via an automatic weighing system, for transport to the site.

All movements of material from the cold feed to the surge site are monitored from the control room with the aid of sensors which monitor conveyor speeds, aggregate weights, temperature, and other critical functions which affect efficient operation.

The main components of a typical *batch hot-mix plant* are shown in Fig. 12.15. A comparison of the first two or three components in the batch and drum facilities shows that the cold feed proportioning systems are generally similar for both types prior to the drying operation.

The dryer in a batch facility is typically of a 'counter-flow' design whereby the aggregates flow in the drum in opposition to the flow of the exhaust gases. The drier, which is mounted at an angle to the horizontal, is essentially a large rotating cylinder that is equipped with a heating unit at the lower end. The heating unit is usually a low pressure air-atomization system using fuel oil, and the hot gases from the burner pass from the lower end up the cylinder and out at the upper end. The cold aggregate is fed into the upper end of the drier, picked up by steel angles or blades set on the inside face of the cylinder, and dropped through the burner flame and hot gases as it moves down the cylinder. The hot aggregate is then discharged from the lower end of the drier onto an open conveyor or an enclosed 'hot elevator' that transports it to screens and storage bins that are mounted at the top of the power plant.

In the preparation of the hot mixtures described earlier, the temperature of the aggregates may be raised to 160°C or more, so that practically all of the moisture is removed.

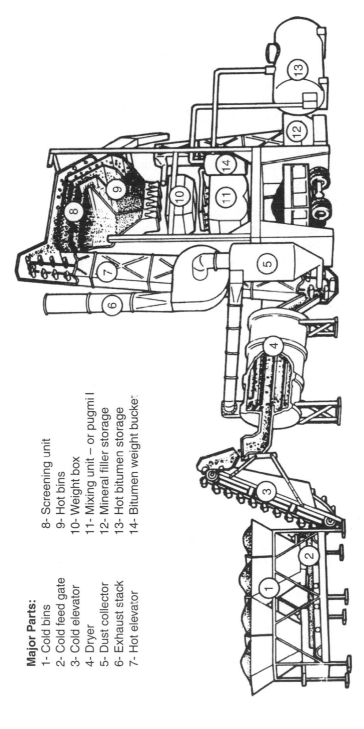

Major Parts:

1- Cold bins
2- Cold feed gate
3- Cold elevator
4- Dryer
5- Dust collector
6- Exhaust stack
7- Hot elevator

8- Screening unit
9- Hot bins
10- Weight box
11- Mixing unit – or pugmill
12- Mineral filler storage
13- Hot bitumen storage
14- Bitumen weight bucket

Fig. 12.15 An asphalt batch hot-mix plant[25] (*Courtesy of the Asphalt Institute*)

The hot dust-laden exhaust gases from the dryer are passed through a dust-collection system to remove dust particles so that emission standards are met. The collected dust is returned to the hot material elevator or filler storage silo for subsequent reintroduction into the mixture as required. Mineral filler that is added to the mixture is not normally passed through the aggregate drier; rather, it is fed by a separate device directly into the mixing unit or into the aggregate batching unit. (Separate feeder units are also used to supply liquefier, other fluxes, or hydrated lime to the mix.)

After the aggregates exit the hot elevator at the top of the tower, they are discharged onto vibrating screens that separate them into a number of sizes which are then stored in hot material bins. The stored aggregates are later proportioned, by the control system, from the bins into a weigh box that is mounted on a set of scales. These dry aggregates are then discharged into the pugmill and the hot bitumen, which has been weighed and stored in a weigh bucket, is sprayed into the pugmill after a few seconds of dry mixing. The pugmill is typically a twin-shaft, counter-rotating mixer designed to coat the aggregate quickly (typically, in about 45 s) with bitumen. After mixing, the bituminous mixture is ready to be discharged into a truck for transport to the site.

Batch plants may be manual, semi-automatic or automatic, depending on the degree of automation. In *manual batch plants*, air cylinders or hydraulic cylinders that are electrically-actuated by the operator, control the bin gates, fines feeders, bitumen supply and spray valves, the weigh box discharge gate, and the pugmill discharge gates. In *semi-automatic batch plants*, the various operations constituting each mixing cycle are under automatic control, i.e. the quantities of bitumen and aggregate introduced into the mixes, the mixing times, the sequencing of the mixing functions, and the operation of the pugmill discharge gate. The *fully automatic batch plant* repeats the weighing and mixing cycle until the operator stops it, or until it stops itself because of material shortage or some other extraordinary event. Automatic plants also normally provide records of the amounts of the component materials in each batch.

12.8.2 Control of hot-mix uniformity

Control of the uniformity of hot mixes is very important, as any appreciable variation in aggregate gradation or bitumen content will be reflected in a change in some other characteristic of the mix. Sampling and testing are therefore among the most important functions in plant control.

Samples are typically obtained at various points in the plant to establish if the processing is in order up to those points. A final extraction test of the mixture is normally carried out to confirm its uniformity, gradation, and bitumen content. The extraction test measures the bitumen content and provides the aggregate for gradation testing. The extraction and gradation results should fall within the job mix tolerance specified; if they do not, corrective measures must be taken to bring the mix within the uniformity tolerance.

12.8.3 Transport of bituminous material

Dump trucks or trailers are normally used to transport the hot-mix material from the plant to the job site. These vehicles should have smooth metal beds that have previously been cleaned of all foreign material. The vehicle bed may be sprayed with a light coat of lime water, soap solution, or some similar substance to prevent adherence of the mixture; fuel oils should not be used as they have a detrimental effect on the mixture. The vehicle is insulated against excessive heat loss in the mixture during hauling, and it is normally covered with a canvas to protect the material from the weather.

12.8.4 Placing hot-mix bituminous material

Surface courses of bituminous materials are frequently placed on new or existing bases that require very little preparation other than a thorough sweeping and cleaning to remove loose dirt and other foreign materials. On absorbent surfaces, e.g. granular bases, a prime coat (typically, a cutback bitumen) is normally applied prior to placing the new mixture. On existing paved surfaces, a tack coat (i.e. a light application of emulsified bitumen) is normally applied. In some cases, a bituminous 'regulating', i.e. levelling, course may be laid to correct irregularities in an existing surface. Placement of a bituminous mixture is permitted only when the underlying layer is dry and weather conditions are favourable. The placement of hot-mix materials is usually suspended when the ambient air temperature is less than 4°C.

Bituminous roadbase, levelling and surface courses are placed and compacted in separate operations. In certain cases, e.g. in roadbases, thick layers composed of the same mixture may be placed in two or more layers.

In the majority of cases, plant mixes (hot or cold) are placed by an 'asphalt paver' that spreads the mixture in a uniform layer of the desired thickness, and shapes or 'finishes' it to the desired elevation and cross-section, ready for compaction. Pavers are widely used with hot mixes, which must be placed and finished rapidly so that they can be compacted while hot. The wheelbases of these machines are sufficiently long to eliminate the need for forms, and to minimize the occurrence of irregularities in the underlying layer. The machines can process thicknesses up to 250 mm over a width of up to 4.3 m; working speeds generally range from 3 to 21 m/min.

As indicated in Fig. 12.16, the mixture that is to be placed by the paver is tipped into a receiving hopper from a transport unit. It is then fed from the hopper towards the finishing section of the machine, and spread and agitated by screws that ensure the uniformity of the spread material over the full processing width. The loose 'fluffed-up' material is then struck off at the desired elevation and cross-

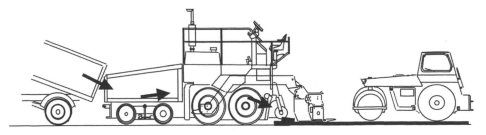

Fig. 12.16 Mode of operation of an asphalt paver

section by one or more oscillating or vibratory screeds that employ a tamping mechanism to strike off and initially compact the material. The screeds are usually provided with heating units to prevent the material from being picked up during the spreading and finishing operations. These pavers are fully adjustable to ensure a uniform flow of material through the machine and to produce a smooth, even layer of the desired thickness and cross-section.

Most bituminous pavers are fitted with electronic screed control systems. These sensors operate on a reference profile, sense changes in the position of the floating screed element of the paver or of the reference profile, and then automatically apply corrections to the angle of the screed so that the surface being laid is continually parallel to the reference profile. Usually the reference profile controls the longitudinal profile of the surface at one side of the machine whilst a slope sensor controls the transverse slope, i.e. the cross-section.

Reference profiles can be either fixed or mobile. A typical fixed reference profile is a taut wire or string line that is stretched between stakes or pins set at close spacings (usually 7 m or less). A moving reference profile is attached to the side of the paver; it may be a long (up to 12 m) rigid or articulated ski that incorporates a taut wire on which the sensor rides, a rolling string line, or a short shoe that rides, for example, on the new surface of an adjacent lane.

When placing the bituminous mixture, special attention must be given to the construction of joints between old and new surfaces or between successive days work. It is essential that a proper bond be secured at longitudinal and transverse joints between a newly placed mixture and an existing bituminous mat. The best longitudinal joint is obtained when the material in the edge being laid against is still warm enough for effective compaction; this means that two pavers working in echelon can produce the most satisfactory joint. Various other procedures are used when laying against a cold edge, e.g. the cold edge may be cut back and painted with bituminous material, or an infrared joint heater may be attached to the front of the finisher to heat the cold material along the edge prior to placing a new mat.

12.8.5 Compaction of the bituminous mat

When the spreading and finishing operations have been completed, and while the mixture is still hot, rolling is begun. Rolling may be carried out by steel wheel or pneumatic tyred rollers or by a combination of the two.

Steel wheel rollers may be of three types: three-wheel rollers of 10 to 12 t in mass, two-axle tandem rollers of 8 to 12 t, and three-axle tandem rollers of 12 to 18 t. Pneumatic-tyred rollers provide a closely knit surface by kneading aggregate particles together; the tyre contact pressures generally range from 276 to 620 kPa. Vibratory steel-wheeled rollers, which provide a centrifugal force of up to 210 kN and up to 3000 vibrations per min, are also increasingly being used to compact bituminous materials.

The bulk of the rolling is done in a longitudinal direction, beginning at the edges and gradually progressing toward the centre – except on superelevated curves where rolling begins on the low side and progresses toward the high side. Rolling procedures vary with the properties of the mixture, thickness of layer, and other factors. In modern practice, rolling is divided into three phases, which follow closely behind one another, namely initial or 'break-down' rolling, intermediate rolling, and finish rolling. The breakdown and intermediate phases primarily provide the compacted density, and the final rolling gives the final smoothness. The practice with some organizations has been to use three-wheel or tandem rollers for breakdown rolling, pneumatic tyred rollers for intermediate rolling, and tandem rollers for finish rolling.

Specifications for the finished surface frequently stipulate that it should be smooth, even, and true to the desired grade and cross-section, with no vertical deviation being permitted that is more that 3 mm from a line established by a 3 m straight edge. While this seems to be a very high standard, these surfaces are readily attainable with most modern construction equipment. Deviations of more than 6 mm from the specified thickness of a wearing course are generally not allowed. The densities to be obtained in the compacted layers are normally stipulated as a percentage of the theoretical maximum density or of the density of the laboratory compacted mixture.

The density of the compacted mixture is determined on samples cored or sawn from the completed mat. In addition to determining the densities of compacted mixtures, many agencies also determine (for control purposes) the percentage of voids and voids filled with bitumen, and obtain large enough samples to enable the gradation and bitumen content of the compacted mixture to be determined. Use is also made of nuclear density gauges to measure density and to control the rolling process.

Rolling completes the construction of most bituminous pavements, and traffic is normally permitted on the surface as soon as the compacted mixture has adequately cooled.

12.9 References

1. Asphalt Institute, *Mix Design Methods for Asphalt Concrete and Other Hot-Mix Types* (MS-2), 6th edn. Lexington, KY: The Asphalt Institute, 1995.
2. Asphalt Institute, *Superpave Level 1 Mix Design*, Superpave Series No. 2 (SP-2). Lexington, Ky: The Asphalt Institute, 1995.
3. *French Design Manual For Pavement Structures*. Paris: Laboratoire Central des Ponts et Chaussées, May 1997.
4. Hopman, P.C., Valkering, C.P. and Van der Heide, J.P.J., Towards a performance related asphalt mix design procedure, *Proceedings of the Association of Asphalt Paving Technologists*, 1992, **61**, pp. 188–216.
5. Mahboub, K. and Little, D.N., An improved asphalt concrete mix design procedure, *Proceedings of the Association of Asphalt Paving Technologists*, 1990, **59**, pp. 138–175.
6. Eckmann, B., EXXON research in pavement design – Moebius software: A case study reduction of creep through polymer modification, *Proceedings of the Association of Asphalt Paving Technologists*, 1989, **58**, pp. 337–350.
7. Gibb, J.M. and Brown, S.F., A repeated load compression test for assessing the resistance of bituminous mixes to permanent deformation. In J.G. Cabrera and J.R. Dixon (eds), *Proceedings of the 1st European Symposium on the Performance and Durability of Bituminous Materials*. London: Spon, 1994, pp. 199–209.

8. Read, J.M. and Collop, A.C., Practical fatigue characterisation of bituminous paving mixtures, *Proceedings of the Association of Asphalt Paving Technologists*, 1997, **66**, pp. 74–108.

9. D'Angelo, J.A., Paugh, C., Harman, T.P., and Bukowski, J., Comparison of the Superpave Gyratory Compactor to the Marshall for field quality control, *Proceedings of the Association of Asphalt Paving Technologists*, 1995, **64**, pp. 611–635.

10. Marvillet, J. and Bougault, P., Workability of bituminous mixes; Development of a workability meter, *Proceedings of the Association of Asphalt Paving Technologists*, 1979, **48**, pp. 91–110.

11. Fordyce, D., Markham, D., Ibrahim, H. and El-Mabruk, H., Measuring the potential compaction performance of bituminous mixtures. In J.G. Cabrera and J.R. Dixon (eds), *Proceedings of the 1st European Symposium on the Performance and Durability of Bituminous Materials*. London: Spon, 1994, pp. 121–136.

12. Cabrera J.G., Hot bituminous mixtures: Design for performance, *Proceedings of the 1st National Conference on Bituminous Mixtures and Flexible Pavements*, University of Thessaloniki, Greece, 1992, pp. 1–12.

13. Zoorob, S.E., Cabrera, J.G. and Suparma, L.B., A gas permeability method for controlling quality of dense bituminous composites. In J.G. Cabrera and S.E. Zoorob (eds), *Proceedings of the 3rd European Symposium on Performance and Durability of Bituminous Materials and Hydraulic Stabilised Composites*. Zurich: AEDIFICATIO Publishers, 1999, pp. 549–572.

14. Cabrera, J.G. and Hamzah, M.O., Aggregate grading design for porous asphalt, In J.G. Cabrera and J.R. Dixon (eds), *Proceedings of the 1st European Symposium on the Performance and Durability of Bituminous Materials*. London: Spon, 1994, pp. 10–22.

15. British Standard Draft for Development BS DD213: 1993, *Method for the Determination of the Indirect Tensile Stiffness Modulus of Bituminous Mixtures*. London: British Standards Institution (1993).

16. British Standards Institution, BS 598 Pt. 111: 1995, *Method for Determination of Resistance to Permanent Deformation of Bituminous Mixes Subject to Unconfined Uniaxial Loading*. London: British Standards Institution, 1995.

17. Lijzenga, J., On the prediction of pavement rutting in the Shell pavement design method. In J.G. Cabrera (ed.), *Proceedings of the 2nd European Symposium on the Performance and Durability of Bituminous Materials*. Zurich: AEDIFICATIO Publishers, 1997, pp. 176–193.

18. BS 598: 1996, Pt. 110, *Methods of Test for the Determination of Wheel Tracking Rate*. London: British Standards Institution, 1996. (Also PrEN 12697–22.)

19. Jacobs, FA, *A30 Winchester Bypass – The Performance of Rolled Asphalts Using the Marshall Test*, TRRL Report LR1082. Crowthorne, Berkshire: The Transport Research Laboratory, 1983.

20. Walsh, I.K., The use of wheel tracking test for wearing course design and performance evaluation. In J.G. Cabrera and J.R. Dixon (Editors), *Proceedings of the 1st European Symposium on the Performance and Durability of Bituminous Materials*. London: Spon, 1994, pp. 210–225.

21. Final draft DD ABF: 1995, *Method for the Determination of Fatigue Characteristics of Bituminous Mixtures Using Indirect Tensile Fatigue*. London: British Standards Institution, 1995. (Also PrEN 12697–24.)

22. BS DD226: 1996, *Method for Determining Resistance to Permanent Deformation of Bituminous Mixtures Subject to Unconfined Dynamic Loading*, London: British Standards Institution, 1996.

23. ASTM D70–97, *Standard Test Method for Specific Gravity and Density of Semi-Solid Bituminous Materials (Pycnometer Method)*. Philadelphia, PA: American Society for Testing and Materials, 1997.

24. Wright, P.H., *Highway Engineering*, 6th edn. London: Wiley, 1996.

25. *Asphalt Plant Manual*, MS-3. College Park, Maryland: The Asphalt Institute, 1986.

26. *Asphalt Paving Manual*, MS-8. College Park, Maryland: The Asphalt Institute, 1987.

27. *The Asphalt Handbook*, MS-4. Lexington, Kentucky: The Asphalt Institute, 1989.

28. BS 598:1990, Part 107, *Sampling and Examination of Bituminous Mixtures for Roads and Other Paved Areas: Method of Test for the Determination of the Composition of Design Wearing Course Rolled Asphalt*. London: British Standards Institution, 1990.

29. Franken, L. (ed.), *Bituminous Binders and Mixes*, Rilem Report 17. London: Spon, 1998.

CHAPTER 13

Concrete pavement construction

C.A. O'Flaherty

13.1 Steps in the construction process

Today, the construction of rigid pavements in major roads in the western world is, to all intents and purposes, mechanized and the hand-spreading of concrete is employed only when it is necessary to lay small and irregular-shaped pavement slabs. The mechanization of the construction process has resulted in reduced labour costs, an increase in the speed of road-making, and, since it allows the use of leaner concrete mixes with low water to cement ratios, an effective saving in the cost of materials.

There are two basic types of mechanized paving equipment used to lay concrete in rigid pavements: fixed-form and slip-form. Nowadays, slip-form paving is used on most major road projects, especially on long continuous stretches of road, whilst fixed-form paving trains are used on smaller sections, often in urban areas, or on complicated interchanges. Both forms of paving require a high standard of subgrade and subbase preparation to ensure uniform slab support, and high-quality finishing and curing to ensure good performance throughout the pavement's design life. The key to successful construction is good site planning and organization.

The basic steps involved in the construction of a rigid pavement can be summarized as follows: (1) prepare the foundation for the slab; (2) place the forms (if fixed-forms are to be used); (3) install the joint assemblies; (4) batch the cement and aggregates and mix and transport them to the site; (5) lay and finish the concrete; and (6) cure the concrete.

13.2 Preparing the foundation

Preparing the foundation for the concrete slab in a rigid pavement involves clearing and grading the site (see Chapter 10), installing any needed subgrade drainage (Chapter 8), stabilizing (if necessary) and compacting and finishing the subgrade (Chapters 6 and 10), and constructing (in most instances) a subbase.

In some instances the subgrade will be composed of a coarse-grained soil or uniform rock and a subbase may not be deemed essential; more often than not, however, it is required. Sometimes, good subgrades have soft spots from which the soft soil has to be ripped, removed and replaced with material of the same quality as the surrounding subgrade, and compacted. If the subgrade as a whole is soft, then a capping layer may have to be installed in order to increase its stiffness modulus and strength and prevent it from 'waving' under construction traffic.

British practice is for a capping layer to have a CBR of at least 15 per cent, and for it to be installed to a depth ranging from 150 mm (when the subgrade CBR is 15 per cent) to 600 mm (when the CBR is 2 or less). Construction of the capping layer may involve removing a layer of weak soil and replacing it with stronger material or, if the soil is cohesive and sufficiently reactive, a lime-stabilization treatment may be an economic option. If the capping layer is not designed as a drainage blanket, drains may have to be placed below the bottom of the capping to ensure drainage of the slab foundation if it is permeable and rain is likely during construction.

When the subgrade (or capping layer) has been thoroughly compacted it will need to be shaped to conform to the specified profile and cross-section. If the pavement is to be constructed over an existing roadway the old pavement will need to be scarified to a depth of at least 150 mm, and then recompacted and shaped to provide uniform support.

The subbase, whether it be a compacted granular or cement-bound material, will need to be laid to a regular and accurate level, and maintained in a clean condition until the concrete is laid; in the interim, its use by construction traffic should be strictly limited. Good control of the subbase surface level improves the uniformity and quality of performance of a concrete pavement, as any unevenness in the finished subbase will be reflected in variations in the thickness of the poured slab.

British practice[1] is for the subbase above a capping layer in a rigid-type of pavement to be 150 mm deep, and for it to be composed of cement-bound granular material or wet lean concrete.

Cement-bound materials are normally plant-mixed using paddle or pan-type mixers and laid by a paving machine similar to those used to place bituminous materials. As with unbound materials the amount of water applied to a mix is normally the optimum moisture content required for satisfactory compaction by smooth-wheeled, rubber-tyred, or vibrating rollers; typically, this is in the range of 5–7 per cent by mass. The mix must not be too dry or the surface will remain loose after compaction; if too wet the material will be picked up by the roller and the surface will be irregular.

If the subbase is to be composed of wet lean concrete the mixing may be carried out in drum mixers. If a slip-form paver is available for slab construction it is normal for the paver to be used to lay the wet lean concrete also; in this case the moisture content of the mix is typically 7–11 per cent and compaction is achieved by vibration. Alternatively, the wet lean concrete is laid using side forms. Whichever method is used the maximum time that may elapse between the admixing of water and the compacting of the lean concrete should meet the requirements in Table 13.1. Note also that Table 13.1 also states that paving is unacceptable in the UK if the concrete temperature at the time of discharge from the delivery truck is in excess of 30 degrees.

Table 13.1 Maximum working times for the laying of concrete slabs[10]

| Concrete temperature at discharge from truck | RC slabs in two layers, without retarding agents: | | All other slabs: | |
	Mixing 1st layer to finishing concrete	Between layers	Mixing 1st layer to finishing concrete	Between layers in 2-layer work
Up to 25°C	3 h	0.5 h	3 h	1.5 h
>25–30°C	2 h	0.5 h	2 h	1 h
>30°C	No paving	–	No paving	–

13.3 Placing the forms for conventional paving trains

After the subgrade and subbase have been compacted and graded, side-forms must be put in place. Typically, these side-forms are custom-built, re-useable, clean and oiled, 3 m long by the depth of the slab, wide-based steel sections that are temporarily fixed to the subbase with steel pins that are driven through holes in their flanges. With conventional fixed-form paving trains the side-forms are usually left in position during the preliminary curing of the concrete slabs: after completion of construction they are carefully removed, avoiding damage to the finished concrete, for cleaning and re-use. Consequently, the contractor must always have on-site enough side-forms for, say, about 3 days work ahead of the train.

Occasionally, the side-forms are permanent edge strips or kerbing that are pre-cast in advance of the paving.

The forms define the width of the slab, contain the concrete as it is being laid, and define the level of its top surface. Side-forms used with wheeled paving vehicles have flat-bottomed metal rails firmly fixed at a constant height below the top of the forms, along which many of the paving machines operate in sequence during construction. Since the equipment has to be supported and guided by the rails, the side-forms must be vertical, rigidly supported and securely fixed to a true line and level, and free from play or movement in any direction whilst spreading and compacting the concrete.

Upon completion of the form-laying and aligning, the foundation (usually the subbase) is checked to ensure that its surface elevations and shape are exactly as specified by the design. If the foundation material is unbound (not British practice) further trimming may need to be carried out by a 'fine-grader' which rides on the rails and employs cutting blades on a rotating drum to achieve the final grade and shape.

The separation membrane is placed on the surface of the foundation ahead of a fixed-form train (but not a slip-form one) prior to placing and fixing any joint crack inducers or assemblies. In the case of unreinforced and jointed reinforced concrete slabs the membrane is normally 125 micron thick plastic sheeting; however, with CRCP and CRCR constructions the waterproof membrane is a bituminous spray.

Too much membrane should not be laid in advance if there is a risk of damage or ponding of rainwater.

13.4 Joint assemblies and reinforcement

At this stage various joint assemblies may be placed in position, depending upon the method of construction employed. Before doing so, however, care should be taken to ensure that any transverse joints in the surface slabs and the subbase are not coincident vertically, but staggered at least 1 m apart.

Irrespective of the method used to lay the concrete, expansion joint assemblies with dowel bars and filler boards are normally prefabricated and securely supported on cradles and fixed to the subbase so that they will not shift when the concrete is being spread and compacted. To work effectively the dowels in each assembly must be aligned parallel to both the surface of the slab and the centreline of the pavement; any misalignment will result in the generation of very high stresses as adjacent slabs expand and contract. Filler boards, which are normally 25 mm thick and have holes that form a sliding fit for the sheathed dowel bars, must be set vertically to ensure that adjacent slab ends are also vertical and that there is no likelihood of one riding up over the other.

The bottom crack inducers of contraction and longitudinal joints are securely fastened to the subbase at the spacings specified in the design prior to any placement of the concrete, if this method of forming the joints is employed. Typically, the crack inducer is a triangular or inverted Y-shaped fillet, with a base width not less than the height, made of timber or of rigid synthetic material.

Dowels are commonly placed on prefabricated metal cradles and fastened to the subbase ahead of the paving train; in the UK, as an alternative to using prefabricated assemblies for contraction joints, the dowels may be vibrated into position as the slab is being formed with the fixed-form process. The tolerances to which dowel bars are specified in Britain are quite stringent, especially for those on cradles, and this onerous need has been queried[2].

Cradles supporting dowel bars should not extend across the joint line.

Tie bars in warping joints and wet-formed longitudinal joints assemblies are normally made up into rigid assemblies with adequate supports, and securely attached to the subbase in advance of the paving train. Alternatively, the tie bars at longitudinal joints may be vibrated into position, using a method that ensures compaction of the concrete about the tie bars. In the case of construction joints the tie bars are often cranked and held at right angles at the fixed-form face until the form is removed and the bars can be bent back; alternatively, the tie bars may be fixed to the side-forms or inserted into the side of the slab and recompacted.

After the joint assemblies have been put in place, then, in the case of single course reinforced construction, any reinforcing steel is securely fastened to supporting metal structures at the specified depth below the finished surface, and distance from the edge, of the slab so as to ensure that the desired cover is achieved. The steel is most commonly either fabric (or mesh) reinforcement that is supplied to the site in large sheets or rolls for placement in front of the paving machinery, or bar reinforcement that has to be assembled manually on-site. The reinforcement, which is fixed higher than tie bars and dowels, is terminated about 125 mm from slab edges and from longitudinal joints with tie bars, and about 300 mm from any transverse joint (excluding emergency construction joints). Lapping in any transverse reinforcement is normally about 300 mm, whilst longitudinal bars are overlapped at least 450 mm; no longitudinal bars should lie within about 100 mm of a longitudinal joint. In all instances the steel used must be clean so as to provide a good bond with the concrete, and the spacing of the bars should not be less than twice the maximum size of aggregate used in the concrete.

The metal support cradles for the reinforcement must be stable and robust enough to carry workmen during construction and not move when concrete is being placed. If two-layer construction is being employed, the JCR, CRCP or CRCR reinforcement, in the form of prefabricated sheets, is placed in or, more usually, on top of the bottom course.

13.5 Preparing the concrete

On large contracts, well-organized trains are able to construct many kilometres of paving per day. For this to happen concrete of a consistent quality must be provided at the site at a steady rate. Steady progress requires the careful and realistic matching of the paving train's productive capacity with the mixing plant's output capability and the availability of enough suitable vehicles to transport the fresh concrete.

In the UK, wet-mixing of the concrete most commonly takes place in a stationary batch type of mixer at a central plant that is either on-site or at a quarry or gravel pit that is conveniently located in relation to the construction site. The use of an existing off-site plant is desirable, where this is an economic and efficient option. The relative permanency of existing plants encourages the production of good quality concrete, i.e. they tend to have more elaborate storage facilities and equipment that can automatically control the flow of materials from the stockpiles, through the batching and mixing plant, to the discharge of the mixed concrete into waiting vehicles for under-cover (to prevent evaporation of water or wetting by rain) transport to the site.

A central plant should be close to the construction site as long hauls require more vehicles to keep the paving train in continuous operation. If the run is through an urban area, the risk of vehicles bunching in traffic congestion increases, with the consequent risk of an irregular supply to the train. If the access haul-road is long and bumpy, agitation is essential to avoid segregation in the wet mixture. If travelling takes too long, the concrete will begin to stiffen and its consistency and workability at the site may be detrimentally affected unless it is agitated enroute. The workability of the concrete at the placement point on-site should be such as to enable it to be fully compacted and finished without undue flow.

The *Compacting Factor test*[3] provides a very useful measure of the on-site workability of C40 concrete (which is used, in the UK, in surface slabs and in continuously reinforced roadbases) when the aggregate is crushed rock or gravel. With this test the freshly mixed concrete is allowed to fall in a controlled way, under gravity, into a steel mould and the measured density is expressed as a ratio of the fully compacted density. Typical Compacting Factor values used in fixed-form work are 0.77–0.80 and 0.80–0.83 for mixes in the bottom and top layers, respectively, of two-course slabs, and 0.80 for single-layer construction; with slip-form operation the target values are 0.88–0.91. The conventional slump test is more commonly used for wet lean concrete of grade C20 or below. Low workabilities ensure that inserted dowel bars are more easily kept in position, whilst higher workabilities allow the use of smaller surcharges, as well as better texturing and finishing.

13.6 Placing and finishing the concrete

The placing of the fresh concrete upon the subbase, or on the lower layer of two-course slabs, should be carried out so as to obtain an even depth, of uniform density, with the minimum of aggregate segregation. The spreading, compacting, and finishing procedures employed will vary according to whether a fixed-form or a slip-form paving train is used.

With a *fixed-form train* (Fig. 13.1) the concrete is usually deposited to a uniform uncompacted density by a traversing box-hopper spreader. This comprises a hopper of 3 or 4.5 m^3 capacity that is mounted on a self-propelled rail-mounted frame so that the box moves across the frame and discharges the concrete at a controlled rate as the frame moves forward. The underside of the hopper then strikes off the deposited concrete to the desired surcharge level – this is the design amount by which the initial depth of spread concrete exceeds the ultimate thickness of the slab – as it returns with the frame remaining stationary. Good surface level control is ensured by the spreader being able to spread the concrete evenly to the correct surcharge; it is bad practice to rely on regulating beams and the diagonal finisher to subsequently achieve the correct levels by major planing.

If the slab is constructed in two layers the spreading of the concrete in the top layer should be carried out within the times given in Table 13.1.

In a conventional fixed-form train, the spreader is followed by a piece of equipment with rotary strike-off paddles and a transverse vibrating compaction beam. The function of the paddles is to trim any minor irregularities in the surface of the surcharged concrete; they are mounted on independently-operated levelling screws at each side of the equipment which allow adjustments to be made for carriageway crossfalls. The compaction beam applies vibration to the surface of the concrete, with the amplitude and frequency used being adjusted to suit the characteristics of the wet mix. The density of C40 concrete without air entrainment should not exceed 3 per cent; with air entrainment the total air voids should not exceed 9.5 per cent or 8.5 per cent for concrete with 20 mm or 40 mm aggregate, respectively. Overall, the aim of the compaction process is to obtain a dense homogeneous slab of concrete that is free from voids, honeycombing and surface irregularities.

After compaction transverse dowel bars and longitudinal-joint tie bars which have not been previously attached to the subbase are vibrated into position. As noted previously, if the design calls for two-course construction with reinforcement, the steel is manually placed on the surface of the compacted bottom layer prior to the placement of the second course.

The initial regulation and finish to the slab surface is carried out by a beam that oscillates transversely or obliquely to the longitudinal axis of the pavement. This beam is a simple oscillating box-section float that is either mounted on the carriage of the compactor machine or is carried on a trailing articulated framework. Only

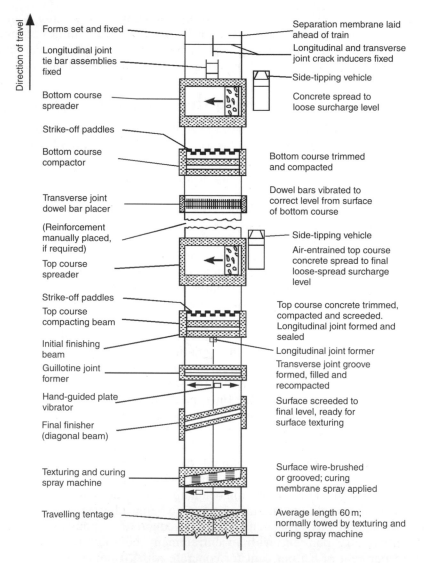

Direction of travel

Forms set and fixed ⎯⎯⎯⎯⎯⎯⎯⎯⎯⎯

Longitudinal joint
tie bar assemblies ⎯⎯⎯⎯⎯⎯
fixed

Bottom course ⎯⎯⎯⎯⎯⎯⎯
spreader

Strike-off paddles ⎯⎯⎯⎯⎯⎯

Bottom course ⎯⎯⎯⎯⎯⎯⎯
compactor

Transverse joint ⎯⎯⎯⎯⎯⎯⎯⎯
dowel bar placer

(Reinforcement
manually placed, ⎯⎯⎯⎯⎯⎯
if required)

Top course ⎯⎯⎯⎯⎯⎯⎯⎯
spreader

Strike-off paddles ⎯⎯⎯⎯⎯⎯
Top course ⎯⎯⎯⎯⎯⎯⎯⎯⎯
compacting beam

Initial finishing ⎯⎯⎯⎯⎯⎯
beam
Guillotine joint ⎯⎯⎯⎯⎯⎯⎯
former

Hand-guided plate ⎯⎯⎯⎯⎯
vibrator

Final finisher ⎯⎯⎯⎯⎯⎯⎯
(diagonal beam)

Texturing and curing ⎯⎯⎯⎯
spray machine

Travelling tentage ⎯⎯⎯⎯⎯⎯

Separation membrane laid
ahead of train

Longitudinal and transverse
joint crack inducers fixed

Side-tipping vehicle

Concrete spread to
loose surcharge level

Bottom course trimmed
and compacted

Dowel bars vibrated to
correct level from surface
of bottom course

Side-tipping vehicle

Air-entrained top course
concrete spread to final
loose-spread surcharge
level

Top course concrete trimmed,
compacted and screeded.
Longitudinal joint formed and
sealed

Longitudinal joint former

Transverse joint groove
formed, filled and
recompacted

Surface screeded to
final level, ready for
surface texturing

Surface wire-brushed
or grooved; curing
membrane spray applied

Average length 60 m;
normally towed by texturing and
curing spray machine

Fig. 13.1 Typical arrangement of fixed-form paving train for two-lane wide two-course concrete pavement construction

used for full-depth or top-course paving, this beam provides a regulated and partially-finished surface that may be adequate as a 'final' finish on minor roads or in industrial estates but requires further treatment for high-quality pavements.

Wet-formed longitudinal joint grooves are formed immediately after the initial oscillating beam has completed its work. The longitudinal joint is formed in one continuous operation using a hollow vertical 'knife' that travels submerged in the plastic concrete, inserts a preformed cellular permanent strip, and recompacts the concrete around the seal; the knife is attached to the underside of a flat plate on which is mounted a small vibrator unit. A bottom crack-inducer is provided at each wet-formed longitudinal joint position.

If transverse and longitudinal joints have not been wet-formed a vibrating blade is used to saw them as soon as the concrete has hardened sufficiently to allow a sharp-edged groove to be produced without spalling, and before random cracks appear in the slab. In practice, sawing is normally carried out between 10 and 20 h after slab construction. Typically, the grooves are between 1/4 and 1/3 of the slab depth and not less than 3 mm wide. Preformed compressed formers that can accommodate the normal expansion and contraction changes in the joint, without allowing the entry of water or grit, are then vibrated into the joints.

Sawed joints (see Fig. 16.5) are generally considered to give better results than wet-formed joints, albeit they are more expensive to produce. However, it is very difficult to saw concrete containing quartzite gravel, especially flint gravel.

The final regulation and finishing of the concrete is usually carried out by a machine that has two oblique finishing beams which oscillate in opposing directions and are mounted on an articulated mobile framework. The leading beam vibrates, makes dense and smooths the surface, whilst the rear beam acts as a float finisher and removes any imperfections prior to texturing. Diagonal rather than transverse beams are preferred as they minimize the area of finisher in contact with a joint former at any given time, thereby reducing the likelihood of damage to the joint. The shearing action of the diagonal screed also gives a more uniform surface finish.

In 1947, to obviate the expense associated with having significant numbers of side-forms on site, highway engineers at the Iowa State Highway Commission in the USA devised a *slip-form paver* which would lay concrete without the need for fixed side-forms. Since then slip-form paving has become a conventional method of concrete road-making throughout the western world.

With slip-form paving much of the conventional paving train is replaced by a single machine frame within which, typically, the main operations prior to the establishment of the transverse joints, namely spreading, trimming, compacting, and finishing, are carried out in a combined and continuous operation. The slip-form paver is self-propelled and mounted on caterpillar tracks that travel outside sliding side-forms attached within the length of the machine; functions such as the forming and finishing of transverse joint grooves, surface texturing and spraying of the curing compound, are done by following equipment (see Fig. 13.2(a)).

The correct alignment and the true surface level of the concrete slab are achieved with the aid of an electronic sensing system that is attached at the four corners of the paving machine; this continuously picks up guidance signals from tensioned guide-wires on stakes on either side of the pavement, and automatically adjusts the machine's operation as necessary. The accuracy to which the vertical and horizontal alignments of the guide-wires is established is absolutely critical to the success of the slip-form paving operation. (More recently, laser guidance systems have also be been used.)

With a conforming plate slip-form paver (Fig. 13.2(b)) the concrete is deposited in a hopper at the front of the moving paver. A hydraulically-adjustable plate attached to the hopper maintains a constant surcharge of concrete above the level of the conforming plate. As the paver moves forward the slab edges are formed by

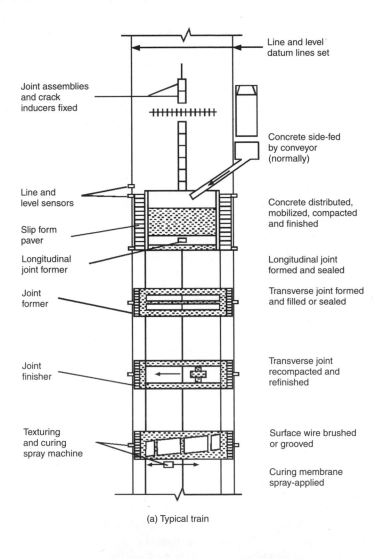

Line and level
datum lines set

Joint assemblies
and crack
inducers fixed

Concrete side-fed
by conveyor
(normally)

Line and
level sensors

Concrete distributed,
mobilized, compacted
and finished

Slip form
paver

Longitudinal
joint former

Longitudinal joint
formed and sealed

Joint
former

Transverse joint formed
and filled or sealed

Joint
finisher

Transverse joint
recompacted and
refinished

Texturing
and curing
spray machine

Surface wire brushed
or grooved

Curing membrane
spray-applied

(a) Typical train

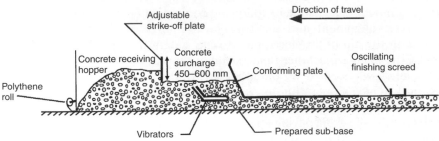

Direction of travel

Adjustable
strike-off plate

Concrete
surcharge
450–600 mm

Oscillating
finishing screed

Concrete receiving
hopper

Conforming plate

Polythene
roll

Vibrators

Prepared sub-base

(b) Conforming plate paver: principle of operation (section)

Fig. 13.2 *continued*

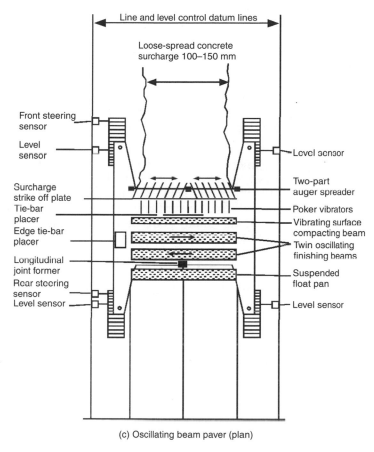

(c) Oscillating beam paver (plan)

Fig. 13.2 Typical slip-form paving plant (based on reference 4)

the moving slip-forms as the concrete is forced under the conforming plate. Immersed vibrators between the strike-off and conforming plates keep the concrete in a fluid state, and help compact it as it is forced into the space between the subbase and the conforming plate. In large pavers the hopper may be fitted with a spreader to help distribute the concrete across its full width.

With the (lesser-used) oscillating beam slip-form paver (Fig. 13.2(c)) the moving side-forms shape the edges of the concrete slab whilst a system of vibrating and oscillating beams does the work of the conforming plate and shapes, compacts and finishes the top of the slab. The fluidity of the concrete, and its initial compaction, are assisted by vibrators in the hopper.

Slip-form pavers are most effectively used when laying plain concrete slabs (up to 14.3 m wide). If reinforcement is required by the design, it is the practice in the UK to attach the steel to cradles fixed to the subbase before placing the concrete. The longitudinal steel tie bars may be welded or spliced on site and positioned so that front loading of concrete to the paver is possible, provided the bars are guided into the correct position in the slab through gates in the paver.

Many slip-form pavers now have the ability to insert transverse contraction-joint dowels, at predetermined positions, into the concrete as it flows from the front of the machine. Crack inducers are not usually used at these points; rather, the required planes of weakness are subsequently created by cutting grooves in the surface of the concrete using a cutting plate. However, the results obtained in this way have not always been completely satisfactory[2].

Waterproof membrane rolls are also fitted to the front of the hopper so that they unroll directly under the slab as the paver moves forward.

Whilst equally good standards have been obtained with both types of paver, the following are the main advantages of slip-form pavers vis-à-vis fixed form ones: (a) slip-form paving is less labour intensive, involves fewer pieces of equipment, and there is no need to provide for the expensive purchase, maintenance, fixing and stripping of large numbers of fixed forms; and (b) higher daily outputs can be achieved with slip-form pavers, e.g. under good summer weather conditions and no joint assembly or plant hold-ups, a slip-form paver in Britain is typically able to lay an 11.2 m wide by 280 mm thick slab (with no expansion joints) at the rate of 65–75 m per hour, as compared with 35–40 m per hour with fixed-form paving[2].

The main disadvantages of slip-form paving are: (a) the slip-form paver is far more complex than the equipment in a conventional fixed-form train, and requires more skilled (and expensive) personnel for its proper operation and maintenance; (b) there is danger of edge slump immediately after the concrete leaves the paver; (c) a minor failure in the control system can cause the whole slip-form paving operation to come to a halt; (d) greater stockpiles of cement and aggregate are required in advance to ensure a high output from the slip-form paver; and (e) the paving contractor is more vulnerable to bad weather hold-ups.

13.7 Texturing of running surfaces

If a carriageway is to have good skid resistance it must have an adequate microtexture and macrotexture. The *macrotexture* allows for the rapid drainage of most of the water trapped between the tyres and the pavement surface, whilst the *microtexture* is expected to penetrate the remaining film of water and maintain the tyre–surface contact. The microtexture at the surface of a concrete slab is ensured by using a fine aggregate in the concrete mix that is more resistant to abrasion than the matrix of the hardened cement paste, i.e. so that the sand grains stand proud of the matrix as the softer paste abrades under traffic. The macrotexture of the running surface is obtained by wire-brushing or grooving the surface, at right-angles to the centreline, upon completion of the final regulation of the slab surface and before the application of the curing membrane.

Wire brushing is carried out manually or mechanically, from a travelling bridge, at right angles to the longitudinal axis of the carriageway. Grooving is carried out mechanically using a profiled vibrating plate which traverses the width of the finished slab and forms grooves with random longitudinal spacings.

From a construction aspect grooving has the advantage that concrete of variable quality can be grooved, even when fairly stiff, as the speed and vibration of the plate can be varied to suit the consistency of the compacted concrete. Further, deep long-lasting texturing is associated with safer roads, and deep grooving (which, unfortunately, is also associated with a higher tyre noise) is hard to achieve by wire-brushing.

Measurements and subjective assessments of the riding quality of concrete surfaces have shown that there is no significant difference between transversely-grooved and brushed surfaces, when both texturing processes are carried out correctly.

13.8 Curing the concrete

Curing aims to ensure that a satisfactory moisture content and temperature are maintained in the newly-placed concrete slab so that hydration of the cement continues until the design strength is achieved. Moisture loss results in drying shrinkage and the development of plastic cracks in the cement paste. Conditions conducive to plastic shrinkage include exposure to the direct rays of the sun, wind, low humidity, high air and concrete temperatures, and concrete with low bleeding tendencies and/or a large surface area in relation to depth.

In practice, curing involves keeping the surface of the concrete damp for at least 7 days using either a resin-based aluminized curing compound, or polythene sheeting, or a sprayed plastic film that can be peeled off before road marks are applied. The method that is now, probably, most used in the UK is to mechanically spray the surface and exposed edges of the concrete slabs with the aluminized compound immediately after the finishing/texturing processes. The spray contains finely-dispersed flake aluminium that completely covers the surface with a metallic membrane that reflects the radiant heat of the sun; the membrane is then worn away by tyre traffic when vehicles are allowed on the carriageway.

Rain damage to the freshly-textured concrete is minimized with the aid of, typically, about 60 m of low-level travelling tentage; this is normally towed behind the curing compound sprayer. The tentage also provides additional protection against drying by the sun and wind.

13.9 Other pavements with concrete

In recent years there has been a rapid growth in the use of two other pavement types involving concrete materials, roller-compacted concrete and concrete block paving.

13.9.1 Roller-compacted concrete

In addition to the cement-bound materials discussed in this chapter and in Chapter 6, mention is made here of the growing use of roller-compacted concrete in road-making in, particularly, the USA, Australia and Spain. Often referred to as simply RCC, roller compacted concrete is a zero slump concrete with a compressive strength of 30 MPa or more that is placed using construction methods that are more akin to those used with bituminous-bound materials rather than the fixed- or slip-form techniques normally associated with concrete pavements. Its use[5,6] is increasing in, mainly, military and industrial pavements that are travelled by vehicles at low operating speeds and, therefore, do not require the same concrete surface evenness as public roads; however, in Spain pavements composed of 200 mm soil cement subbase, 220–250 mm RCC roadbase, and 100 mm bituminous surface course are used in public roads carrying 800–2000 commercial vehicles per day over subgrades with CBRs > 20 per cent.

Air-entraining agents are not used with roller compacted concrete and, hence, its use is confined to locales with mild climates.

The proportions in an RCC mix typically comprise 81–84 per cent aggregate, 12–14 per cent cement, and 4–6 per cent mixing water by mass. The maximum aggregate size is normally limited to 20 mm to avoid a harsh or bony concrete surface, whilst the grading is similar to that for a well-graded crushed rock. Mixing is usually carried out on-site using continuous pugmill mixing plants. Whilst wet-mix batch plants have been successfully used, the low water:cement ratio (typically 0.3–0.4) and lack of fluidity of the mix often leads to long cycle times and excessive build-up within the mixer.

The RCC material must be placed, compacted and finished within 60 min of mixing. Placement is carried out using purpose-built equipment, similar to bituminous pavers, with tamping bars that carry out the initial compaction. Further compaction is carried out by steel-drum vibrating rollers. The maximum lift thickness cannot exceed about 250 mm if good density is to be achieved, whilst the edges must be rolled first to provide confinement for the material in the interior of the pavement. The surface texture can be improved if a pneumatic-tyred roller, travelling behind the vibratory one, is used to work some fines to the top to help close any voids, cracks or tears in the surface. Any marks left by the pneumatic tyres are removed by a non-vibrating, steel wheel, finish roller.

Because of its low moisture content the initial curing of rolled compacted concrete is carried out with a wet process such as a water spray, followed by a curing compound after 2–3 days. Traffic is generally allowed on the RCC pavement after 7 days.

RCC pavements used in industrial applications are sometimes allowed to crack naturally in a random way. However, many of these cracks are ragged and subsequently ravel. More often, sawn contraction joints, one-quarter to one-third the slab depth, are provided to ensure desired planes of weakness. The sawing is carried out as soon as the RCC can be cut without ravelling. Joint spacings are, typically, 40 times the slab depth; this spacing compares with about 20 times the depth for conventional plain concrete slabs.

13.9.2 Block paving

World-wide, the use of coloured concrete and brick paving blocks in roadways has mushroomed in the past 25 years. Much of this growth has been due to a trend towards beautifying urban areas, a greater governmental emphasis on controlling traffic through local-area management schemes, and the development of block-making equipment to mass-produce blocks with high compressive strengths and close dimensional tolerances for use in those schemes. Typical applications include residential roads, precincts for pedestrians only or shared between pedestrians and vehicles, car parks, building forecourts, and footways and landscaped areas.

Advantages attributed to block paving are their: (a) attractiveness of appearance, (b) simplicity of construction, (c) immediate accessibility by traffic upon completion of laying, (d) use of contrasting colours to differentiate areas of usage and/or the location of underground services, (e) ease of access to underground services, and (f) the ability for trench reinstatement with almost complete re-use of the old paving blocks and without visible patching (assuming additional 'spare' blocks are available).

The general approach used with concrete blocks in road pavements is to substitute the block construction for the conventional surface course in a flexible pavement. The block construction is typically composed of 50–80 mm thick blocks – these may be rectangular or shaped to provide lateral interlock – on a bedding of 30–50 mm compacted sand. The sand, which does not contain more than 3 per cent by mass of silt and/or clay and has a maximum size of 5 mm, is spread to a uniform height of about 65 mm in the case of a 50 mm thick layer. The blocks are laid between edge restraints and vibrated down to the desired level using plate vibrators, and additional sand is brushed into the joints.

The optimum joint width is 2 to 5 mm; wider joints result in a loss of structural efficiency. The sand serves a number of functions[7]: (a) it provides a separation cushion which ensures that there are no high local stresses which might cause spalling of the blocks, (b) it caters for needs resulting from variations in block plan dimensions and in laying, and (c) it partially restricts the ingress of surface water which could damage the foundation. Experience has shown that, over time, natural detritus will seal the sand voids so that the joints become practically impervious; thus block pavements are usually constructed to normal longitudinal and crossfall requirements to avoid ponding of rainfall on the surface.

Detailed advice regarding the design of block pavements is available in section 16.12 and in the literature[8,9].

13.10 References

1. *Design Manual for Roads and Bridges, Volume 7, Section 2, Part 2 – HD 25/94: Foundations*. London: The Department of Transport, 1994.
2. Carroll, L.J., Concrete pavement construction. In A.F. Stock (ed.), *Concrete Pavements*. London: Elsevier Applied Science, 1988, pp. 103–47.
3. BS 1881: Part 103: 1983, *Methods of Testing Concrete: Determination of Compacting Factor*. London, The British Standards Institution, 1983.
4. Bannon, C.A., *Concrete Roads: An Overview of Design, Specification and Construction Issues*. Paper presented at a joint meeting of The Institution of Engineers of Ireland and The Irish Concrete Society on 4 February 1991.
5. Rollings, R.S., Design and construction of roller compacted concrete pavements, *Proceedings of the 14th Australian Road Research Board Conference*, 1988, **14**, Part 8, pp. 149–63.
6. Petrie, R.E. and Matthews, S.C., Roller compacted concrete pavements: Recent Australian developments and prospects for the 90's, *Proceedings of the 15th Australian Road Research Board Conference*, 1990, **15**, Part 2, pp. 66–79.
7. Lilley, A., Shaped versus rectangular paving blocks in flexible pavements, *Highways and Transportation*, 1994, **41**, No. 1, pp. 24–9.

8. Knapton, J. and Cook, I.D., An integrated design method for pavements surfaced with concrete blocks or clay pavers, *Highways and Transportation*, 1993, **40**, No. 3. pp. 4–14.
9. BS 7533: 1992, *British Standard Guide for Structural Design of Pavements Constructed with Clay or Concrete Block Pavers*. London: The British Standards Institution, 1992.
10. Highway Agency, *Specification for Highway Works*. London: HMSO, 1998.

CHAPTER 14

Current British thickness design practice in relation to new bituminous and concrete pavements

C.A. O'Flaherty

14.1 Introduction

The design methods for new trunk roads (including motorways) given in this chapter are based on the governmental publications HD 25/94[1], HD 24/96[2] and HD 26/94[3]. These, in turn, are based on a number of Transport and Road Research Laboratory publications, particularly references 4 (for flexible and flexible composite construction), 5 (for rigid and rigid composite construction), and 6 and 12. The design methods also take account of research on new pavement materials.

Reference 4 is based on observations and measurements of full-scale bituminous road experiments obtained over a 20 year period, supplemented by structural analysis to rationalize and extend the data. The analyses used the elastic stiffness modulus of the various pavement and foundation layers to calculate the strains developed within the structure, and these strains were then related to pavement life.

Reference 5, which is mainly empirical, is based on outputs from full-scale concrete road experiments. However, the performance data available for continuously reinforced concrete (CRC) pavements were limited, so that the designs were extrapolated from a comparison between jointed unreinforced (URC) and reinforced (JRC) concrete slabs. For rigid composite pavements, an allowance was made for the structural contribution and thermal insulation effected by the bituminous surfacing.

The review reported in reference 12 was prompted by the fact that, at the time it was initiated, the most heavily trafficked motorways and trunk roads had carried in excess of 100 million standard axles, whereas the traffic levels had not exceeded 20 msa at the time that the original design curves were developed. In addition, research has resulted in more information becoming available on the fundamental behaviour of road pavements which has helped to explain the observed performances of road pavements.

The basic governmental approach to the design of new roads involves three main steps:

Step 1. Design the foundation for construction traffic loading.
Step 2. Assess the anticipated commercial vehicle traffic that will use the design carriageway lane over the design life.
Step 3. Determine the pavement thicknesses of roadbase and surfacing required to cope with the traffic demand.

Before discussing this approach it must be emphasized that the design recommendations illustrated in this chapter were devised for conditions and materials applicable to the UK, and care should be exercised before directly applying them to environmental and construction conditions elsewhere that are not akin to those experienced in the UK.

14.2 Foundation design

The main function of the foundation, i.e. the subbase and any capping layer beneath it (see Fig. 9.3), is to distribute applied vehicle loads to the underlying subgrade without causing distress to the foundation layers or to the layers overlying them, during construction and during the design life of the pavement. The foundation also acts as a construction platform for the laying and compaction of the overlying layers, as well as protecting the subgrade from the adverse effects of bad weather during construction.

Whilst the number of stress repetitions from construction vehicles is relatively low when a pavement is being built, the stresses that they cause to the foundation can be high. Thus, the approach that is normally adopted is to design the foundation to withstand construction traffic stresses, using Fig. 14.1. The curves in this figure limit foundation deformation caused by construction vehicles to a maximum of 40 mm for 1000 passes of a standard axle; this is the maximum depth that can be tolerated if the subbase surface is to be effectively reshaped and recompacted.

Use of Fig. 14.1 requires determination of the California Bearing Ratio (CBR) of the subgrade; this is most commonly measured in the laboratory on recompacted specimens or, if that is not feasible, estimated from established relationships (e.g. see Table 14.1). The laboratory CBR tests are carried out over a range of conditions of moisture content and density that are likely to be experienced during construction and in the finished pavement. Cohesive soils are compacted to not less than 5 per cent air voids, to reproduce the likely conditions on site. The equilibrium moisture content can be deduced from measurements on a suction plate[7]. From a construction aspect it is not desirable for the foundation design to often vary along the road and, consequently, tests are carried out only where there are significant changes in subgrade properties; in practice, this means that design changes rarely occur at intervals of less than about 500 m.

The thicknesses of the subbase and capping layers are obtained from Fig 14.1. Note that the subbase is omitted only if the subgrade is a hard intact rock, or if it is granular and has a CBR greater than 30 per cent and does not have a high water table. If the subgrade CBR is more than 15 per cent, the subbase thickness is 150 mm; this is the minimum practicable thickness for spreading and compaction.

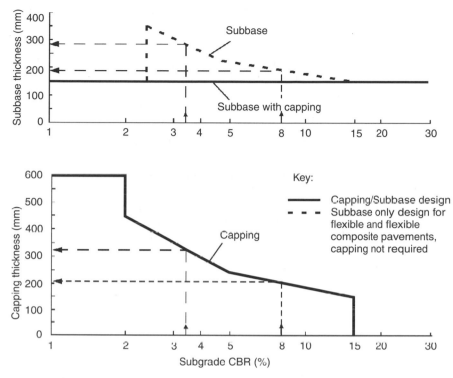

Fig. 14.1 Capping and subbase thickness design for new trunk roads[1]

If the subgrade CBR is between 2.5 and 15 per cent two options are available for flexible (including flexible composite) pavements: 150 mm of subbase on a thickness of capping or an increased thickness of subbase and no capping, e.g. if the subgrade CBR is 3.5 per cent the options are 150 mm of subbase on top of 330 mm of capping or 280 mm of subbase and no capping. For all pavements on subgrades with CBRs below 2.5 per cent, and for rigid (including rigid composite) pavements on CBRs less than 15 per cent, 150 mm of subbase on a capping layer is required.

Table 14.1 Estimating the equilibrium subgrade CBR from the plastic index[1]

Type of soil	PI	Predicted CBR, %
Heavy clay	70	2
	60	2
	50	2
	40	2–3
Silty clay	30	3–4
	20	4–5
Sandy clay	10	4–5
Sand (poorly graded)		20
Sand (well graded)		40
Sandy gravel (well graded)		60

In all cases the California Bearing Ratio of the material used in a capping layer must be at least 15 per cent.

Granular and cement-bound subbases can be used with flexible and flexible composite pavements, but only cemented ones with rigid and rigid composite pavements.

A Type 1 granular material is most commonly used in subbases in flexible and flexible composite pavements; however, a Type 2 material may be used instead if the pavement has a design traffic loading of less than 5 msa when the road is opened. The CBRs of all unbound granular subbases must be at least 30 per cent, and the material's grading must enable it to be compacted to a dense impermeable mass that will shed rainwater during construction and have a relatively high stiffness modulus. Alternatively, a weak cement-bound material, CBM1 or CBM2, or a weak wet lean mix, grade C7.5, can be used in these subbases.

Some or all of an unbound subbase in a flexible-type pavement can be replaced with a bituminous material, provided that a substitution rate of 30 mm of bituminous roadbase is used per 100 mm of Type 1 subbase. However, this technique cannot be applied to the lowest 150 mm of subbase if it lies directly on a subgrade with a CBR of less than 5 per cent at construction time.

The stress generated in a concrete slab partly depends on the stiffness ratio between the slab and its underlying support. Consequently, cement-bound subbases are required beneath rigid and rigid composite constructions to (a) minimize the risk of erosion and weakening of the subbase from water that might subsequently penetrate joint sealants and cracks in slabs, and (b) enable better compaction of the overlying concrete. These subbases are normally composed of either a strong cement-bound material, CBM3, or a grade C15 wet lean concrete; however, if the initial design life of the pavement is less than 12 msa a CBM2 or C12 material is permitted. An impermeable membrane is required over these subbases to stop the suction of water from the overlying concrete and, in the case of jointed concrete slabs, to act as a slip layer.

The gradations of the unbound Type 1 and Type 2 materials, and of the cement-bound materials CBM1, CBM2 and CBM3, are given in Tables 6.2 and 6.4, respectively. Details of these materials, as well as of the wet lean concrete materials, are specified in reference 8.

If the subgrade CBR is much below 2 per cent, the soil may deform and 'wave' under construction traffic so that, typically, a 0.5 to 1 m thick layer of the subgrade may have to be removed and replaced with a stronger material. Even though the replacement material is of good quality the subgrade is still assumed to have a CBR of less than 2 per cent and to require 600 mm of capping; this may mean that a total construction thickness of around 1.5 m will be required. If the subgrade soil is reasonably permeable the installation of a deeper than normal drainage system may be appropriate; if it is a cohesive soil a lime, lime-cement, or lime-PFA treatment may be an economic option.

Drainage of the subbase can only be omitted if the underlying capping and subgrade are more permeable than the subbase, and the water table is never closer than 300 mm to the formation.

Normally, all material within 450 mm of the road surface must not be frost-susceptible. However, this requirement may be reduced to 350 mm in locales where (e.g. in coastal areas) the mean annual frost index of the site is less than 50.

The frost index, *I*, is the number of days of continuous freezing multiplied by the average amount of frost (°C) on those days; it is related to the depth of frost penetration by

$$H = 4I^{0.5} \qquad\qquad (14.1)$$

where *H* = the depth of frost penetration (cm).

14.3 Traffic assessment

When designing a new road the total flow of commercial vehicles in one direction per day at the road's opening, and the proportion of these vehicles in the Other Goods Vehicle (OGV2) class (see Subsection 9.3.12), are normally required in order to determine the cumulative design traffic over the design life, using the *standard method* of traffic assessment. In the UK a commercial vehicle is defined as one whose unladen weight is more than 15 kN. For new road schemes the traffic flow is determined from traffic studies using the principles described in the *Traffic Appraisal Manual*[9] except that, for Scotland, the principles are set out in the *Scottish Traffic and Environmental Appraisal Manual*[10]. If the traffic studies suggest that the flow at the road opening is not a good basis for forecasting the cumulative flow over the design life (e.g. if an adjacent link is to be built after a few years) an adjustment must be made to reflect this situation.

If the traffic flow figures available are for a two-way flow, the directional split is assumed to be 50:50 unless traffic studies show otherwise.

Either Fig 14.2 or Fig. 14.3 is then used to obtain the cumulative design traffic in millions of standard axles (msa) for flexible- and rigid-type pavements, respectively. A minimum cumulative design traffic of 1 msa is assumed for lightly-trafficked roads.

Note that the figures are based on the high growth prediction from the National Road Traffic Forecasts (NTRF)[11], and that average growth rates have been assumed for PSV + OGV1 and for OGV2 vehicle classes. The average design life flow divided by the flow at opening (i.e. the proportional flow increase) is calculated on the basis that the year of opening is 1995.

The curved shapes of the relationships shown at Figs 14.2(b) and 14.3(b) are due to the greater proportion of traffic carried by the outer lane(s) of a dual carriageway at high vehicle flows.

The distribution of commercial vehicle traffic between lanes can be expected to vary at particular points along a road, e.g. where lanes leave/join a carriageway, or at traffic signals or at roundabouts. Nonetheless, the design of new roads is based on the traffic distribution away from junctions, and all lanes – including hard shoulders, so that they can be used by traffic during subsequent maintenance activities – are designed to carry the heavily-trafficked left-hand lane traffic load.

Note also that the figures assume a 20 year life for flexible composite pavements, 20 *or* 40 years for fully-flexible pavements, and 40 years for rigid and rigid composite pavements. Recent research work[12] has indicated that, from a whole-life cost aspect, it is more cost effective to design flexible pavements, especially those at heavily-trafficked locations, for at least 40 years so as to obviate the need for an interim (after an initial 20 years) structural strengthening/major maintenance and its associated traffic delays. Consequently, an option to design a thicker fully-flexible pavement with a 40 year design life is permitted for (i) road locations where

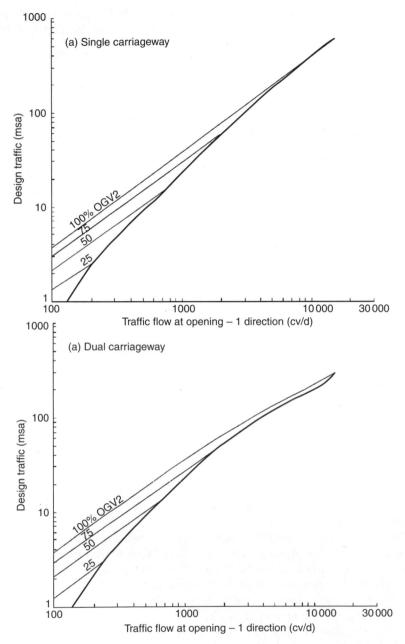

Fig. 14.2 Cumulative design traffic for flexible (20 year life) and flexible composite pavements[2]

the design traffic is heavy in relation to the capacity of the layout, and (ii) in all cases where whole life costing is required to be taken into account. In normal circumstances, however, the first stage design for flexible (and flexible composite) pavements assumes that investigatory conditions (see Subsection 9.3.3) will arise

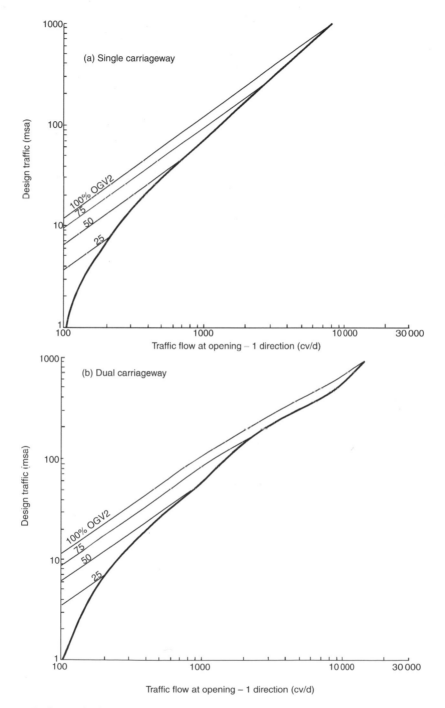

Fig. 14.3 Cumulative design traffic for rigid, rigid composite and flexible (40 year life) pavements[2]

after a 20 year design life, at which time it can be expected that major maintenance will need to be carried out to ensure a further 20 years of design life.

The steps involved in carrying out a *full traffic assessment* (see Table 14.2) follow. It should be noted that this approach is not normally applied to new roads but is primarily intended for use in structural assessments and in the design of maintenance measures. It is summarized here for completeness.

Step 1.   Estimate the present one-way commercial vehicle flow, or the traffic flow at the opening in the case of a new road design, F, for each of the seven classes shown in Fig. 9.13. In many instances, however, it will be sufficient to group the traffic into the same two categories, PSV+OGV1 and OGV2, as in the standard method.

Step 2.   Select the initial design period, Y.

Step 3.   Determine the 'growth factor', G, for each category of vehicle (see Fig. 9.15). In practice, the grouped PSV+OGV1 and OCV2 growth lines are commonly used unless actual growth rates are known for specific vehicle groups.

Step 4.   Determine the approximate national average wear factor, W, to be used with each vehicle class (from Fig. 9.13). Alternatively, weighted average wear factors of 0.6 and 3.0 may be assumed for the grouped categories PSV+OGV1 and OGV2, respectively; these are based on the proportion of vehicles in each class (as determined from the 1992 statistics).

Step 5.   Calculate the cumulative design traffic in each vehicle class for the design period, using Equation 9.1. In the case of a two-way single carriageway pavement the total design traffic, T, is the summation of the cumulative design traffic in each

Table 14.2 Example of carrying out a full traffic assessment for a 20 year design and NTRF growth[2]

Vehicle class	AADF (F)	Design period (Y)	Growth factor (G)	Wear factor (W)	Cumul. traffic (T)*
Either					
PSV:					
Buses and coaches	398	20	1.0	1.3	3.8
OGV1:					
2-axle rigid	2084	20	1.0	0.34	5.2
3-axle rigid	196	20	1.0	1.70	2.4
3-axle articulated	95	20	1.0	0.65	0.5
OGV2:					
4-axle rigid	209	20	1.5	3.0	9.2
4-axle articulated	912	20	1,5	2.6	36.0
5-axle articulated	743	20	1.5	3.5	28.5
Or					
OGV1+PSV				0.6	
OGV2				3.0	

Total cv/day	4637	
Total traffic load (all lanes)		85.6 msa
Proportion of commercial vehicle traffic in left hand lane (P)		79%
Design traffic load in left hand lane (=P× Total/100)		67.6 msa

*$T = 365F \times Y \times G \times W \times 10^{-6}$ msa as defined by Equation 9.1

category in a given direction. In the case of a dual carriageway road the proportion of vehicles in the most heavily-trafficked lane is normally obtained (from Fig. 9.14) and applied to the total accumulation to derive the design traffic; note, however, that if the flow is greater than 30 000 cv/day, the proportion is assumed to be 50 per cent. When carrying out maintenance, it is sometimes necessary to estimate the traffic load in the other lanes separately. Thus, for three-lane carriageways it is assumed that all commercial vehicles not in the left-hand lane are taken by the middle lane and, for a four-lane carriageway, the number of commercial vehicles not in the left-hand lane is assumed to be evenly distributed between the two middle lanes (unless, in either instance, specific data indicate otherwise). In each such case, however, the outer lane is designed to carry the same traffic as the middle lane(s).

In the case of maintenance or widening work to areas about junctions, consideration will need to be given to non-standard traffic distributions when determining the design load.

14.4 Thickness design of pavements

The thickness design procedures are more easily described by presenting the flexible types of pavement separately from the rigid types. However, it should be noted that, in most instances, pavement designs are prepared for all flexible- and rigid-type options.

14.4.1 Flexible and flexible composite pavements

In the UK the surfacing and roadbase materials used in fully-flexible pavements in trunk roads are bound with a bituminous binder. Permitted roadbase materials are dense bitumen macadam (DBM), hot rolled asphalt (HRA), dense tarmacadam (DTM) with a tar binder, DBM plus 50 penetration bitumen (DBM50), and heavy duty macadam (HDM). Designs for this pavement type are available, based on the predicted traffic flow over a 20 or 40 year life; Fig. 14.4 is used to obtain the combined thickness of bituminous material for this type of pavement for either design life.

Only the surfacing and upper roadbase materials in flexible composite pavements are bound with a bituminous binder and (as is indicated in Fig. 9.3) the lower roadbase is composed of a cement-bound material (CBM). In this case, designs (see Fig. 14.5) are based on a 20 year life.

Both (fully) flexible and flexible composite pavements are deemed suitable for all normal road construction applications bar those where large differential movements or settlements associated with compressible ground, or considerable subsidence caused by mining, are expected.

In the case of a *flexible pavement*, the wearing course thicknesses and materials permitted are either 45 mm or 50 mm of hot rolled asphalt (HRA) or 50 mm of porous asphalt (PA). If HRA is used, its contribution to the pavement's thickness design is assumed to be the full wearing course depth; if a PA wearing course is used, its contribution to the thickness is only considered to be 20 mm.

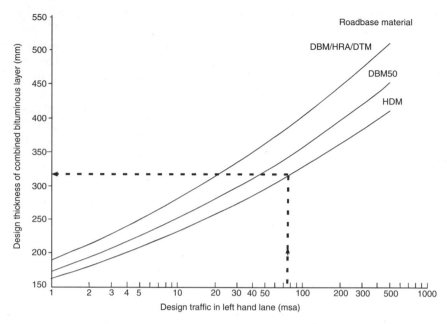

Fig. 14.4 Design thicknesses for flexible pavements[3]

A basecourse is optional beneath an HRA wearing course; if used it must be at least 50 mm thick and composed of any permitted material, except that basecourses over HDM or DBM roadbases must be HDM and DBM, respectively. A 60 mm thick basecourse of either HRA, DBM, DBM50, or HDM is always required beneath a PA wearing course; if the roadbase is HDM or DBM50 the basecourse must also be HDM and DBM50, respectively, unless HRA is used as an alternative in either case.

Dense bitumen macadam (DBM) roadbases and basecourses must contain 100 pen binder, whilst hot rolled asphalt (HRA) roadbases and basecourses must contain 50 pen binder. Dense tar macadam (DTM) roadbases and basecourses must contain C50 or C54 grade tar or, if gravels containing flints are used, C58 tar.

Bituminous basecourse materials, and roadbase materials when the traffic load exceeds 80 msa, must normally contain crushed rock or slag coarse aggregate. For DTM basecourses the fine aggregate must also be crushed rock or slag.

A DBM, HRA or DTM thickness design for traffic loads in excess of 80 msa must include a 125 mm thick lower roadbase layer of HRA, with DBM or DTM as the remainder of the roadbase material.

The combined thickness of bituminous-bound roadbase and surfacing in a flexible pavement, as read from Fig. 14.4, is rounded up to the nearest 10 mm. Note that DBM, HRA, and DTM roadbases are considered to be of similar quality in relation to light and medium traffic loads, i.e. their differing strengths and weaknesses are assumed to cancel out; hence, a single design line is used for all three materials in this figure. However, when the traffic loads exceed 80 msa the designs in Fig. 14.4 incorporate a 125 mm lower roadbase layer of HRA, which is easier to lay and compact than either DBM or DTM above an unbound granular subbase,

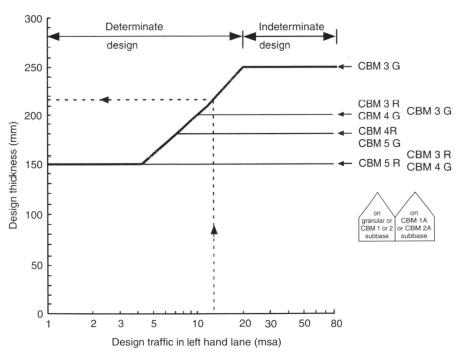

Fig. 14.5 Design thicknesses for flexible composite pavements[3]
(*Note:* R = roadbase of crushed rock aggregate, with a coefficient of thermal expansion <10 × 10⁻⁶ per deg C; G = roadbase with gravel aggregate or roadbase with crushed rock aggregate and a coefficient of thermal expansion >10 × 10⁻⁶ per °C)

and is also better able to sustain the horizontal tensile strains present at this level. DBM and DTM roadbases, both of which have slightly better deformation characteristics than HRA, are retained for the upper roadbase to give the best resistance to deformation in the zone which experiences higher temperatures and shear stresses. Studies have shown that the inclusion of an HRA lower roadbase gives no benefit when DBM50 or HDM roadbases are used.

Example design   As an example[3], consider a design traffic loading of 75 msa, using an HDM roadbase. From Fig. 14.4 the combined thickness of roadbase and surfacing that is required is 320 mm. Options then available are: (a) 45 mm HRA wearing course + 55 mm HDM basecourse + 220 mm HDM roadbase; (b) 45 mm HRA wearing course + 275 HDM roadbase; and (c) 50 mm PA wearing course + 60 mm HDM basecourse + 240 mm HDM roadbase.

In the case of a *flexible composite pavement* the design thickness of the combined bituminous-bound layers and of the cement-bound lower roadbase are obtained from Fig. 14.5(a) and (b), respectively, with each thickness being rounded up to the nearest 10 mm. The only permitted wearing course material is HRA and it must be either 45 mm or 50 mm thick. A basecourse of any permitted material is optional beneath the HRA wearing course; however, if used, the basecourse must be at least 50 mm thick. Where a bituminous roadbase is required, it may be of any permitted material.

Bituminous basecourse materials normally contain crushed rock or slag coarse aggregate. The fine aggregate used in a DTM basecourse must also be crushed rock or slag.

HRA roadbase and basecourse materials must contain 50 pen bitumen, and DBM types must contain 100 pen bitumen. DTM roadbases and basecourses must contain C50 or C54 grade tar, except that C58 grade tar must be used if the mix contains gravels with flints.

The cement-bound material used in the lower roadbase is most commonly CBM3. The minimum thickness of CBM3 or CBM4 used in Scotland is 175 mm rather than 150 mm for the rest of Great Britain.

It might be noted here that, early in the life of a flexible composite pavement, a CBM roadbase tends to exhibit transverse shrinkage cracks that can be reflected through the bituminous layers above and become visible at the surface. However, whilst they may be unsightly, research has shown that these primary transverse cracks do not normally have a significant effect upon the structural performance of the pavement unless the soil beneath the subformation is moisture susceptible.

The combined effects of traffic and temperature eventually result in a gradual deterioration of the CBM roadbase from both secondary transverse cracking and longitudinal cracking, typically in the wheelpaths. Designs where this gradual deterioration is expected are said to have a 'determinate life' and are designed on the same basis as fully flexible 20 year pavements, i.e. after 20 years some reconstruction and replacement of the roadbase may be required in conjunction with a strengthening overlay, to extend the pavement's life in a predictable manner. If the strength and thickness of the CBM roadbase are both sufficient to withstand the combined effects of traffic and temperature, the design is said to be 'indeterminate' and the roadbase should not normally need replacement at the end of the first stage of its design life. Indeterminate life designs are used for traffic loadings greater than 20 msa and/or where the traffic delays associated with future maintenance are likely to be unacceptable.

The risk of longitudinal cracking induced by combined stresses is minimized if individual construction widths of CBM roadbase that are designed for an indeterminate life do not exceed 4.75 m. Determinate life designs (i.e. thinner constructions) are more likely to deteriorate by general cracking and, consequently, restricting the individual laid width does not necessarily lead to improved performance.

If a cement-bound subbase is used, it should be checked to ensure that no longitudinal cracks are present before the roadbase is laid. If CBM1A or CBM2A subbases are used they should comply with the requirements for CBM1 or CBM2, respectively, except that the minimum compressive strength must be appropriate to a CBM3 material.

Example design As an example[3] of the use of the design process for a flexible composite pavement, note (Fig. 14.5(a)) that a traffic load of 13 msa requires a total thickness of 140 mm of bituminous material above the lower roadbase. The surfacing options available are therefore (a) 45 mm HRA wearing course+95 mm DBM, DTM or HRA basecourse or upper roadbase, or (b) 50 mm HRA+90 mm DBM, DTM or HRA basecourse or upper roadbase. The lower roadbase options (Fig. 14.5(b)) are (i) 220 mm CBM3G on granular, CBM1 or CBM2 subbase, (ii) 200 mm CBM3R on granular, CBM1 or CBM2 subbase, (iii) 200 mm CBM3G on CBM1A or CBM2A subbase, (iv) 200 mm CBM4G on granular, CBM1 or CBM2 subbase, (v) 180 mm CBM4R on granular, CBM1 or CBM2 subbase, (vi) 180 mm CBM5G on granular, CBM1 or CBM2 subbase, (vii) 150 mm CBM3R on CBM1A or CBM2A subbase, or (viii) 150 mm CBM4G on CBM1A or CBM2A subbase, or (ix) 150 mm CBM5R on granular, CBM1 or CBM2 subbase.

14.4.2 Rigid and rigid composite pavements

Factors taken into account in the design of rigid and rigid composite pavements in the UK are the (i) magnitude and position of the wheel loading, (ii) underlying support for the concrete slab, (iii) temperature and moisture changes, (iv) strength of the concrete, (v) fatigue behaviour of the concrete, (vi) thickness of the concrete slab, and (g) amount of reinforcement.

The concrete used in all rigid-type pavements must be pavement quality concrete (PQC) that is manufactured, laid and cured in accordance with the *Specification for Highway Works*[8]. As the load-induced stresses at the slab corners and edges are greater than in the slab centre, dowel bars must be provided between jointed slabs to distribute loads and, normally, the concrete slab is extended at least 1 m beyond the outer edges of lanes carrying commercial vehicles.

The design thicknesses of slab to be used in *unreinforced (URC)* and *jointed reinforced (JRC) concrete pavements* are obtained from Fig. 14.6(a). The thickness determined in either case is rounded up to the nearest 10 mm. In the case of JRC pavements four design alternatives must be prepared, one for each of the design curves, for different quantities of longitudinal reinforcement.

The maximum transverse contraction joint spacings permitted with URC pavements are 4 m and 5 m for slab thicknesses up to 230 mm, and for 230 mm and over, respectively.

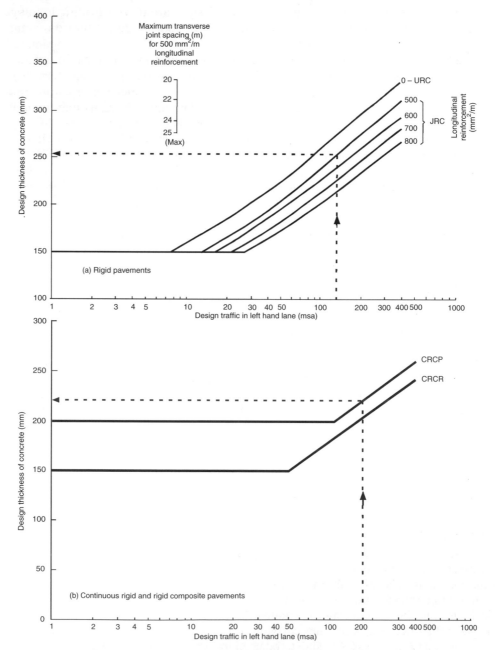

Fig. 14.6 Design thicknesses for rigid and rigid composite pavements[3]

The maximum transverse joint spacing allowed in a JRC pavement is 25 m, except for slabs having 500 mm^2/m of longitudinal reinforcement when the maximum spacing must be read from the insert in Fig. 14.6(a). (500 mm^2/m is the minimum amount of longitudinal reinforcement permitted in JRC pavements.)

The figure can also be used to interpolate intermediate values of maximum transverse joint spacing, longitudinal reinforcement area, and slab thickness.

With both URC and JRC pavements, all transverse joint spacings may be increased by 20 per cent if limestone coarse aggregate is used throughout the depth of slab.

A separation membrane is required between the slab and the subbase for both URC and JRC pavements. Whilst this is mainly required in order to reduce the loss of water from the fresh concrete, it also helps to reduce the friction between the slab and the subbase and, consequently, the formation of mid-bay cracking is inhibited.

Example design   As a JRC thickness design example[3], consider a pavement with 500 mm^2/m of longitudinal reinforcement that is required to carry a traffic load of 130 msa. From Fig. 14.6(a) the design thickness is 260 mm and the transverse joint spacing is 25 m.

Design thicknesses for *continuously reinforced concrete pavement (CRCP)* surface slabs, and *continuously reinforced concrete roadbase (CRCR)* slabs with a bituminous surfacing, are obtained from Fig. 14.6(b).

A major advantage associated with the use of CRCP and CRCR pavements is the elimination of transverse joints with their continuing maintenance liability; these transverse joints are essential in unreinforced and jointed reinforced pavements. A CRCP pavement can be considered as part of a staged construction, as it can be strengthened with a bituminous surfacing or a concrete overlay at a later time; this (more expensive) type of pavement therefore tends to be used when it is desired to reduce the traffic disruption associated with future maintenance works.

The bituminous surfacing in a CRCR pavement helps to limit the amount of water penetrating into the concrete slab and, hence, the potential for corrosion of the reinforcement is reduced; it also provides the concrete roadbase with thermal protection from rapid temperature changes.

The wearing course in a CRCR pavement must be either hot rolled asphalt (HRA) or porous asphalt (PA). If an HRA wearing course is used the surfacing must be at least 100 mm thick, and comprise either 45 mm or 50 mm of HRA over 55 mm or 50 mm, respectively, of dense bitumen macadam (DBM), dense tar macadam (DTM) or HRA basecourse. If a PA wearing course (which must use a modified binder) is used it must comprise either 50 mm of PA over either 90 mm or 60 mm of HRA basecourse; however, if only 60 mm of HRA basecourse is employed the CRCR slab thickness must be increased by 10 mm. If an HRA wearing course is used the material must meet specified[8,13] stability and flow values related to traffic loading.

Normally, all bituminous basecourse materials must contain crushed rock or coarse slag aggregate; in the case of DTM basecourses the fine aggregate must also be crushed rock or slag. HRA and DBM basecourses must contain 50 and 100 pen bitumen, respectively, whilst DTM basecourses must contain C50 or C54 grade tar or, when gravels containing flints are used, the tar must be grade 58.

The longitudinal reinforcement used in CRCP pavements must comprise 16 mm diameter deformed steel bars and amount to 0.6 per cent of the area of the slab cross-section; in CRCR pavements it must comprise 12 mm diameter bars and be 0.4 per cent of the slab cross-section area. If transverse reinforcement is required it must comprise 12 mm diameter deformed bars at 600 mm spacings, for both CRCP and CRCR pavements; if transverse steel is omitted, the spacing between longitudinal joints in both types of pavement is limited to 4.2 m (or 5 m, when limestone aggregate is used).

Unlike for jointed slabs, a separation membrane is not required with CRCP construction as the restraint provided by the higher level of friction between the slab and the subbase reduces the movement at the ends of the pavement, as well as encouraging a desirable crack pattern. (The initial crack spacing in CRCP slabs is about 3 or 4 m, and the extent to which further crack propagation occurs is closely related to the proportion of steel and the strength of the concrete.) The insertion of discontinuities, e.g. gullies and manholes, in the main slab should be avoided as these encourage the development of closely-spaced cracks and the risk of subsequent spalling. The continuous longitudinal reinforcement is intended to hold the cracks tightly closed; this minimizes corrosion of the steel and ensures that load transfer is by aggregate interlock.

The ends of CRCPs must be restrained by ground beam anchorages, or movement accommodated within wide flange steel beam expansion joints, to ensure that expansion of the slab does not result in forces being transmitted to structures or adjacent forms of pavement construction. However, ground beam anchorages cannot be used when the subgrade strength is poor, especially on high embankments where consolidation may be insufficient to restrain the movement of the beam downstands.

No anchorages are required in a CRCR pavement with at least 100 mm surfacing, as the bituminous surfacing protects the slab from large thermal stresses. The use of only a wearing course as a surfacing material is not permitted as experience suggests that the durability of a wearing course laid directly on concrete is insufficient to achieve a maintenance-free life of more than a few years; also anchorages would be required as the wearing course would be insufficient to protect the slab from large thermal stresses.

Note that the design graphs in both Fig. 14.6(a) and (b) assume that there is a 1 m hardstrip or a hard shoulder beside the most heavily-trafficked (left) lane. If a road, e.g. an urban road, does not have an adjacent 1 m strip or hard shoulder, then the slab thickness determined for the left-hand lane will need to be increased by an amount determined from Fig. 14.7.

In the UK unreinforced concrete (URC) and jointed reinforced concrete (JRC) pavements are considered suitable for all normal road construction applications, except in areas where differential movement, subsidence caused by mining, or appreciable settlement caused by compressible ground, are anticipated. CRCP and CRCR pavements are also suitable in all applications, especially when large differential movements are expected, as they can accommodate significant strains whilst remaining substantially intact. Both CRCP and CRCR must be considered as alternative pavement constructions when the design traffic loading exceeds 30 msa; they may also be prepared as options for less heavily-trafficked schemes when there are advantages associated with lower maintenance.

Example design   As a thickness design example[3] consider a pavement that is being designed to carry a traffic load of 170 msa. If a continuous rigid pavement is to be used the rounded up CRCP slab's thickness will be 230 mm. If a rigid composite

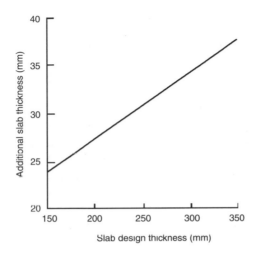

Fig. 14.7 Additional concrete slab thickness for pavements without 1 m hard-strip[3]

pavement is being considered, options available are (a) 45 mm HRA wearing course + 55 mm DBM, DTM or HRA basecourse + 210 mm CRCR; (b) 50 mm PA wearing course + 90 mm HRA basecourse + 210 mm CRCR; and (c) 50 mm PA wearing course + 60 mm HRA basecourse + 220 mm CRCR.

14.5 References

1. Department of Transport, *Foundations*, HD 25/94. London: The Department of Transport, 1994.
2. Department of Transport, *Traffic Assessment*, HD 24/96. London: The Department of Transport, 1996.
3. Department of Transport, *Pavement Design*, HD 26/94. London: The Department of Transport, 1994 – plus Amendments No. 1, Mar 1995, No. 2, Dec 1995, and No. 3, Feb 1998.
4. Powell, W.D., Mayhew, H.C. and Nunn, M.E., *The Structural Design of Bituminous Roads*, TRRL Report LR1132. Crowthorne, Berkshire: The Transport and Road Research Laboratory, 1984.
5. Mayhew, H.C. and Harding, H.M., *Thickness Design of Concrete Roads*, TRRL Report RR87. Crowthorne, Berkshire: The Transport and Road Research Laboratory, 1987.
6. Robinson, R.G., *Trends in Axle Loading and Their Effect on Design of Road Pavements*, TRRL Report RR138. Crowthorne, Berkshire: The Transport and Road Research Laboratory, 1988.
7. Black, W.P.M. and Lister, N.W., *The Strength of Clay Fill Subgrades: Its Prediction in Relation to Road Performance*. Crowthorne, Berkshire: The Transport and Road Research Laboratory, 1979.
8. *Specification for Highway Works*. London: HMSO, 1998.

9. *Traffic Appraisal Manual (TAM)*. London: The Department of Transport, 1984.
10. *Scottish Traffic and Environmental Manual (STEAM)*. Edinburgh: Scottish Office Industry Department, 1986.
11. *National Road Traffic Forecasts (Great Britain)*. London: The Department of Transport, 1989.
12. Nunn, M.E., Brown, A., Weston, D. and Nicholls, J.C., *Design of Long-life Flexible Pavements for Heavy Traffic*, TRL Report 250. London: HMSO, 1997.
13. BS 594:1992, *Hot Rolled Asphalt for Roads and Other Paved Areas*: Part 1- *Specification for Constituent Materials and Asphalt Mixtures*. London: British Standards Institution, 1992. (Annex B: Table B1.)

CHAPTER 15

Analytical design of flexible pavements

J. McElvaney and M.S. Snaith

15.1 Introduction

A flexible pavement combines layers of generally different materials in a structural system designed to withstand the cumulative effects of traffic and climate to the extent that, for a predetermined period, the foundation or subgrade is adequately protected and the vehicle operating costs, safety and comfort of the road user are kept within tolerable limits.

Basic inputs required for pavement thickness design include some measure of subgrade strength, the range in climatic conditions, the characteristics of available materials and an estimate of cumulative traffic loading. Selection of the most economic combination of layer materials and thicknesses requires evaluation of various design strategies developed from consideration of a wide range of factors. Analysis of life-cycle costs and benefits provides a basis for comparing alternative strategies. However, the economic analysis is meaningful only if the design method used can predict accurately the performance of the alternative constructions being considered.

It is convenient to group current methods of pavement thickness design into two broad categories: empirical and analytical. The analytical term, as used here, includes the structural analysis of candidate pavements and the prediction of their performance from the computed parameters. Empirical methods are those that have evolved from observation of the performance of experimental pavements laid either on public roads, and hence subjected to normal road traffic, or on test tracks where loading was strictly controlled. The majority of current methods of flexible pavement design are in the empirical category.

The analytical approach to design involves: (i) characterization of the pavement layers and subgrade to meet the requirements of a theoretical model; (ii) calculation of parameters considered to have a primary influence on selected aspects of pavement performance; and (iii) utilization of these parameters in performance models to evaluate the structural adequacy of the pavement under consideration. The approach is illustrated in the flow diagram shown on Fig.15.1.

In analytical pavement design it is assumed that pavements deteriorate due to repetitions of the stresses, strains and deflections generated by traffic loads, ultimately

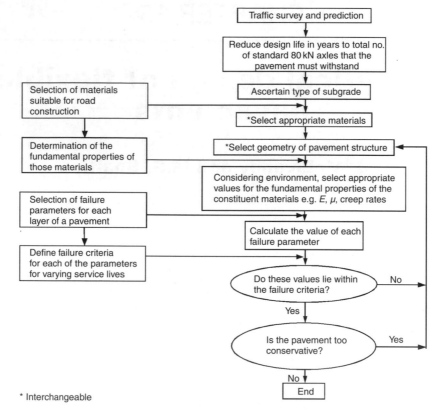

Fig. 15.1 Flow diagram of the analytical approach to road pavement design[1]

reaching a terminal condition that necessitates strengthening. Performance models attempt to relate those parameters associated with a particular mode of deterioration to the number of repetitions of these that can be sustained before a terminal condition is reached. In the more advanced design methods the models have been derived by correlating the results of laboratory and theoretical studies with empirical data.

In the majority of analytical design methods it is assumed that the pavement and subgrade can be modelled as a system of elastic layers of finite thickness supported by a semi-infinite elastic mass, and that the stresses, strains and deflections generated in the system, by traffic, can be determined by solving the general elastic equations governing the behaviour of multilayer linear elastic systems[2]. A material that is linear elastic, homogeneous and isotropic needs only to be characterized by two properties, the modulus of elasticity *(E)* and Poisson's ratio (μ), for purposes of structural analysis, and the pavement and subgrade materials are assumed to possess these attributes. Both pavement and subgrade are usually considered to be infinite in horizontal extent and it is usual to assume full friction at the interfaces between layers. A simplified pavement system modelled for purposes of analysis is illustrated on Fig. 15.2.

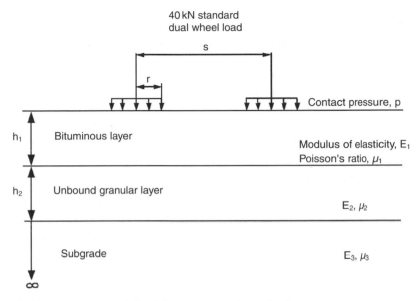

Fig. 15.2 Pavement modelled for structural analysis

Whilst a theoretical basis for calculating stresses and displacements in pavements modelled as two-layer and three-layer elastic systems has been available since the 1940s[3,4], the complexity of the theory and the time then required to perform calculations precluded its general application to flexible pavement design for a number of years. Rapid advances in computer and materials testing technology subsequently accelerated the development of the analytical approach, and interest was further stimulated by papers on the topic presented at the 1962 *International Conference on the Structural Design of Asphalt Pavements* and by the 1963 Shell design charts[5] whose derivation was based on structural analysis.

A number of agencies have now developed analytical design procedures and published the designs in chart format. This method of presentation imposes constraints on the design process in that it is unlikely to cover all of the site-specific factors that can influence design decisions, e.g. the particular climatic and subgrade conditions that exist at a site, characteristics of the anticipated loading, materials available for construction, and local construction techniques and practices. Where resources are available and the importance of the project warrants it, a rigorous analysis that seeks to take account of these factors is desirable. Such analyses are now feasible with the availability of appropriate software for use on personal computers. However, in order to make effective use of any analytical design method, it is necessary to understand how the pavement and subgrade layers are characterized and the background to the development of the performance models.

15.2 Pavement design period and design loading

The ability of a pavement to carry load and serve the road user is progressively reduced as a result of the separate and interactive effects of climate and traffic.

Reduction in the structural integrity and serviceability of a flexible pavement ultimately becomes apparent in deformation of the longitudinal and transverse profiles and cracking and disintegration of the surfacing. Various types of deterioration and possible causes are tabulated in reference 6.

The extent to which a pavement has deteriorated at any particular time is usually described in terms of its 'condition' at that time. The design period is the number of years from the time that the road is opened to traffic until a terminal condition, however defined, is reached; the pavement is designed to withstand the loading that will be applied during this period.

15.2.1 Pavement condition

Pavement condition may be categorized as either functional or structural: the former is indicative of how well the pavement is serving the road user at any particular time, the latter relates to the pavement's ability to sustain load and protect the subgrade.

The Present Serviceability Index (PSI)[7] provides a quantitative rating of a pavement's *functional condition*. First used at the AASHO Road Test in the USA in the 1958–60 period, the PSI was originally calculated from a regression equation that included measurements of wheeltrack rutting, cracking and patching, and of longitudinal unevenness, with the latter being by far the dominant parameter. The regression equation was derived by correlating road user opinion of pavement adequacy with objective measurements of the above parameters. A PSI-value in the 4–5 range indicates a pavement in 'very good' condition, whilst a pavement in 'very poor' condition has a PSI-value in the range 0–1. Flexible pavement test sections constructed at the AASHO Road Test had an average initial (i.e. immediately after construction) value of 4.2.

Another widely-used measure of functional condition, as indicated by 'roughness', is the International Roughness Index (IRI)[8]. IRI-values that are indicative of different levels of road serviceability are shown in Table 15.1. The corresponding Bump Integrator (BI) values are also indicated in this table.

The skidding resistance provided by a pavement surface is also an important factor in evaluating its functional condition.

A pavement's *structural condition* is normally evaluated from measurements of surface deflection under a moving wheel load (using the Benkleman Beam or Deflectograph meaurement methods), a vibratory load (the Dynaflect and Road Rater methods) or a falling weight (the Falling Weight Deflectometer method). Another readily-quantifiable measure of pavement structural condition is wheeltrack rutting which also influences surface roughness and safety and, hence, functional condition. The rutting parameter, particularly when evaluated in combination with cracking, correlates well with deflection as an indicator of the onset of a terminal structural condition.

15.2.2 Pavement performance

The curve linking functional or structural condition measurements made over a period of time characterizes *pavement performance*. Factors influencing pavement performance include initial structural capacity, quality of construction, load magnitude and

Table 15.1 International Roughness Index values for different road serviceability levels[9]

Paved roads			Unpaved roads		
Serviceability description	m/km, IRI	mm/km, BI	Serviceability description	m/km, IRI	mm/km, BI
Ride comfortable over 120 km/h. Undulation barely perceptible at 80 km/h in range 1.3 to 1.8. No depressions, pot-holes or corrugations are noticeable: depressions <2 mm/3 m. Typical high quality asphalt = 1.4–2.3. High quality surface treatment = 2.0–3.0	1.5-2.5	1000–2000	Recently bladed surface of fine gravel, or soil surface with excellent longitudinal and transverse profile (usually found only in short lengths)	1.5-2.5	1000–2000
			Ride comfortable up to 80–100 km/h. Aware of gentle undulations or swaying. Negligible depressions (e.g. <5 mm/ 3 m) and no pot-holes	3.5–4.5	2500–3500
Ride comfortable up to 100–200 km /h. At 80 km/ h, moderately perceptible movements or large undulations may be felt. Defective surface: occasional depressions, patches or pot-holes (e.g. 5–15 mm/ 3 m or 10–20 mm/5 m with frequency of 1–2 per 50 m) or many shallow pot-holes (e.g. on surface treatment showing extensive ravelling). Surface without defects: moderate corrugations or large undulations	4.0–5.5	3000–4000	Ride comfortable up to 70–80 km/h but aware of sharp movements and some wheel bounce. Frequent shallow moderate depressions or shallow pot-holes (e.g. 6–30 mm/3 m with frequency 5–10 per 50 m). Moderate corrugations (e.g. 6–20 mm/0.7–1.5 m)	7.5–9.0	6000–7000
Ride comfortable up to 70–90 km/h, with strongly perceptible movements and swaying. Usually associated with defects: frequent moderate and uneven depressions or patches (e.g. 15–20 mm/3 m or 20–40 mm/ 5m with a frequency of 5–3 per 50m) or occasionally pot-holes (eg 1–3 per 50 m). Surface without defects: strong undulations or corrugations	7.0–8.0	5500–6500	Ride comfortable at 50 km/h (or 40–70 km/h on specific sections). Frequent moderate transverse depres- sions (e.g. 20–40 mm/ 3–5 m at frequency of 10-20 per 50 m) or occasional deep depressions or pot- holes (e.g. 40–80 mm/ 3m with frequency less than 5 per 50 m). Strong corrugations (e.g. <20 mm/0.7–1.5 m)	11.5–13.5	9500–11500

(*Continued on p. 400*)

Table 15.1 (*continued*)

Paved roads			Unpaved roads		
Serviceability description	m/km, IRI	mm/km, BI	Serviceability description	m/km, IRI	mm/km, BI
Ride comfortable up to 50–60 km/h, frequent sharp movements or swaying. Associated with severe defects: frequent deep and uneven depressions and patches (e.g. 20–40mm/3m or 40–80 mm/5 m with a frequency of 5–3 per 50 m) or frequent pot-holes (e.g. 4–6 per 50 m)	9.0–10.0	7000–8000	Ride comfortable at 30–40 km/h. Frequent deep transverse depressions and/or pot-holes (e.g. 40–80 mm/1–5 m with frequency less than 5 per 50 m) with other shallow depressions. Not possible to avoid all the depressions except the worst	16.0–17.5	14000–15500
Necessary to reduce velocity below 50 km/h. Many deep depressions pot-holes and severe disintegration (e.g. 40–80 mm deep with a frequency of 8–16 per 50 m)	11.0–12.0	9000–10000	Ride comfortable at 20–30 km/h. Speeds higher than 40–0 km/h would cause extreme discomfort, and possible damage to the car. On a good general profile: frequent deep depressions and/or pot-holes (e.g. 40–80 mm/1–5 m at a frequency of 10-15 per 50 m) and occasional very deep depressions (e.g. <80 mm/0.6–2 m). On a poor general profile: frequent moderate defects and depressions (e.g. poor earth surface)	20.0–22.0	18000–20000

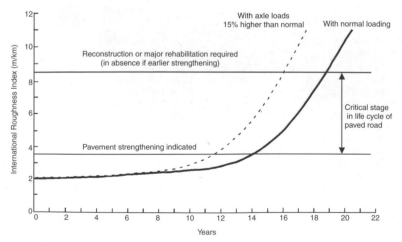

Fig. 15.3 Deterioration of paved roads over time[10]

repetitions, drainage conditions, climate and maintenance policies and practices. A functional performance curve based on the International Roughness Index (IRI) is shown on Fig.15.3.

A feature of performance curves, whether functional or structural, is the lengthy initial stage during which pavement condition remains relatively stable. There then follows a stage during which condition rapidly worsens. During this stage, often termed the 'critical' stage, strengthening of the pavement by overlay is necessary; if deterioration is allowed to continue beyond the critical stage, a failure condition is reached that normally requires reconstruction of the pavement.

15.2.3 Design period

Selection of the optimum initial design period requires economic evaluation of alternative design strategies over a fixed analysis period. For example, one strategy would be to design the pavement to last for the duration of the analysis period, i.e. a single-stage strategy. Alternatively, a stage-construction (planned rehabilitation) approach might be considered in which two or more design periods are included in the analysis period. At the end of each design period, the pavement is strengthened by a structural overlay designed to enable the pavement to carry the loading estimated for the following design period.

A range of strategies should be evaluated and the optimum identified[11]. However various constraints, e.g. availability of funds, disruption to traffic, etc. normally limit the number of feasible options. The accuracy with which future traffic volume and loading can be estimated is a further constraint.

A clear statement on terminal condition is absent from many pavement design procedures. In those procedures that specify a terminal functional condition, pavement roughness is often the primary criterion, possibly because of its direct impact on the road user and transportation costs and its relative ease of measurement. The extent of surface cracking and wheeltrack rutting are criteria used to specify a terminal structural condition.

The 1982 Asphalt Institute analytical design procedure[12] incorporates structural performance characteristics (fatigue cracking of the bituminous layer and wheel-track rutting) as well as functional characteristics (PSI). Current flexible pavement design standards in the UK are based on an analysis[13] of the performance of 144 sections of experimental road as indicated by wheeltrack rutting, cracking and deflection. Pavements were considered to have reached a terminal structural condition when wheeltrack rutting exceeded 10 mm or when structural cracking was observed.

Few design procedures address directly the important question of overall design reliability. There are two main uncertainties, one relating to the prediction of traffic loading, the other to prediction of pavement performance. Whilst uncertainty due to the former can be progressively reduced by systematic monitoring of traffic volume and axle loading and feedback of data into the design process, the latter is a more intractable problem. A 'reliability design factor' is included in the *AASHTO* design equation[14] to allow the designer to take discretionary account of these uncertainties. The design curves[13] developed for flexible pavements in the UK incorporate a probability of 85 per cent that a pavement constructed to the layer thicknesses and compositions indicated will not reach a terminal condition before it has carried the estimated design loading. Another procedure for dealing with design uncertainty takes account of variability in subgrade modulus. Depending on traffic volume, the modulus selected is that which is exceeded by a specified percentage of the measured values; for heavily trafficked roads, 87.5 per cent is typically specified[12].

15.2.4 Design loading

Traffic load is normally considered in terms of repetitions of axle load and during the design period the pavement is subjected to axle loads that vary in magnitude, frequency and configuration. The latter includes single, tandem and tridem or tri-axle configurations. A distinction may also be made between axles with single wheels and those with dual wheels[15]. In empirical design methods it is normal practice to convert, by means of load equivalency factors, the estimated spectra of axle loads of different configurations to the number of repetitions of a standard axle that causes an equivalent amount of damage. Thus, design loading is quantified as cumulative equivalent standard axle loads (ESALs). The practice has continued in analytical design, possibly for convenience, but also because the designs were generally presented in chart format. However, the complex interaction of wheel loads and pavement temperatures on pavement performance warrants a more rigorous analysis that considers separately the effect of individual axle loads and configurations on design criteria over the full range of pavement temperatures.

Application of the ESAL concept requires definition of a 'standard axle' and the derivation of a load equivalency factor for each axle configuration and load. The equivalency factor for the load carried by a particular axle configuration equals the number of passes of the standard axle that will cause the same damage as that caused by one pass of that axle configuration and load. Internationally, the most widely used factors are those determined from analysis of the AASHO Road Test data[16]. The axle selected as standard was an 18 000 lb (80 kN) dual-wheel single axle which, at that time (late 1950s), was the maximum legal single axle in the majority of states in the USA. Load equivalency factors were derived for single and tandem axle configurations; more recently, AASHTO[14] has published factors for a tridem axle configuration.

For purposes of structural analysis, it is normally assumed that wheel loads are applied through circular areas over which contact stress is uniform and equal to tyre pressure. Neither of these assumptions is strictly accurate but the errors introduced are relatively insignificant. Critical stresses, strains and deflections are usually calculated for the standard 40 kN dual-wheel load acting normal to the pavement surface and approximated by two circular loaded areas. Horizontal loads at the surface are generally ignored. The radius of each loaded area, the contact stress and the centre-to-centre spacing between the loaded areas are specified (see Fig. 15.2).

15.3 Structural analysis of layered elastic systems

Analysis of a one-layer system is based primarily on equations developed by Boussinesq[17] for the particular case of a point load acting at the surface of a semi-infinite mass characterized as elastic, homogeneous and isotropic. In the case of a circular load acting at the surface, the stresses, strains and deflections at any point in the mass can be determined by summation.

Many theories have been proposed for the analysis of two-layer and three-layer systems (see reference 18 for an excellent summary of each theory, including the assumptions made and the stresses and deflections that can be determined). Burmister[3,4] was the first to provide a theoretical framework for the analysis of two-layer and three-layer systems and, in a 1962 paper[19] dealing with the application of layered system analysis to the problems of design and construction, he discussed concepts 'of fundamental importance in understanding and dealing with layered pavement system performances'.

In the last 30 years or so, analytical models have been developed[20–23] which allow the determination of a number of parameters under a variety of conditions. The advent of these models has permitted the development of the more comprehensive analytical design procedures now in use. However, it should be borne in mind that calculation of stress, strain and deflection is but one component of the design process. The calculated values are only relevant when the pavement system has been modelled realistically and, in this regard, the specification of modulus values for cement-bound and unbound granular materials is a particular problem. The design process is also critically dependent on the adequacy of the models that utilize the calculated parameters to predict pavement performance. In a discussion in reference 14 on the use of analytical design procedures the point is made that the most important step in their implementation is the field testing and calibration of the performance models.

15.4 Design criteria used in analytical methods

Throughout the design period a pavement is subjected to repetitions of a complex system of stresses, strains and deflections whose cumulative effect is to weaken progressively the pavement and subgrade, and create the conditions for a rapid increase in deterioration. Cracking of the bituminous layer and wheeltrack rutting are the primary modes of traffic-induced pavement structural deterioration; thus, the design problem is to identify the stresses, strains and deflections, including their location, that are the major determinants of these modes of deterioration and to put limiting values on these. It should be recognized that cracking is also caused by non-traffic-related factors such as thermal stresses and volume change in the pavement foundation.

In the development of the 1963 Shell design charts[5], strains considered to have a major influence on pavement performance were (i) the horizontal tensile strain at the bottom of the bituminous layer, ε_t and (ii) the vertical compressive strain at the surface of the subgrade, ε_c. These are the primary design criteria currently used in the majority of analytical design methods. The tensile strain criterion is used to control cracking of the bituminous layer whilst the compressive strain criterion is intended to limit permanent deformation of the subgrade and, hence, at the pavement surface.

Cracking under repeated loading is attributed to fatigue, and early laboratory investigations[24] indicated that the fatigue behaviour of bitumen and bituminous mixes could be fully characterized in terms of tensile strain.

An analysis of 23 pavement sections from the AASHO Road test[2] indicated good correlation between calculated vertical stress at the surface of the subgrade and the weighted number of axle load repetitions that reduced the Present Serviceability Index to a value of 2.5. For convenience, Shell adopted vertical compressive strain rather than stress as a control parameter in the development of the 1963 thickness design charts; however, the use of either criterion results in designs that are very similar[2].

Using these two design criteria, the design objective is to ensure that the cumulative effects of repetitions of ε_t and ε_c do not cause critical levels of cracking and wheel-track rutting, respectively, to occur before the end of the design period. For pavements incorporating a cemented layer, either tensile stress or tensile strain is used as the design criterion, depending on the performance model.

Using compressive strain in the subgrade as the sole criterion for control of wheeltrack rutting does not take account of the permanent deformation that occurs in the unbound and bitumen-bound layers of the pavement. In the current Shell design method[25], candidate structures are identified from consideration of the primary design criteria and for each candidate structure an estimate is made of the permanent deformation that will occur in the bituminous and unbound layers during the design period. However, most analytical design methods consider the subgrade strain criterion to be adequate if the pavement materials are well designed and good construction practice is followed. Procedures for the design, placement and quality control of materials are specified as supplementary design criteria.

Deflection of the pavement surface has been proposed, both in combination with other design criteria[1] and as the sole criterion[26], for the design of new pavements. However, it is rarely used for this purpose and the importance attached by Burmister[19] to limiting the deflection and, hence, the magnitude of the shear stresses developed in a pavement system has received little attention.

When a dual-wheel load is applied normal to the surface of the pavement the maximum tensile strain usually occurs at the bottom of the bituminous layer, either at the centreline or edge of one of the wheels or mid-way between the wheels; in certain circumstances the maximum tensile strain occurs higher up in the layer[25]. Cracking is considered to be initiated at the bottom of the layer and to propagate upwards. However, cracking is frequently observed to have initiated at or near the layer surface where thermal and flexure-induced tensile stresses and strains are present, together with those generated by horizontal forces.

Radial-inward shear stresses develop at the tyre–pavement interface when a vertical load is applied by a pneumatic tyre, and where there is an increase in uni-directional shear during, for example, acceleration or deceleration. An analysis[27] indicates that the vertical stress calculated for a combination of vertical and radial-inward shear loading is greater than that calculated for vertical loading alone, and that the contribution of the shear loading to total stress is more significant at a shallow depth. The pattern of vertical normal and horizontal shear stresses generated by a unidirectional shear load acting at the surface is described in terms of their lateral distribution and the variation with depth of the maximum values observed. Both, but particularly the shear stress, are of significant magnitude at a shallow depth. Analysis of a combination of vertical and radial-inward loading at the surface of a three-layer pavement system indicates significant levels of surface tensile strain at the edge of the loaded area[28].

A possible explanation for the initiation of cracking at the surface is that oxidation of the bitumen at the layer surface lowers significantly the fatigue resistance of the material in the top few millimetres of the layer, thereby exacerbating the effects of repetitions of thermal and traffic loading. This explanation is given credence by the fact that initiation of cracks at the surface is observed most widely in the tropics and at high altitudes where the rate of oxidation of bitumen is relatively high. Whilst the more 'traditional' design criterion (ε_t) to deal with cracking is given prominence in this discussion, there is no reason why the analytical design process should not include tensile strain in the top fibres of the bituminous layer, modelled to reflect the bitumen hardening process, as an additional criterion.

15.5 Pavement material and subgrade properties required for structural analysis

Whilst it is reasonable, for practical purposes, to represent the pavement and its foundation as an essentially elastic system[18,25], it is necessary when characterizing individual system layers for purposes of analysis to take account of deviations from linear elastic behaviour.

Non-linearity of the dynamic stress–strain relationship is a feature of the majority of materials used in flexible pavement construction, and it is particularly evident in the case of unbound granular and cohesive materials. Furthermore, the response to load is not purely elastic. The deformation of unbound granular and cohesive materials includes an irrecoverable or permanent component; deformation of a bituminous material can comprise elastic, delayed elastic and viscous (irrecoverable) components, depending on the duration of loading and the temperature at which it is applied. In recognition of these deformation patterns, various terms are in use to characterize the stress–strain relationships of pavement materials. Thus, terms used for a bituminous material include 'dynamic' modulus, 'stiffness' modulus and 'resilient' modulus. The resilient modulus term is also used when characterizing unbound granular and cohesive materials. In the discussion that follows, the term *modulus of elasticity*, denoted by E, is used for all layers in the system – except when discussing data published by Shell where the term 'stiffness modulus', as used by that organization to characterize the stress–strain relationship for a bituminous material, is retained.

15.5.1 Bituminous materials

A bituminous mixture is a multiphase system in which the volume percentages of air, bitumen and aggregate have a significant influence on mechanical properties. Additionally, for any given combination of air, bitumen and aggregate, mechanical behaviour is time- and temperature-dependent, indicating the dominant effect of bitumen, a viscoelastic material, on the response of the mixture to loading.

The general properties required of bituminous mixtures and methods of mixture design are discussed in Chapter 12. From the viewpoint of structural analysis and design, the properties of primary interest are the modulus of elasticity and fatigue resistance.

Various types of test (e.g. flexural, direct axial and indirect tensile) have been used in the laboratory to determine the modulus of elasticity of bituminous materials, and procedures that have been standardized for this purpose are described in references 29 and 30. In the absence of suitable laboratory equipment to perform these tests the modulus can be estimated using models based on the volumetric composition of a mixture and the properties of its constituents. Models of this type have been developed by Shell, The Asphalt Institute and the University of Nottingham.

In Shell publications, the stress–strain relationship for bitumen and bituminous mixtures is described in terms of a 'stiffness' modulus, denoted by $S(t, T)$. The concept was introduced in early Shell investigations[31,32] to emphasize the dependency of the relationship on temperature and time of loading, and the stiffness modulus was defined as:

$$S_{(t, T)} = \left[\frac{\sigma}{\varepsilon_{(t)}} \right]_T \qquad (15.1)$$

where $\varepsilon_{(t)}$ = time-dependent strain under constant stress σ applied at temperature T.

Analysis of the results of an extensive testing programme enabled a nomograph to be constructed for estimating the bitumen stiffness modulus, S_{bit}, over a range of temperatures and loading times. To use the nomograph in its present form it is necessary to know the penetration index (*PI*) of the bitumen and the temperature at which a penetration of 800 (in units of 0.1 mm) is obtained under standard test conditions (T_{800} pen). The stiffness modulus of a bituminous mixture, S_{mix}, was found to be dependent primarily on S_{bit} and on C_v, the volume concentration of aggregate (defined as the ratio of the compacted aggregate volume to the combined volumes of aggregate and bitumen, with the volumes expressed as percentages of the total volume of the mix). Further research[33] resulted in the derivation of a mathematical relationship between S_{mix}, S_{bit} and C_v. This relationship, subsequently modified, is the basis for the University of Nottingham's equation[34] for predicting S_{mix}. Inputs required for a more recent Shell model[35] for estimating S_{mix} are S_{bit} and the volume percentages of aggregate and bitumen. S_{mix} can be read from a nomograph or, alternatively, can be calculated using a computer program (MODULE) which includes a subroutine (POEL) for calculation of S_{bit}.

The properties of bitumen change during the process of mix production, transportation and laying, largely because of bitumen oxidation and volatilization of its lighter fractions. When deriving S_{mix} from the Shell and the University of Nottingham models, properties determined on bitumen recovered from the mix must be used in the estimation of S_{bit}.

A constraint on using the above models is that they are valid only for values of $S_{bit} > 5$ MPa. In this high bitumen stiffness region, a plot of S_{mix} versus S_{bit} will result in curves that are almost coincident for mixes having approximately the same volume concentration of aggregate C_v, indicating that S_{mix} is primarily a function of S_{bit} and C_v. At values of S_{bit} less than 5 MPa, mixes having approximately the same C_v will yield individual relationships if differences exist in aggregate shape, grading, texture and interlock; in this case also the degree of confinement becomes significant.

The Asphalt Institute's pavement design programme DAMA[36] incorporates a regression equation for estimating the modulus of a bituminous mixture. In the equation the bitumen is characterized in terms of the absolute viscosity at 70°F (21.1°C) of the bitumen as supplied, and the time of loading in terms of frequency.

Providing that due consideration is given to their limitations, the above models can generally be considered to provide estimates of bituminous mixture stiffness or modulus that are of acceptable accuracy for practical purposes.

Laboratory determination of Poisson's ratio can be made from either static or dynamic tests in which lateral and longitudinal deformations are measured. Ratio values are influenced by temperature and stress state, and range from about 0.25 at 5°C to a value approaching 0.5 at 60°C. However, changes in the Poisson's ratio of flexible pavement materials have a relatively minor effect on calculations of maximum tensile strain in the bituminous layer and compressive strain in the subgrade[25]. Modulus and ratio values that have been determined at different temperatures for a variety of bituminous materials are given in reference 6.

Fatigue refers to the initiation and propagation of cracks in a material subjected to repetitive loading. *Fatigue life* (i.e. the number of load repetitions a material can withstand before failure) has two components: the number of load repetitions to initiate cracks, and the additional number of repetitions to propagate the cracks until a failure condition is reached.

Factors governing the initiation and propagation of cracks in a bituminous material subjected to repeated cycles of stress/strain have been extensively investigated and various models have been proposed to describe the phenomenon. However, the interpretation of laboratory fatigue data (which are the basis for many of the performance models used to predict in-service cracking of a bituminous layer) is complicated by the fact that the test type and mode have a significant influence on laboratory fatigue behaviour. Types of test used in the laboratory include flexural tests, direct and indirect tensile tests, and rotating-bending tests. The test mode may be either controlled-stress or controlled-strain (except for rotating-bending tests which are conducted in the controlled-stress mode only).

In controlled-stress tests the stress amplitude is kept constant for the duration of the test. The strain amplitude increases as the modulus reduces, and failure is defined as the number of stress repetitions required either to fracture the sample completely or to increase the initial strain amplitude by a specified amount. In this test mode the major proportion of fatigue life is the number of stress repetitions to crack initiation; once cracking is initiated, it propagates rapidly and failure quickly follows.

In the controlled-strain mode the strain amplitude is kept constant throughout. The stress amplitude reduces as the test progresses and failure is defined as the number of strain repetitions at which the stress amplitude has reduced to a specified percentage of its initial value. In this test mode crack-propagation time is predominant.

As with all fatigue testing, bituminous materials tested under cyclic loading exhibit scatter in the test results (i.e. nominally identical samples tested under nominally identical conditions give a range of fatigue lives). When plotted on logarithmic scales, the relationship between stress/strain and the number of repetitions to failure at different levels of probability of failure/survival is linear. Unless otherwise stated, relationships presented in the literature are generally those corresponding to a 50 per cent probability of failure/survival (i.e. the relationship is the 'best-fit' line established for the logarithmic means).

When a bituminous mixture is tested in controlled-stress at various temperatures and/or loading times, and the resultant data are plotted in the format 'log stress vs. log repetitions to failure', fatigue life at a given stress level increases with reduction in temperature and/or loading time (i.e. with increase in modulus). The reverse is true for the mixture when tested in controlled-strain; at a given strain level, fatigue life increases with decrease in modulus.

If controlled-stress data are presented in the format 'log initial strain in the mixture vs. log repetitions to failure' the influence of temperature and loading time and, hence, modulus is generally reflective of the pattern described above for controlled-strain testing. Fatigue behaviour is then characterized by an equation of the form:

$$N = K (\varepsilon_i)^{-m} (E_i)^{-c} \qquad (15.2)$$

where N=number of repetitions of constant stress amplitude σ that will cause failure; ε_i=initial strain in the mixture $(=\sigma/E_i)$; E_i=initial mixture modulus; K and m are material constants for the particular type of test; and c is a constant indicative of the influence of the modulus.

Mixture composition factors that influence significantly the fatigue characteristics of bituminous material include the volume and properties of the bitumen and the volume of air voids. The laboratory fatigue model which is the basis for the in-service fatigue performance model used in the Asphalt Institute design procedure has the form:

$$N = 4.32 \times 10^{-3} \times \varepsilon_i^{-3.291} \times E_i^{-0.854} \qquad (15.3)$$

where E_i=initial mixture modulus, psi.

The relationship described by Equation 15.3 was established[37] from controlled-stress tests made on a bituminous mixture of the same composition as that used at the AASHO Road Test. It may be applied to other bituminous materials by multiplying the right-hand side of the equation by a factor C, which is defined as

$$C = 10^M \qquad (15.4)$$

where

$$M = 4.84 \left(\frac{V_b}{V_v + V_b} - 0.69 \right)$$

and V_b=bitumen volume, %; and V_v=volume of air voids, %.

Research at the University of Nottingham[38] on a wide range of bituminous mixtures indicated that the fatigue lines determined for the mixtures tended to intersect at some common point or focus, and that the volume of bitumen and its softening point temperature were primary factors influencing fatigue performance. The focus was found to occur at N=40 and ε_i=630 $\mu\varepsilon$. The fatigue life at 100 $\mu\varepsilon$ was selected as the basis for defining the relationship between fatigue performance and bitumen volume and type. The relationship is expressed as

$$\log N = 4.13 \log V_b + 6.95 \log T_{R\&B} - 11.13 \qquad (15.5)$$

where N=fatigue life at initial strain ε_i= 100 $\mu\varepsilon$; V_b=volume of bitumen, %; and $T_{R\&B}$=Ring and Ball softening point temperature, °C, of the bitumen as supplied.

These two points, the focus and the fatigue life calculated for $\varepsilon_i = 100\ \mu\varepsilon$, enable the laboratory fatigue line for any mix composition to be determined.

Fatigue tests at Nottingham[38] were conducted in the controlled-stress mode using rotating-bending test equipment. Tests carried out over a range of temperatures and loading frequencies showed the relationship between initial strain in the mixture, ε_i, and the number of repetitions to failure, N, to be independent of the modulus, i.e. the fatigue characteristics of a mixture were completely described by the relationship;

$$N = K(\varepsilon_i)^{-m} \tag{15.6}$$

where K and m are mixture constants.

Two laboratory fatigue models have been developed by Shell, one based on direct interpretation of laboratory fatigue measurements[39] and the other on interpretation of the same measurements using a dissipated energy concept[40,41]. The model based on direct interpretation was used in deriving the 1978 Shell design charts and is also used in the NAASRA design method[15]. Based on controlled-strain test data, it has the form:

$$\varepsilon = (0.856 V_b + 1.08) \times S_{mix}^{-0.36} \times N^{-0.2} \tag{15.7}$$

where ε = permissible strain for a given number of repetitions, N; V_b = volume of bitumen, %; and S_{mix} = stiffness modulus of the mix, Pa.

15.5.2 Unbound granular materials

The description 'unbound' denotes the absence of a binding agent, e.g. bitumen or Portland cement, from the granular materials used in the subbase and (now less frequently) in the roadbase of flexible pavements. The mechanical properties of unbound layers that are of interest in pavement design are resistance to shear failure and to elastic and permanent deformation. For this discussion it is the elastic characteristics that are of primary interest.

Repeated loading of unbound granular material in a triaxial compression test results in a total deformation per cycle that has an elastic component and a permanent or irrecoverable component. The elastic response is characterized in terms of a modulus of elasticity, Eg, defined as:

$$Eg = \frac{\sigma_d}{\varepsilon_e} \tag{15.8}$$

where σ_d = repeated deviator stress; and ε_e = elastic axial strain.

The latest AASHTO procedure[42] for the laboratory determination of Eg supersedes two earlier procedures, and was adopted from the Strategic Highway Research Programme (SHRP) Protocol P46. As noted previously the stress–strain relationship is non-linear and various equations have been proposed[43] to model the stress-dependency of Eg. The equation most widely used at present considers the influence of confining stresses only:

$$Eg = K_1(\theta) K_2 \tag{15.9}$$

where θ = bulk stress or first stress invariant = $\sigma_1 + \sigma_2 + \sigma_3$ ($= \sigma_1 + 2\sigma_3$ in a triaxial compression test); and K_1 and K_2 are experimentally determined coefficients.

Equation 15.9 is valid provided the applied stresses are not sufficiently large to cause failure[44]. AASHTO[14] lists values of the coefficients K_1 and K_2 for roadbase and subbase materials when they are 'dry', 'damp' or 'wet' and suggests values for the stress state (θ) in a roadbase layer for a range in values of bituminous layer thickness and subgrade modulus. Material characteristics that influence *Eg* include its density and degree of saturation, particle shape and surface texture and the amount of material finer than 75 µm. At a given stress level, *Eg* increases with increase in density and increased angularity and surface roughness of the aggregate particles; an increase in saturation causes a reduction in *Eg* as does an increase in the amount of material finer than 75 µm[44].

Repeated load triaxial compression tests indicate that the principal stress ratio (σ_1/σ_3) has a major influence on Poisson's ratio which increases[44] with a decrease in σ_3 or an increase in σ_1. Density, degree of saturation and the amount of material finer than 75 µm have a slight influence. At high stress ratios, the Poisson's ratio of granular material may be as high as 0.6 or 0.7, indicating an increase in volume[43].

15.5.3 Cemented materials

Cemented materials are those produced by the addition of Portland cement to either naturally occurring or processed material in quantities sufficient for the development of significant tensile strength.

The modulus of elasticity of a cemented material is very dependent upon the type and properties of the untreated material, the cement content and mix density, and the method of testing. Included in the latter are flexural tests, direct and indirect tensile tests and direct compression tests. The method preferred by NAASRA[15] for direct measurement of the modulus is a test involving the three-point flexural loading of beam specimens having a span/depth ratio greater than 3; the modulus is obtained from the linear portion of the load–deflection relationship using simple-beam theory. Beams are also used to determine dynamic modulus of elasticity from measurements of resonance frequency; the modulus thus determined is in reasonable agreement with that determined on in-situ material using the wave velocity technique[45].

Correlations have been established between elastic modulus and compressive strength, the strength parameter most widely used in the routine design and quality control of cemented material[14,15]. Correlations are also available between flexural strength and modulus of elasticity[46,47].

Precise determination of the Poisson's ratio of cemented materials is not of particular importance for pavement analysis due to its second-order effect on the design criteria. Resonance frequency and pulse velocity measurements made on cemented granular materials indicate typical values of 0.2, while compression tests on these materials gave values of the order of 0.15[46].

The fatigue behaviour of cemented material is typically characterized in terms of ratios (e.g. stress ratio or strain ratio). In some fatigue models account is taken of thermal or shrinkage stresses and strains, whilst in others they are ignored.

One model[48] that considers thermal effects and characterizes fatigue behaviour in terms of strain ratio is:

$$N_f = 10^{9.1(1 - \varepsilon_s/\varepsilon_b)} \tag{15.10}$$

where N_f=number of repetitions of ε_s to crack initiation; ε_s=modified traffic-induced strain; and ε_b=tensile strain at break measured by static beam flexure. The modified strain ε_s is obtained by increasing the traffic-induced strain by a factor to take account of thermal cracking. The factor ranges from 1.1 to 1.4, depending on the material strength, the severity of cracking and the layer thickness.

15.5.4 Subgrade soils

The subgrade ultimately supports the dead and live loads imposed by the pavement and by traffic, respectively. The support that the subgrade can provide determines the structural capacity required of the pavement to carry the design traffic loading.

The modulus of elasticity as defined earlier for granular material (Equation 15.8) is used also to characterize the elastic properties of the subgrade. Laboratory procedures for determining the modulus of elasticity of granular and cohesive subgrade soils are described in reference 42.

If the subgrade soil is granular, its modulus is influenced primarily by those factors discussed earlier for unbound granular material, and confining stresses are of particular importance.

In the case of cohesive soils, it is the deviator or shearing stress which has a primary influence on the modulus. At low levels of deviator stress, the modulus of a cohesive material reduces rapidly with increase in deviator stress; outside this range, an increase in deviator stress results in a modest change in modulus. The influence of deviator stress on measurements of elastic (resilient) modulus made after 100 000 stress repetitions is shown in Fig. 15.4, where it can be seen that the variation in modulus is greatest over the range in stress levels likely to occur at the surface of a subgrade supporting a structurally adequate pavement. Hence, it can readily be understood why the pavement failure process, once initiated, can be very rapid (Fig. 15.3). Compacted soil characteristics that influence the modulus of elasticity of a cohesive soil include dry density, moulding moisture content and soil structure or particle arrangement, with the last of these being of particular importance when preparing samples for laboratory evaluation.

The relatively limited data available on the influence of stress state on the Poisson's ratio of cohesive soils indicate that the ratio increases slightly with increase in deviator stress but is relatively unaffected by variation in mean normal stress[50].

15.6 Layer characterization for purposes of structural analysis

Calculation of stresses, strains and deflections in layered elastic systems requires modulus and Poisson's ratio values to be assigned to each layer of the system. From the earlier discussion of factors influencing modulus and Poisson's ratio, it will be apparent that it is not easy to assign values that are representative of the wide range of conditions that occur in practice.

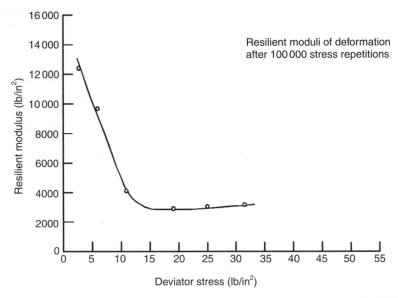

Fig.15.4 Effect of stress intensity on resilience characteristics – AASHO Road Test subgrade soil[49]

15.6.1 Bituminous layers

Temperature and duration of loading are the primary factors considered when assigning a modulus value to a bituminous layer; the progressive increase in modulus due to in-service hardening of the bitumen, whilst significant, is not usually taken into account.

Identifying a range of in-service operating conditions that is comprehensive but manageable presents a significant design problem. Ambient temperature can vary considerably in the course of a day, and from day to day, over the annual cycle of climatic conditions, with consequences for the temperature gradients that develop in a bituminous layer. Load duration varies from about 0.01s for relatively fast-moving traffic to about 0.1s for slow-moving traffic. In the range of operating conditions from slow-moving traffic/high ambient temperature to fast-moving traffic/low ambient temperature, there is wide variation in modulus.

Information on *mean monthly air temperature (MMAT)* is readily available at most locations and a common procedure is to relate a representative bituminous layer temperature to this parameter. A representative temperature may be derived from structural analysis of pavements modelled to reflect the daily range in temperature (modulus) gradients associated with a particular value of MMAT[25]. In the Asphalt Institute design procedure[12], the *mean monthly pavement temperature (MMPT°F)* is related to MMAT (°F) as follows:

$$MMPT = MMAT \left[1 + \frac{1}{z + 4} \right] - \frac{34}{z + 4} + 6 \qquad (15.11)$$

where z = depth below pavement surface, inches. The temperature at one-third of the depth, which is calculated using Equation 15.11, is chosen as the representative temperature for the bituminous layer. In a pavement comprising upper and lower bituminous layers of thicknesses h_1 and h_2, respectively, z is taken to be $h_1/3$ for the upper layer and $(h_1 + h_2/3)$ for the lower layer. The corresponding modulus values can be input directly to the DAMA design programme[36] if laboratory data on the 'modulus–temperature' relationships for the materials are available. Alternatively, the values are assigned using modulus–temperature relationships generated by the modulus prediction model incorporated in the programme. The 1978 Shell designs[39] are presented in chart form and the concept of a weighted mean annual air temperature (w-MAAT) is used to present designs for a wide range of climatic conditions. Development of the concept is described in references 25 and 41.

Loading time at the underside of the bituminous layer may be estimated from the following relationship[51]:

$$t = \frac{1}{V} \tag{15.12}$$

where t = loading time, s; and V = vehicle speed, km/h. This relationship applies for layer thicknesses of 150 to 300 mm.

Poisson's ratio values of 0.35 or 0.4 are typically used. The Asphalt Institute[12] and Shell[25] use a value of 0.35, the latter stating that this value applies at short times of loading, independent of stress. The University of Nottingham[51] and NAASRA[15] use a value of 0.4.

15.6.2 Unbound granular layers

Characterization of an unbound granular layer for purposes of structural analysis is complicated by the stress dependency of its modulus and by its limited tensile strength[52]. Equation 15.9 is not normally used directly in structural analysis as it involves a lengthy iterative-type of solution and unbound layers are characterized by more manageable procedures that take account of factors influencing the stress state that can develop in the material.

Shell[25] has approached the problem primarily by recognizing the dependence of the granular material modulus on the subgrade support. The approach assumes that a granular layer significantly stiffer than its supporting layer will develop tensile stresses sufficiently large to cause decompaction, leading ultimately to an equilibrium condition. Hence, there is a limit to the extent to which the modulus of an unbound granular layer can exceed that of its supporting layer. This limit is expressed in the form of a modular ratio as follows:

$$\frac{E_g}{E_{sg}} = k \tag{15.13}$$

where E_g = modulus of the unbound granular layer; and E_{sg} = modulus of the subgrade. Shell defines k as:

$$k = 0.2 \, (h_2)^{0.45} \text{ and } 2 < k < 4 \tag{15.14}$$

where h_2 = thickness of unbound granular layer (roadbase plus subbase), mm. In an 1985 addendum[53] to the Shell design manual, an alternative approach for calculating the modulus of the granular layer is described.

The Asphalt Institute considers degree of confinement as represented by Equation 15.9 to be the primary influence on granular layer modulus and, in its DAMA programme[36], the modulus is calculated using a regression equation that reflects the relative effect on confinement of the following pavement characteristics: (i) the thickness and modulus of the overlying bituminous layer; (ii) the support provided by the underlying subgrade layer; (iii) the thickness of the granular layer; and (iv) the quality of the granular material.

In the NAASRA design procedure[15], the granular layer is divided into a number of sublayers. Sublayer thickness is required to be in the range 50–150 mm and the modular ratio between adjacent sublayers must not exceed 2. Only the modulus of the top sublayer and of the subgrade need be specified, and the modulus of the other sublayers is determined from the modular ratio, R, which is constant between sublayers and is calculated from

$$R = \left[\frac{E_{TSL}}{E_S}\right]^{1/n} \tag{15.15}$$

where E_{TSL} = modulus of the top sublayer, MPa, a function of the bituminous layer thickness and temperature; E_S = modulus of the subgrade, MPa; and n = number of sublayers. A Poisson's ratio value of 0.35 is typically assumed for the unbound granular layer.

15.6.3 Cemented layers

Design of a pavement incorporating a cemented layer requires analysis of the combined effects of thermal and traffic-induced stresses. In the UK, for example, lean concrete roadbases typically develop lateral cracks at 5–8 m intervals[47,54]. These cracks are most likely to develop in the interval between laying the roadbase and the provision of the surfacing, when there is a large difference between day and night temperatures. The development of a more general pattern of cracking occurs under the combined action of traffic and thermal stresses.

The thickness of the overlying bituminous layer is governed largely by the need to insulate the cemented layer in order to minimize thermal stresses and from considerations of reflection cracking (i.e. propagation through the bituminous layer of cracks originating in the cemented layer). Given the pattern of crack development outlined above, a question that arises when characterizing a cemented layer for purposes of structural analysis is whether the modulus assigned to the layer should represent the 'cracked' or 'uncracked' condition. If a two-stage analysis is undertaken, the further question arises as to when it is appropriate to use the cracked modulus and what value should be assigned for this condition. Suggested correlations between modulus and unconfined compressive strength for various types of cemented material, before and after extensive cracking, are given in Table 15.2.

15.6.4 Subgrade

The subgrade modulus is best determined either from in situ measurements of dynamic deflection made on the soil at a representative moisture content and using

Table 15.2 Moduli of cemented materials[48]

Unconfined compressive strength (MPa) (Pre-cracked phase)	Pavement materials cemented	Pre-cracked phase		Post-cracked phase	
		Modulus range (GPa)	Recom-mended modulus (GPa)	Under bound materials (GPa)	Under a cracked or untreated material (GPa)
6–12	Crushed stone	7–30	14	1.5	1.2
3–6	Stone or gravel	4–14	8.5	1.0	0.75
1.5–3	Gravel	3–10	6.0	0.75	0.50
0.75–1.5	Gravel	2–7	3.5	0.50	0.30

Poisson's ratio = 0.35

loads of a magnitude likely to occur in practice, or from a laboratory test procedure such as that specified in reference 42. Alternatively, it may be estimated from the California Bearing Ratio (CBR):

$$E_{sg} (\text{MPa}) = 17.6 \, (CBR)^{0.64} \qquad (15.16)$$

$$E_{sg} (\text{MPa}) = 10 \, CBR \qquad (15.17)$$

Equation 15.16 (see reference 13) is applicable to the 2 to 12 per cent range of CBR values. Equation 15.7 is based on reference 52.

Poisson's ratio is assumed by Shell[25] to be 0.35 whilst the University of Nottingham[51] considers values in the range 0.4 to 0.5. Representative values suggested by NAASRA[15] are 0.45 and 0.35, respectively, for cohesive and granular soils.

15.7 Damage computations and performance models

15.7.1 Damage computations

Fatigue life, as discussed earlier, is normally defined for simple-loading conditions (i.e. in which stress or strain amplitude is maintained constant for the duration of a test). In practice, however, materials are usually subjected to a wide spectrum of stress/strain amplitudes (i.e. compound-loading conditions exist) and the problem then is to estimate fatigue life under compound loading. A number of models have been proposed that utilize simple-loading fatigue data to predict fatigue life under compound loading. Of these, the simplest and most widely used model is based on the linear summation of cycle ratios[55], and laboratory investigations[56] have shown that it predicts compound-loading fatigue life with reasonable accuracy. The model states that failure will occur when the summation of cycle ratios equals unity, i.e. when

$$\sum_{i=1}^{r} \frac{n_i}{N_i} = 1 \qquad (15.18)$$

where n_i = number of repetitions of strain amplitude ε_i; N_i = number of repetitions of strain amplitude ε_i that will cause failure; and r = number of strain amplitudes in the spectrum.

If the application of a particular spectrum of strain amplitudes does not cause failure but is repeated a number of times until failure occurs, then the number of repetitions (F) of the strain amplitude spectrum that the material can sustain before failure occurs is given by:

$$F = 1 / \sum_{i=1}^{r} \frac{n_i}{N_i} \qquad (15.19)$$

The steps involved in the application of Equation 15.19 to pavement design can be summarized as follows:

Step 1.   Identify intervals during the year when climatic conditions, and hence layer modulus values, are relatively constant. (Convenient intervals are the individual months of the year or individual seasons and the transition periods between seasons.)

Step 2.   For the particular structure being evaluated, assign appropriate layer modulus values for each climatic condition, and calculate the maximum tensile strain in the bituminous layer when the standard wheel load is applied at the surface.

Step 3.   For the strain amplitude characteristic of each climatic condition, use the fatigue performance model to determine fatigue life if that strain amplitude were applied until failure occurred.

Step 4.   Determine the design loading (i.e. the cumulative ESALs during the design period) and calculate the number of ESALs carried during each climatic condition in a single year. For example, if the pavement is expected to carry W ESALs during a design period of t years, and monthly variations in climatic condition are being considered, then the relevant number is given by $W/12t$. (Note the assumptions that the strain amplitude characteristic of a particular climatic condition remains constant throughout the design period, and that the damage caused by a particular strain amplitude remains constant.)

Step 5.   The number of repetitions of each characteristic strain amplitude during a single year is divided by the number of repetitions to failure at that strain amplitude. Summation of all such cycle ratios gives the proportion of fatigue life consumed during the year. As this is assumed to be constant for each year of the design period, and since failure is assumed to occur when the summation of cycle ratios equals unity, the design period in years is obtained by dividing unity by the summation of cycle ratios determined for a single year.

The above outlines a procedure for estimating the fatigue life of the bituminous layer under strain amplitudes of varying magnitude. The same procedure can be applied to estimating the 'permanent deformation' life under varying amplitudes of subgrade compressive strain.

15.7.2 Performance models

Bituminous material fatigue models developed in the laboratory normally provide a reference point for in-service fatigue performance models. Since laboratory models, derived under relatively simplistic conditions, greatly underestimate fatigue performance in the field, it is necessary to adjust laboratory models to take account of this disparity. Some factors contributing to the disparity, and taken account of, as necessary, in its adjustment, are discussed below.

In conventional laboratory fatigue tests, a bituminous material is subjected to continuous repetitions of a constant amplitude of stress/strain until failure occurs. In the field, however, loading is intermittent rather than continuous (i.e. between successive loading pulses there are rest periods whose duration depends upon vehicle speed, axle spacing, and traffic density). Furthermore, wheel loads are distributed transversely across the design lane, i.e. all wheels moving along the pavement do not follow exactly the same line. Thus, with reference to the centre-line of the wheelpath, maximum strain at the bottom of the layer is generated by those wheel loads that pass over the centreline, while wheel loads distributed about the centreline generate strains of lower magnitude at the centreline. Crack propagation time is a further consideration. Laboratory models developed from controlled-stress tests predict essentially the number of repetitions to crack initi-ation; crack propagation time in this test mode is relatively short. When this type of model is applied to in-service conditions, and the areal extent of cracking at the surface is a measure of pavement terminal condition, the model needs to be adjusted to allow for the number of repetitions required for upward propagation of cracks through the layer.

The Asphalt Institute used its laboratory fatigue model (Equation 15.3) in com-bination with layered elastic theory and the linear summation of cycle ratios concept to predict the initiation of cracking at the underside of the bituminous layer in various pavement sections of the AASHO Road Test[37]. The time interval and, hence, the number of load repetitions between predicted crack initiation and the observation of cracking at the pavement surface was noted, and the model was then adjusted for use in service by applying a shift factor consistent with the extent of observed cracking. Shift factors of 13.4 and 18.4 were determined for up to 10 per cent and greater than 45 per cent cracking, respectively, in the wheelpath area. The shift factors were found to be the same for bituminous layer thicknesses of 4 in (102 mm) and 6 in (152 mm). Equation 15.3, multiplied by a shift factor of 18.4 and incorpor-ating the mix composition factor C (Equation 15.4), is the fatigue performance model that is used in The Asphalt Institute DAMA programme[36]. Because of the method of derivation of the shift factor, it was not necessary to adjust the laboratory model to take account of rest periods and the transverse distribution of wheel loads.

In contrast with the Asphalt Institute approach, the Shell laboratory fatigue model (Equation 15.7) was adapted to field conditions using structural analysis to calculate the adjustments needed to account for temperature gradients in the bitu-minous layer and the transverse distribution of wheel loads[41]. Laboratory tests were performed to determine an appropriate adjustment for intermittent loading. In the computer version of the Shell design procedure that is now available[57], the fatigue life indicated by Equation 15.7 is increased by a factor of 10 to account for these effects. As noted earlier, the model is based on controlled-strain test data and an adjustment for crack propagation was not considered necessary[40]. The fatigue life, N, is the number of repetitions of the permissible strain (ε) that will result in the development of 'real cracks' in the material.

In-service fatigue models developed[13] for roads in the UK with a bituminous roadbase are presented in the form:

for rolled asphalt: $$\log N = -9.78 - 4.32 \log \varepsilon_t \qquad (15.20a)$$

for dense bitumen macadam: $$\log N = -9.83 - 4.16 \log \varepsilon_t \qquad (15.20b)$$

where N = number of repetitions of the standard 40 kN wheel load; and ε_t = horizontal tensile strain at the bottom of the bituminous layer under the standard wheel load.

The models are applicable for a single 'equivalent' pavement temperature of 20°C and the modulus is therefore not a variable. The designation 'equivalent' means that if the bituminous layer is characterized in terms of the modulus at 20°C, fatigue damage under repetitions of the standard wheel load is equivalent to that computed by considering the full range of combinations of pavement temperatures and wheel loads representative of conditions in the UK. Regarding reliability, there is a probability of 85 per cent that pavements designed to these criteria will survive the design life.

The interactive effects of climate and traffic make it highly unlikely that realistic field performance models can be derived solely from theoretical analysis. Future design procedures will develop from a combination of the analytical and empirical approaches, and the methodology described in references 13 and 58 is a good illustration of how this can be achieved.

In respect of fatigue of cemented material, Equation 15.10 predicts the number of strain repetitions that will initiate cracking at the underside of a cemented layer. To allow for crack propagation throughout the layer, the number of repetitions is adjusted by a shift factor whose value depends on layer thickness and road category. Roads are categorized as A, B or C in order of decreasing importance. The shift factors for a 100 to 200 mm thick layer in road categories A and C, for example, are 1.2 and 3.0, respectively; the corresponding factors when layer thickness exceeds 200 mm are 3 and 7.5. An alternative to using fatigue models for cemented layers is to calculate the combined thermal and traffic stresses at the underside of the cemented layer for various combinations of bituminous and cemented layer thicknesses, using the flexural strength of the cemented material as a control parameter. Based on an analysis of this type, a 200 mm bituminous surfacing and a 250 mm lean concrete roadbase has been recommended[13] for roads in the UK designed for long but indeterminate lives.

The parameter most widely used in analytical design procedures for control of wheeltrack rutting is the vertical compressive strain at the surface of the subgrade and the performance model has the form:

$$\varepsilon_c = a \left(\frac{1}{N} \right)^b \qquad (15.21)$$

where ε_c = permissible vertical compressive strain at the surface of the subgrade for N repetitions; and a and b are regression coefficients, generally determined for the 40 kN standard wheel load.

A number of models of the type described by Equation 15.21 have been proposed since the concept was first introduced by Shell[5] and were derived from back-analysis of pavements that performed satisfactorily in practice. The Shell model, derived from an analysis of a representative selection of AASHO Road Test pavement sections, relates permissible strain to the number of repetitions of the standard axle load that will reduce Present Serviceability Index to a value of 2.5. The Asphalt Institute model was obtained from back-analysis of pavements designed according to the method specified by the State of California Department of Transportation (Caltrans). These models are described as follows:

Shell: $$\varepsilon_c = 2.8 \times 10^{-2} \times (N)^{-0.25} \qquad (15.22)$$

The Asphalt Institute: $$\varepsilon_c = 1.05 \times 10^{-2} \times (N)^{-0.223} \qquad (15.23)$$

Given the method of derivation, no adjustment is needed for transverse distribution of wheel loads.

Using the methodology referred to earlier to develop a fatigue performance model for the bituminous layer, a performance model[13] developed in the UK for wheeltrack rutting was presented in the form:

$$\log N = -7.21 - 3.95 \log \varepsilon_c \qquad (15.24)$$

where N = road design life in standard 80 kN axles; and ε_c = vertical compressive strain at the subgrade surface under a standard 40 kN wheel load. As with the fatigue model, the equivalent temperature of the bituminous layer is 20°C (i.e. the model is applicable for the pavement temperatures and wheel loads found in the UK).

15.8 Concluding comments

Observation of the performance of full-scale pavements is essential for checking the reliability of, and calibrating, a particular design procedure. The limitations of its usefulness as the sole basis for developing a design procedure are readily apparent, particularly in its inability to respond rationally and quickly to new materials and material processing procedures, improved construction techniques and practices and apparently ever-increasing and varied traffic loading.

It will be apparent that the analytical approach offers a rational process for the design of flexible pavements. It has the potential to cater for hitherto unforeseen traffic loadings and a variety of environmental conditions and materials. This chapter sets out to draw attention to some of the better-known methods currently available and the problems inherent in the process due, for example, to incomplete knowledge of material behaviour or lack of clarity with regard to failure criteria. The intention was not to describe in detail how to use any particular method – this information is readily available in the literature – but rather to enable the reader to decide which method is appropriate to a particular set of conditions and how the method selected may be adapted to provide a realistic design within the constraints of current knowledge.

15.9 References

1. Kirwan, R.W., Glynn T.E. and Snaith, M.S., A structural approach to the design of flexible pavements. *The Irish Engineer*, 1975, **28**, No. 8.

2. Peattie, K.R., A fundamental approach to the design of flexible pavements, *Proceedings of the International Conference on the Structural Design of Asphalt Pavements*, The University of Michigan, 1962.

3. Burmister, D.M., The theory of stresses and displacements in layered systems and applications to the design of airport runways, *Proceedings of the Highway Research Board*, 1943, **23**.

4. Burmister, D.M., The general theory of stresses and displacements in layered systems, *Journal of Applied Physics*, 1945, **16**.

5. *Shell Design Charts for Flexible Pavements*. London: The Shell International Petroleum Co., 1963.

6. Evdorides, H.T, and Snaith, M.S., A knowledge-based analysis process for road pavement condition assessment, *Proceedings of the Institution of Civil Engineers -Transport*, 1996, **117**.

7. Carey, W.N., and Irick, P.E., The pavement serviceability – performance concept, *Highway Research Board Bulletin 250*, 1960.

8. Sayers, M., Gillespie, T.O., and Queiroz, C.A.V., *The International Road Roughness Experiment; Establishing Correlation and a Calibration Standard for Measurements*, Technical Paper 45. Washington DC: World Bank, 1986.

9. *A Guide to Road Project Appraisal*, Overseas Road Note 5. Crowthorne, Berkshire: The Transport and Road Research Laboratory, 1998.

10. *Road Deterioration in Developing Countries – Causes and Remedies*. Washington DC: World Bank, 1988.

11. Kerali, H.R., Odoki, J.B and Wightman, D.C., The new HDM-4 analytical framework. Paper presented at the *Joint 18 ARRB Transport Research Conference and Transit New Zealand Transport Symposium*, Christchurch, New Zealand, 1996.

12. The Asphalt Institute, *Research and Development of the Asphalt Institute's Thickness Design Manual (MS-1)*, 9th edn, Research Report No. 82–2. Lexington, Ky: The Asphalt Institute, 1982.

13. Powell, W.D., Potter, J.F., Mayhew, H.C. and Nunn, M.E., *The Structural Design of Bituminous Roads*, LR1132. Crowthorne, Berkshire: The Transport and Road Research Laboratory, 1984.

14. AASHTO, *Guide for Design of Pavement Structures*. Washington, DC: American Association of State Highway and Transportation Officials, 1986.

15. NAASRA, *Pavement Design: A Guide to the Structural Design of Road Pavements*. Sydney: National Association of Australian State Road Authorities, 1987.

16. Liddle, W.J., Application of AASHO Road Test results to the design of flexible pavement structures, *Proceedings of the International Conference on the Structural Design of Asphalt Pavements*, The University of Michigan, 1962.

17. Boussinesq, V.J., *Application des potentiels a l'étude de l'équilibre et du mouvement des solides élastiques avec les notes etendues sur divers points de physique, mathématique et d'analyse*. Paris: Gautlier – Villais, 1885.

18. Whiffin, A.C., and Lister, N.W., The application of elastic theory to flexible pavements, *Proceedings of the International Conference on the Structural Design of Asphalt Pavements*, The University of Michigan, 1962.

19. Burmister, D.M., Application of layered system concepts and principles to interpretation and evaluation of asphalt pavement performances and to design

and construction, *Proceedings of the International Conference on the Structural Design of Asphalt Pavements*, The University of Michigan, 1962.

20. Warren, H. and Dieckmann, W.L., *Numerical Computations of Stresses and Strains in a Multi-layer Asphalt Pavement System*. Unpublished Internal Report, Chevron Research Corporation, 1963.
21. Peutz, M.G.F., van Kempen, H.P.M. and Jones, A., Layered systems under normal surface loads, *Highway Research Record 228*, 1968.
22. De Jong, D.L., Peutz, M.G.F. and Korswagen, A.R., *Computer Programme Bisar: Layered Systems Under Normal and Tangential Surface Loads*, External Report AMSR. 0006.73. Amsterdam: Koninklijke/Shell Laboratorium, 1973.
23. Snaith M.S., McMullen, D., Freer-Hewish, R.J. and Shein, A., *Flexible Pavement Analysis; Final Technical Report*. Birmingham: Department of Transportation and Environmental Planning, University of Birmingham, 1980.
24. Pell, P.S., Fatigue characteristics of bitumen and bituminous mixes, *Proceedings of the International Conference on the Structural Design of Asphalt Pavements*, The University of Michigan, 1962.
25. Claissen, A.I.M., Edwards, J.M., Sommers, P. and Uge, P., Asphalt pavement design – The Shell method, *Proceedings of the Fourth International Conference on the Structural Design of Asphalt Pavements*, The University of Michigan, 1977.
26. Snaith, M.S. and Hattrell, D.V., A deflection based approach to flexible pavement design and rehabilitation in Malaysia, *Proceedings of the Institution of Civil Engineers: Transport*, 1994, **105**.
27. Barber, E.S., Shear loads on pavements, *Proceedings of the International Conference on the Structural Design of Asphalt Pavements*, The University of Michigan, 1962.
28. Molenaar, A.A.A., Fatigue and reflection cracking due to traffic loads, *Proceedings of the Association of Asphalt Paving Technologists*, 1984, **53**.
29. *Dynamic Modulus of Asphalt Mixtures*, ASTM D3497–85. Philadelphia: American Society for Testing and Materials, 1991.
30. *Resilient Modulus of Asphalt Concrete from Diametral Strain*, ASTM D4123–82. Philadelphia: American Society for Testing and Materials, 1991.
31. Van der Poel, C., A general system describing the visco elastic properties of bitumen and its relation to routine test data, *Journal of Applied Chemistry*, 1954, **4**, Pt 5.
32. Van der Poel, C., Time and temperature effects on the deformation of asphaltic bitumens and bitumen mineral mixtures, *Journal of the Society of Petroleum Engineers*, 1955.
33. Heukelom, W. and Klomp, A.J.G., Road design and dynamic loading, *Proceedings of the Association of Asphalt Paving Technologists*, 1964, **33**.
34. Brown, S.F., Stiffness and fatigue requirements for structural performance of asphaltic mixes, *Proceedings of Eurobitume Seminar*, London, 1978.
35. Bonnaure, F., Gest, G., Gravois, A. and Uge, P., A new method of predicting the stiffness of asphalt paving mixtures, *Proceedings of the Association of Asphalt Paving Technologists*, 1977, **46**.
36. *Computer Programme DAMA; User's Manual*. Lexington, Ky: The Asphalt Institute, 1983.

37. Finn, F.N., Saraf, C., Kulkarni, R., Nair, K., Smith, W. and Abdullah, A., The use of distress prediction subsystems for the design of pavement structures, *Proceedings of the Fourth International Conference on the Structural Design of Asphalt Pavements*, The University of Michigan, 1977.

38. Pell, P.S. and Cooper, K.E., The effect of testing and mix variables on the fatigue performance of bituminous materials, *Proceedings of the Association of Asphalt Paving Technologists*, 1975, **44**.

39. *Shell Pavement Design Manual – Asphalt Pavements and Overlays for Road Traffic*. London: The Shell International Petroleum Company, 1978.

40. Van Dijk, W., Practical fatigue characterization of bituminous mixes, *Proceedings of the Association of Asphalt Paving Technologists*, 1975, **44**.

41. Van Dijk, W. and Visser, W., The energy approach to fatigue for pavement design, *Proceedings of the Association of Asphalt Paving Technologists*, 1977, **46**.

42. *Interim Method of Test for Resilient Modulus of Unbound Granular Base/Subbase Materials and Subgrade Soils*, AASHTO T294–92I. Washington, DC: American Association of State Highway and Transportation Officials, 1992.

43. Uzan, J., Witczak, M.W., Scullion, T. and Lytton, R.L., Development and validation of realistic pavement response models, *Proceedings of the Seventh International Conference on Asphalt Pavements* held at the The University of Nottingham, 1992.

44. Hicks, R.B. and Monismith, C.L., Factors influencing the resilient response of granular materials, *Highway Research Record 345,* 1971.

45. Ros, J., Pronk, A.C. and Eikelboom, J., The performance of highway pavements in the Netherlands and the applicability of linear elastic theory to pavement design, *Proceedings of the Fifth International Conference on the Structural Design of Asphalt Pavements*, The Delft University of Technology, 1982.

46. Williams, R.I.T., *Cement-treated Pavements: Materials, Design and Construction*. London: Elsevier Applied Science Publishers Ltd., 1986.

47. Croney, D., *The Design and Performance of Road Pavements*. London: HMSO, 1977.

48. Freeme, C.R., Maree, J.H. and Viljoen, A.W., Mechanistic design of asphalt pavements and verification using the heavy vehicle simulator, *Proceedings of the Fifth International Conference on the Structural Design of Asphalt Pavements*, The Delft University of Technology, 1982.

49. Seed, H.B., Chan, C.K. and Lee, C.E., Resilience characteristics of subgrade soils and their relation to fatigue failures in asphalt pavements, *Proceedings of the International Conference on the Structural Design of Asphalt Pavements*, The University of Michigan, 1962.

50. Chaddock, B.C.J., *Repeated Triaxial Loading of Soil; Apparatus and Preliminary Results*, Supplementary Report 711. Crowthorne, Berkshire: The Transport and Road Research Laboratory, 1982.

51. Brown, S.F., Pell, P.S. and Stock, A.F., The application of simplified, fundamental design procedures for flexible pavements, *Proceedings of the Fourth International Conference on the Structural Design of Asphalt Pavements*, The University of Michigan, 1977.

52. Heukelom, W. and Klomp, A.J.G., Dynamic testing as a means of controlling pavements during and after construction, *Proceedings of the International*

Conference on the Structural Design of Asphalt Pavements, The University of Michigan, 1962.

53. *Addendum to the Shell Pavement Design Manual*. London: The Shell International Petroleum Company, 1985.

54. Lister, N.W. and Kennedy, C.K., A system for the prediction of pavement life and design of pavement strengthening, *Proceedings of the Fourth International Conference on the Structural Design of Asphalt Pavements*, The University of Michigan, 1977.

55. Miner, M.A., Cumulative damage in fatigue, *Journal of Applied Mechanics, Transactions of the American Society of Mechanical Engineers*, 1945, **66**.

56. McElvaney, J. and Pell, P.S., Fatigue of asphalt under compound-loading., *Journal of Transportation Eng.Div., ASCE*, 1974, **100**.

57. Valkering, C.P. and Stapel, F.O.R., The Shell pavement design method on a personal computer, *Proceedings of the Seventh International Conference on Asphalt Pavements*, **I**, The University of Nottingham, 1992.

58. Lister, N.W., Powell, W.D. and Goddard, R.T.N., A design for pavements to carry very heavy traffic, *Proceedings of the Fifth International Conference on the Structural Design of Asphalt Pavements*, The Delft University of Technology, 1982.

CHAPTER 16

Analysis of stresses in rigid concrete slabs and an introduction to concrete block paving

J. Knapton

16.1 Introduction

This chapter deals primarily with the development of stresses in rigid pavements but includes a section on concrete block paving since this is a form of concrete pavement which behaves in a different way from rigid concrete. Stresses in rigid concrete slabs result from applied load and from restraint to slab movement induced by moisture loss and temperature changes. This chapter explains how those stresses can be calculated and how the calculated stresses can be compared with concrete strength in order to proportion slab thickness. In other words, this chapter presents a mechanistic design method for rigid slabs. In particular, tensile and flexural tensile stresses are considered since it is these stresses which can lead to concrete cracking and to subsequent deterioration of the pavement. Compressive stresses are usually so low as to be negligible.

A rigid concrete pavement comprises the concrete slab supported by its foundation which may be a subbase constructed directly over the subgrade. In the case of low strength subgrades, the foundation may include a capping layer separating the subbase from the subgrade. Capping layer material usually comprises locally available low-cost material with a California Bearing Ratio (CBR) of 15 per cent or more. It may be crushed concrete, hardcore or crushed rock of insufficient strength or stability to function as a subbase. When a capping is specified, for mechanistic design purposes it is assumed that the subgrade has been strengthened so that its CBR is 5 per cent and the effect of the capping is otherwise ignored. The stresses in a rigid slab are influenced by the properties of the subgrade and the subbase, as now described.

16.2 Subgrade

The subgrade is the naturally occurring ground or imported fill at formation level. Homogeneity of the subgrade strength is particularly important and avoiding hard and soft spots is a priority in subgrade preparation. Any subgrade should be suitable material of such grading that it can be well compacted. Material containing variable piece sizes often proves difficult to compact, giving rise to settlement and early failure of the pavement. On very good quality subgrades, such as firm sandy gravel the subbase material may be omitted.

16.3 Subbase

For most types of subgrade, a subbase is essential. This layer usually consists of an inert, well-graded granular material although a cement-treated subbase, such as lean concrete or cement-bound granular material may, be employed. In-situ cement stabilization may prove an economic means of improving a poor subgrade. In the case of wheel loading, the subbase assists in reducing the vertical stress transmitted to the subgrade. Where a distributed load is present the pavement slab achieves very little load spreading and the bearing capacity of the underlying subgrade may limit the maximum load applied to the pavement. Subgrade is often specified primarily according to its CBR and values of between 2 per cent and 4 per cent are commonly used. Other factors such as frost resistance, grading and plastic fines are often included in highway authority specifications.

16.4 Modulus of subgrade reaction

In considering the value of the stresses induced in a slab under loading, the influence of the subbase and subgrade is often treated as that of an elastic medium whose strength is defined by its modulus of subgrade reaction, K. The modulus of subgrade reaction characterizes the deflexion of the ground and/or the foundation under the pavement slab. California Bearing Ratio (CBR) tests and plate bearing tests can be used to establish K-values. In many instances subgrades are variable and results obtained from in-situ tests can often show scatter. CBR and plate bearing tests induce a shallow stress bulb and may not reflect the influence of deeper material which might become stressed beneath a loaded slab. Assumed values of the modulus of subgrade reaction are shown in Table 16.1. The values in Table 16.1 may be assumed if no plate bearing test or CBR test results are available. Note that the

Table 16.1 Assumed modulus of subgrade reaction, K, for typical British soils.

Soil type	Typical soil description	Subgrade classification	Assumed K (N/mm^3)
Coarse grained soils	Gravels, sands, clayey or silty gravels/sands	Good	0.054
Fine grained soils	Gravely or sandy silts/ clays, clays, silts	Poor	0.027
		Very poor	0.013

Note: This is a coarse assessment and Table 16.4 may provide more accurate data

units of K are N/mm^3 which can be considered as the vertical pressure in N/mm^2 to produce a deflexion of 1 mm.

When a lean concrete subbase is specified beneath a pavement quality concrete, the K-value of the subgrade material is used to calculate the required thickness of the concrete slab. This calculated thickness is then apportioned between the pavement quality concrete slab thickness (the higher strength concrete) and the lean concrete subbase thickness. This relationship is shown in Table 16.3 when a C40 concrete is used for the slab and a C20 lean concrete is used for the subbase. (The prefix C is used commonly to denote the characteristic compressive strength of the concrete measured on a 150 mm cube.)

Chandler and Neal[1] suggest that the sub-base can be taken into account by enhancing the effective modulus of subgrade reaction, K, of the subgrade as in Table 16.2.

16.5 Plate bearing and CBR testing

The plate bearing test procedure is to load the ground through a steel disk, usually mounted on the back of a vehicle, and to record load and corresponding deflexion. The value of K is found by dividing the pressure exerted on the plate by the resulting vertical deflexion and is expressed in units of N/mm^3, MN/m^3 or kg/cm^3. K is normally established by plate bearing tests with a plate diameter of 750 mm. A modification

Table 16.2 Enhanced value of K when a subbase is used

K-value of subgrade alone	Enhanced value of *K* when used in conjunction with: A cement-bound subbase thickness (mm)				A granular subbase of thickness (mm)			
	150	200	250	300	100	150	200	250
0.013	0.018	0.022	0.026	0.030	0.035	0.050	0.070	0.090
0.020	0.026	0.030	0.034	0.038	0.060	0.080	0.105	–
0.027	0.034	0.038	0.044	0.049	0.075	0.110	–	–
0.040	0.049	0.055	0.061	0.066	0.100	–	–	–
0.054	0.061	0.066	0.073	0.082	–	–	–	–
0.060	0.066	0.072	0.081	0.090	–	–	–	–

Table 16.3 The modified thickness of a pavement quality concrete slab with a C20 lean concrete subbase

Calculated thickness of slab (mm)	Modified thickness of pavement quality concrete slab required (mm) when used in conjunction with a lean concrete subbase of thickness:		
	100 mm	130 mm	150 mm
250	190	180	–
275	215	200	–
300	235	225	210

is needed if a different plate diameter is used, e.g. for a 300 mm diameter plate K is divided by 2.3 and for a 160 mm diameter plate it is divided by 3.8.

Alternatively, the CBR can be measured and CBR values expressed as percentages are obtained. Table 16.4 shows the relationship between CBR and modulus of subgrade reaction, K, for a number of common soil types.

16.6 Fibre-reinforced pavement quality concrete

During recent years pavement construction methods involving the addition of polypropylene or steel fibres into a concrete mix have become common in Europe. For centuries, man has attempted to reinforce construction mortars and concretes with various types of fibres. Pharaoh ordered his foremen to cease supplying reinforcing straw to Israelite brick makers (Exodus 5) and later the Romans used hair fibres in structural mortars.

With the advent of fast-track systems in the construction industry, concrete pavements have had to meet quicker construction programmes. For example, with the use of laser guided screeding machines, steel or polypropylene fibres are often specified instead of conventional mesh because of the inconvenience in positioning individual mats of mesh immediately in front of the laser-guided screeding machine as the placing of the concrete progresses. Laser-guided screeding machines cannot construct conventional mesh reinforced pavements efficiently because the mesh impedes the machine. Tests from a manufacturer of polypropylene fibres revealed the change in strength characteristics shown in Table 16.5.

Compressive strength tests conducted in accordance with BS 1881[2] indicated that the fibres, when used at the recommended dosage rate of $0.9 kg/m^3$, slightly increase the early strength gain of concrete. The fibres have no significant effect on

Table 16.4 Modulus of subgrade reaction values for a number of common subgrade and subbase materials

Material	CBR (%)	Modulus of subgrade reaction (N/mm^3)
Humus soil or peat	<2	0.005–0.015
Recent embankment	2	0.01–0.02
Fine or slightly compacted sand	3	0.015–0.03
Well compacted sand	10–25	0.05–0.10
Very well compacted sand	25–50	0.10–0.15
Loam or clay (moist)	3–15	0.03–0.06
Loam or clay (dry)	30–40	0.08–0.10
Clay with sand	30–40	0.08–0.10
Crushed stone with sand	25–50	0.10–0.15
Coarse crushed stone	80–100	0.20–0.25
Well compacted crushed stone	80–100	0.20–0.30

the 28 day compressive strength of concrete cubes, nor do they have any substantial effect on the flexural strength of concrete.

Steel fibres are now commonly used in place of mesh reinforcement. The stresses occurring in a pavement slab are complex and depending on the type of load, tensile stresses can occur at the top and at the bottom of the slab. There are, in addition, stresses that are difficult to quantify, arising from a number of causes such as sharp turns, moisture loss, thermal effects and impact loads. The addition of steel wire fibres to a concrete slab results in a homogeneously reinforced slab achieving a considerable increase in flexural strength and enhanced resistance to shock and fatigue.

A manufacturer of anchored steel fibres commissioned TNO, Delft to undertake flexural strength tests using fibres embedded in C30 concrete. These tests have resulted in values of flexural strength of up to 4.5 N/mm^2, depending on dosage, type and size of fibre. Partly from these results and partly from work undertaken at the UK Cement and Concrete Association the flexural strength values shown in Table 16.6 have been developed.

16.7 Thermal and moisture-related stresses in concrete slabs

The basic premise underlying most concrete pavement design methods is that stresses developed as a result of the concrete slab changing temperature or moisture

Table 16.5 Test results comparing the strength of polypropylene fibre-dosed concrete and conventional plain concrete

Measurement	Strength of polypropylene fibre-reinforced concrete	Strength of unreinforced concrete
Compressive strength (N/mm^2) (equivalent cube method)		
1 day	16.5	16.0
3 days	28.5	24.5
7 days	34.0	35.0
28 days	43.5	39.5
Cube compressive strength (N/mm^2)		
1 day	16.0	14.5
3 days	28.0	27.5
7 days	34.0	36.0
28 days	48.5	44.5
Flexural strength (N/mm^2)		
1 day	2.3	2.1
3 days	4.0	3.7
7 days	4.2	4.8
28 days	4.6	6.2

Table 16.6 Concrete design flexural strengths

Concrete grade and dosage	Flexural strength (N/mm^2)
Plain or polypropylene reinforced C30 concrete	2.0
20 kg/m^3 steel fibre C30 concrete	2.4
30 kg/m^3 steel fibre C30 concrete	3.2
40 kg/m^3 steel fibre C30 concrete	3.8
Plain or polypropylene reinforced C40 concrete	2.4
20 kg/m^3 steel fibre C40 concrete	2.8
30 kg/m^3 steel fibre C40 concrete	3.8
40 kg/m^3 steel fibre C40 concrete	4.5

content are contained by the provision of stress relieving joints whereas stresses developed by traffic and other applied loads are controlled by proportioning the thickness of the slab and its underlying supporting courses. The exception is in continuously reinforced concrete pavements (CRCP) in which case temperature and moisture loss stresses are contained by the composite action of the reinforcement and the concrete.

Whether temperature or moisture loss stresses are predominant depends upon many factors which are difficult to calculate. Moisture-related stresses are potentially greater than temperature-related stresses by an order of magnitude. However, it is often the case that a highway pavement retains much of its moisture throughout its life. Also, moisture-related stresses develop slowly so creep often reduces their effect significantly. Temperature stresses, on the other hand, are often at their most severe immediately following construction as the setting concrete cools. Furthermore, temperature-related stresses are usually diurnal, so creep has little mitigating effect. For this reason, temperature-related effects are the ones of most concern in most concrete highway pavement projects.

Although moisture-loss effects are usually less important than temperature-related effects, they need careful consideration in the case of concrete roads constructed in a dry climate. Both temperature and moisture can cause a slab to shrink uniformly, to curl upwards at its perimeter or to curl downwards at its perimeter. The way in which these three conditions impart stress into the slab is now considered.

16.7.1 Uniform shrinkage

As a result of uniform temperature fall or moisture loss, a concrete slab will shrink uniformly about its centre on plan. Theoretically, the centre will remain stationary. At a distance from the centre the slab will attempt to displace horizontally and this displacement will increase uniformly towards the edge of the slab. Frictional restraint between the underside of the concrete slab and the surface of the subbase will inhibit or prevent this movement and so generate tensile stress within the slab. This stress in the concrete resulting from frictional restraint to shrinkage can be calculated. The force required to overcome the frictional resistance is given by the

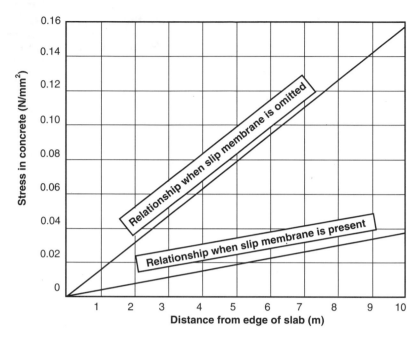

Fig. 16.1 Relationship between friction-induced stress developed in a slab and distance from the edge of the slab, for slabs with and without membranes

expression $F_f = w\mu$, where F_f=friction force, w=weight of the concrete (calculated by assuming a density of $24\,kN/m^3$) and μ=coefficient of friction. As the weight of concrete generating frictional restraint increases with distance from the slab edge the stress gradually increases to a maximum at the slab centre (there is zero stress at the edge of the slab or at the joints). Assuming the values for the coefficient of friction between concrete and subbase and polythene are 0.65 and 0.15 respectively, the theoretical stresses which result from uniform shrinkage friction are shown in Fig. 16.1. Even without a slip membrane, the stresses are low, attaining a value of less than $0.1\,N/mm^2$ in a 6 m long slab.

Figure 16.1 illustrates why the provision of a slip membrane is not a crucial issue. Indeed, in an unreinforced concrete pavement with closely spaced joints at say 5 m spacings, the provision of a slip membrane may be detrimental in that it may concentrate movement at one joint, which can then become a maintenance problem, while other joints never operate. (It may be preferable to provide a layer of polyethelene to prevent concrete water loss into the subbase.)

16.7.2 Slab perimeter curling downwards (hogging)

As a result of the underside of the concrete slab cooling or drying faster than the top, non-uniform shrinkage develops throughout the slab with the lower concrete shrinking more than the upper. The result of this non-uniform shrinkage will be hogging of the slab. Assuming the hogging slab can be represented by a simply supported beam of length L with a uniformly distributed load equal to the concrete weight then the maximum moment $M=wL^2/8$ where $L=$ length of the slab (the distance between joints) and $w=$ dead weight of the concrete (assumed to be $24\,\text{kN/m}^3$). The stress is calculated from the equation $\sigma/y=M/I$, where $\sigma=$ stress, $y=$ depth to the neutral axis, $M=$ bending moment, and $I=$ second moment of area. In the extreme case, L would be the distance between joints allowing rotational freedom.

Figure 16.2 shows stresses which would develop in the hogging situation for a 200 mm thick slab and demonstrates that such behaviour would crack each bay. In fact, the temperature fall or moisture loss is usually insufficient to cause the slab to separate from its subbase so this extreme condition rarely occurs.

16.7.3 Slab perimeter curling upwards (curling)

The result of the upper side of the slab cooling or drying faster than the underside will be that the slab will attempt to curl upwards at its edges. This curling can be represented by a cantilever of length L equal to the curled length. Assuming this cantilever carries a uniformly distributed load generated by the weight of the concrete (based upon an assumed density of $24\,\text{kN/m}^3$), the bending moment at the slab's point of contact with the ground is given by the expression, $M=wL^2/2$. As in

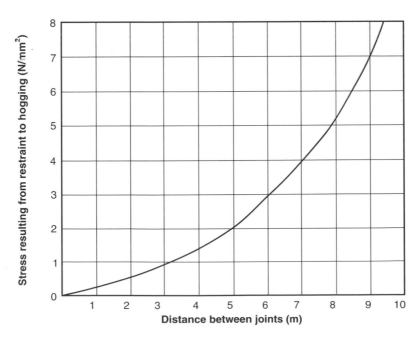

Fig. 16.2 Relationship between joint spacing and stress developed as a result of restraint to hogging

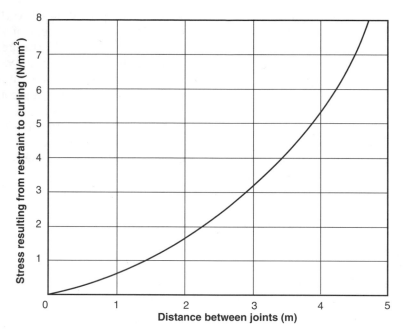

Fig. 16.3 Relationship between joint spacing and stress developed as a result of restraint to curling

the case of hogging, the stress can then be calculated from the expression $\sigma/y = M/I$. Figure 16.3 shows curling stresses calculated for a 200 mm thick slab.

16.7.4 Calculation of slab temperature changes

In order to determine temperature-related stresses through the depth of a slab, the temperature profile at the time of set needs to be known. From that time forward, whenever the 'at set' temperature profile is replicated, temperature stresses will disappear. The 'at set' temperature profile has been investigated by several researchers and they have concluded that there can be no standard profile of value to the designer. The profile depends upon the type of concrete, the curing regime, the weather during concreting and the time of day at which the concrete set.

Figure 16.4 illustrates differing temperature profiles at set and at subsequent times in different climatic conditions. It shows that concrete which sets during a warm mid-afternoon may have a locked in profile as shown in Fig. 16.4(b). If this slab were in a warm climate, then it might subsequently be subjected to a profile as in Fig. 16.4(e) during the night. In such a case, the temperature of the slab surface has fallen by, say, 20° and the temperature of the underside of the slab has increased by 17°. This will cause the slab to attempt to curl upwards at its perimeter (curling). The self-weight of the slab together with applied loading will attempt to keep the slab in contact with its subbase so tensile stresses will develop at and near the

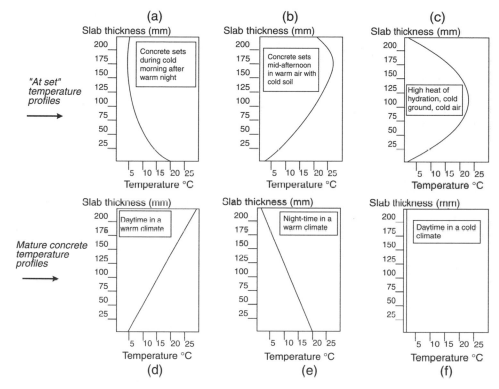

Fig. 16.4 Concrete slab temperature profiles at set and at subsequent times

upper surface of the slab. Their value will depend upon the relative elastic properties of the slab and the underlying subbase material.

The opposite effect would occur for slabs which developed their initial set during a cold morning following a relatively warm night, in which case the 'as set' profile shown in Fig. 16.4(a) would apply. If subsequently this slab were subjected to the temperature profile as shown in Fig. 16.4(d), the surface temperature rises by, say, 25° and the underside temperature falls by 15°. This causes the slab to attempt to curl downwards at its perimeter (hogging).

The above two cases represent extremes which slabs might be expected to sustain in normal situations. In the first case, the upwards perimeter curling is caused by a temperature differential of 37° and, in the second case, the downwards perimeter curling temperature differential is 40°. In order to gauge the magnitude of the stresses which might be generated, consider the extreme case in which the slab is fully restrained against hogging. The concrete might have a coefficient of thermal expansion of 0.000 009 per °C. Therefore, the maximum tensile strain would be 40 times 0.000 009 or 0.000 36. Concrete with a Young's Modulus of 20 000 N/mm^2 would develop a tensile stress of approximately 7 N/mm^2 which is sufficient to crack the concrete. This value is never attained in the concrete because full restraint is never achieved and because curing regimes ensure that the 'as set' profiles shown in Fig. 16.4 are not attained in well-controlled projects. In practice, it is found that the provision of transverse warping joints, i.e. joints which permit rotation at spacings of between 5m and 30m (depending upon the level of reinforcement) is sufficient to control temperature stresses. Tables 16.7(a) and (b) show joint spacings which have been found to be sufficient to control temperature and moisture related cracking.

Difficulties can occur when different concretes are used in two layer construction. The use of low-heat concrete below pavement quality concrete can lead to excessive tensile stresses and difficulties were experienced, for example, on the London Orbital Motorway (M25) in which transverse cracking occurred.

Moisture-related shrinkage occurs as the concrete slab loses its free water through evaporation. Typically, pavement quality concrete will have a water/cement ratio of 0.45. Approximately two-thirds of this water is combined chemically with the cement during hydration and the remainder acts as a lubricant, creating workability. All of this free water has the opportunity to evaporate and the volume which it occupied is no longer present so shrinkage occurs. There may be $50\,l/m^3$ of water evaporating over a period of several months following construction. This represents 5 per cent of the volume of the concrete so significant shrinkage can occur. In fact, creep and precipitation reduce the effect of shrinkage to a controllable level and the joint spacings shown in Table 16.7 are usually sufficient to prevent distress.

16.8 Crack control methods

Most concrete pavements include joints to reduce the effect of temperature and moisture-related stress. The joints are characterized by the type of movement which they permit. Full movement joints permit expansion and contraction but not

Table 16.7 Transverse joint spacings for concrete pavements

(a) Conventionally-reinforced pavements

Weight of conventional mesh reinforcement (kg/m^2)	Spacing of joints (m)
2.61 (6 mm Ø wires at 100 mm c/c) – C283	16.5
3.41 (7 mm Ø wires at 100 mm c/c) – C385	21
4.34 (8 mm Ø wires at 100 mm c/c) – C503	27.5
5.55 (8 mm Ø wires at 80 mm c/c) – C636	35

(b) Plain and steel fibre-reinforced pavements

Type of concrete	Spacing of joints (m)
Plain or polypropylene reinforced C30 concrete	6
20 kg/m^3 hooked steel fibre reinforced C30 concrete	7
30 kg/m^3 hooked steel fibre reinforced C30 concrete	8
40 kg/m^3 hooked steel fibre reinforced C30 concrete	10
Plain or polypropylene reinforced C40 concrete	6
20 kg/m^3 hooked steel fibre reinforced C40 concrete	8
30 kg/m^3 hooked steel fibre reinforced C40 concrete	10
40 kg/m^3 hooked steel fibre reinforced C40 concrete	12

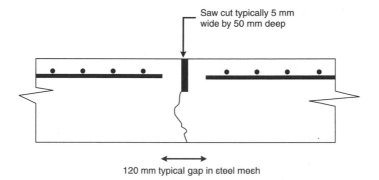

Fig. 16.5 A typical induced joint that is formed by the saw-cut technique

rotational movement. Contraction joints permit contraction but not expansion. Warping joints permit rotation and a little contraction but no expansion. Usually, nearly all the joints in a concrete pavement are either contraction joints or warping joints.

A cost-effective way of constructing such joints is to cast the concrete continuously through the joint and then to saw through part of the concrete before stresses become severe. In some rigid pavements, all of the joints are formed in this way (except for the end-of-day construction joints). The saw-cut acts as a line of weakness which is incorporated into the slab at the position of the joint such that the slab will crack at that point owing to an increase in the tensile stress in the remaining depth of slab. Saw-cuts are formed when the concrete has gained sufficient strength to withstand the weight of a concrete saw but not so much that the effect of sawing would damage the pavement. Sawn joints are particularly durable. They are expensive and can cost as much as ten times the cost of wet formed joints. The saw-cut has a depth of between 1/4 and 1/3 that of the slab (see Fig. 16. 5). As well as the joint being the most durable crack induction form, a sawn joint is also very serviceable, with no difference in level at each side of the cut.

Joints should be cut when the concrete has gained sufficient strength to support the weight of the cutting equipment but before it has gained sufficient strength that sawing might loosen or pull out aggregates and fibre reinforcement. A suggested sawing time scale is 24 to 48 h after initial concrete set. This leaves a time window for sawing, as follows:

Assuming that – the concrete is mixed 1 hour before placing;
               – the first concrete is mixed at 7.00 am;
               – the last concrete is mixed at 4.00 pm;
               – the concrete is placed between 8.00 am and 5.00 pm on day 1;
and   – the concrete takes 6 h to reach initial set;
then   – earliest initial set = 1.00 pm and latest initial set = 10.00 pm.

If the concrete is to be sawn between 24 and 48 hours after the initial set, the first cut can be performed 24 h after the latest initial set so as to ensure that all of the concrete has gained sufficient strength. All saw-cutting must be finished 48 h after the earliest initial set. In this example, saw-cutting can commence at 10.00 pm on day 2 and must be finished by 1.00 pm on day 3. This gives 15 h sawing time.

16.9 Spacing of joints

Table 16.7 shows transverse joint spacings for various types of concrete. The joint spacings in this table have been used successfully for many years in Europe. Some consider that spacings can be greater than 12 m for steel fibre-reinforced concrete pavements. Whilst 14 m joint spacings will probably be acceptable for most concrete roads, the additional movement which would occur at joints might lead to loss of aggregate interlock and joint degeneration in harsh climates.

In the case of conventional mesh, longitudinal joint spacing is usually dictated by mesh size and a mesh length of 4.8 m is common. In the case of fibre-reinforced concrete, no such considerations apply and longitudinal joints can be located at transverse joint spacings. This provides the designer with the opportunity to locate longitudinal joints at lightly loaded positions such as lane boundaries.

16.10 Mechanistic rigid pavement design

The design method presented here is based upon assessing the stresses developed in rigid concrete slabs and comparing such stresses with those known to be suitable in different categories of concrete. It uses Westergaard equations[3] in the case of patch loading, and has been formulated into a set of Design Charts to facilitate use. Examples are presented. Providing conventional joint details are specified, stresses induced by moisture related shrinkage or by temperature effects will remain minimal.

Loading of highway pavements will be one or a combination of the following: (1) uniform distributed load (UDL); (2) patch loading from vehicle tyres (PL); and (3) horizontal loading (HL).

Pavement design comprises assessing the loading regime which the pavement is predicted to sustain and selecting the materials, thicknesses and joint configuration which will sustain those loads whilst at the same time satisfying flatness, durability, abrasion and riding quality requirements. An essential part of the design process is the assessment of the anticipated load regime. Horizontal loading may be introduced into the pavement if vehicles undertake turning manoeuvres. In such cases, special care is needed when detailing joints which must be able to transmit tension between neighbouring slabs. Small slabs may need to be restrained. The design of the restraint can be undertaken conservatively by assuming the pavement slab spreads the applied horizontal load through a projected 90° path.

The design procedure set out in this section comprises calculating stresses resulting from the loading regime and ground conditions and comparing those stresses with the strength of the concrete. Thus, the following factors have to be taken into account in rigid pavement design: (a) loading regime, (b) strength of concrete, and (c) strength of existing ground and effect of the subbase.

16.10.1 Loading regime

Fatigue often leads to pavement distress and in some cases a fatigue factor is built into the design method by the load safety factor. A load safety factor of 2.0 is recommended for an infinite number of load repetitions and may therefore be used conservatively in all cases. The position of the load relative to the slab edge is critical and in the case of patch loads from vehicle wheels, three alternative cases may need to be considered: internal loading (greater than 0.5 m from edge of slab); edge loading; and corner loading. With internal or edge loading the maximum stress occurs beneath the heaviest load at the underside of the slab. Corner loading creates tensile stress at the upper surface of the slab a distance away from the corner. This distance can be calculated from:

$$d = 2[(2^{0.5})rl]^{0.5} \tag{16.1}$$

where r = radius of loaded area (mm), l = radius of relative stiffness (mm), and d = distance from slab corner to position of maximum tensile stress (mm).

16.10.2 Strength of concrete

Road pavement slabs are frequently constructed using C30 or C40 concrete with a minimum cement content of $300 \, \text{kg/m}^3$ with a slump of 25 mm or less. Design is based upon comparing material flexural strength with calculated flexural stresses, and specification is by characteristic compressive strength. Table 16.6 shows flexural strength values for a range of commonly-used concretes.

16.10.3 Strength of existing ground, and effect of subbase

The design method requires a value for the modulus of subgrade reaction, K, which defines the deformability of the material beneath the pavement. The following four values of K are used in the design procedure:

$K = 0.013 \, \text{N/mm}^3$ – very poor ground
$K = 0.027 \, \text{N/mm}^3$ – poor ground
$K = 0.054 \, \text{N/mm}^3$ – good ground
$K = 0.082 \, \text{N/mm}^3$ – very good ground (no sub-base needed)

The beneficial effect of a granular subbase is taken into account by increasing K according to the thickness and strength of the subbase as shown in Table 16.3.

16.10.4 Stress in the concrete slab

The stress in a pavement slab depends upon:

(a) the properties of the subgrade;
(b) the loading regime
    (i)   uniformly distributed load (UDL)
    (ii) patch loads;
(c) the thickness of the pavement slab; and
(d) the strength of the subbase.

16.10.5 Design method for uniformly distributed loading

Whilst most road pavements will be required to withstand only patch loads, some may need to be designed to withstand distributed loading. This will be the case for example when industrial roads are used for storage. A common loading system comprises alternate unloaded areas and loaded storage areas. The maximum negative bending moment (hogging), $M_{(hog)}$, occurs within the centre of the unloaded areas:

$$M_{hog}=-q/2\lambda^2.(B\lambda a'-B\lambda b') \tag{16.2}$$

where a=width of the unloaded area, b=width of the loaded area, $a'=a/2$, and $b'=a/2+b$.

The maximum positive bending moment (sagging), M_{sag}, occurs beneath the centre of the loaded area:

$$M_{sag}=q/2\lambda^2.B_{(1/2)\lambda b} \tag{16.3}$$

where q=UDL(i.e. the characteristic load×load factor)(N/mm^2), $B_x=e^{-x}.\sin x$, $\lambda=(3K/Eh)^{1/4}$, K=modulus of subgrade reaction (N/mm^3), E=concrete modulus (i.e. 10 000 N/mm^2 for a sustained load), and h=concrete slab thickness (mm).

The two moment equations can be simplified into a single conservative equation for any combination of unloaded and loaded width:

$$M_{max}=-0.168q/\lambda^2 \tag{16.4}$$

The corresponding maximum flexural stress, σ_{max} (N/mm^2), is given by:

$$\sigma_{max}=6.M_{max}/h^2$$
$$=1008\,q/(\lambda^2 h^2) \tag{16.5}$$

where the maximum flexural strength cannot exceed the relevant value from Table 16.6. To ease calculation, Table 16.8 shows values of $\lambda^2 h^2$ for common combinations of modulus of subgrade reaction, K, and slab thickness.

Uniformly distributed load example. Consider a 150 mm thick pavement slab carrying a distributed load of 50 kN/m^2 between unloaded areas on very poor ground. Using Equation 16.5

$$\sigma_{max}=1008q/(\lambda^2h^2)$$

and $\qquad q=0.05\,\text{N/mm}^2\times2$ (i.e. a load safety factor)

$$=0.10\,\text{N/mm}^2$$

For very poor ground, $K=0.013\,\text{N/mm}^3$.

Table 16.8 Values of $\lambda^2 h^2$ for combinations of slab thickness and modulus of subgrade reaction

Modulus of subgrade reaction, K (N/mm^3)	$\lambda^2 h^2$ for slab thickness (mm) of:				
	150	175	200	225	250
0.082	0.061	0.066	0.070	0.074	0.078
0.054	0.049	0.053	0.057	0.060	0.063
0.027	0.035	0.038	0.040	0.043	0.045
0.013	0.024	0.026	0.028	0.029	0.031

From Table 16.8:

$$\sigma_{max} = 1008 \times 0.10/0.024$$

$$= 4.2\,\text{N/mm}^2$$

This stress can be withstood by a C40 concrete incorporating $40\,\text{kg/m}^2$ of 60 mm long 1 mm diameter anchored steel fibre which can be used for any combination of aisle and stacking zone width.

16.10.6 Design method for patch loading using Westergaard equations

The maximum flexural stress occurs at the bottom of the slab under the heaviest wheel load. The maximum stress under the wheel can be calculated by Westergaard and Timoshenko[3] equations:
(a) patch load in mid-slab (i.e. more than 0.5 m from slab edge)

$$\sigma_{max} = \frac{0.275(1 + \mu)P\log(0.36Eh^3/Kb^4)}{h^2} \tag{16.6}$$

(b) patch load at edge of slab

$$\sigma_{max} = \frac{0.529(1 + 0.54\mu)P\log(0.20Eh^3)/(Kb^4)}{h^2} \tag{16.7}$$

(c) patch load at slab corner

$$\sigma_{max} = \frac{3P[1 - 1.41\{12(1 - \mu^2)\}^{1/4}]Kb^4/Eh^3}{h^2} \tag{16.8}$$

where σ_{max} = flexural stress (N/mm^2); P = patch load (N), i.e. a characteristic wheel load $\times$ load factor; μ = Poisson's ratio, usually 0.15; h = slab thickness (mm); E = elastic modulus, usually 20 000 N/mm^2; K = modulus of subgrade reaction (N/mm^3); b = radius of tyre contact zone (mm) = $(W/\pi.p)^{1/2}$; W = patch load (N); and p = contact stress between wheel and pavement (N/mm^2).

Twin wheels, bolted side by side, are assumed to be one wheel transmitting one-half of the axle load to the pavement. In certain cases, wheel loads at one end of an axle magnify the stress beneath wheels at the opposite end of the axle, at distance S away (where S is measured between load patch centres). To calculate the stress magnification, the characteristic length (i.e. the radius of relative stiffness, l) has to be found from Equation 16.9:

$$l = [Eh^3/12(1 - \mu^2)K]^{1/4} \tag{16.9}$$

Once Equation 16.9 has been evaluated, the ratio S/l can be determined so that Fig. 16.6 can be used to find M_t/P, with M_t being the tangential moment. The stress under the heaviest wheel has to be increased to account for a nearby wheel, and this is calculated by adding an appropriate flexural stress calculated from

$$\sigma_{add} = (M_t/P)(6/h^2)P_2 \tag{16.10}$$

where P = greatest patch load, and P_2 = other patch load.

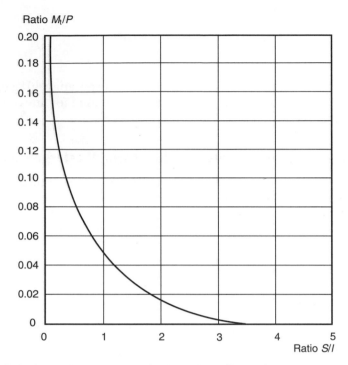

Fig. 16.6 Relationship used to calculate the effect of two patch loads in close proximity

Next, sum the stresses and verify (Table 16.6) that the flexural strength has not been exceeded for the prescribed concrete mix.

Road vehicle example.   Consider a road vehicle with a rear axle load of 11 000 kg and assume a slab thickness of 200 mm on good ground ($K=0.054$ N/mm^3). The load safety factor is 2.0. The design axle load is 220 000 N (i.e. 110 000 N×load safety factor of 2.0) so the design wheel patch load is 110 000 N.

p=contact stress between wheel and pavement=0.7 N/mm^2
b=radius of contact area=$(W/\pi p)^{1/2}$
  $=(110\,000/\pi 0.7)^{1/2}$
  $=224$ mm

Substituting known values into Equation 16.7:

$$\sigma_{max} = 0.529(1 + 0.54 \times 0.15)\left(\frac{110000}{200^2}\right) \times \log\left(\frac{0.2 \times 20000 \times 200^3}{0.054 \times 200^4}\right)$$

$=3.73$ N/mm^2 beneath one wheel

Assuming that the wheel at the other end of the axle is at a distance of 2.7 m (i.e. $S=2700$ mm), the radius of relative stiffness, l, using Equation 16.9 is

$$l=[20,000 \times 200^3/12(1-0.15^2) \times 0.054]^{1/4}$$

$$= 709 \, \text{mm}$$

and $\quad S/l = 2700/709$

$$= 3.8$$

From Fig. 16.6

$$M_t/P = 0.005$$

Therefore, using Equation 16.10

$$\sigma_{add} = (M_t/P) \times \frac{6}{h^2} \times P_2$$

$$= 0.005 \times \frac{6}{200^2} \times 110\,000$$

$$= 0.08 \, \text{N/mm}^2$$

Total stress $= 3.73 + 0.08$

$$= 3.81 \, \text{N/mm}^2$$

Therefore, a C40 concrete incorporating $40 \, \text{kg/m}^2$ 60 mm long, 1 mm dia, anchored steel fibres is adequate for this design.

16.10.7 Design method for concrete slabs subjected to patch loading, using Design Charts

The design procedure is as follows:

Step 1: Assess the existing conditions.   Determine the actual point load, *APL*, and the modulus of subgrade reaction, *K*. The *K*-values at Tables 16.1 and 16.4 can be used to confirm the category of subgrade.

Step 2: Verify the distance between point loads and determine σ_{add}.   If the distance between loads is greater than 3 m, the *APL* can be used directly (depending on the radius of contact zone – see Step 6) to calculate the thickness of the slab using the relevant Design Chart. If the distance between loads is less than 3 m, the radius of relative stiffness, *l*, has to be calculated from Equation 16.9. Table 16.9 shows values of radius of relative stiffness for different *K*-values and slab thicknesses.

Determine σ_{add} using Equation 16.10.

Step 3: Select a proposed concrete mix from Table 16.6, hence σ_{flex}.

Step 4: Calculate σ_{max}.   When two patch loads are acting in close proximity (i.e. less than 3 m apart), the greater patch load, *P*, produces a flexural strength σ_{max} directly beneath its point of application. The smaller point load, P_2, produces additional stress σ_{add} beneath the larger load. Then

$$\sigma_{max} = \sigma_{flex} - \sigma_{add} \tag{16.11}$$

Table 16.9 Radius of relative stiffness values for different slab thicknesses and support conditions

Slab thickness (mm)	Modulus of subgrade reaction, K (N/mm³):			
	0.013	0.027	0.054	0.082
150	816	679	571	515
175	916	763	641	578
200	1012	843	709	639
225	1106	921	774	698
250	1196	997	838	755
275	1285	1071	900	811
300	1372	1143	961	865

Note: Elastic modulus, $E = 20\,000$ N /mm² and Poissons Ratio, $\mu = 0.15$

Step 5: Calculate the equivalent single point load, ESPL, which, when acting alone, would generate the same flexural stress as the actual loading configuration:

$$ESPL = APL(\sigma_{flex}/\sigma_{max}) \qquad (16.12)$$

where APL = actual point load.

Step 6: Prior to using the Design Charts, modify the actual point load, APL, to account for contact area as well as wheel proximity to obtain the equivalent single point load, ESPL. Design Charts 1 to 8 apply directly when patch loads have a radius of contact between 150 and 250 mm. Multiply the patch load by a factor determined from Table 16.10 prior to using the Design Chart.

Step 7: Use the corresponding Design Chart for the mix selected in Step 3 to determine the slab thickness, and return to Step 3 if an alternative concrete mix is required.

Design example for multiple patch loading. Assume that two patch loads are applied to a concrete pavement, such that a 60 kN patch load is situated 1 m away from a 50 kN patch load. The 60 kN patch load has a contact zone radius of 100 mm and the 50 kN point load has a 300 mm radius. The existing ground conditions are poor, so that $K = 0.027$ N/mm³.

Table 16.10 Point load multiplication factors for loads with a radius of contact outside the range 150 mm to 250 mm

Radius of contact (mm)	Modulus of subgrade reaction, K (N/mm³):			
	0.013	0.027	0.054	0.082
50	1.5	1.6	1.7	1.7
100	1.2	1.2	1.3	1.3
150	1.0	1.0	1.0	1.0
200	1.0	1.0	1.0	1.0
250	1.0	1.0	1.0	1.0
300	0.9	0.9	0.9	0.9

Assume a thickness of slab $= 200$ mm.

The radius of relative stiffness, l is given by Table 16.9. From this table

$l = 843$ mm

The distance apart $= 1$ m (i.e. $S = 1000$ mm). Thus

$$S/l = 1000/843$$

$$= 1.186$$

From Fig. 16.6, $M_t/P = 0.053$

so, from Equation 16.10, $\sigma_{add} = (0.053)(6/200^2)(50\,000)$

$$= 0.4 \text{ N/mm}^2$$

Try polypropylene fibre reinforced C40 concrete with a flexural strength of 2.4 N/mm^2 (see Table 16.6), i.e.

$$\sigma_{flex} = 2.4 \text{ N/mm}^2$$

From Equation 16.11

$$\sigma_{max} = \sigma_{flex} - \sigma_{add}$$

$$= 2.4 - 0.4$$

$$= 2.0 \text{ N/mm}^2$$

This is the maximum flexural stress which the 60 kN load can be allowed to develop.

Calculate the equivalent single point load using Equation 16.12:

$$ESPL = APL(\sigma_{flex}/\sigma_{max})$$

$$= 60 \times 2.4/2.0$$

$$= 72 \text{ kN}$$

From Table 16.10, the modification factor to be applied to the *ESPL* is 1.2. Therefore

$$72 \times 1.2 = 86.4 \text{ kN}$$

From the static loading curve in the Design Chart in Fig. 16.7 the thickness of slab $= 200$ mm.

A 200 mm thick C40 concrete slab with 0.9 kg/m^3 polypropylene fibre reinforcement is therefore adequate for this design.

16.11 Some research results

The writer has carried out a site investigation into the performance of the joints in two ground-supported concrete industrial pavements. The two pavements were Unit 114/14 and Unit 114/10 of the Boldon Business Park, Tyne & Wear, UK. Each 1000 m^2 pavement was constructed by laser screeding, using polypropylene fibre

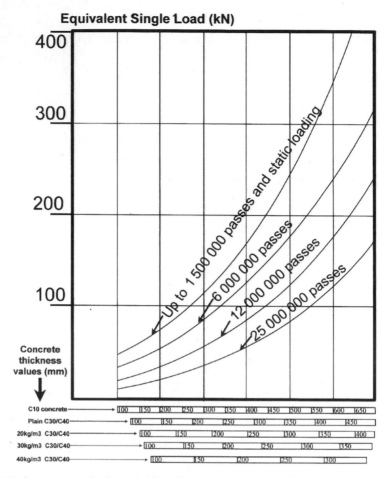

Equivalent Single Load (kN)

Fig. 16.7 A concrete industrial hardstanding Design Chart

reinforced concrete. Once the pavement slabs had been concreted and the saw-cut joints made, the arrangement of studs shown in Fig. 16.8 was established on each concrete surface to permit joint movements to be monitored.

These examples illustrate the behaviour of rigid slabs when diurnal temperature effects are eliminated so that only drying shrinkage takes place. Whilst this is an unusual situation for an in-service road, it illustrates the maximum levels of moisture-related shrinkage which a road pavement might sustain in a dry climate. Using an extensometer with an accuracy of one-hundredth of a millimetre, weekly measurements were taken during a period of 9 months commencing December 1993.

16.11.1 Interpretation of results

Figures 16.9(a) and (b) show the movement of each of the measured joints in Fig. 16.8, from week to week. The larger movements at the beginning of the slabs' life represent joints cracking, and the influence of cracked joints on neighbouring

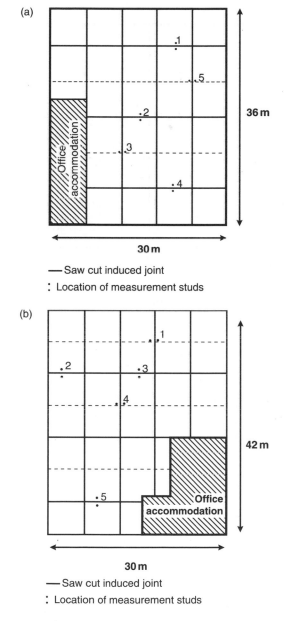

(a)

.1

.5

36 m

Office accommodation

.2

.3

.4

30 m

— Saw cut induced joint

: Location of measurement studs

(b)

1

.2

.3

4

42 m

.5

Office accommodation

30 m

— Saw cut induced joint

: Location of measurement studs

Fig. 16.8 The locations of five measurement points in each of the floors of Units 114/10 (a) and 114/14 (b) at Boldon Industrial Estate, Tyne and Wear, UK

joints can be seen. When all the joints had cracked and were working, a more uniform movement became evident throughout the slabs.

The data were used to calculate the cumulative movement of each joint with time in both pavements. Note that there were some sudden larger movements during the

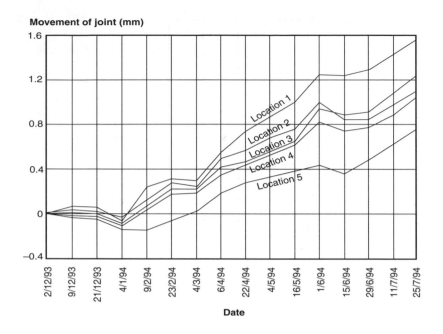

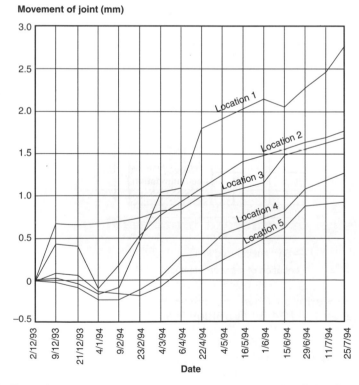

Fig. 16.9 Cumulative movements of five joints in each of the floors in Units 114/10 (a) and 114/14 (b) at Boldon Industrial Estate, Tyne and Wear, UK

early stages of each slab's life which corresponded with the active joint movement period. Once all the joints had cracked and were working, the movements become more uniform. The cumulative movements were of the same order demonstrating uniform slab movement.

In another research programme, twelve concrete slabs, each containing the same basic concrete mixture but with different quantities (0, 20, 25 and 30 kg/m^3) of hooked 1.0 mm diameter by 60 mm long steel fibres, were prepared. Each slab was 1 m × 1 m × 50 mm thick, and was simply supported on four edges and loaded in the centre.

The following data were obtained for the fibre-reinforced concrete slabs, and were used in developing the values used in the design procedures presented in this chapter:

Fibre dosage (kg/m^3)	0	20	25	30
First-peak load (kips)	2.09	2.14	2.47	2.47
First-peak stress (psi)	638	652	754	754
Maximum load (kips)	2.11	2.79	2.97	3.10
Plateau load (kips)*	0	1.96	2.36	2.47

* Plateau load corresponds to a state of increasing deflection under constant load and represents the collapse load of the slab.

In a second set of tests, the following data were recorded:

- nine slabs were tested (each 3 m × 3 m × 150 mm);
- $K - 0.35$ N/mm^3;
- each slab was loaded until failure with a point load in the slab centre;
- the load at the first visible crack was also recorded.

Steel fibre type	None	Length = 60 mm; diameter = 1 mm	Length = 60 mm; diameter = 1 mm	Length = 60 mm; diameter = 0.8 mm	Length = 60 mm; diameter = 0.8 mm
Quantity kg/m^3	0	20	30	20	30
Load at first visible crack, kips	40.5	47.2	53.9	58.4	65.2
Maximum load, kips	45.0	73.1	76.4	87.7	77.6

16.12 Design of concrete block paving

During the last 50 years, concrete block paving has become a significant surfacing material for many categories of roads. World-wide, over 600 000 000 m^2 are now installed annually. There are many reasons for this.

Concrete block paving was first introduced by the City of Rotterdam shortly after World War 2 as a temporary, expedient, substitute for the brick pavers that were, previously, used widely throughout the city centre. A post-war shortage of coal led to a shortage of brick pavers, and all of the bricks that could be manufactured were needed to rebuild the damaged city buildings. Engineers then found that the concrete pavers had superior engineering properties, particularly in relation to dimensional stability and skidding resistance. This led to a significant Dutch paver market which attained a volume of one square metre per capita per annum by

1970. Because Dutch pavers were introduced as a replacement for paving bricks, rectangular pavers predominated and, today, in many parts of the world, rectangular pavers are known as Holland Stones.

German building block manufacturers introduced pavers in 1962 following a collapse in their market for their traditional product. This led to Germany becoming the world's largest consumer of pavers with over 100 million m^2 being installed annually. In Germany, pavers of non-rectangular shape have become popular, with each city having its preferred shape.

The worldwide adoption of pavers in the 1970s was a result of German manufacturers of paver manufacturing equipment marketing their products internationally alongside a general perception that roads and streets should have a more interesting appearance. The introduction of pedestrianization schemes and traffic calming measures in the 1980s further stimulated concrete paver usage.

The technology of concrete block paving has evolved since the early 1970s and is recorded in the proceedings of a series of international conferences and workshops which have been held since 1980[4-8].

Usually, concrete block roads have a flexible pavement comprising the pavers bedded in sand over a lean concrete or bituminous roadbase. The road foundation usually comprises a crushed rock or cement stabilized subbase designed as for a flexible pavement. In the main, this type of pavement has proved successful for most types of road; the exceptions relate to heavily-trafficked city centre streets which have channelized traffic which, in turn, has led to unacceptably high levels of rutting and surface irregularity. Thus, for example, during the 1980s, questions were asked regarding the suitability of pavers as a surfacing material for roads that were trafficked principally by buses in city centres. Such concerns led to the development of improved specifications for paver bedding sands and to a change from flexible pavements to rigid ones.

In the UK, concrete block pavements are designed according to a British Standard publication[9]. The design of new pavements is based upon the method given in reference 10.

16.12.1 Components of a concrete block pavement

The cross-section of a classical block pavement is illustrated in Fig. 16.10. Note that not all of the component layers need be present in every pavement.

16.12.2 The design process

In the UK the design normally proceeds according to the flow chart shown in Fig.16.11. The first steps involve assessing the strength of the subgrade, and the amount of traffic expected to use the pavement during the design period, expressed in millions of standard axles (msa).

The design California Bearing Ratio (CBR) of the subgrade is most commonly obtained by direct measurement; alternatively, the CBRs of cohesive subgrade soils may be estimated from measurements of the plasticity index or of the equilibrium suction index (see reference 10). In situations where it is possible that the subgrade will become saturated during part or all of the life of the pavement, the test method that employs the soaking procedure should be used to determine the California Bearing Ratio.

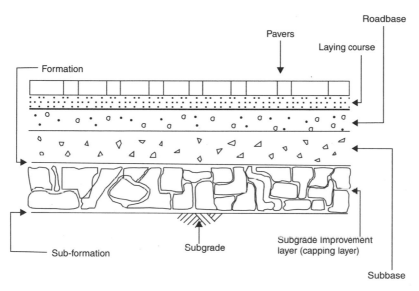

Fig. 16.10 Typical cross section of a concrete block pavement

The thickness of the subgrade improvement (capping) layer is selected according to the subgrade CBR value, as shown in Fig. 16.11. For a CBR of 2 or 5 per cent, the designer is given a choice of thickness of subgrade improvement layer. This permits engineering judgement to be used according to when, where and how the CBR was measured, local knowledge regarding the performance of the subgrade, the extent to which the subgrade improvement layer and subbase will be trafficked, the proximity of subgrade drainage, and the likely time of year for construction.

A subgrade improvement layer material should provide a CBR in excess of 15 per cent on which to lay the subbase, when designed according to the values given in Table 16.2.

If the subbase is not to be used as a site access road, then its thickness should be either 150 mm if a subgrade improvement layer is provided, or 225 mm if there is no subgrade improvement layer. If a subgrade improvement layer is provided and the road is to be used as an access road, then the subbase thickness may have to be increased depending upon the amount of traffic which will travel directly over the subbase (see Fig. 16.11). This traffic flow may be determined in terms of either the number of standard axles trafficking the subbase, or the number of dwellings being constructed, or the equivalent size in square metres of the industrial or commercial property being constructed. In the case of sites that cannot be categorized in one of these three ways, it should be assumed that the subbase will serve as an access road to a large development trafficked by 5000 standard axles.

Engineering judgement will normally be needed when assessing the amount of construction traffic to which the subbase will be subjected. If the roadbase to be provided is unbound a Type 2 subbase should not be used. If the projected traffic is likely to exceed 2 msa the subbase strength should be least equivalent to a CBR of 30 per cent. The detailed preparation of the subbase should be in accordance with the recommendations in BS 6717: Part 3 or BS 6677: Part 3 as appropriate. When rainfall is expected, it is expedient to cover the subbase as quickly as possible to

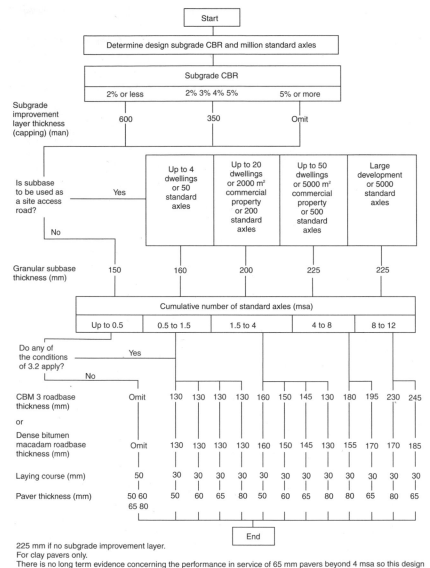

Fig. 16.11 Design procedure for a new concrete block pavement, used in the UK

prevent saturation and to protect the underlying materials. Remedial work is often required to the surface of a subbase that is fairly heavily used as a construction access road. Excessive trafficking of the subbase can cause rutting of the subgrade or contamination of the subbase by subgrade material; in these circumstances removal of the subbase and remedial works to the subgrade will be necessary.

Figure 16.11 also shows the recommended roadbase thicknesses for cement-bound material (CBM) and dense bitumen macadam (DBM) roadbases, as defined in Clauses 1038 and 908 of the UK Department of Transport *Specification for High-way Works*. Note that in the case where the anticipated traffic flow is less than 0.5 msa, there is no need to provide a roadbase. The minimum nominal roadbase thickness of 130 mm has been included in Fig. 16.11 for practical construction reasons. The alternative roadbase thicknesses recommended for traffic flows in excess of 1.5 msa vary according to the relevant paver thickness.

The laying course (bedding sand) should be in accordance with the recommendations in BS 6717: Part 3 or BS 6677: Part 3. The nominal laying course thickness should be either 50 mm, if placed over an unbound roadbase, or 30 mm, when placed over a bound roadbase. In the case of streets that are subjected to severely channelled traffic, particularly buses, the bedding sand should be further constrained so that it comprises a naturally occurring uncrushed silica-based material, and the fraction passing the 75 micron sieve is less than 0.3 per cent by weight.

Generally, the paver thickness should be either 60 mm, 65 mm or 80 mm. There has been considerable debate regarding the preferred shape and size of paver. The author has found 100 mm × 200 mm × 80 mm rectangular units to be suitable for all trafficked situations. Pavers of other sizes and shapes should be used with care. In some situations, permeable pavers may be considered, albeit care is needed in dealing with the water which will then enter the bedding material. Limiting the fine material in the laying course can lead to a sand which remains stable when saturated. Draining the bedding material can be a counterproductive measure.

16.13 References

1. Chandler, J.W.E. and Neal, F.R., *The Design of Ground Supported Concrete Industrial Floor Slabs*, BCA Interim Technical Note 11. Crowthorne, Berkshire: The British Cement Association, April 1988.
2. BS 1882: Part 116, *Method of Determination of Compressive Strength of Concrete Cubes*. London: British Standards Institution, 1986.
3. Westergaard, HM(1947) New formula for stress in concrete pavements of airfields, *Proceedings of the American Society of Civil Engineers*, 1947, **73**, No. 5, pp. 687–701.
4. *Proceedings of the First International Conference on Concrete Block Paving*. Newcastle: University of Newcastle upon Tyne, 1980.
5. *Proceedings of the Second International Conference on Concrete Block Paving*. Delft: Delft University of Technology, 1984.
6. *Proceedings of the Third International Conference on Concrete Block Paving*. Treviso, Italy: Pavitalia, 1988.
7. *Proceedings of the Fourth International Conference on Concrete Block Paving*. Auckland, NZ: University of Auckland, 1992.
8. *Proceedings of the Fifth International Conference on Concrete Block Paving*. Haifa, Israel: Haifa Technion, 1996.
9. BS7533: Part 2, *Guide for the Structural Design of Pavements Constructed with Clay or Concrete Block Pavers*. London, British Standards Institution, 1992.
10. Powell, W.D., Potter, J.F., Mayhew, H.C. and Nunn, M.E., *The Structural Design of Bituminous Roads*, TRRl Report LR1132. Crowthorne, Berkshire: The Transport and Road Research Laboratory, 1984.

CHAPTER 17

Basic road maintenance operations

J.E. Oliver

Design standards for new road construction are based on the expectation that necessary maintenance will be carried out periodically to deal with the inevitable deterioration caused by traffic loading, climatic effects and other deleterious influences. Consequently, road maintenance is a fundamental necessity, as important as the original road provision. The maintenance of the roadway asset must, additionally, be planned, designed and carried out in the knowledge that the road is there to provide a high level of service to users who, rightly, expect their needs to be met even when activities are being carried out on the network. Moreover, all maintenance has to be carried out with full regard to the requirements of governmental legislation which affects many aspects of the work involved.

This chapter emphasizes the need for good information and a thorough understanding of the technologies that determine the construction and condition of the road before deciding about the need for, and nature of, maintenance treatments. Road pavements normally perform very well and road maintenance engineers are offered a stimulating challenge to maintain their network within professional aims and without wasting scarce, valuable, resources. Understanding analytical approaches and innovation are essential prerequisites for successful maintenance management.

17.1 Importance of maintenance

Road maintenance is an essential activity for a number of reasons. First, various Acts of Parliament place legal obligations on road authorities to maintain their roads in a safe condition, and to ensure that maintenance operations are carried out safely. Second, roads are very often the 'vehicle' for carrying the apparatus of Statutory Undertaker's, e.g. electricity, gas, water and telephones, and work on the provision and maintenance of this equipment is also controlled by statute. Third, well-maintained roads support national and local economies by ensuring that freight and businesses can move efficiently and safely. Fourth, the way of life in developed countries now depends substantially on the availability of the road network, e.g. the vast majority of trips to schools, shops, hospitals and leisure activities are made via the road network.

17.1.1 Legal duty to maintain roads

Over 99 per cent of the roads in England and Wales, for example, are maintainable at the public expense and the duty to maintain these safely, for the benefit of all the users, is normally vested in a road authority under the requirements of *The Highways Act 1980*[1]. Road authorities in England are usually county, metropolitan district or London borough councils who are responsible for the maintenance of the roads in their own administrative areas; the national motorways and trunk roads are the responsibility of the Highways Agency, an agency of the Department of Environment, Transport and the Regions. Over recent years, an increasing amount of maintenance work and highway management has been contracted to the private sector, but this does not relieve the road authorities, their agents or the private sector operators of the duties contained within the Act.

Other legislation also exists which affects the planning, design, operations and management activities carried out by road authorities and those working for them. An important example of this is health and safety issues under *The Health and Safety at Work etc. Act 1974*[2]. There are many subsequent regulations under this Act, an increasing number of which now recognize the requirements of European health and safety legislation. The requirements of the *Construction, Design and Management Regulations under the 1974 Act* play a particularly important part in the field of maintenance.

Whilst the maintenance engineers are charged with providing a safe and efficient road for its users, they must always be conscious that there are others who experience the effects of the road's operation and maintenance. Some of these effects can be beneficial in terms of improved access, but others are intrusive. For example, roads carry a great deal of industrial traffic which, unwittingly, may discharge pollutants onto the carriageway which, in turn, are washed into the drainage system and discharge points which can affect other parts of the community; the road authority and its agents are required to provide appropriate controls to avoid pollution in such situations. Requirements to control such events also cover noise, litter, visual intrusion, disposal of materials, working hours, and other intrusive effects.

17.1.2 Utility apparatus in roadways

In the UK, a great deal of utility apparatus is placed within roadways (except motorways) for a variety of good reasons. Customers usually live close to roads so that supply of services is straightforward, and the maintenance of equipment can be carried out efficiently. Utility customers require their services to be maintained at all times, which usually means that immediate action is called for when a service breaks down. However, the interests and objectives of road authorities and utility companies can be quite different and therefore disruptive of road network operations.

As there are well over a hundred road authorities and operators in, for example, England and Wales, and even more utility companies operating within boundaries which do not match those for the roads, the interaction between these parties is both substantial and complex. Therefore, in England and Wales, the activities of the two parties are controlled by the requirements of *The New Roads and Street*

Works Act 1991[3]. The controls within the Act are designed to ensure that the works carried out by the two parties are well coordinated, managed safely, completed to required standards and generally managed to meet the needs of all users.

17.1.3 Highways as a national asset

An effective and efficiently managed transport network is an essential requirement of any developed or developing trading nation. The transport network includes air, rail, road and water, but by far the major proportion of freight movement in the UK, and in virtually all major trading nations, is carried by road. This is likely to continue for the foreseeable future.

Because of the extensive utilization of road networks by both private and commercial users, the costs of delay on congested roads can be considerable. If, in addition to the inherent congestion caused by large volumes of traffic, roads are also disrupted by the maintenance works of road authorities and utility operators, the delay costs can be substantially higher. The Confederation of British Industry (CBI)[4] has estimated that congestion costs the economy £15 billion a year, or in excess of £10 a week for every household. As roadworks are clearly an impact element in this cost, the CBI considered that their effect was a major problem, and pointed to the need for innovation and imagination in the planning, design and management of all maintenance works to reduce this cost to the economy as far as possible.

17.2 Scope of road maintenance

Maintenance of a roadway takes on many forms because of the duties placed on the various parties and because of the diverse range of issues encountered. These duties and issues relate to the road and its underlying strata and structures, to the influences of the weather and natural conditions, to the effects of its use and the impacts of its users and, finally, the effect that all of these may have on those affected by the road, i.e. almost everyone. Figure 17.1 illustrates a range of factors which contribute to the deterioration of a road and to its consequent condition at any given time.

Roads are constructed across a very wide range of ground conditions. For example, poor-grade land and reclaimed landfill sites are often likely locations for new roads because of their lower costs and easy availability. The behaviour of these sites, and the emissions from them, are less predictable than those from undisturbed ground, and call for careful monitoring and carefully-developed maintenance plans. Or, the roads may be constructed over old mine workings or other structures which should be (but may not be) known to the design engineer. Consequently, road maintenance issues and influences can extend well below and beyond the carriageway width.

For a modern road to operate efficiently and effectively for the benefit of all users, it is required to meet defined customer requirements. At the present time, with technological capability increasing rapidly and the pressure on roads intensifying, a major contribution can be made through the well-considered use of modern technology, e.g. in the form of real-time driver information, dynamic traffic control

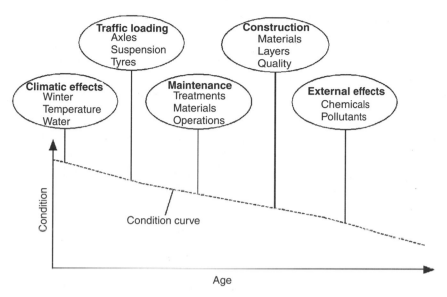

Fig. 17.1 Typical factors affecting pavement condition

units, network management systems, fixed traffic regulation measures, and high-speed monitoring equipment. Such systems can utilize both land-based and satellite technologies, each of which offer powerful tools to help meet the challenges of operating modern roads.

Maintenance operations and technologies themselves are evolving rapidly to meet the demands of modern road networks. Pavement design and rehabilitation is to some extent an empirical science supported by long-term observation, research and development. Because pavements are designed to last for many years before strengthening is needed, the observation and understanding of their whole-life performance is a continuous process. The understanding of pavement performance must, additionally, be carried out as an integrated road maintenance study because of the interaction of the various parts of the roadway and the influence that one part of the infrastructure can have on another, e.g. road drainage performance plays a vital role in ensuring the efficient structural performance of a pavement.

17.2.1 Importance of maintenance in relation to road purpose

Before discussing the detail of highway maintenance operations, it is important to emphasise the essential purposes of a road and its operation. Almost all roads are paid for through public funds for the benefit of users, whether they be industry, public services or the general public. The service to these users is enhanced through a well-structured and thought-through programme of maintenance that is designed to keep the roadway safe and working efficiently. In addition, the maintenance operations must be planned to meet the wide-ranging environmental requirements and meet the reasonable expectations of those living near to, or otherwise affected by, the roadway and its maintenance and/or operation.

17.3 Maintenance management systems

Efficient and effective maintenance management is most simply expressed as doing the correct thing at the correct time and in the correct place. Remembering that a typical road network represents the largest in-place asset component of the national publicly-financed transport infrastructure, it is necessary and morally correct to manage and, importantly, to be seen to manage the asset as efficiently and effectively as possible. A structured approach to this task with the assistance of an appropriate management system should meet both of these objectives. The following discussion therefore touches on the development of maintenance management systems, describes their position today in the engineer's armoury and, finally, offers some thoughts about their future development and value.

17.3.1 Development of pavement management

Whilst this chapter focuses on maintenance operations, the management of maintenance needs to be considered within a wider framework to ensure the best service to the user. To this end, management tools take the form of pavement management (rather than maintenance) systems, to ensure that full account is taken of all the issues which influence the state and performance of the modern road. Roads vary enormously, from high capacity, high standard, motorways to modest local roads; each type, however, fulfils an important function and must be accorded appropriate consideration. Thus, any management system used to support these requirements must be flexible enough to meet the needs of all road classes.

The principles underlying the design and construction of road pavements have been the subject of evolution and refinement rather than major change; this can be seen clearly when a Roman pavement of 2000 years ago is compared with a Macadam pavement of the early 19th century and, in turn, with a modern pavement. The uses for which these pavements were designed and constructed have, however, varied enormously and the loadings and other conditions to which they were subjected bear little resemblance to each other. With change being certain in the modern lifestyle, it is clear that future road management must be responsive to whatever influences impact upon travel and performance.

17.3.2 Fundamentals of pavement management systems

Pavement management systems are most effective if they fulfil a number of essential requirements in relation to the roads and road network to which they are applied. First, and most important, they can be a powerful tool when used in predictive mode to persuade governments and other decision-makers of the need for a sustained programme of investment. Second, a well-structured system will assist in the prioritization of maintenance works across a road network, and between roads of different types and categories. Finally, pavement management systems can assist the engineer in identifying the most appropriate treatment on selected sections of the road network through the use of economic analysis, predictive models and time-series information.

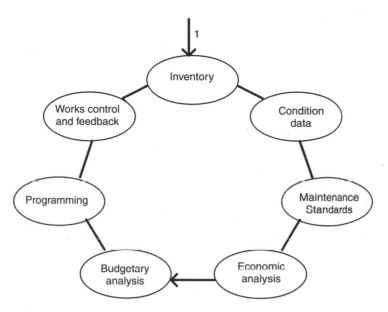

Fig. 17.2 Maintenance management planning stages

Many different forms of pavement management system are in use in various countries. These usually reflect particular conditions and requirements as found by the national road administrations which adopt their usage. Most, however, can be adapted for use on many road networks and some are used widely, e.g. the World Bank model known as HDM IV has been enhanced over a number of years from a model that was initially calibrated for use in developing countries and then changed so as to meet the wider requirements of other countries. (An advantage associated with using the HDM IV model is that it can be used to support the claims of user countries for investment by the World Bank or by other sources.) Such models are capable of being adapted for use on the networks of developed countries, but more bespoke systems are often seen as being more focused and cost-effective.

Whilst an effective management system must meet a number of core or critical requirements, too much sophistication should be viewed with caution and additional modules should be justified incrementally. Fundamental requirements of most systems include those illustrated in Fig. 17.2.

1. *Inventory data*   These data are essential. It is impossible to maintain an asset without a full understanding of the 'what, where, when, etc.' of all of it.
2. *Condition data*   It is vital to know the condition of the asset and, just as importantly, changes in condition over time. Deterioration rates, even of similar road components, can be quite variable.
3. *Maintenance standards*   The carrying out of maintenance works should be specified against appropriate and consistent investigatory standards.
4. *Economic analysis*   Decision-making regarding alternative maintenance treatments should consider various options, performance modelling, implementation timing, and pay particular attention to traffic delay costs.

5. *Budgetary analysis*   Treatment priorities are invariably affected by the known availability of funds.
6. *Programming*   Good programming is vital in ensuring that maintenance works are carried out at the correct time, and that unnecessary disruption is avoided whilst they are being carried out.
7. *Works control and feedback*   This is the final step required to complete the management process and to ensure that the system's records are up-to-date.

17.3.3 Selected system modules

It has already been emphasized that road maintenance must be regarded as an integrated management activity and not as a series of disparate tasks. Thus, the following discussion should be considered alongside other descriptions of maintenance activities in this chapter and elsewhere in this text.

The importance of an up-to-date, reliable and comprehensive *inventory* cannot be overstated. Modern technological and information handling capabilities enable the maintenance engineer to hold extensive information on the road asset. Much information is still collected manually and logged electronically, but to an increasing extent new capabilities, e.g. ground penetrating radar and laser technology, provide detailed information at lower cost and at higher survey speeds which result in less traffic disruption during the collection process.

The information gathered should include the location, nature, age, purpose and provider of each road component. The reasons for holding these items include: (1) location is needed to ensure that condition information and works details are accurately correlated against a referenced network; (2) the nature of each component needs to be understood so that its behaviour and subsequent treatment can be defined correctly; (3) the age of each component enables the engineer to check the validity of condition information against expectations; (4) the purpose or reason for a particular type of construction helps the engineer to ensure that proposed treatments are consistent with historic requirements or undertakings; and (5) knowledge of the provider of a component is essential in the event of failure, so that liability can be determined and redress can be pursued.

Maintenance standards have historically tended to be of a prescriptive nature in the UK, with the product and the construction process often being specified in some detail. The prescriptive approach has had some benefits, but it is now widely regarded that the specification of performance requirements is a much more satisfactory way of securing the desired result, albeit it may not always be the best approach when, say, performance is hard to measure over limited periods. The prescriptive approach, it is argued, stifles innovation on the part of the contractor and/or consultant who are often very experienced in providing the end product; also, the client is contractually in a stronger position if the maintenance provider is simply required to guarantee a level of performance rather than having to work to a specified method which can be later claimed to be the cause of any subsequent failure.

Time is a very important consideration in relation to the application of a standard to maintenance works. Pavements and other highway components deteriorate at rates which are not necessarily predictable; in fact, they are usually variable.

The correct point in time for intervention is therefore a critical consideration for the engineer, and good supporting time series data are needed to assist in the decision process. Thus, maintenance standards often need to be viewed in the context of being 'investigatory' rather than 'interventional'; in other words, when condition information shows that a component is approaching a stage where treatment appears necessary, the correct action is to first consider whether more investigation is needed and not to decide on a treatment precipitately. The additional investigations can in some cases be quite simple, whereas in others they may be extensive.

In order to discharge their duties properly, road authorities adopt nationally (and often internationally) recognized maintenance standards and codes of practice. In the UK, local authorities[5] and the national highway agencies[6] have published codes of practice. There are also nationally recognized design standards[7] which are largely adopted by all road authorities and agencies. These documents draw upon a wide range of published work and developments from academic, research, industry and road authority sources.

To carry out maintenance treatments in the correct place at the correct time, it is necessary to understand the nature and likely future behaviour of the component in question. Whilst pavement deterioration is not necessarily a uniform and, therefore, an easily predictable phenomenon, the benefits to be gained from reliable *performance modelling* are considerable and much research continues to be carried out to develop this capability, e.g. via the Long Term Pavement Performance (LTPP) studies under the American Strategic Highway Research Program (SHRP).

Principles underlying pavement performance modelling, which are also applicable to other forms of infrastructure behaviour, are shown in Fig. 17.3.

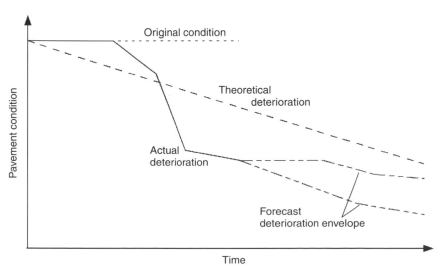

Fig. 17.3 Pavement performance modelling

17.4 Maintenance rating systems for bituminous roads

The most common pavement construction and road surface in the UK is bituminous in nature; these are referred to as flexible pavements and flexible construction (see Subsection 9.2.1). Modern roads may have up to three or four different bituminous layers on top of an unbound subbase and, depending on the strength of the underlying subgrade, possibly an unbound capping layer.

If the large budgets that are now available for road maintenance are to be spent wisely, the mechanisms used to determine deterioration and treatments must be objective, robust, and consistent, so that the needs on one part of the network can be accurately compared with those on other parts. To ensure this end, rating systems and objective road inspection regimes have been developed from research and experience. Whilst these systems take various forms, all depend on objective and consistent condition survey programmes which rely on teams of trained inspectors to collect information on condition – still based, largely, on visual observation – at predetermined intervals and in prescribed formats. An increasing amount of information is now, however, being collected via machine surveys; these offer benefits which include speed, consistency, lower cost, and repeatability (which ensures that successive surveys on the same road section can be confidently compared, due to the absence of human bias).

17.4.1 Performance principles for bituminous roads

The design of bituminous pavements in the UK[7] is currently based on research reported in *LR1132*[8]. Pavements that were designed prior to the publication of reference 8, at times of much lower traffic loadings, may require specific analysis to ensure that the most appropriate maintenance is applied.

The most common causes of flexible pavement deterioration arise from the actions of traffic and climate, and result in surface defects and/or structural deterioration. Such deterioration typically includes: (a) polishing of the stone in the surfacing, thereby reducing skidding resistance; (b) loss of surface texture, also reducing skidding resistance; (c) deformation of the surface due to traffic loading; (d) oxidation of the binder, resulting in cracking and surface deterioration; and (e) fatigue strain of the foundation, which causes structural deterioration.

The accurate identification of a defect, i.e. including its cause, is crucially important because there are often similarities between the visual appearances of pavements with quite different defects. For example, rutting of a flexible pavement can occur because of either surface deformation or structural failure of the foundation, and the nature of the appropriate remedial treatments and their costs are substantially different.

17.4.2 Condition survey strategies

Nowadays, it is not only necessary to monitor highway condition effectively, but they must be carried out in ways which are not excessive and minimize disruption to the travelling public. Thus, condition strategies should initially utilize *high-speed, first-pass, survey techniques* to obtain an overview of pavement condition. The frequency of these initial surveys are usually determined on the basis of known information about the pavement's construction, age, and (previously-determined) condition; in other words, well-built, newer, pavements normally need less frequent surveys than older pavements for which a modest level of deterioration has already been recorded. The high-speed survey results determine the need for, and frequency of, other more detailed surveys to establish whether and what type of maintenance treatment is required, e.g. the further investigations may include material analysis to accurately establish the correct treatment.

The high-speed surveys are, as far as possible, carried out with the aid of equipment which can travel in the normal flow of traffic without causing significant disruption. In the UK machines using laser technology have been used for over a decade to obtain an overview picture of the carriageway profile on major roads at travel speeds of up to 60 mph. The latest model is known as Traffic Speed Condition Surveys (TRACS). Skidding resistance is measured by a Sideways-force Coefficient Routine Investigation Machine (SCRIM)[10] which travels at about 30 mph. These machine surveys must, however, be backed up by detailed visual surveys and, very often, by pavement material analyses to establish precisely the nature(s) of the deterioration. For many years the UK primarily used two visual inspection systems known as CHART (Computerised Highway Assessment Ratings and Treatments) and MARCH (Maintenance Assessment Rating and Costing of Highways), but these have now been replaced by the UKPMS (United Kingdom Pavement Management System) coarse and detailed visual inspections known as CVI and DVI respectively.

The CHART system remains for the time being the basis of the National Road Maintenance Condition Survey (NRMCS). The NRMCS monitors, on an annual cycle, the condition of a range of categories of road in England and Wales, and is likely to be extended to other parts of the United Kingdom. CHART has been used to record defects both on and off the carriageway.

Visual inspection surveys are based on the knowledge that roadway components deteriorate gradually and that, at various stages in their lives, there will be a certain priority for treatments to be carried out. The UKPMS CVI was originally designed as a rapid low-cost network survey from which lengths for more detailed assessment would be identified. It has, however, become the default network monitoring tool for Non-Principal roads in most of the UK. Wherever possible the CVI is carried out from a moving vehicle. DVI, on the other hand, was intended as a scheme-specific visual inspection carried out on foot at targeted locations. In practice, CVI is being used widely by rural highway authorities whereas urban highway authorities tend to favour DVI on their more congested networks. Visual condition surveys identify priorities for treatment in relation to levels of defectiveness recorded. A schematic deterioration curve is illustrated in Fig. 17.4.

In practice, the defectiveness rating principles illustrated in Fig. 17.4 apply to a number of different modes of deterioration for the various components of a typical roadway (including its pavement). Inspection records are established for referenced sections whose position and length are determined in relation to important physical features on the network. The following are a typical range of measures of defectiveness for a bituminous pavement.

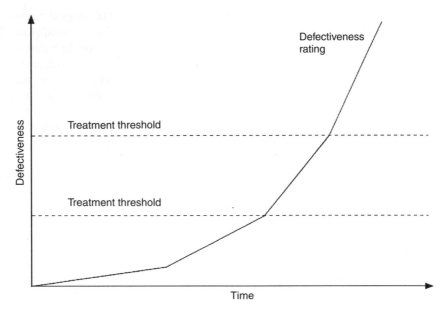

Fig. 17.4 Schematic illustration of a component defectiveness profile

1. *Whole carriageway minor deterioration.* Includes potholing, fine crazing, surface fretting and permeability, loss of chippings and excess bitumen at the surface (i.e. 'fatting-up') of surfaced dressed roads. The level of defectiveness is expressed as a percentage of the section area.
2. *Whole carriageway major deterioration.* Includes cracking, coarse crazing, loss of aggregate or serious fretting. Again, the level of defectiveness is given in terms of percentage of the section area, albeit potholes, being a source of potential danger, would be recorded separately and treated as necessary.
3. *Loss of skidding resistance.* Can have significant safety implications, and is noted for prompt action.
4. *Wheel track rutting.* Can be caused by either plastic deformation at the surface or failure of the foundation; the former defect usually requires only a surface treatment, whereas the latter is likely to require major strengthening works. The measure of rut defectiveness is its depth; however, this is considered alongside other indicators, e.g. cracking in the wheeltrack, which assist in defining the problem and/or any further investigation that is required prior to decision-making regarding the treatment.
5. *Surface irregularity.* Affects driver comfort and, in severe cases, vehicle stability. This defect can be identified quickly and consistently with the aid of the HRM, which uses laser technology to measure both transverse and longitudinal profiles and surface texture.

Rating systems also allows for ratings of condition that are based on 'off-pavement' inspections made over the same sections as for the pavement defects. In this case particular features, which can be on or off the pavement, are inspected against a common network referencing system, so that the results are capable of direct comparison to ensure the efficient operation of the whole roadway. Reference 11 illustrates the range of features to be inspected, but the following highlight some important issues.

1. *Footways and cycleways* are for the benefit of vulnerable travellers, e.g. pedestrians, including the elderly, the disabled, the young, those carrying shopping, and cyclists. Defects noted for these facilities include potholes and trip-points in paving slabs and utility ironwork. Modern laser-based systems can also be used to record these defects quickly and accurately.

2. *Electrical systems* should be operating correctly at all times (e.g. failure of traffic lights can result in hazardous situations developing on the network) and, hence, the road authority is required to ensure their proper operation. It is normally more efficient for the monitoring of electronic systems such as traffic lights to be done remotely through a central facility. Authorities normally have additional arrangements with other public service bodies, e.g. the police, to provide reports on problems that may exceptionally occur.

3. *Road signs* must convey clear messages at all times. All road signs reduce in clarity over time, and poorly-located signs, in particular, are easily dirtied by spray. Sign inspections, therefore, need to be carefully timed to ensure they identify problems of reduced legibility or reflectivity.

4. *Road markings* are subject to considerable wear where they are run over by vehicles, especially heavy lorries. Laser technology is now available which can measure the effectiveness of road markings whilst the vehicle carrying the measuring equipment is travelling at normal traffic speed. Thus, even over a large road network, the need and priority for replacement of these markings can be now relatively easily established.

5. *Street lighting* provides a sense of safety and security to, especially, those on foot. Road users indicate quite clearly that they welcome facilities which offer security in what is often seen as a threatening environment. Thus, inspections check on the adequacy of lamps, columns and electrical infrastructure to ensure the continued provision of the service.

17.5 Maintenance rating systems for concrete roads

Roads with concrete, or rigid, pavements (see Subsection 9.2.2) account for only about 15 per cent of the major road network in England. However, they form a greater percentage of the local roads, particularly in urban residential areas.

Both the nature and the performance of concrete roads are quite different from those for bituminous roads and, consequently, a *Manual for the Maintenance and Repair of Concrete Roads* was published in 1988 to provide an appropriate inspection regime for these road types in the UK. This manual[12] provides a visual inspection methodology that has been developed by road authorities for computerized condition recording.

Condition surveys for concrete roads are best carried out during the colder months of the year when cracks have opened and are therefore easier to define. The three essential features of the advice in the manual are as below.

1. *Ensure the integrity of the road.*   In this respect, particular attention is paid to the state of the joints between concrete slabs, as these are usually the greatest source of potential weakness, e.g. they may allow water to penetrate to the foundation if they are not well maintained.
2. *Identify defects and ensure that they are treated at the appropriate time.*   Deterioration of concrete roads tends to be in terms of slab cracking, edge deterioration, and the spalling of joints. These defects, if undetected and left untreated, can lead to very serious problems.
3. *Ensure that concrete roads are continually able to behave as intended at the design stage.*   Jointed roads, for example, are constructed so that the 'rigid' concrete slabs can expand or contract as the temperature changes, and this need requires that joints be well-maintained and kept clear of detritus. This maintenance operation has important benefits in helping to prevent 'blow ups' whereby, in hot weather, joints that are blocked with detritus prevent slab expansion so that adjacent slabs buckle and lift.

It has not been possible to develop ratings for defectiveness in concrete roads to the same degree as for bituminous roads due to the differences in behaviour of a rigid pavement compared with one which is relatively flexible. Hence, the need for treatment of a defective concrete pavement is largely based on the extent of such defects as cracking, joint defects and spalling.

17.6 Maintenance of bituminous pavements

The maintenance of all types of pavement consists of a structured programme of investigation and intervention to safeguard the fabric of the road and to ensure that its economic performance is optimized. This objective is achieved by minimizing whole life costs, which involves careful consideration of the condition of the road, the likely future deterioration, the expected performance of the various treatment options and the costs, including user costs, of the works. Whole life costing techniques have been developed by a range of organizations and authorities and will form an increasingly important part of highway maintenance techniques in the future.

To meet the various treatment needs of different pavement types, industry and research institutions have developed a range of materials and treatments to offer the engineer a wide variety of effective solutions. The materials are variously required to meet a number of criteria including strength, resistance to deformation, impermeability, good skidding resistance, low noise and spray generation, and value for money. Many modern materials offer a good range of these qualities, but it is fair to say that the perfect material has yet to be developed. The principal maintenance treatments for bituminous roads are described in the following paragraphs.

17.6.1 Surface dressings and thin surfacings

Surface dressing (see Chapter 19), which is a treatment usually employing up to two layers of binder and either coated or uncoated chippings of specific grading, can be a very important and cost-effective maintenance treatment for bituminous and concrete pavements. The treatment can be unpopular with road users because of the risk of loose chippings causing minor damage to cars. As with maintenance operations in general, care needs to be exercised by the maintenance planner and the contractor to ensure that surface dressing applications are completed safely and successfully. If a high level of control is provided, surface dressing will provide a very good result for all categories of road, including motorways.

Surface dressing is a very effective treatment for restoring skidding resistance, treating small cracks in roads and providing an impermeable surface. It does not, however, restore the profile of a rutted or uneven road. To meet this need cost-effectively the industry has developed a range of *thin surfacings* which are thicker than surface dressings, but thinner than a full resurfacing treatment; these too are discussed in Chapter 19.

17.6.2 Resurfacing

Resurfacing usually involves providing a new 40 to 50 mm thick bituminous layer to either: (a) replace an old surface layer, i.e. inlaying; (b) add a new surface layer, i.e. overlaying; (c) recycle the existing surface, i.e. where proportions of old and new materials are mixed and recompacted; or (d) apply a surface with special properties, e.g. to reduce noise levels from traffic.

Resurfacing is most commonly carried out where the existing surface shows signs of visual distress and the materials in the pavement surfacing have deteriorated significantly. These defects are normally identified through a visual inspection, e.g. using UKPMS and laboratory analyses of the surfacing materials. Resurfacing is not, however, appropriate for the treatment of pavements where major structural and/or foundation problems are identified. Replacing an old surface is normally more expensive than simply adding an extra layer. However, overlaying may be impractical or uneconomic in urban areas, where utility ironwork and footway kerbs will need to be lifted in order to ensure an acceptable carriageway profile.

Resurfaced pavements on major roads can give many years of excellent performance, but it must be remembered that they are laid at various times of year under different circumstances which call for care in both the design and works phases. Mix design, material specification, temperature and wind chill can all have a major impact on the success of the resurfacing operation. When the resurfacing material is being placed and laid by the paving machine, it is particularly important that it be at a temperature that is suitable for successful rolling for a reasonable time afterward.

For both inlays and overlays the resurfacing materials to be used are often determined by the location and nature of the road to be treated. Hot rolled asphalt[13] is probably the most commonly used surfacing material on major roads in England, albeit a number of different alternatives are considered in specific situations. For example, the need to reduce spray and noise levels would suggest the use of an open-textured material like porous asphalt, whereas tar-based surfacings rather than bitumen-based ones might be used at locations where there is a likelihood of chemical or fuel spillages.

Open-graded materials reduce the effect of noise and facilitate the rapid drainage of surface water from the road. Because of their open texture they are less stiff than conventional hot rolled asphalts and other dense surfacing materials, and therefore provide a reduced structural contribution to the pavement. A resurfaced open-textured wearing course must be laid on top of a dense basecourse to ensure that it functions as expected under the design requirements. The materials are popular with road users because of the improved travelling comfort that it offers; however, it can be more expensive, as well as presenting other issues over durability and winter maintenance which need to be considered at the planning stage.

A growing number of materials, binders and modifiers offer the maintenance engineer a wide range of resurfacing options for pavements. Some of these are proprietary materials, the compositions of which, for reasons of commercial confidentiality, are closely-guarded secrets; however, they offer benefits of performance through being effective in a wide range of situations and extreme conditions.

17.6.3 Pavement strengthening

When the investigations indicate that the defects are affecting the structural strength of the pavement rather than simply its surface, the repairs identified are likely to suggest pavement strengthening as being appropriate. *Strengthening* is carried out by reconstructing all or part of the pavement, or by applying a structural overlay to the existing pavement, or by carrying out a combination of the two. *Reconstruction* may involve replacement of the complete pavement or replacement of only certain layers; in the latter case this would normally include the roadbase. Pavement deterioration is seldom a consistent phenomenon either along or across a carriageway, so that an optimized pavement strengthening scheme will very often include a combination of overlay and reconstruction both along and across the carriageway.

Research has resulted in the development of detailed guidance on the optimized design of strengthening schemes[14]; this advice is now available in computerized form. Detailed theoretical advice of this type must, however, also take account of the need for proposals to be refined into efficient works proposals which do not impose uneconomic working requirements.

Notwithstanding that the application of a strengthening overlay usually costs less than reconstruction, as well as needing less (scarce) natural resources, a careful assessment of pavement condition and of the available solutions is normally required to ensure an optimized combination of treatment in a given situation. Also, the preferred type of strengthening treatment will sometimes be determined by the location and nature of the road to be treated; for example, urban roads which, typically, have inherent constraints such as in-ground utility apparatus, footways, and accesses, are often unsuitable for the application of strengthening overlays because of the implications of adding an additional thickness.

Also, it is now being recognized that many of the stronger-built major road pavements are demonstrating virtually indeterminate structural lives, and may 'never' need strengthening through reconstruction or a thick overlay (see Subsection 9.3.3).

It is again emphasized that pavement assessment and treatment design should be based on information from an appropriately wide range of information on pavement condition. Detailed advice, based on work[15] by the Transport Research Laboratory, is available to the maintenance engineer in relation to the design of structural overlays; this methodology is based on measurement and analysis of pavement deflection. Knowing the condition history of the pavement and its deflection performance, the maintenance engineer using the TRL approach is able to determine the remaining life in the pavement and the type and thickness of strengthening overlay that is needed. The design takes account of the existing pavement structure, the traffic loading (in equivalent standard axles) that it has carried, the traffic loading it is forecast to carry, and the deflection corrected to a standard level depending on temperature.

Pavements can deteriorate to a condition past which the application of a strengthening overlay is unlikely to be successful because of the poor condition of the lower layers. It is usually uneconomic to allow a pavement to deteriorate to this extent and, when it does, the pavement can only be restored to an acceptable structural condition by reconstruction. If this reconstruction has to be delayed for a considerable time through, for example, lack of resources, interim periodic maintenance in the form of crack sealing and patching will be required as the pavement will have rutting and cracking in the wheeltracks.

When reconstruction is decided upon, the treatment selected should be such that the new road will perform as if it were designed anew to carry the traffic forecast over its new design life. Current design methodology for reconstructed roads takes account of the traffic forecast to be carried up to a point where a pre-emptive strengthening is necessary (typically after 20 years); this future strengthening would normally be a structural overlay and the original design and specification would allow for this and ensure there is, for example, the necessary cross-sectional freedom to permit the future application of an overlay.

17.7 Maintenance of concrete pavements

Concrete pavements behave quite differently from flexible pavements and also deteriorate in a different and, in some ways, less predictable manner. The normal indicators of deterioration are spalling of the concrete at joints, surface scaling or crazing, and cracks of various types and sizes. Priorities for treatment are determined in the light of the degree and extent of these defects.

Concrete pavements can be formed of unreinforced, jointed reinforced or continuously reinforced slabs. Concrete pavements cannot absorb temperature movements in the same way as bituminous ones, and the slabs require either joints or reinforcement to do so. Continuously reinforced pavements require anchors at specified points to absorb the movement stresses, whereas the unreinforced and jointed reinforced pavements require preformed joints at longitudinal spacings of, typically, 5–6 m in order to accommodate the temperature movements. Continuously reinforced concrete pavements tend to give good long-term performances with little need for periodic maintenance; they are, however, an expensive form of construction and are therefore normally used in specific circumstances where their extra cost is justified.

Jointed slabs require regular maintenance and, because of the slab movements, this maintenance is often in the form of treatment to the slab edges and/or the joints, including replacement of the joint sealants. The effectiveness of the joints is crucial to both the safety and serviceability of the pavement, i.e. malfunctions not only lead to surface problems but also to water penetration into, and deterioration of, the subbase and subgrade and result in the need for subsequent extensive (and expensive) repairs. Whilst subgrade and subbase problems can sometimes be economically dealt with by pressure and vacuum-grouting techniques, it is often necessary to remove the slab(s) to effect more substantial repairs.

In order to minimize traffic disruptions, the tendency in recent years has been to use rapid hardening cements for small-scale remedial works, so that slabs can be repaired or replaced overnight and roads opened to vehicles the following morning. Non-shrink concretes are available for thin-bonded repairs that reduce the risk of warping, and consequent de-bonding, due to high temperatures that might arise soon after placement. Care needs to be taken, however, when carrying out these small and rapid repairs, to ensure that the work is effective and that the repair lasts the required length of time; it is wasteful and disruptive to have to return to replace defective treatments.

When major repairs are required to even a single concrete slab, the resulting works can be extremely disruptive to traffic and other road users.

17.8 Maintenance of unsurfaced soil–aggregate pavements

Whilst unbound soil–aggregate roads are very rare as public highways in the UK and in most European countries, they are quite common in overseas countries, e.g. in the United States and Australia, and in many developing countries. Depending upon the composition and volumes of traffic, climatic conditions, the nature of materials available locally, and the quality of the engineering expertise involved in their construction, these roads can perform very effectively. By their nature, however, soil–aggregate roads require regular maintenance at frequent intervals.

The essential requirements of an unbound pavement is that it be structurally sound, have adequate crossfall (typically 4–5 per cent) to ensure that water disperses quickly, and have a longitudinal profile that is not so steep that it exacerbates natural surface erosion. Maintenance work, therefore, has to be carried out at frequent intervals to ensure that the riding quality is satisfactory, and that the profile facilitates the dispersion of water, and minimizes the possibility of ponding and consequential deterioration.

Maintenance works on soil–aggregate pavements usually involve: (1) filling and patching, to deal with small local defects; (2) light grading, to deal with general surface defects; (3) heavy grading, to deal with defects which extend deeper into the surface; (4) reprocessing, to rectify deteriorated surfaces; (5) light reshaping, to restore a badly rutted or potholed road; or (6) heavy reshaping, to rectify defects in a pavement which would be virtually impassable. Very often, a certain amount of reworking of the surface (including trimming, cutting, ripping and/or reshaping) is required before a soil–aggregate pavement can be regraded to the correct profile. The trimming/regrading process may also include the addition of coarse and/or fine materials and water prior to compaction by rollers.

A wide range of generally understood maintenance principles, which are based on practical experience, has been brought together in a maintenance handbook commissioned by the Permanent International Association of Road Congresses

(PIARC)[16]. Detailed advice on the maintenance of soil–aggregate pavements tends to be tailored to specific national requirements because of the variability of the local climatic and other conditions.

17.9 Maintenance of other roadway features

The maintenance and management of a modern road involves the upkeep of an asset of massive replacement value, and extends to many more features than the pavement. It is not possible in this short chapter to cover in detail the full range of infrastructure which has to be maintained by a road authority, but the following categories offer a general indication of those features which are of significant importance, and are normally included in maintenance schedules: features that (1) are of structural significance, (2) are of statutory importance, (3) meet environmental needs, (4) improve the road's appearance, (5) are provided as user services, and (6) are for pedestrians and cyclists. However, it should be borne in mind that these categories are neither exhaustive nor definitive and that some features may fit into more than one category.

All road authorities have codes of practice and maintenance plans which set out their policies in relation to the above works.

17.9.1 Features of structural significance

Road drainage falls into this category because of its importance in relation to road-side slope stability and pavement performance.

In practice, significant roadside earthwork failures rarely occur in the UK; in fact, those that do happen usually occur during construction rather than after the road is opened to traffic. Minor slope slips do, however, occur and their cause must be identified so that the appropriate remedial treatments can be programmed. These slips are often attributable to the penetration of water into the slopes of embankments or cuttings. Remedial measures usually require some replacement of the slip areas and the provision of some forms of counterfort drain.

The most important non-structural feature contributing to the performance of a road pavement is its drainage system. The subgrade and the subbase are both designed to perform at a specified moisture content and if, through the poor performance of the drainage system, the moisture content rises above the design value at particular locations then these parts of the pavement structure will deteriorate more rapidly and, probably, fail prematurely. Regular monitoring and good maintenance is therefore critical to ensuring that drainage systems perform as required, and they should always be safeguarded during other highway works so that damage does not occur to hidden items of infrastructure.

Drainage systems form a continuous network from the point of initial collection to outfall, and a failure at any point is likely to result in water-induced damage somewhere within the roadway. Formal inspections and/or observations during general maintenance activities should therefore seek to identify the extent to which the drainage system is performing properly. Simple observations to identify whether water is not running when it should, help to monitor the condition of the drainage. Detailed inspections may require a more thorough examination where ditches or filter drains are involved, e.g. a specialist closed circuit television (CCTV) survey of piped drainage systems may be desirable when possible defects are not visible.

Drainage systems not only carry rainwater away from the carriageway, but they also carry the results of pollutant spillages. It is not only hazardous chemicals which present a threat; spillage from a milk tanker is just as unwelcome if it reaches a local water company's aquifer. The road authority has a duty in law to take all reasonable steps to prevent pollutants arising on its land from causing a nuisance elsewhere and, consequently, it works closely with environmental agencies and emergency services such as the fire service, to ensure that appropriate interceptors are installed within the road drainage systems. These interceptors must be marked and their locations known to the emergency services; the locations must also be kept clear so that, in the event of a spillage, the spillage material is both contained but removable.

17.9.2 Features of statutory importance

Public roads allow the virtually free movement of all members of the public and industry carrying a wide range of goods and commodities in vehicles of many shapes and sizes in a variety of states of repair. It is therefore inevitable that numerous laws and regulations apply to and affect the operations of roads, albeit they should not unduly affect the right of the user to enjoy reasonably unrestrained travel. Whilst the following illustrate some of the issues that are of concern in this respect, it must be stated that it is increasingly important for the maintenance engineer to be alert to the need for specialist legal advice in appropriate instances.

There is a fundamental requirement in law for road authorities to maintain their highways *safely*. Thus, for example, they have a responsibility to be aware of tree conditions on land close to a road and to require the owner to take action where an adjacent tree presents a potential threat to road users, e.g. from falling branches. Maintenance activities such as the trimming of roadside foliage, e.g. where tall grass obscures a driver's view at a junction or where overhanging tree branches may be dangerous if struck by passing lorries, have safety as well as environmental benefits.

Research has shown consistently that accident rates for roads on which works are being carried out are consistently higher than for the same stretches of road in the absence of works. To reduce this accident rate as far as possible, it is normal for speed restrictions to be imposed on roads on which works are being carried out. These speed controls, which should be a carefully considered part of the planning process for the maintenance activity, must be imposed through an Order under the appropriate legislation. Orders are also required to implement traffic controls such as contraflow operation.

Feedback from road users has consistently shown that *personal security* (as distinct from road safety) is a major concern, particularly in the evenings and at night, not so much because of 'road rage' but because of the physical presence of large volumes of high speed and/or heavy traffic. Survey respondents have consistently indicated the value of road lighting and signing communications in offering comfort in what is clearly regarded as a threatening environment. Lighting using low pressure sodium lighting (which minimizes glare and light spread) to provide greater driver comfort and safety is therefore often supplied on major non-urban roads as well as on roads in urban areas, albeit it may not be liked by some rural residents who regard it as visually intrusive. The maintenance of these facilities is carried out within the normal works programme with increasing emphasis being given to their importance in providing a good standard of user service.

There are a significant number of regulations which implement requirements of governmental legislation; in recent years these have been increasingly enshrined within European legislation. The most important regulation for the road authority in Great Britain is the Construction, Design and Management Regulations[17]; these specify the responsibilities of the various parties within road-works contracts and need to be considered early in the planning process for maintenance works as they can significantly influence the manner in which they are carried out and controlled.

On occasions maintenance engineers may be taken to task by the public for not having performed certain remedial activities. In such instances, the 'defence of reasonableness' may well be appropriate; e.g. it is usually considered reasonable for maintenance work not to be immediately carried out if funds and other resources are not available because of higher priorities elsewhere. In such instance, however, it should be demonstrable that the need has been recognized and that appropriate remedial work was programmed after careful assessment of the relative priorities across the network. If the deferred work involves rectifying a potential hazard in the roadway, road users should be alerted to the potential danger and advised of the need for caution via, say, a prescribed warning sign.

17.9.3 Features that meet environmental needs

Environments that need to be considered from a maintenance aspect are those of (1) the users of the roadway, (2) the living world within it (i.e. roads are very often rich wildlife habitats), (3) persons who live and/or work adjacent to a road, and (4) persons elsewhere who are indirectly affected by road activities.

The *Environmental Protection Act 1990*[18] requires public places such as roadways to be reasonably free of litter. Litter is dropped by the public and/or blown from elsewhere and, once the accumulations have reached a certain level, they must be removed; if litter is not removed, members of the public can apply to a Court to obtain an order to have it removed.

Litter removal should, however, be differentiated from maintenance involving debris and detritus. Items of debris can be quite large and present a hazard to road users, in which case they must be removed or made safe so as to ensure the road's safe operation. Detritus is usually composed of silt which can block drains and service ducts; it is normally swept away or otherwise removed within a maintenance operation.

Landowners (including road authorities) have responsibilities with regard to the prevention of the spread of pests from their own land onto adjoining plots. Examples which may need to be controlled on a regular basis via the maintenance activities of road authorities are injurious weeds at the roadside which may prove deadly if eaten by cattle, and pests such as rabbits. When controlling such pests, authorities have to ensure that the methods adopted do not cause other problems, e.g. through the use of toxic chemicals.

The wildlife within a roadway are often the subject of special maintenance provisions which must be handled with sensitivity and imagination. For example, some major roads have badger tunnels beneath embankments which must be maintained carefully, as the human scent may deter the badger from using these essential facilities.

17.9.4 Features that are designed to improve a road's appearance

The aesthetic appearance of a highway is important if its impact on the landscape is to be minimized. As well as enhancing the appearance of a road, horticultural provisions (often referred to as the 'soft estate') can be used to screen it, and traffic, from nearby homes of residents. It is normal for new roads to be approved only after soft landscaping has been approved, following a comprehensive environmental assessment. Landscaping schemes typically include the provision of new trees, bushes, flowers and other indigenous species. It is the responsibility of the maintenance engineer to ensure that the young flora specimens are well established and managed in order to achieve the desired landscaping effects.

17.9.5 Features that are provided as road user services

Examples of these features are motorway communication systems and driver information systems.

Road users continually seek information about such matters as conditions on the network, guidance on preferred routes, and how to summon assistance when required. Nowadays, major roads are increasingly equipped with sophisticated control networks that draw upon the extensive range of electronic equipment that are operated, in conjunction with the police, to supply real-time advice and information to road users; these also help the police to deal with incidents that require immediate attention, e.g. road accidents.

Electronic developments now enable traffic conditions to be assessed from closed-circuit television cameras, detection loops, vehicle detection devices, and global positioning systems, so that up-to-date information can be provided to motorists on network efficiency via radio broadcasts, dynamic signing systems, and in-car information units. The electronic technologies underlying these services require specialist maintenance skills that are different from those traditionally provided by road maintenance engineers; these are therefore provided under specific maintenance contracts or in the form of service provision arrangements whereby specialist companies are commissioned to design, provide and maintain the service.

17.9.6 Features specifically provided for pedestrians and cyclists

Road users such as cyclists and pedestrians are normally more vulnerable than vehicle users, and require appropriate consideration in maintenance policies and practices. These users may include people who are young, elderly and/or infirm, as well as those who are disabled and have hearing and/or visual difficulties.

Uneven surfaces, unprotected works, utility reinstatements, poorly maintained barriers, etc., present special challenges to these road users. The road authority must therefore make reasonable provision for these persons in its maintenance planning; consequently, it is now common practice for representatives of these groups to be asked to advise on how appropriate provision can be made for their needs, within the constraints with which the road maintenance authority has to work.

17.10 Winter maintenance operations

Winter maintenance is carried out to provide, as far as is practically possible, a safe road which will allow the efficient movement of traffic under snow and ice conditions. In order to carry out this task effectively it is necessary to have a clear policy on what is to be included in the winter maintenance plan, accurate weather forecast information, and a well-resourced network of equipment that is ready to respond to sudden changes in weather conditions.

17.10.1 Winter maintenance plan

A road maintenance authority must have a well-defined plan to state which roads will be given priority of treatment during periods of severe winter weather, as it may be impossible for all roads to be kept clear under such conditions, e.g. it is likely that, in periods of severe weather, only major roads and those of special significance to, say, hospitals can be kept open and snow- and/or ice-free. These plans should be made public so that people in general, and travellers in particular, are aware of which roads will be treated and which will not. Also, as treated carriageways do not offer the same levels of grip between the surface and the tyre as dry roads, the public needs to be advised that caution is required if these roads are used under such conditions.

17.10.2 Weather forecast information

If roads are to be maintained safely, and economically, it is necessary to know when ice and/or snow is expected and in the case of ice, where it is likely.

Roads have different temperature characteristics depending on altitude, topography, tree cover and other local factors. The roads within a network can, therefore, be said to have a thermal profile or thermal map. Current practice is for the maintenance authority to have a network of sensors placed within pavements at a small number of critical locations within the road network over which it has control. These serve as reference points from which the road authority, on the basis of experience and analysis, is able to estimate the temperature along the whole length of road (in relation to the thermal map) and decide whether and where treatment is necessary.

The winter maintenance network is normally divided into individual lengths over which the authority can complete a treatment within a reasonable period, e.g. typically, about 2 h. These are known as 'salting routes', and are selected to reflect weather conditions; on some occasions, one salting route may not need treatment because of its temperature profile, whereas another one will.

Road authorities normally have arrangements with weather forecasting organizations whereby information from pavement sensors is directly fed to the forecaster, who combines it with information backing the general weather forecast to provide the maintenance body with a specific ice-prediction forecast for its road network. These forecasts are in real time so that the authority can respond, at very short notice, if conditions change; such changes occur frequently in those parts of the UK where winter weather conditions are marginal or close to freezing.

Road authorities may also employ inspectors for local monitoring in particular locations, where unusual conditions might be expected to develop. Monitoring is also undertaken by the police as part of their patrolling roles.

17.10.3 Winter maintenance operations

Winter maintenance operations in much of England and Wales usually emphasizes the prevention of ice formation; however, in the north of England, in Scotland, and over higher ground in Wales, more emphasis is placed on the removal of snow. Whilst the public often considers these operations as 'de-icing' it is more correct to describe them in terms of 'anti-icing' or the prevention of ice formation. Prevention normally requires only about 25 per cent of the mass of salt (the usual anti-icing chemical) required to remove ice that has already formed. Also, a road that is already iced is very slippery road whilst it thaws, which is a slow process.

To prevent ice from forming, salt is normally spread on the road at concentrations of about $10\,g/m^2$; this spread rate is usually enough to cope with normal winter temperatures, but it is less effective at lower temperatures. *Salt* (*sodium chloride*) is normally spread in dry form in Great Britain; a wetted salt – where wetting usually takes place at the point of discharge of the salt to the spreader – tends to be preferred in mainland Europe. Wetted salt is more efficient in that it goes into solution more quickly (which is the essential state for the salt to act); this is especially important at times of low humidity when dry salt may not go into solution.

Whilst salt is a very economic anti-icing material, there are significant disadvantages associated with its use, e.g. it is corrosive to steel in vehicles and structures, and it is harmful to roadside vegetation. Much work has been carried out to develop alternatives, some of which have their own disadvantages and others are prohibitively expensive. For example, *urea*, which is used on some roads with major structures where the corrosivity of salt would be unacceptable, can be harmful to aquatic life and is less efficient as an anti-icing agent. *Calcium magnesium acetate*, which has few side effects, is as effective as salt but is about 30 times more expensive.

As noted above, salt is harmful to vegetation, and developments in recent years have therefore placed emphasis on better control of the spreading process. Nonetheless, the actions of traffic cause spray to reach roadside vegetation, with consequent damage to trees and shrubs; e.g. see reference 19 in which are highlighted the problems caused to tree roots as a result of manual salt-spreading practices on footways in urban areas. Wetted salt has a reduced detrimental effect in that there is less likelihood of the spread salt being blown off the carriageway.

Because of the problems associated with the use of salt, it can be expected that research will continue until a more-generally acceptable, efficient, economic, anti-icing alternative is found.

17.11 External influences on maintenance operations

Maintenance operations are carried out within an environment which affects, and is affected, by a large proportion of the population. It is important, therefore, to identify these impacts throughout the preparation, planning and programming, design and works stages of a road maintenance programme.

Roads are located immediately adjacent to land owned by others, and this places a duty on the maintenance engineer to take full and proper account of the needs of neighbouring owners during the *preparation stage*. For example, some maintenance operations require access across neighbouring land for which formal easements or wayleaves are required. These need to be legally binding in perpetuity so that change of ownership, etc. will not result in the highway authority being presented with insurmountable problems when necessary maintenance is initiated.

Maintenance planning, as used here, refers to the implications for maintenance programming of the statutory planning processes controlled by the local authority, which may or may not be the same body as the road authority. The local authority administers planning control over a variety of activities, including noise levels, anti-social working hours, disposal of materials, impacts on the local area, as well as various other statutory considerations. Because of the need to consider their impacts thoroughly, many of these considerations can take time to analyse and complete and must therefore be programmed into the maintenance schedule.

Central to the proper programming of maintenance works is the impact that the works will have on both the travelling public and the public at large. On busy road networks such as those in the UK, maintenance works are normally planned and programmed to minimize the effects of congestion and delay to the travelling public, and this planning principle is applied irrespective of whether the maintenance involves minor routine works or a major pavement reconstruction contract. Most works can normally be timed with a degree of flexibility which takes account of major events taking place in locales which might be impacted upon by maintenance operations, e.g. it would have been invidious to carry out major maintenance works on, say, the North Circular Road near Wembley on the days when FA Cup Finals were being staged.

In practice, it is necessary to programme maintenance works precisely only when special issues arise, e.g. when safety issues are involved.

Routine works include such mundane tasks as road sweeping, gully emptying, grass cutting, replacement of road markings, and the sealing of cracks in the pavement surface. These typically involve short-term activities which last only a few hours or, at most, a few days, and seldom require a long-term carriageway or lane closure. Timing is crucial to reducing the impact of these works, and should be selected to avoid periods when the reduced road capacity due to, for example, a lane closure is less than the traffic demand when the works starts and ends.

Other procedures used to reduce the impact of routine maintenance works on busy roads involve timing them to be carried out at night or on weekends, when traffic flows are low. Re-timing works to combine them with other proposed road-works is another useful procedure.

Even when *maintenance works* have been designed and are taking place, they cannot take priority over the normal requirements for safe and efficient road operation. Unforeseeable events, e.g. traffic accidents and severe weather conditions on other parts of the road network, may well require changes to the programmed activities within the maintenance contract,

It might be noted here that the simple traffic cone is one of the most useful and effective pieces of safety equipment available to use in maintenance operations; its use is normally essential to the security of both the road user and the maintenance worker. However, the cones and related signs required for safe lane-closures take time to set out before work can begin and to recover after the works end. In order to minimize this 'dead time', therefore, a system of *mobile lane closures* has been developed for use on some projects; these involve the use of a number of lorries carrying easily-legible prescribed signs behind them, which move in a controlled manner behind slow-moving maintenance works and provide directions to following vehicles and protection for the worker.

Major maintenance works inevitably take a long time and very often they require a long-term (termed 'permanent') lane or carriageway closure to provide a road environment which is reasonably safe, e.g. to guard an excavation or other obstruction which constitutes an unacceptable hazard to the road user. As these closures are long-term, their maintenance contracts are normally planned so that the following conditions are met.

1. *The number of closures is kept to a minimum.*   This can be achieved by combining works as far as it is economical and sensible to do. If, for example, roadworks and bridgeworks are planned for the same section of road but several months apart, it would normally be indefensible to award two contracts with the associated disruption; preferably the two projects would be combined into one contract and, ideally, carried out simultaneously.
2. *The traffic delays during a closure are minimized.*   This can be achieved by careful assessment of the delays expected from various traffic management options, using tools that are available to assist with these assessments, e.g. the computer program QUADRO (*QU*eues *A*nd *D*elays at *RO*adworks)[20]. These assessments include the number of lanes to be closed, the 'tidality' of the traffic flows (i.e. the different flows on a stretch of road during, for example, the morning and evening peak flows), the variations in hourly flow and the practicability of diverting some or all traffic onto other roads.
3. *The contract durations are kept to a minimum.*   This can be done by planning the works to ensure that the operations, techniques and materials required are provided on time to assist the speedy, satisfactory, completion of the works. It is also beneficial to offer the contractor the opportunity to propose innovative alternative methods to meet the contract requirements in shorter time. In this latter context, the works duration can often be kept to reasonable minimum through the use of incentive contracts. For example, *Lane Rental contracts*, which has been in use in the UK since the mid-1980s, are designed to offer a bonus to contractors who complete their contracts in less than the agreed time, but require them to pay a charge for late completion. The bonus and the charge are at the same level, and determined in relation to the road user cost savings associated with the shorter contract period.

As noted previously, statutory undertakers and private contractors have legitimate claims to carry out roadworks. The road authority has the power, and the responsibility, to ensure that these other activities are planned and controlled so that they are carried out safely and with the minimum of disruption to road users.

17.12 Future directions of road maintenance

There is no doubt but that the external influences which affect road operators and authorities will continue to play a major part in shaping the development of road maintenance. Indeed, it is probable that the most successful road maintenance will be that which is noticed least in terms of its impacts on the road user, the environment, and those living and working close to the carriageway. Some general indicators of future challenges for the road maintenance engineer are given below.

Good quality information about road conditions is an essential prerequisite for sound decision-making about the need for road maintenance, and type of treatment that is subsequently applied. This information-collection requirement is being increasingly met through the introduction of new technologies, and it can be expected that this will continue as the ability to handle large amounts of data also increases. Technologies already in wide use include laser systems, ground-penetrating radar, image-recognition systems, and electronics. Increased interest is likely to be paid to the concept of *smart roads*, which use sensors within road components to feed real-time information regarding their condition and performance.

A greater emphasis is also likely to be placed on improving *communication with road users* to provide them with up-to-date and reliable advice about the state of the road, including weather, traffic, safety and other conditions. Current technologies provide information to the motorist through electronic signing systems, the media, and in-car guidance systems. Car technology has already developed a virtually driverless car, and capabilities of this type could eventually form the basis of better highway maintenance control if the demand for limited roadspace grows excessively.

These and other material and technical developments will be introduced for the benefit of all road users, and not just vehicle users. The very wide use of roads, and their impact upon the locales which they serve, offer great challenges to road maintenance engineers to ensure that the assets for whose upkeep they are responsible are maintained for the benefit and convenience of all.

17.13 References

1. *Highways Act*. London: HMSO, 1980.
2. *Health and Safety at Work etc. Act*. London: HMSO, 1974.
3. *New Roads and Street Works Act*. London: HMSO, 1991.
4. Confederation of British Industry, *Trade Routes to the Future*. Hurstpierpoint, Sussex: Gwynne Printers Ltd, 1989.
5. Local Authority Associations, *Highway Maintenance – A Code of Good Practice*. London: Association of County Councils, 1989.
6. Department of Transport, *Trunk Road Maintenance Manual, Vol. 2 – Routine and Winter Maintenance Code*. London: Highways Agency, 1992.

7. Department of Transport; *Design Manual for Roads and Bridges – Vol. 7. Pavement Design and Maintenance*. London: HMSO, 1994.

8. Powell, W.D., Potter, J., Mayhew, H., and Nunn, M., *The Structural Design of Bituminous Roads*, TRRL Report LR1132. Crowthorne, Berkshire: The Transport and Road Research Laboratory, 1984.

9. Cooper, D.R.C., *The TRRL High-speed Road Monitor: Assessing the Serviceability of Roads, Bridges, and Airfields*, TRRL Research Report 11, Crowthorne, Berkshire: The Transport and Road Research Laboratory, 1985.

10. Hosking, J.R. and Tubey, K.W., *Measurements of Skidding Resistance: Pt. 5, The Precision of SCRIM Measurements*, TRRL Report SR642. Crowthorne, Berkshire: The Transport Research Laboratory, 1981.

11. *CHART5: Illustrated Site Manual for Inspectors*. London: Department of Transport, 1986.

12. Mildenhall, H.S. and Northcott, G.D.S., *A Manual for the Maintenance and Repair of Concrete Roads*. London: Department of Transport and The Cement and Concrete Association, 1986.

13. BSI 594:1992, *Hot Rolled Asphalt for Roads and Other Paved Areas; Pt. 1; Specification for Constituent Materials and Asphalt Mixtures*, and *Pt. 2; Specification for the Transport, Laying and Compaction of Rolled Asphalt*. London: British Standards Institution, 1992.

14. Pynn, J., *Pavement Strengthening at Minimum Cost*, TRL Report RR73. Crowthorne, Berkshire: The Transport Research Laboratory, 1987.

15. Kennedy, C.K. and Lister, N.W., *Prediction of Pavement Performance and the Design of Overlays*, TRRL Report LR833. Crowthorne, Berkshire: The Transport Research Laboratory, 1978.

16. *International Road Maintenance Handbook: Vol. 2, Maintenance of Unpaved Roads*. Crowthorne, Berkshire: The Transport Research Laboratory, 1994. (Revised English edition.)

17. *Construction, Design and Management Regulations*. London: HMSO, 1996.

18. *Environmental Protection Act 1990*. London: HMSO, 1990.

19. Dobson, M.C., *De-icing Salt Damage to Trees and Shrubs*, Forestry, Commission Bulletin 101. London: HMSO, 1991.

20. Department of Transport, *QUADRO – QUeues And Delays at ROadworks*. London: HMSO, 1991.

CHAPTER 18

Wet skid resistance

A.R. Woodside and W.D.H. Woodward

18.1 Introduction

This chapter considers the wet skid resistance of bituminous highway surfacing materials. It first asks the question what is wet skid resistance and why should highway engineers consider this. It then considers the measurement of wet prediction. The measurement of wet skid resistance on the road is followed by considering the methods used to predict this value in the laboratory. Current UK specification limits are outlined. The chapter concludes with details of research to improve its prediction and measurement, including reappraisal of the PSV test method and identification of the dynamic stresses that occur within the contact patch as a rolling tyre passes over a road surface.

18.2 What is wet skid resistance?

To the typical road user, the idea of skid resistance is, normally, will the vehicle stop if urgently required to do so? Can an accident be avoided or will the vehicle skid uncontrollably into another oncoming vehicle? Road users also know that if they try to brake quickly on a wet surface they may aquaplane. These basic concepts must be considered when designing a highway surface for skid resistance, i.e. can the likelihood of a serious life-threatening skidding-related accident be reduced, particularly when the surface is wet?

The term *skidding resistance* refers to the extent to which the road contributes to friction and typically refers to wet conditions. It is measured and predicted in a number of different ways. These all involve sliding some form of rubber over a wet surface and measuring the forces developed under a known load. Frictional resistance is used to describe these forces that resist motion when the two surfaces are in contact. A *coefficient of friction* is defined as the force resisting motion divided by the vertical load. For a road surface, the force resisting motion is developed between a vehicle tyre and the road surface[1].

Therefore, for any given situation, the amount of friction depends on a complex interaction of factors including the properties of the tyre, the properties of the road surface material, and the presence of contaminants or lubricants, e.g. oil or water, and the ability of either the tyre or road surface to remove them or reduce their effect. Consider the example of a drunk 18-year-old male driver in a sports car with bald tyres, driving on a smooth road in heavy rain – statistically, this is probably the worst possible combination and the greatest threat to life.

This chapter concentrates primarily on the bituminous surfacing conditions and their effect on the skidding resistance of vehicles, i.e. those conditions that can be controlled by the road engineer through proper selection of raw materials and choice of surfacing bituminous mixture.

The reason for the use of the title *Wet skid resistance* is that most types of bituminous surfacing can provide an adequate level of skid resistance in the dry. Compare the phenomenal road-holding performance of a Formula 1 racing car with smooth tyres on a smooth racetrack during dry weather conditions, with its performance on the same surface after a sudden rain shower; in the latter case the action of the smooth tyre passing over the smooth surface will cause a lubricating layer of water to form between the tyre and road surface leading to increased risk of aquaplaning, poor road holding, increased braking distances, and a probable eventual crash.

The presence of water which acts as a lubricant between the road surface and tyre, and how this may be controlled, is therefore the key aspect that must be considered in relation to understanding wet skid resistance. To avoid the conditions that increase the risk of a wet skidding accident occurring, there must be some underlying process whereby the film of water is reduced to the extent that it is broken, so that the rubber of the tyre comes in contact with the aggregate used to construct the road surfacing. This concept is currently the topic of interest in a number of countries around the world. It may be described as the *total effective texture* encompassing all of the main controllable factors involved in wet skid resistance, i.e. the tyre, road surface and aggregate. As the term implies, this considers the different degrees of texture involved in the removal or reduction of lubricating water films. These are shown in Fig. 18.1 and consist of: (1) a highway surface profile and geometry, particularly crossfalls, that enable the removal of surface water and which are resistant to permanent deformations that may lead to standing water in the wheel paths; (2) an adequate tread on the vehicle tyre to remove excess water, particularly at higher speeds and where the highway surface has poor macrotexture; (3) an adequate tread or texture on the road surface to remove excess water – known as texture depth, this is essentially the same as tread depth on a vehicle tyre; and (4) the use of aggregate with a microscopic surface roughness or micro-texture that will break through the water film that develops between the road surface material and the tyre tread.

18.3 Why should the highway engineer consider wet skid resistance?

When designing, constructing or maintaining a road, the engineer must ensure that the resulting structure is safe, economic and durable. This is a difficult balance of three sometimes-conflicting factors. However, there is no point in providing a cheap surfacing that lasts for ever if it results in unacceptable outcomes for its users. In terms of human suffering, there were some 150 road deaths and 13 294 people injured in road traffic accidents in Northern Ireland during 1998/99; in terms of financial community cost, these causalities cost an estimated £452 000 000.

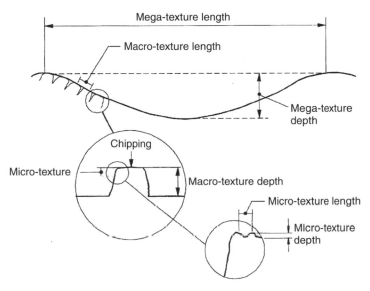

Fig. 18.1 Total effective texture – explanation of micro, macro and megatextures and lengths

Such figures clearly indicate the need for a safe road network brought about by continuous improvement through safety performance programmes.

The phenomenon of wet skid resistance has been studied in the UK for almost 100 years. The results of these studies have been summarized[2] and show the development of highway surfacing and aggregate skid resistance standards. Wet skid resistance data, road surface texture, and accidents have been correlated[3] to clearly show the close relationship between these factors.

These studies have all had the intention of preventing or reducing serious accidents by identifying those sections or locations within the highway network that have low or deteriorating levels of wet skid resistance. The data collected have formed the basis of pavement management programmes that include maintenance or rehabilitation activities, evaluation of material types, and new construction practices.

A summary of the current types of surfacing option is given in reference 4. Table 18.1 shows the current UK requirements for newly constructed wearing courses. Note that the different locations within the road network are given site categories and site definitions, and that these are further categorized based on commercial vehicles per lane per day at design life. Depending on site and traffic conditions a minimum polished stone value (PSV) is specified for the aggregate used. Finally, as the wearing course loses wet skid resistance when trafficked, investigation limits are given whenever they fall to potentially unacceptable values.

These are areas where the road engineer and researcher has input, i.e. understanding the risk of an accident occurring in relation to speed and volume of traffic, turning and stopping movements, and road geometries. Accident statistics and road surface measurements have clearly shown that accident risk increases with decreasing skid resistance, with the predominant cause of traffic accidents being wet weather skidding, poor texture, and high speed[3].

Table 18.1 Minimum PSV of chippings or coarse aggregate in unchipped surfaces for new wearing courses (HD36/99)[4]

IL Band	Default IL	Site Categories	Site Definitions	Traffic (cv/lane/day) at design life									
				0–250	251–500	501–750	751–1000	1001–2000	2001–3000	3001–4000	4001–5000	5001–6000	Over 6000
I	0.35	A, B	Motorway (mainline), dual carriageways (non-event)	50	50	50	50	50	55	60	60	65	65
Ia	0.35	A1	Motorway mainline, 300 m approaches to off-slip roads	50	50	50	55	55	60	60	65	65	65
II	0.40	C, D	Single carriageways (non-event), dual carriageways approaches to minor junctions	50	50	50	55	60	65	65	65	65	68+
III	0.45	E, F, G1, H1	Single carriageways minor junctions, approaches to and across major junctions, gradients 5–10% >50 m (dual, downhill only), bends <250 m radius >40 mph	55	60	60	65	65	68+	68+	68+	68+	70+
IV	0.50	G2	Gradients >50 m long >10%	60	68+	68+	70+	70+	70+	70+	70+	70+	70+
V	0.55	J, K	Approaches to roundabouts, traffic signals, pedestrian crossings, railway level crossings and similar	68+	68+	68+	70+	70+	70+	70+	70+	70+	70+
VI	0.55 (20 km/h)	L	Roundabouts	50–/70+	55–/70+	60–/70+	60–/70+	60–/70+	65–/70+	65–/70+			
VII	0.60 (20 km/h)	H2	Bends <100 m	55–/70+	60–/70+	60–/70+	65–/70+	65–/70+	65–/70+	65–/70+			

Notes:
1 Where '68+' material is listed in this table, none of the three most recent results from consecutive tests relating to the aggregate to be supplied shall fall below 68.
2 Throughout this table '70+' means that specialised high-skidding resistance surfacings complying with MCHW1 Clause 924 will be required.
3 For site categories L and H2, a range is given and the PSV should be chosen on the basis of local experience of material performance. In the absence of other information, the highest values should be used.
4 Investigatory Level (IL) is defined in Chapter 3 of HD 28 (DMRB 7.3.1).

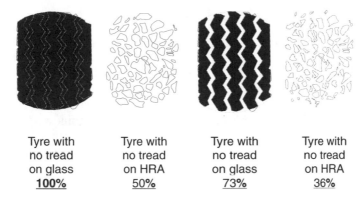

Tyre with no tread on glass	Tyre with no tread on HRA	Tyre with no tread on glass	Tyre with no tread on HRA
100%	**50%**	**73%**	**36%**

Fig. 18.2 Illustration of tyre/aggregate contact within the contact patch for a hot rolled asphalt surface

18.4 Development of methods to measure wet skid resistance

Early interest in measuring skid resistance concentrated on methods to measure road surfaces and, in the UK, led to the development of the Sideway-Force Coefficient Routine Investigation Machine (SCRIM) (see Subsection 18.7.1). This was introduced in the early 1970s to measure wet skid resistance of the road network. The Mean Summer SCRIM Coefficient (MSSC) data obtained with this machine are used as the basis for the national skid resistance specification limits shown in Table 18.1. Below a certain limit of MSSC the risk of a skidding-related accident for a given location is such as to warrant resurfacing, to renew skid resistance to a higher level.

Due to the requirement for a test method to assess aggregate in the laboratory prior to use, research[5] in the early 1950s resulted in a machine being designed that could assess the polishing characteristics of different aggregates. This original method, which has changed relatively little since then and is now known as the Polished Stone Value (PSV) test method, has been accepted as the Euro Norm to measure the property of aggregate wet skid resistance[6].

Despite a considerable background of research and specification implementation, there is now renewed interest around the world in the differing test methods involved and their ability to measure and/or predict wet skid resistance. It has been found that some aggregates and types of surfacing perform better than expected, whilst others appear to polish more and thereby give a lower in-service skid resistance. There have also been important changes in relation to the types of bituminous mixtures being laid as surfacing layers.

Criteria such as noise characteristics, negative texture, spray generation, layer thickness, and the availability and cost of limited sources of high PSV aggregates, have resulted in a shift towards thinner, smoother and quieter surfacings. It should also be recognized that surfacing aggregate is now used in ways that were never considered in the historical development of the skid resistance standards and specifications that were originally developed in the UK.

18.5 What is a wet road surface?

Technically, a road surface becomes wet when precipitation intensity exceeds 0.1 mm/h. This precipitation creates a water film that varies in thickness from a few microns to a few millimetres. The film formed acts as a lubricant between tyre and the surface which decreases skid resistance. This state of surface wetness is a major factor when considering the degree of skid resistance possessed by a surface.

The presence of a very thin film will have limited effect on low-speed skid resistance, as the tyre has time to penetrate the film and make contact with the aggregate microtexture. However, as the film thickness increases, the tyre–surface contact time affects the ability of excess water to be expelled; this is particularly important as vehicle speeds increase and contact times decrease. This illustrates the complex interaction between tyre (tread depth and pattern, contact patch area), surface texture (type of surfacing mixture), aggregate microtexture (rock type, degree of polish), speed (contact time) and environmental conditions (water film thickness). It has been shown by analysis of wet weather accidents that these accidents occur where wet skid resistance is low, and that they may be reduced with resurfacing maintenance.

18.6 The mechanics of wet skidding

The mechanics of skidding are related to energy losses. If one considers a car skidding on a wet road surface, due to its weight and speed it will possess a considerable amount of momentum and will only stop once that energy is dissipated, i.e. the car's momentum energy must be transferred to the road surface through the interaction of tyre and surface. The surface water that is present acts as a lubricant. So, in terms of energy losses, there are two main components: (a) a friction component between the tyre and the road surface, causing energy to be dissipated as heat; and (b) a hysteresis component that relates to a tyre's ability to deform its shape around the aggregate particles in the road surface, and so cause loss of energy.

In terms of engineering controls there are further components which are of importance:

- the tread on the vehicle tyre to remove excess water;
- the tread or texture on the road surface to remove excess water;
- the use of aggregate with a microscopic texture to break through the water film.

These different factors or degrees of texture, which are illustrated in Fig. 18.1, have been termed collectively as *total effective texture*. (Note: the tyre tread is not shown in Fig 18.1.)

18.6.1 Tread on the vehicle tyre to remove excess water

One of the most effective mechanisms for removing water from between the tyre and surface is the tread on the vehicle tyre. This is typically ensured by a national requirement for a legal minimum tyre tread, e.g. a minimum tread depth of 1.6 mm is specified in the UK, with 4 mm in Germany. Another important vehicle

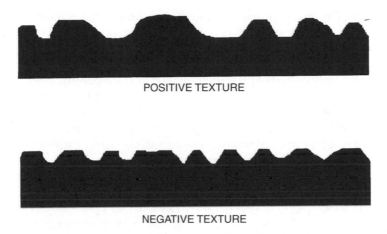

POSITIVE TEXTURE

NEGATIVE TEXTURE

Fig. 18.3 Positive and negative texture at the road surface

development is the anti-lock ABS braking system, which is now fitted to most modern cars throughout the world.

Recent research (see Fig. 18.2) has indicated that the actual amount of tread rubber in contact with a hot rolled asphalt surfacing may be as little as 36 per cent of the total contact patch area, as compared with 53 per cent for a stone mastic asphalt surfacing. It was found that the value obtained is dependent upon the type of surface and its texture, tyre tread pattern, tyre pressure, and the tyre loading. Initial findings would indicate a possible dilemma, where lower tyre/surface contact areas are expected to provide greater removal of water or superior wet skid resistance.

18.6.2 Tread or texture on the road surface to remove excess water

The realisation that a road surface should possess a degree of texture has not been as widely accepted around the world as tyre tread. However, in the UK there is a minimum national texture requirement for newly-laid surfacings; a new surface must have a texture depth of at least 1.5 mm prior to trafficking. The purpose of this texture is similar to tyre tread, i.e. to dispose of excess water during wet conditions and to reduce the amount of lubricant between tyre and road surface.

This requirement for a certain value of highway surface texture (commonly referred to as *texture depth*), or lack of it, has had a profound influence on the development and use of bituminous materials around the world. For example, in the UK the need to construct a road surface with texture has, historically, favoured the use of materials such as surface dressing or hot rolled asphalt with 20 mm chips applied to its surface to ensure that the aggregate particles protrude above the bituminous matrix. In other countries, e.g. the USA, there has traditionally been no requirement for this property; rather, smooth surfaces of asphaltic concrete have been the norm which, when wet, offer very poor levels of skid resistance. However, a major change in road surfacing requirements is now occurring around the world. For example, the issue of safety is one of the fundamental ideals of European harmonization. This has resulted in many European countries having to recognize that their road surfaces must provide a minimum level of wet skid resistance, resulting in a change from the use of a smooth surface to one which possesses a high degree of texture and, thus, an improved ability to remove water. The technical term used to describe this property is *macrotexture* (i.e. texture depth); it is formed by the general shape of, and the spaces between, the particles or grooves in a road surfacing.

Whilst a range of terms are used to describe surface texture (e.g. see reference 1), probably the main ones are positive and negative texture (see Fig. 18.3), and these may be used to explain most of the types of surfacing material now in use: (a) those with a positive texture, i.e. where the coarse aggregate content protrudes above the plane of the surface, as with a surface dressing or 20 mm chippings applied to the surface of a hot rolled asphalt; (b) those with a negative texture, i.e. where the texture largely comprises voids between particles whose upper surfaces form a generally flat plane, e.g. a stone mastic asphalt or a thin propriety surfacing such as ULM; and (c) those with a porous texture, i.e. where a coarse grading provides a high void content mixture such as a porous asphalt. (With porous asphalt the water can penetrate into, and flow through, the surfacing mixture, thus providing a means of removing the water.)

The engineer can therefore decide upon the most suitable means of water removal. However, there are other important issues that need to be considered that include rolling resistance, fuel efficiency, generation of road noise, use of aggregate resources, etc. For example, a high texture surface dressing may have very good wet skid resistance but its greater rolling resistance will result in greater tyre wear, road noise and fuel inefficiency. In terms of a heavily-trafficked urban environment this type of texture, although safe, may not be suitable.

The principal method of measuring macrotexture or depth involves obtaining an average texture depth. There are two main methods:

- The *sand patch (SP)* method (reference 7) – this is a simple volumetric method (Fig. 18.4) which uses sand to physically fill the hollows in the road surface up to the peaks with a known volume of sand. The surface area covered is then measured and the average texture depth is calculated.
- The *sensor-measured texture depth (SMTD)* method – this uses a 'texture meter' laser apparatus to make a sequence of displacement measurements along the line of the surface profile, and then the root-mean-square of the texture is calculated to determine the average texture depth.

18.6.3 The use of aggregate with a microscopic texture to break through the water film

The microscopic texture found on the surfaces of aggregate particles in the road surfacing is known as *microtexture*, and relates to the aggregate's geological properties and its condition or state when used. For example, rounded particles of uncrushed river gravel will normally offer poor skid resistance due to their smooth surface; however, if the same gravel is crushed and the fresh surfaces exposed to the traffic, then its skid resistance will be much greater due to the rougher microtexture of the crushed surfaces.

(i) Known volume of fine sand of uniform particle size poured on road

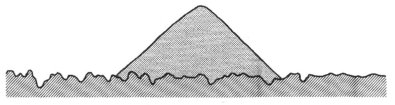

(ii) Sand spread to form circular patch with 'valleys' filled to level of 'peaks'

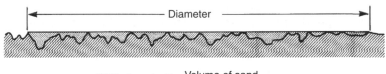

(iii) Texture depth = $\dfrac{\text{Volume of sand}}{\text{Area of patch}}$

Fig. 18.4 Sand patch method for measuring texture depth[7]

Figure 18.5 shows how skid resistance values measured by means of the British Pendulum Skid Tester (BPST) can produce a range of pendulum values (PVs) for an aggregate. In a freshly crushed dry condition, almost all types of aggregate have a similar level of skid resistance, as is shown by the dry unpolished pendulum values (DUPV). However, this will reduce significantly, as is shown by the wet unpolished pendulum value (WUPV) data, when the surfaces are wet and a lubricating water film is allowed to form between the rubber and aggregate surface. Then, as each aggregate is subjected to simulated trafficking in the Polished Stone

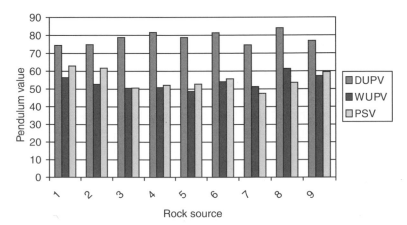

Fig. 18.5 Comparison of skid resistance for nine different aggregates, under dry unpolished (DUPV), wet unpolished (WUPV), and polished (PSV) conditions

Value (PSV) test, the results change yet again as each aggregate reacts to the test conditions and different aggregates react differently.

The considerable lowering of skid resistance values, due to the presence of water and the polishing action of the PSV test, clearly indicates the importance of considering wet skid resistance, particularly when a surfacing is to be subjected to heavy trafficking and/or high stresses during braking or turning movements.

The values obtained for wet skid resistance, particularly the polished wet skid resistance as measured by the PSV test, are strongly linked to rock type. Research has shown that not all types of rock make a good surfacing aggregate. Some may be too weak, with very low aggregate crushing values and, thus, prone to break-up due to construction or in-service trafficking and environmental conditions. Of those that are suitable, a number of different groupings are possible, based on aggregate surface texture, strength and geological properties, e.g. the hard fine-grained rocks such as quartzite, the hard coarse-grained rocks such as granite, and the softer fine and medium-coarse arenacous rocks such as sandstone and greywacke. Figure 18.6 shows the distribution of PSV data with respect to rock type for aggregate sources available in the UK.

For the naturally-occurring aggregates shown in Fig. 18.6, it can be seen that the gritstone or arenaceous types of rock provide the better values of wet skid resistance. This is because they are composed of grains held together by a finer-grained softer-bonding medium which produces a sandpaper type of microtexture. Under trafficking, the grains are plucked from the binding medium to expose a new grain beneath; this mechanism results in a renewable microtexture and, hence, a greater value of wet skid resistance. However, the advantage of this renewable texture must be balanced against other essential properties, particularly resistance to abrasion (AAV), i.e. the aggregate must not wear away too quickly and neither must it

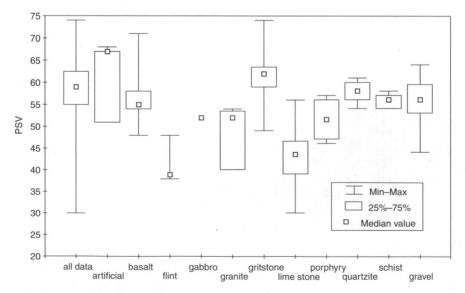

Fig. 18.6 Distribution of PSV data with respect to rock type[8]

remain intact to become polished. This aspect of polishing is the basic reason why most types of hard aggregates – defined as those with a very low 'aggregate crushing value' or a very high 'ten per cent fines value' – have poorer levels of skid resistance, irrespective of whether they are fine-, medium- or coarse-grained. Note that diamond is an extreme example of a very hard aggregate which has a very high degree of polish. Geologically, all of the component minerals are of similar hardness and so there is no differential wear leading to the renewal of the microtexture.

It must be stressed that skid resistance is only one of many requisite properties of a good surfacing aggregate. Research has shown that for most types of aggregate, increasing values of skid resistance are usually achieved at the expense of almost every other property, e.g. strength, abrasion and soundness. It is therefore recommended that high values of PSV should always be considered carefully, in case premature in-service failure should occur. This is particularly important when dealing with new sources or new types of aggregate for which historical in-service data do not exist.

18.7 The measurement and prediction of wet skid resistance

In order to provide a safer road it is necessary for the highway engineer to be capable of measuring and predicting the wet skid resistance. A range of techniques are available in order to do this, i.e. methods that either (i) measure the wet skid resistance of the in-situ road surface, or (ii) predict wet skid resistance in the laboratory.

18.7.1 Measurement of the wet skid resistance of the highway surface

Measurement of the highway surface in relation to its wet skid resistance may be assessed using different methods ranging from the low-cost and simple to the expensive and technical. Selection of the most appropriate method depends upon the reason for the assessment, e.g. whether it is to assess a local 'black spot' with a high incidence of accidents or to form part of a national pavement maintenance programme.

The British Portable Skid Tester (BPST) was developed[9] in the 1950s to measure the skid resistance of road surfaces. It works on the same principle as a person sliding a foot along a surface to assess its slipperiness. The apparatus (Fig. 18.7) involves a pendulum supported on a stand, with a spring-loaded rubber slider projecting from the underside of the pendulum bob. When in use the stand is adjusted and levelled so that as the pendulum swings the rubber slider makes contact with the surface for a distance of 125 mm. The final height reached by the pendulum is indicated on a scale calibrated in skid-resistance values (SRVs) that are related to the energy lost to friction during the slide.

The BPST is very useful in that it is portable, easily operated, and may be used both to measure in-situ road surfaces and for the laboratory determination of PSV. Testing is always carried out on a wetted surface. The equipment can be used to determine wet skid resistance up to equivalent vehicle speeds of 50 km/h; the relevance of the results may be questioned at speeds in excess of this, due to the effect of texture depth on the retardation of vehicles.

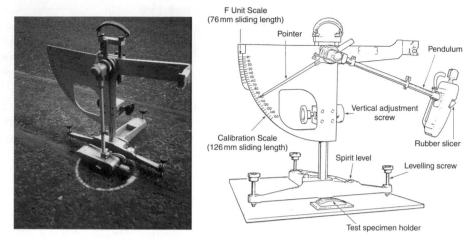

Fig. 18.7 The British Portable Skid Tester (BPST)

The principal piece of equipment used in the UK to measure the in-service wet skid resistance of road surfaces is the Sideway-force Coefficient Routine Investigation Machine (SCRIM). The SCRIM apparatus evolved from motorcycle-based testing machines in the 1930s when research discovered that the force exerted on a wheel that is angled to the direction of travel, and held in a vertical plane with the tyre in contact with the road surface, could be correlated with the wet skid resistance of that surface. The sideways force coefficient (Sfc) determined in this way was defined as the force at right angles to the plane of the inclined wheel, expressed as a fraction of the vertical force acting on the wheel.

The SCRIM apparatus (Fig. 18.8) is capable of giving a continuous record of Sfc in the wheel path at speeds up to 100 km/h. It basically consists of a lorry with a water tank and an inclined solid rubber test tyre mounted on an inside wheel-track.

Fig. 18.8 The SCRIM-measuring wheel assembly

Fig. 18.9 The GripTester

(Note: It is possible to have two test wheels on the same vehicle, one in each of the wheel-tracks.) Water is sprayed in front of the tyre to give a film thickness of constant depth. The investigatory limits for different site categories are shown in Table 18.1.

The *GripTester*, shown in Fig. 18.9, has been available since 1987 and is now in use in more than thirty countries around the world. It is a highly versatile surface friction tester comprising a three-wheel trailer of 85 kg weight that is towed behind a small van. A constant film of water is sprayed in front of the tyre, depending on the test speed. A simple transmission system brakes a measuring wheel by 15 per cent giving a continuous measurement of both drag and load. The friction coefficient, known as the GripNumber, is recorded and may be easily analysed for inspection using a laptop computer.

The GripTester equipment is easy to handle, operate, calibrate and maintain. It may be towed at any speed up to 130 km/h, and may also be operated manually in order to measure pedestrian areas. Whilst SCRIM has a long established history in the UK, its high running costs and limited availability make difficult its widespread usage. As a result, the use of GripTester is now becoming more popular in the UK and world-wide. It successfully took part in the 1992 PIARC project for the international harmonization of road friction measurement[10]. The Transport Research Laboratory (TRL) has also established a strong correlation between the results obtained from the GripTester and from SCRIM.

The main problem with the SCRIM equipment is that although it can assess skid resistance at speeds up to 100 km/h, the resulting slip speed between the test tyre and the road is relatively slow because it uses an angled wheel. At higher speeds, e.g. up to 160 km/h, a *Braking-force Trailer*[1] is typically used; this equipment comprises a towing vehicle and a purpose-built trailer. The trailer has two wheels, one of which is fitted out as the test wheel. The test wheel axle is fitted with a two-axis force transducer that measures the vertical force (load) and horizontal force (drag) on the wheel. Shaft encoders allow the speed of both wheels to be measured. During the test, water is pumped onto the surface of the road in front of the test wheel, with the pump rate increasing with speed to maintain a nominal water depth of 0.5 or 1.0 mm. The test uses a locked-wheel principle to measure friction values.

Fig. 18.10 Test specimens used for the GRAP, Flat Bed, and PSV test methods

18.7.2 Predicting wet skid resistance in the laboratory

The laboratory prediction of wet skid resistance is vital to avoiding the use of aggregates that easily polish and become dangerously slippery due to trafficking. There have been a number of laboratory prediction methods developed, of which the most popular throughout the world is the British PSV test. Other methods include the French/Canadian GRAP test and the British Flat Bed Test. The test specimens used for the three methods are shown in Fig. 18.10; note that the large flat mould is for the GRAP test, the small flat mould is for the Flat Bed Test, and the small slightly-curved mould is for the PSV test.

The *Polished Stone Value (PSV)* test[6], which was first introduced as a British Standard in 1960, is the main laboratory method used throughout the world to measure an aggregate's resistance to skidding. Apart from some minor modifications, the original method as devised in 1952 is still in use today, and has been accepted as the CEN European Norm for measuring the polishing characteristics of an aggregate. The PSV test is carried out in two parts. With the first part, slightly curved test specimens of cubic-shaped 10 mm sized aggregate chippings (Fig. 18.10) are subjected to a wet polishing action using the accelerated polishing machine shown in Fig. 18.11. This part of the test lasts for 6 h, during which time each specimen is subjected to 115 200 passes of a solid tyre under a force of 725 N. Coarse emery abrasive is used as the polishing medium for the first three hours, followed by fine emery flour for the remaining 3 hours. With the second part of the test, the state of polish reached by each specimen is measured using the (BPST) pendulum friction tester shown in Fig. 18.7. The result is expressed as a laboratory-determined polished stone value (PSV), the higher the value obtained (between 30–80) denoting an aggregate with better resistance to polishing.

The *GRAP test* was developed in France from an idea that had originated in Canada as an alternative to the PSV test. With the GRAP test, flat test specimens

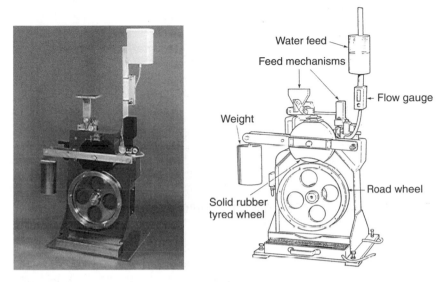

Fig. 18.11 The accelerated polishing machine

(as shown in Fig. 18.10) are prepared and then polished using a jet of water sprayed onto the exposed aggregate surface; the degree of wet skid resistance is then measured using the BPST. Comparison of the results obtained for the GRAP and PSV tests on identical aggregates showed poor correlation between the two methods (Fig. 18.12).

The British *Flat Bed test* was developed as a means of measuring the skid resistance of concrete and clay paving bricks that are primarily used in lightly trafficked and pedestrian areas. The method adapts the Dorry abrasion apparatus which was essentially developed to measure the British Standard Aggregate Abrasion Value, i.e. a rubber pad is placed on the grinding lap and, using a water and emery abrasive, flat test specimens which have been cut from the surface of the paver are

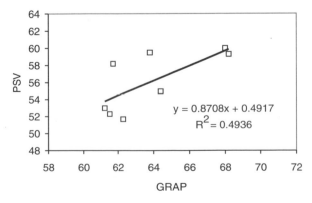

Fig. 18.12 Correlation of PSV and GRAP

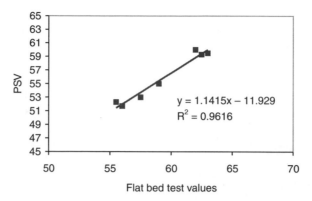

Fig. 18.13 Correlation of PSV with flat bed test values

polished face down on the rubber pad. The wet skid resistance is then measured using the BPST apparatus. As shown in Fig. 18.13, the results from flat test specimens made with 10 mm sized aggregate correlate very well with PSV test results.

18.8 A critical evaluation of the PSV test method

The British PSV is now widely used to predict aggregate skid resistance in the laboratory, and has recently been accepted as the CEN Euro Norm to measure the degree of polishing. However, when it is critically evaluated[8,11,12] one must conclude that, in contrast with modern traffic conditions which have changed significantly, this 50 year old machine has changed relatively little during its lifetime.

In the UK, the desire for very high PSV values has favoured the use of one type of aggregate, i.e. gritstone, as the principal type of surfacing aggregate. Based on reported research[8] small increases for higher PSV aggregates correspond to large decreases in terms of wear and other test properties. In the example shown in Fig. 18.14 the PSV and AAV plots for arenaceous rocks clearly show that with small increases in PSV this type of aggregate is susceptible to an increasing amount of wear due to abrasion.

Another fundamental problem is that the PSV test assesses single size 10 mm aggregate whilst other British Standard reference tests assess 14 mm size particles. By contrast, bituminous mixtures typically contain a wide range of aggregate sizes. Although the PSV test attempts to simulate in-service factors such as water, traffic loading and the presence of detritus, it cannot adequately simulate real in-service conditions where there are tremendous variations in traffic conditions, stresses, types of bituminous mixtures, etc.

Although the use of a solid tyre may have helped to improve the PSV test, real trafficking is done by a pneumatic tyre that has different conditions within the contact patch area. It is therefore emphasized that the 6 h standard PSV test is not a true reflection of the performance of an aggregate, but rather it offers a ranking mechanism under simplified laboratory conditions where end-use, climate and trafficking conditions are not considered. This must always be borne in mind when considering the use of the values obtained by the PSV test.

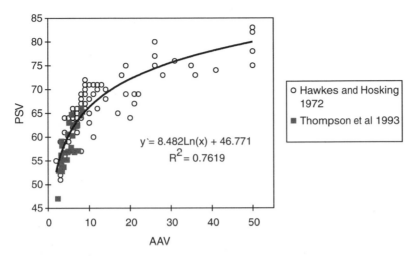

Fig. 18.14 Plot of PSV and AAV for arenaceous rocks

The problem of how to improve the accuracy of predicting an aggregate's wet skid resistance is currently of considerable importance in the UK. The Transport & Road Assessment Centre (TRAC) at the University of Ulster, has been considering this problem in detail for many years, and unique testing protocols based on the PSV test apparatus have been developed to better simulate in-service conditions. These have resulted from assessments of extended polishing cycles, the repetitive usage of coarse and fine emery polishing cycles, different types of polishing agents, use of low-temperature conditioning cycles, and sideways polishing to increase the amount of polish experienced during testing. The results from these methods were compared with those obtained using the French GRAP and the British Flat Bed tests. The overall conclusion by TRAC was that *the standard PSV test remains the best laboratory test to measure wet skid resistance.*

However, the use of the standard 6 hour test does not give the lowest possible value of skid resistance for an aggregate, i.e. compared with its ultimate state of polish. Rather, it was found that individual aggregates react differently to the different protocols assessed. Some continued to lose a significant amount of skid resistance whilst others remained relatively unaffected.

Many of the experiments involved extended polishing cycles where some of the aggregates were subjected to 50 hours of polishing. However, a 50 hour test is not viable for routine testing, so a method involving only 3 hours additional offset polishing was developed as a way of achieving the ultimate levels of polishing such as are experienced at in-service sites that are subjected to heavy trafficking or in-service stressing.

18.9 The contact patch: understanding what happens

In the UK, road surfacings are designed to have wet skid resistance provided by high-quality aggregates and a minimum level of texture depth or surface roughness.

However, there has been a noticeable increase in the occurrence of premature failure of surfacings due to loss of wet skid resistance. This phenomenon is difficult to explain, particularly as aggregates were specified with very high levels of skid resistance as measured by the standard PSV test. Possible causes have been attributed to a range of factors including increased trafficking, axle loads, and the use of super singles tyres.

It is proposed that greater awareness be required of the conditions experienced within the contact patch during the pavement design processes. The contact patch is that area of road surface which is in contact with a tyre as it rolls over the road surface. Research[13,14] involving static and dynamic measurements of a different type has endeavoured to understand the processes within the contact patch. This research has shown that dynamic contact stress is probably the most important characteristic that exists at the interface between tyre and road surface.

Stress was found to be influenced by many factors including load, tyre inflation pressure, speed and surface macrotexture, with dynamic contact stress existing in three directions, i.e. vertical, longitudinal and lateral. During the contact time a vertical contact stress is generated in the z-direction, whilst longitudinal and lateral contact stresses occur in the x- and y-directions, respectively.

The research involved the development of a test platform to measure dynamic contact stress using twelve specially-designed, inverted T-shaped transducers positioned across the width of the contact patch to simulate exposed surfacing aggregate. A cover plate allowed the tips of the inverted T-shaped transducers to protrude to simulate differing values of texture depth. Strain gauges attached to each transducer allowed data to be recorded in the x-, y- and z-directions as the tyre transversed over them. By adjusting the transducer height the effect of load and tyre inflation pressure was assessed, and it was found that transducer height, i.e. texture depth, had the greatest effect on the generation of vertical contact stresses. This has a significant effect on the contact action between tyre and road surface, particularly where the surfacing has a positive texture. During a contact action, loading from the tyre is transferred to the road structure through the surface chippings.

When the transducer height is small, i.e. a smooth road surface or minus texture, lower contact stresses are produced. However, for higher transducer heights, loading is concentrated on the top of exposed chippings, resulting in much higher levels of contact stresses. In practice, these stresses may induce premature chipping loss or cause levels of aggregate wear or polishing that will cause unpredicted loss of skid resistance, particularly in wet conditions. Increased loading was shown to increase the level of contact stress generated.

It was also found that an increase in tyre inflation pressure resulted in an increase in contact stress. This can also be explained by the mechanism of contact action between tyre and road surface. When the tyre inflation pressure is low and at a constant load, the tyre surface deflects to give a wide contact area that results in low contact stress. When the tyre inflation pressure is increased it results in a smaller area of contact and a higher contact stress.

It has been shown that longitudinal contact stress changes direction during a contact action, suggesting that a single chipping suffers two types of stressing as a tyre passes over it, i.e. compressive and tension. Combining the different directions of stressing indicates that a screwing type of action occurs. This will have a greater effect on surfaces with higher texture, causing the harder aggregates to polish and softer aggregates to wear away. It may also influence the shearing mechanism of the aggregate-bitumen bond, thus causing a loss of integral strength and cohesion.

It has been shown that during the contact between a tyre and the road surface the maximum vertical contact stress occurs for approximately 75 per cent of the contact duration within the contact patch. Maximum vertical contact stresses have been shown to increase significantly towards the edge of the contact patch, and this is probably due to the transferral of load through the stiff sidewall or shoulder of the tyre. This indicates that instead of the assumed constant level of contact stressing within the contact patch, chippings or the road surface located at the edge of a contact patch may be damaged much more rapidly than those located within the contact patch. Given the tendency for channelized heavy traffic the data suggest that this may be one of the major causes of premature loss of wet skid resistance.

18.10 Conclusions

This chapter has considered the wet skid resistance of road surfacings, i.e. what is it, how may it be measured in-service, and how may it be predicted, and some current research being carried out to improve its prediction has been summarized. It cannot be over-stressed that the phenomenon of wet skid resistance is the combination of many factors, with some being more important than others, particularly in terms of their location on the road surface.

It is crucial that the different types of texture involved in wet skid resistance be understood, i.e. the microtexture of the aggregate surface, the macrotexture of the road surface and its ability to displace water either with positive texture or negative texture (minustexture) as vehicle speeds increase, and the megatexture of the road surface and its ability to remove water with adequate crossfalls and to minimize standing water particularly in permanently-deformed wheel tracks. Finally, there is the role played by the tyre tread and its ability to remove water as vehicle speeds increase.

The in-service measurement of wet skid resistance will become more important as additional countries around the world become more aware that this is an issue that relates to road deaths and the immense cost of these, and the need for data to implement improved pavement management systems. Equipment such as the GripTester, which offers a reliable and cheap capability, is likely to become more readily available and adopted.

In terms of predicting performance, the laboratory measurement of wet skid resistance using methods such as the standard PSV test offer limited ability to simulate the wide variation in conditions now experienced by aggregates in the many types of bituminous mixes available in modern roads under modern traffic conditions. Testing protocols that increase the amount of polish have been shown to affect different aggregates in differing ways. Typically, all aggregates will polish to a higher degree than is predicted by the PSV test. Simulation of turning traffic using an offset-angled polishing wheel appears to offer the most potential to assess aggregates for higher stress sites.

Research has shown that the use of bituminous mixtures to create surfacings with high texture depths will result in aggregates experiencing higher levels of dynamic vertical contact stress than smooth- or negative-textured surfacings. The levels found have been greater than expected, and is closely related to axle loading, tyre inflation pressure and texture depth. It is suggested that these are the principal factors which contribute towards the loss of skid resistance and other types of premature failures.

It is recommended that consideration of the contact stresses to be induced on the road surface should be incorporated into the surfacing design process, whether for a simple surface dressing or a high-performance thin surfacing. The research shows that in order to minimize contact stress, road surfacings should be designed to have a smooth-textured surface or one possessing a negative texture. Whilst this offers a challenge in terms of bulk water removal, it has been shown that these surface textures will suffer less dynamic vertical contact stressing and, consequently, will be less susceptible to premature loss of skid resistance and permanent deformation.

Finally, despite almost 100 years experience of wet skid resistance in countries such as the UK, and in other countries with little such experience, there is considerable scope for research and development into this phenomenon. The aim should be to improve the fundamental understanding of the mechanisms involved, to develop new improved predictive techniques, and to design bituminous surfacing materials that will result in safer roads in the future.

18.11 References

1. Roe, P.G., Parry, A.R. and Viner, H.E., *High and Low Speed Skidding Resistance: The Influence of Texture Depth*, TRL Report 367. Crowthorne, Berkshire: Transport Research Laboratory. 1998.
2. Hosking, R., *Road Aggregates and Skidding*, State-of the-art Review 4. Crowthorne, Berkshire: Transport Research Laboratory. 1992. ISBN 011 5511156.
3. Roe, P.G., Webster, D.C. and West, G., *The Relation Between the Surface Texture of Roads and Accidents*, TRRL Research Report 296. Crowthorne, Berkshire: Transport and Road Research Laboratory, 1991.
4. *Design Manual for Roads and Bridges, Vol 7 – Pavement Design and Maintenance, Section 5 – Pavement Maintenance Assessment, Part 1 – Skidding Resistance*, HD 36/99. London: British Standards Institution, 1999.
5. Maclean, D.J and Shergold, F.A., *The Polishing of Roadstone in Relation to the Resistance to Skidding of Bituminous Road Surfacings*, DSIR Road Research Technical Paper No. 43. London: HMSO, 1958.
6. BS EN 1097–8, *Tests for Mechanical and Physical Properties of Aggregates, Determination of the Polished Stone Value*. London: British Standards Institution. 1999.
7. BS 598–105: 2000, *Sampling and Examination of Bituminous Mixtures for Roads and Other Paved areas, Methods of Test for the Determination of Texture Depth*. London: British Standards Institution, 2000.
8. Woodward, W.D.H. *Laboratory Prediction of Surfacing Aggregate Performance*, DPhil thesis. Newtownabbey, Northern Ireland: University of Ulster. 1995.

9. Giles, C.G., Sabey, B.E. and Cardew, K.H.F., Development and performance of the portable skid resistance tester, *Rubber Chemistry Technology*, 1965, **33**, pp. 151–7.

10. Wambold, J.C., Antle, C.E., Henry, J. and Rado, Z., *International PIARC Experiment to Compare and Harmonize Texture and Skid Resistance*, Chapter 4. Paris: Association Internationale Permanente des Congres de La Route, 1995.

11. Woodside, A.R., *A Study of the Factors Affecting the Polishing of Aggregates and Skidding Resistance*, MPhil thesis. Newtownabbey, Northern Ireland: Ulster Polytechnic, 1981.

12. Perry, M., *A Study of the Factors That Influence the Polishing Characteristics of Gritstone Aggregate*, DPhil thesis. Newtownabbey, Northern Ireland: University of Ulster, 1997.

13. Liu, G. X., *The Area and Stresses of Contact Between Tyres and Road Surface and Their Effects on Road Surface*, DPhil thesis. Newtownabbey, Northern Ireland: University of Ulster, 1993.

14. Siegfried, *The Study of Contact Characteristics Between Tyre and Road Surface*, DPhil thesis. Newtownabbey, Northern Ireland: University of Ulster, 1998.

CHAPTER 19

Design and use of surface treatments

H.A. Khalid

19.1 Surface treatment types and purposes

When a road surface quality deteriorates with increase in traffic use, the resulting condition can be improved by applying a surface treatment of some type, provided that the road is structurally sound. Many kinds of surface treatment measures exist and have been used in the UK and Europe to improve the skid resistance of surfacings, seal surface cracks and correct the longitudinal profiles of roads. This chapter will discuss various types of surface treatments with special emphasis on surface dressing, acknowledged to be one of the most common and cost-effective measures used to improve the quality of road surfacings.

Surface dressing has been practised in the UK for over 90 years, and longer still in other countries like France and Australia. However, the development of specifications for the process only started in the 1930s with full-scale road experiments conducted by the (then) Road Research Laboratory[1]. Its effectiveness as an economic road maintenance measure has led to an increased appreciation and use by road engineers. Over the past five years, an annual average of 60 million m^2 of surface dressing has been undertaken by member companies of the Road Surface Dressing Association[2]; this accounts for around half of the total dressing carried out in the UK.

In its simplest sense, surface dressing comprises spraying a thin layer of binder, usually an emulsion or cutback, onto an existing road surface, i.e. the 'substrate', followed by spreading a layer of chippings which is then rolled. Rolling helps initiate and enhance the bond between the binder and the stone chippings and achieve their embedment in the underlying substrate. Surface dressing is mainly used as a maintenance tool to perform designated functions which can be summarized as follows:

1. to provide a non-skid wearing surface;
2. to seal the entire road surface against the ingress of water; and
3. to suppress the disintegration and fretting of the road surface.

Note that a surface dressing does not restore the riding quality of a deformed road.

The selection of materials and their application rates for surface dressing design to meet traffic and environmental conditions encountered in the UK is detailed in Road Note 39[3].

Other surface treatments used in the UK and Europe include thin surfacings which can either be laid cold, e.g. *slurry seals*, or hot. e.g. *very thin surface layers (VTSL)* and *ultrathin hot mix asphalt layers (UTHMAL)*. VTSLs are laid in 20–30 mm thick layers, whereas UTHMALs are about 10–20 mm thick. These thin surfacings are single layers that are laid and compacted using machines and rollers; they are considered as a cross between thin wearing courses and surface dressings. Their uses include all those of surface dressings, as well as regulating longitudinal road profiles without having to raise manholes or lose headroom at overbridges[4]. They do not require the after-care service necessary with surface dressing, and they can be laid in weather conditions outside those suitable for surface dressing works.

19.2 Surface dressings recommended in Road Note 39

There are several types of surface dressing systems which vary according to the number of layers of chippings and binder applied. Figure 19.1 shows the main types of surface dressings as depicted in Road Note 39, excluding the resin-based high skid-resistant systems. Figure 19.2 summarizes the procedure utilized to select each surface dressing type.

A *single surface dressing* consists of a single application of a binder followed by a single application of chippings. This system is recommended mainly for low shear stress sites.

A *pad coat* is essentially a single surface dressing system with small (usually 6 mm) size chippings. It is usually applied to porous, hungry, hard, or uneven roads to produce a more uniform surfacing which can be subsequently surface-dressed. The pad coat and single dressing construction is now referred to as *inverted double surface dressing*.

The *racked-in system* consists of a single application of binder followed by two applications of chippings of different sizes. The first application consists of larger chippings, usually 10 to 14 mm, and constitutes about 90 per cent cover which leaves gaps that are filled by the second application of smaller size (3–6 mm) chippings. This system is recommended for sites of heavy or fast traffic. On very hard or hard surfacings, like concrete, the racked-in system is now preferred to pad coats in reducing the effective hardness of the substrate.

The *double surface dressing* is similar to the single system, but uses two applications of binder and chippings. The first application contains larger chippings while the second contains smaller ones. This system, which is not common in the UK, is recommended for road surfaces that are binder lean.

The *sandwich surface dressing* is mainly used in situations where the substrate is binder rich. It consists of one application of binder interposed between two applications of chippings. It is similar to the double system but without the first binder application[5]. The double and sandwich systems are characterised as having a

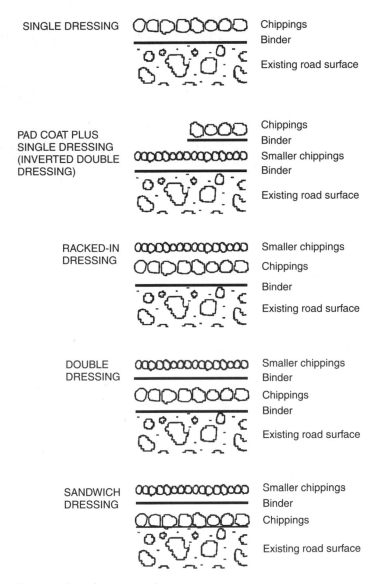

Fig. 19.1 Types of surface dressing

higher texture depth than that produced with a single system with the same size chippings[6].

In some instances, where sites are classified as most difficult, special surface dressing systems are used. These sites often occur in urban areas, e.g. at round-abouts, approaches to traffic signals and pedestrian crossings, and other locations where there are turning and braking movements. In such situations, proprietary high performance systems, e.g. 'Shell Grip' and 'Spray Grip'(which use bitumen-extended

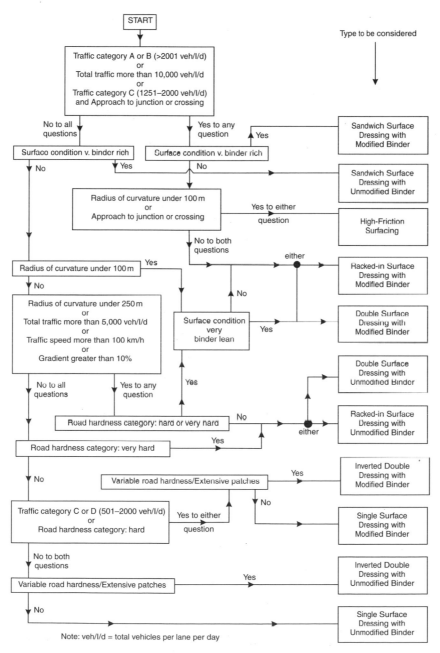

Fig. 19.2 Road Note 39 method of selection of surface dressing type[3]

epoxy resin binders and very high PSV aggregates such as calcined bauxite) are used[5,6]. The exceptionally high durability and wear-resistant properties of these systems justify their use despite their higher initial cost compared with conventional systems.

19.3 Factors affecting the use of surface dressing systems

The selection of an appropriate surface dressing system is a function of a number of parameters. The selection process involves the choice of chipping size and their spread rate, embedment and binder type and application rate. The factors affecting the choice of surface dressing systems are categorized under the following headings: (1) traffic categories; (2) hardness and surface condition; (3) location and geometry of the site; (4) site requirements; and (5) seasonal and weather conditions.

19.3.1 Traffic categories

Embedment of chippings is related to the volume of traffic and the percentage of commercial vehicles; these are considered to cause most of the embedment.

To assess the probable amount of embedment using Road Note 39, traffic levels have been divided into eight categories, ranging in severity from 0 to over 3250 commercial vehicles/lane/day. Traffic speed is also accounted for in the sense that the higher the speed the less the embedment that is caused; however, this also means that there is more chance of loose chippings that may cause windscreen damage.

19.3.2 Hardness and surface condition of the existing road surface

The *hardness* of the road is a very important factor in determining the size of the chippings. Road Note 39 divides surface hardness into five categories ranging from very hard, e.g. concrete roads, to very soft. The hardness of the road is measured with a hardness probe[7] when the surface temperature is, preferably, between 15 and 35°C.

The *condition* of the existing road surface is important in determining the most appropriate type of surface dressing. The binder content of the road surface affects the amount of binder required for the initial retention of the chippings prior to their long-term embedment. Based on binder content, the surface condition is therefore divided in Road Note 39 into five categories ranging from very binder rich to very porous and binder lean.

19.3.3 Location and geometry of site

Included under location and geometry of site are parameters that are likely to vary along the entire length of the road being considered. These parameters are radius of curvature and superelevation of the road, longitudinal gradient, altitude and shade. They can affect the stresses imposed on the road surfacing by vehicles, the properties and quality of the binder required to retain the chippings, and road hardness.

Road Note 39 divides sites into categories in respect of the above-mentioned parameters and considers these categories in the selection of the optimum surface dressing system that suits the prevailing conditions.

19.3.4 Site requirements for skid resistance

As stated in Section 19.1, the main function of a surface dressing is the restoration of the diminished skid resistance of a surfacing in an otherwise structurally-sound pavement structure. The polished-stone values (PSV) and aggregate abrasion values (AAV) of chippings affect the skid resistance of a road surface and are used in Road Note 39 as criteria for the selection of chipping type for the particular site conditions and traffic intensity.

19.3.5 Seasonal and weather conditions

The weather is considered to be second only to traffic in its disruptive action against surface dressing. The two most important weather conditions that affect surface dressing are temperature and humidity. Road Note 39 uses temperatures and humidity criteria in the selection of binder and chipping types.

Binders are visco-elastic and are, therefore, temperature susceptible, some more than others. *Temperature effects* are significant during both construction and in service. Chipping embedment into the substrate is also affected by temperature, as the hardness of the (bituminous) surfacing will vary with the temperature, becoming soft at high temperatures and hard at low temperatures. Persistent high temperatures will likely cause considerable embedment leading to bleeding or fatting up.

The presence of *moisture* on the chippings and/or road surface affects the rate of build-up of strength of a newly-laid surface dressing. Dampness on the chippings may reduce or delay the bond formation between the chippings and the binder. This situation is not so critical with bitumen emulsions as with cutbacks because emulsions are more tolerant to moisture.

In *humid weather* conditions, emulsions break slowly and agents which speed up their breaking may become necessary. Cutbacks favour the use of coated chippings if sprayed at low (below 15°C) temperatures, as it becomes very difficult to establish good adhesion between uncoated chippings and binders.

19.4 Theory underlying the design of surface dressing

There are two primary aims involved in the design of any surface dressing system: (a) to determine the type and amount of chippings; and (b) to determine the type and amount of binder required. Achievement of these aims enables the proper construction of the dressing and ensures satisfactory performance.

The means of achieving these objectives depends on the design approach adopted. There are currently two design methodologies, namely, the Engineering (or Rational) approach originally devised by Hansen[8] and subsequently developed by other workers (see references 9–12), and the Road Note 39 method.

19.4.1 The Engineering (Rational) approach

The theory behind the Rational approach is that the quantity of binder and chippings used depends on the *average least dimension (ALD)* of the chippings and the

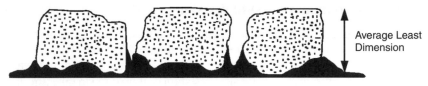

(a) The average least dimension after trafficking

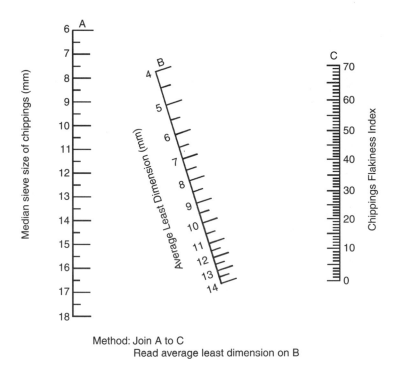

Method: Join A to C
Read average least dimension on B

(b) Normograph for determining the average least dimension

Fig. 19.3 The determination of the average least dimension (ALD)[8]

percentage voids available between the chippings when they are spread and rolled. The ALD concept (see Fig. 19.3(a)), involves obtaining the average of the least dimensions of the chippings forming a single surface dressing coat.

After full compaction by traffic, chippings tend to adopt a position in which their least dimension is vertical and their compacted depth is related to their ALD. The ALD can be measured with the use of callipers as the least dimension of each of a sample of 200 chippings from a batch, with the average representing the ALD. Alternatively, the median size and flakiness of the chippings may be used to evaluate the ALD from the nomograph shown in Fig. 19.3(b)[13].

With the Rational method, the chipping size is selected to provide the desired texture depth and skid resistance level, taking into account any chipping embedment expected to occur into the underlying layer. The factors considered in the

Table 19.1 Factors used in the Rational method for designing surface dressings[12]

Traffic level Category (veh/day)	Value	Type of chipping Category	Value	Existing surface Category	Value	Climate Category	Value
0–100	+3	Round	+2	Untreated	+6	Wet and cold	+2
100–500	+1	Dusty	+2	Primed base	+6	Wet and hot	+1
500–1000	0	Cubical	0	Very lean	+4	Temperate	0
1000–3000	−1	Flaky	−2	Lean	0	Dry and hot	−1
3000–6000	−3	Pre-coated	−2	Average	−1	Very dry and	
6000+	−5			Rich	3	very hot	2

selection of the chippings are: type of treatment, nature and volume of traffic, surface condition, life expectancy of treatment, climatic condition, and site geometry. The chipping spread rate can be obtained by considering the effect of some of these factors and by allocating numerical values to them for each category, as shown in Table 19.1. The chipping spread rate, which assumes that 60 to 75 per cent of the voids in the compacted layer will be filled with binder, is then obtained by use of these values, the ALD and the design chart in Fig. 19.4. The binder spread rate can also be obtained from the diagram in Fig. 19.4 or, alternatively, it can be calculated from the percentage of embedment of the chippings into the binder[8,9,14].

The design binder and chipping application rates are determined by summing up the values obtained for the four factors in Table 19.1 and entering this value in Fig. 19.4. The intersection point of the ALD and the factor line gives the design binder application rate (bottom scale). The intersection point of the ALD and the line AB gives the chipping application rate (top scale). Thus, for an ALD of 14 mm and a total factor of +4, binder and chipping application rates of 1.3 l/m^2 and 19 kg/m^2, respectively, are obtained.

19.4.2 The Road Note 39 approach

The Road Note 39 design approach is empirical and is based entirely on experience gained from extensive road trials.

The methodology begins with the initial selection of the surface dressing type (Section 19.2) based on traffic intensity, location and geometry of site and the surface hardness (Section 19.3). The next step is the selection of the size of the chippings; this is achieved by relating the number of commercial vehicles and the road hardness to the size of chippings required. The selection of the size of chippings is followed by the application rate; this must be sufficient to cover the binder film.

The selection of the type of binder is based on the traffic category, time of year, weather conditions and the likely stresses to be encountered. Finally, the binder application rate is selected; this is related to the size and shape of the chippings, nature of the existing road surface, and the envisaged degree of embedment of chippings by traffic.

For the regular users of the Road Note 39 design method, there now exists a computer program which carries out the design process. The computer program is

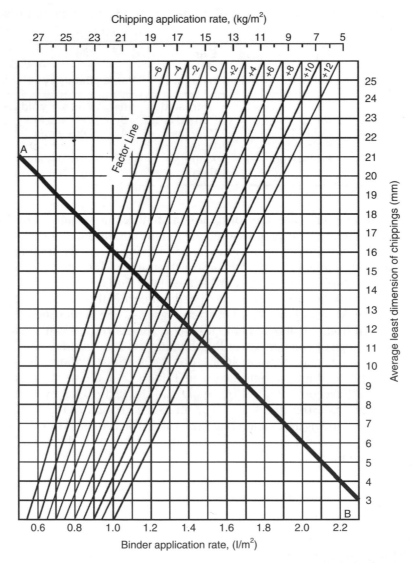

Fig. 19.4 Chart for the determination of the design binder and chipping application rate using the Engineering (Rational) design approach for surface dressings[12]

intended only to supplement, and not to replace, the Guide. The program is supported by and can be obtained from the Transport Research Laboratory.

19.4.3 Comparison of Road Note 39 and Engineering (Rational) methods

As with all empirical design techniques, the Road Note 39 approach is applicable to sites and conditions similar to those experienced previously. Identifying design parameters such as road hardness and surfacing condition tends to reduce the method's efficiency because any given site has to be related to another with similar traits. Although the most recent edition of Road Note 39 includes detailed recommendations for the use of modified binders, it is the writer's opinion that not enough experience has been gained to-date with these materials for them to be used in a general recommendation, whilst new ones continue to be developed. It is also felt that important design considerations like existing road surface texture, variation in chipping compaction with aggregate type and shape, and the need to account for the resulting texture depth are still not considered directly.

One advantage of the Road Note 39 approach, however, is that once the site and its prevailing conditions have been classified, the design process becomes fairly easy as the application rates can then be obtained from preset tables.

The Engineering approach, on the other hand, is a more flexible method that allows the engineer to design for any particular condition. It enables the designer to take into consideration the factors affecting the performance of surface dressings in the design process. It has also been found[15,16] to give greater confidence and a more reliable and consistent performance, thus achieving a more cost-effective solution.

The one major shortcoming in both design approaches, however, is the failure to include assessment tests, simulative or otherwise, to check the designs before implementation. Simulative evaluation tests would greatly reduce the risk of failure by highlighting, in advance, any weakness areas[17]. The need for performance-measuring methods cannot be over-emphasized, especially with the increasing use of newly developed polymer-modified binders.

19.5 Applying surface dressing and avoiding failure

The majority of surface dressing failures are predominantly associated with failures to implement the basic principles of surface dressing operation. These basic principles include the dependency of the size of the chippings and the spread-rate of the binder on the hardness of the road and traffic volume, especially the heavy goods vehicles component. Road Note 39 uses temperatures and humidity criteria in the selection of binder and chipping types.

However, there are other parameters more related to the planning, organization and management – these are largely taken for granted – that are equally critical to the success of surface dressing operations.

In relation to design and performance-related characteristics, a surface dressing is classified as having 'failed' when it ceases to perform any of its stated functions satisfactorily. Common modes of failure of surface dressings are most easily discussed under four main headings: (1) chipping loss; (2) chipping embedment; (3) bleeding and fatting up; and (4) wearing of chippings.

19.5.1 Failure due to chipping loss

Chipping loss, known as *scabbing*, usually occurs either in the very early stages of surface dressing or after the first winter following placement. It is caused by the action of traffic which breaks down the adhesive bond between chipping and binder or the bond between binder and substrate.

This type of failure is normally attributed to a binder failure, i.e. due to the use of the wrong binder type or the wrong application rate. Scabbing can occur as a result of a combination of very little or no embedment of the chippings into the substrate due to the surfacing being too hard, or there is insufficient time between laying the dressing and the onset of cold weather, coupled with too brittle a binder[18].

19.5.2 Failure due to chipping embedment

Failure due to chipping embedment is normally due to the substrate being too soft. However, this failure mode can also be attributed to the selection of the wrong chipping size and an increase in traffic intensity.

On a soft substrate, the action of traffic forces the chippings into it, which causes the binder to rise. This leads to a loss of texture depth and a consequent reduction in the skid resistance of the surfacing. Research has shown[19] that the rate of embedment depends on, inter alia, the substrate temperature, influence of heavy goods vehicles, and the loading duration.

19.5.3 Failure due to bleeding of the binder and fatting up

Bleeding of the binder often occurs within a few days or weeks on dressings laid early in the season, or during the following summer on dressings laid late in the previous year. It results from the use of the wrong type of binder in respect of temperature susceptibility, or the wrong rate of binder application.

A highly temperature-susceptible binder has the tendency to become very fluid at high ambient temperatures. This causes the binder to flow under the action of traffic so that it rises to the surface of the road and flushes the chippings, resulting in a loss of texture depth. During warm summer months, the binder may rise to the surface of the road as the chippings become embedded into the substrate.

Fatting up is thought to be due to either too heavy a rate of binder application or the use of smaller size chippings than is necessary for a given traffic condition and road hardness[20]. It may also be caused by subsequent embedment of chippings into the substrate, crushing of the chippings, or the absorption of dust by the binder leading to its increased effective volume[21].

19.5.4 Wearing of the chippings

Most chippings will wear out with time and action of traffic. Even though the overall structure or stability of a surface dressing may still be intact, it may have lost its ability to provide adequate skid resistance.

Wearing out of chippings can be the result of use of chippings of inadequate AAV. A good surface dressing design should ensure that, for the chipping selected, its design life has been exceeded by the time the traffic has worn the chippings to below acceptable levels.

19.6 Slurry seals

Slurry seals are cold-mixed, thin, surface treatments that are used as a maintenance measure to improve the unevenness of roads and produce a good ride quality. They can also fill up surface depressions to a certain extent, and provide a seal against the ingress of moisture.

Two grades of slurry seal, i.e. with 1.5 mm and 3.0 mm finished thicknesses, are used in the UK. These are specified in the *Specification for Highway Works*[22] and BS 434[23]. However, recent developments using modified binders have enabled the successful use of larger aggregate sizes and layer thicknesses of up to 20 mm[5].

The composition of a slurry seal comprises graded aggregates, bitumen emulsion and an additive, usually ordinary Portland cement or hydrated lime. The binder is predominantly a cationic, class K3, bitumen emulsion that (a) is sufficiently stable to allow mixing with aggregate without breaking during mixing and laying, and (b) is capable of producing a mix which develops, upon laying, early resistance to traffic and rain. The emulsion, which is used at the rate of approximately 180–250 litres per tonne of dry aggregate, produces a fully cured mix with a bitumen:aggregate ratio, by weight, of about 1:4. The amount of additive, e.g. cement, is not more than 2 per cent by weight of dry aggregate; its function is mainly to control consistency, setting rate, and mix segregation. The precise proportions of each mix constituent are normally selected with the aid of laboratory tests, the most common of which is the Wet Track Abrasion test[23]. The maximum amount of loss in this test, expressed in g/m^2, is usually specified to enable the selection of optimum material proportions.

A cationic emulsion tack coat is applied, where required, prior to laying of a slurry seal. Depending on the type and condition of the existing surfacing, the tack coat is applied at rates of 0.15 to 0.6 l/m^2.

Normally, slurry sealing operations are carried out using mobile mixer-spreader vehicles. These purpose-built vehicles are equipped with separate hoppers for the aggregates, emulsion, cement and water, from which metered quantities of materials are continuously fed to a mixer located at the rear of the vehicle. The mixer feeds the slurry to a spreader-box, drawn behind the vehicle, which controls the laying process. These mixer-spreader vehicles can apply up to 8000 m^2 of slurry seal per day[5].

The use of slurry seals in roads is predominantly restricted to lightly trafficked areas, e.g. in residential estates. As the slurry seal is laid in thin layers on an existing substrate, it is not seen as contributing to the overall strength of the pavement structure. Also, the resultant texture depth is too low, due to the fine aggregate grading used. It is the writer's opinion that slurry seals' main effect is cosmetic at this time, in that it results in a smooth even appearance being produced on roads which have developed an irregular surface texture. However, ongoing research into cold-mix technology and the development of novel polymer modified binders and synthetic fibres may have a major impact on the role of slurry seals in road maintenance operations in the longer-term future.

19.7 Thin surface treatments

The advantages of surface dressings are well recognized by those involved in their design, specification and use. It is also well known, however, that surface dressings cannot be used to reshape the longitudinal profile of roads and that they produce high noise levels[5]. Moreover, their success is critically dependent on the weather conditions during their application, as well as during the first months of trafficking.

The need for additional surface treatments that would act as an intermediate solutions between conventional thin wearing courses and surface dressings has been identified in relatively recent years. Their performance requirements would be to eliminate the shortcomings of surface dressings, reduce the chance of flying chippings, and maintain the major qualities of impermeability and safety.

In the late 1970s and early 1980s, thin and very thin hot-applied asphalt surfacings were developed in France[24,25]; these were designed mainly to meet the rehabilitation requirements of old road networks. The thickness of these surfacings ranged between 30 and 40 mm for thin hot-mix materials, and between 20 and 30 mm for very thin surface layers (VTSLs). More recently[26], ultra thin hot-mix asphalt layers (UTHMALs), have been developed in France to be laid in 10–20 mm courses; these have experienced increased interest and rapid growth in France and in other parts of the world, including the UK, USA and Austria[4,27,28].

The best known UTHMAL is the French process ‘NOVACHIP’, developed in the mid-1980s by Screg Routes and the Laboratoire Central des Ponts et Chaussées[24], which has since then been marketed as a proprietary product. In the UK, an UTHMAL by the name of ‘Safepave’ has been marketed by Associated Asphalt, and there are several sites where it has been laid. A VTSL technique called ‘UL-M’ was developed in France by Enterprise Jean Lefebvre[29] and has been recently acquired[4] under Licence for the UK market.

19.7.1 Mix design

In France, the use of VTSLs and UTHMALs requires the pre-application of a tack coat using an emulsion, which is often polymer modified, containing 70 per cent residual bitumen. The application rates depend on the condition of the substrate, but are normally between 0.4 and 0.7 l/m^2 for VTSLs and 0.8 to 1.0 l/m^2 for UTHMALs. This tack coat is important as it provides the bonding and sealing that the treatment alone cannot provide, due to the presence of a large volume of voids (about 10 per cent).

The most common aggregate grading is a 10 mm nominal size mix with a gap between the 2 mm and 6 mm size fractions. The main difference between VTSLs and UTHMALs lies in the percentage of coarse aggregate and binder content. For VTSLs, on average, there would be 65 to 70 per cent 6/10 mm size fractions and approx. 5.5 to 6.0 per cent by weight of binder; for UTHMALs, there would be 75 to 80 per cent 6/10 mm stones and 5.2 to 5.6 per cent by weight of binder[26]. The rest of each material is made up by a combination of fine aggregates and filler, with UTHMALs being predominantly coarser.

The binder is most often a 60/70 or a 80/100 pen grade bitumen, often modified by polymers or fibres to improve the mechanical properties and temperature susceptibility of the material.

19.7.2 Placement

After the application of a tack coat, VTSLs are laid with a normal paving plant which spreads both binder and aggregate. Compaction is carried out with a steel-wheeled roller.

A specially designed paving machine is used to lay UTHMALs. This machine performs three operations, i.e. it spreads the binder, applies the hot mix, and smooths the layer in a single pass within a short time. A paving speed of up to 1.5 km/h has been reported[26] with this specialist equipment. Final compaction is achieved with two or three passes of a steel-wheeled roller.

19.7.3 Outline of performance assessment

A number of studies have monitored the performances of VTSL and UTHMAL courses in France, enabled by the extensive use of these increasingly popular maintenance techniques[25,26]. Measurement of texture depth, as determined by the Sand Patch test[30], has shown that adequate values are obtained with both types of surfacing, with better results being obtained with UTHMAL materials, due to their more mono-granular nature, e.g. average texture depth values of 1.0 mm for VTSLs and 1.6 mm for UTHMALs (both of 10 mm nominal size gradings) were achieved after 1 year of service on a heavily trafficked road in France[26]. In the UK, initial texture depth values for UTHMALs of above 1.5 mm were reported on high-speed roads, which satisfy governmental requirements[22]. VTSL values of 1.4 mm were reported on a test section which was not on a high-speed road[4].

In France, the skid resistance of UTHMALs as measured by the coefficient of longitudinal friction between 40 and 120 km/h was reported to be very good. In the UK, however, even though good SCRIM values were obtained, there was some concern about the reduction in skid resistance with increasing speed.

Good improvements in respect of both road unevenness and riding quality have been reported in France and the UK[4,26] for both VTSLs and UMTHMALs. A limited improvement to the ride quality was witnessed in the US after laying UTHMAL courses in Texas[27].

Noise measurements in France have indicated a net reduction of 3 dB(A) in favour of UTHMALs when compared with a surface dressing of similar size aggregate[26]. In the UK, the spray and noise properties of UTHMAL are reported to be at least as good as conventional hot rolled asphalt and surface dressing surfacings[4].

19.8 References

1. Road Research Laboratory, *Bituminous Materials in Road Construction*. London: HMSO, 1962.
2. Bracewell, E., *Surface Dressing Statistics* (private communication). Bristol, Road Surface Dressing Association, 1998.
3. Nicholls, J.C., *Design Guide for Road Surface Dressing*, Road Note 39 (4th edition). Crowthorne, Berkshire: Transport Research Laboratory, 1996.
4. Mercer, J., Nicholls, J.C. and Potter, J.F., Thin surfacing material trials in the UK, *Transportation Research Record 1454*, 1994, pp. 1–8.

5. Hoban, T.W.S., Surface dressings and other surface treatment. In *Bituminous Mixtures in Road Construction*. London: Thomas Telford, 1994, pp. 316–51.

6. Whiteoak, D., *The Shell Bitumen Handbook*. Chertsey, Surrey: Shell Bitumen, 1990, pp. 281–318.

7. Wright, N., *Surface Dressing: Assessment of Road Surface Hardness*, TRRL Report SR573. Crowthorne, Berkshire: Transport Research Laboratory, 1980.

8. Hanson, F.M., Bituminous surface treatment of rural highways, *Proceedings of the New Zealand Society of Civil Engineers*, 1935, **21**, pp. 189–220.

9. Kearby, J.P., Tests and theories on penetration surfaces, *Proceedings of the Highway Research Board*, 1953, **32**, pp. 232–37.

10. Benson, F.J., Seal coats and surface treatments, *Proceedings of the 44th Annual Road Research Meeting*. Purdue University, 1958, pp. 73–84.

11. Mackintosh, C.S., Rates of spread and spray in bituminous surface dressing of roads, *Civil Engineer of South Africa*, 1961, **3**, pp. 183–85.

12. Jackson, G.P., *Surface Dressing*. London: Shell International Petroleum Co., 1963.

13. Country Roads Board, *Bituminous Surfacing of Roads*. Victoria, Australia, Country Roads Board, 1954.

14. Mcleod, N.W., Basic principles for the design and construction of seal coats and surface treatments, *Proceedings of the Association of Asphalt Paving Technologists*, 1960, Supplement to Vol. **29**, pp. 1–15.

15. Heslop, M.F.W., Elborn, M.J. and Pooley, G.R., Recent developments in surface dressing, *The Highway Engineer*, 1982, July, pp. 6–19.

16. Heslop, M.F.W., Elborn, M.J. and Pooley, G.R., Aspects of surface dressing technology, *Journal of the Institute of Asphalt Technology*, 1984, **34**, pp. 24–34.

17. Khalid, H., Scratching the surface, *Highways*, 1992, **60**, pp. 18–21.

18. Robinson, D.A., Surface dressing – Variations upon a theme, *Journal of the Institution of Highway Engineers*, 1968, April, pp. 17–21.

19. Abdulkarim, A.J., *Investigation of the Embedment of Chippings for an Improved Road Surface Design Procedure*, PhD thesis, Heriot-Watt University, 1989.

20. Wright, N., Recent developments in surface dressing in the UK, *Eurobitume*, 1978, pp. 156–61.

21. Southern, D., Premium surface dressing systems, *Shell Bitumen Review*, 1983, **60**, pp. 4–8.

22. Department of Transport, *Specification for Highway Works*. London: HMSO, 1998.

23. BS434: 1984, Part 2, *Bitumen Road Emulsions (Anionic and Cationic): Code of Practice for Use of Bitumen Road Emulsions*. London: British Standards Institution, 1984.

24. Samanos, J., Novachip: New concept of chip seal system. Paper presented at the *International Symposium on Highway Surfacing* held at the University of Ulster, 1990.

25. Serfass, J.P., Bense, P., Bonnot, J. and Samanos, J., New type of ultrathin friction course, *Transportation Research Record 1304*, 1991, pp. 66–72.

26. Bellanger, J., Brosseaud, Y. and Gourdon, J.L., Thinner and thinner asphalt layers for maintenance of French roads, *Transportation Research Record 1334*, 1992, pp. 9–11.

27. Estakhri, C.K. and Button, J.W., Evaluation of ultrathin friction course, *Transportation Research Record 1454*, 1994, pp. 9–18.
28. Litzka, J.H., Pass, F. and Zirkler, E., Experiences with thin bituminous layers in Austria, *Transportation Research Record 1454*, 1994, pp. 19–22.
29. Soliman, S., Sibaud, C. and Potier, J.P., The situation of ultrathin bituminous concrete surfacing, *Eurasphalt & Eurobitume Congress*, Paper No. 7.010, 1996.
30. BS 598: 1990, Part 105, *Sampling and Examination of Bituminous Mixtures for Roads and Other Paved Areas: Methods of Test for the Determination of Texture Depth*. London: British Standards Institution, 1990.

CHAPTER 20

Structural maintenance of road pavements

D. McMullen and M.S. Snaith

20.1 Introduction

This chapter describes the techniques that are used to decide when a road pavement needs structural maintenance, and what strengthening measures may be appropriate. The usefulness of deflection measurements within the process is considered, and several overlay design systems are described. It is not appropriate to 'mix and match' elements from different design procedures, and the latest manuals or other sources should be consulted when adopting a particular procedure.

20.2 Concept of pavement strengthening

20.2.1 Pavement deterioration

Road pavements generally deteriorate gradually under normal traffic loading. Evidence of structural deterioration (i.e. that which affects the structural integrity of the pavement) in flexible pavements is generally seen as rutting or cracking in the wheel tracks. These defects may develop at the same time but, in a properly designed pavement, neither their extent nor severity should be excessive before the anticipated design traffic has been carried.

The rate of pavement deterioration depends on several factors including the volume and type of traffic, pavement layer material properties, the environment and the maintenance strategy adopted[1]. Deterioration will progress more rapidly if there is an increase in the volume of heavy vehicles or if the vehicle wheel loads increase. In hot climates, particularly, wheel-track rutting of bituminous road pavements is common on uphill road sections which carry slow-moving heavy wheel loads, due to the viscoelastic properties of bitumen. Additionally in hot climates, pavement cracking is often initiated at the surface due to age hardening of the bitumen[2].

Pavement deterioration may also be initiated or accelerated by poorly designed or maintained drainage. Drainage problems account for a significant number of premature pavement failures, generally caused by water which is trapped within the pavement layers or subgrade[3].

Table 20.1 Road surface conditions which are broadly comparable with the condition of the whole pavement[5]

Wheel-track cracking	Wheel-track rutting under a 2 m straight edge			
	<5 mm	**5 to <10 mm**	**10 to <20 mm**	**≥20 mm**
None	Sound	Sound	Investigatory	Failed
Less than half width* or single crack	Investigatory	Investigatory	Investigatory	Failed
More than half width*	Failed	Failed	Failed	Failed

* Half width is likely to be in the range 0.5–1.0 m. If there is no rutting to define the wheel tracks, use 0.5 m as half the wheel track width.

20.2.2 Investigatory and failure conditions

It is helpful to identify two levels of pavement condition which trigger a response – *critical* and *failure*. Ideally failure should never be reached, while the critical level should only occur near the end of the initial service or design period of the road[4]. The 'critical' condition may be more appropriately referred to as 'investigatory'[5]. This implies that, as the end of the service life is approached, it may be necessary to investigate any signs of failure with a view to establishing a timely and economic remedial works programme, thereby avoiding failure with its consequent direct costs to the Road Authority and indirect costs to road users.

As an example of how this may be effected, the investigatory and failure levels in the UK are given in Table 20.1.

20.2.3 Timing of pavement strengthening

Typical treatments which might be applied should investigatory surveys show signs of incipient failure would be a programme of structural overlays or, more immediately, resealing works to preserve the structural integrity of the pavement. It is known from experience that these will only be effective, from an economic and engineering standpoint, if effected at the investigatory deterioration level. Clearly, as a road reaches the end of its designed service life, be it from new or after remedial works, the need for monitoring condition surveys will increase. Where investigatory levels of deterioration are detected, further detailed investigations will be required at project level to determine the most appropriate rehabilitation strategy[6].

20.3 Structural assessment procedure

The aim of structural assessment is to determine the ability of a pavement to carry the anticipated future traffic loading for a given design period. The results of the assessment will usually include an estimate of the pavement's remaining life (expressed in terms of the cumulative number of standard axles) to a pre-defined pavement condition, such as is defined in Table 20.1. The assessment should also produce recommendations for any necessary remedial works (e.g. repairing the

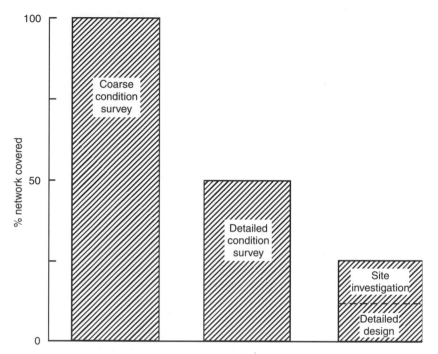

Fig. 20.1 The stepped approach to structural assessment

surface or drainage defects) and, additionally, for any strengthening works (e.g. overlaying, edge haunching). It is important that the causes of the observed defects are identified, to ensure that the recommended remedial and strengthening works be appropriate and compatible.

Structural assessment usually includes a deflection survey. The surface deflection is regarded as a quantitative indicator of the strength of flexible pavements. With an appropriate analysis of deflection measurements (see Section 20.4), surface deflection should give a reasonable estimate of the pavement's remaining life.

The structural assessment procedure generally follows a stepped approach (see Fig. 20.1) comprising a coarse (functional) condition survey of the full network (or road under study), followed by a more detailed survey of those road sections (targeted by the coarse survey) where remedial or strengthening works are anticipated (i.e. where investigatory levels are breached). The detailed survey may be undertaken at selected representative test lengths.

20.3.1 Structural assessment procedure for UK trunk roads

As an example of such a system, the structural assessment procedure currently recommended for trunk roads in the UK is summarized in Fig. 20.2. The procedure is sequential, with progression to the detailed investigation dependent on the extent and severity of the pavement distress. The full process outlined in Fig. 20.2 may not be necessary depending on the needs of the project and the availability of

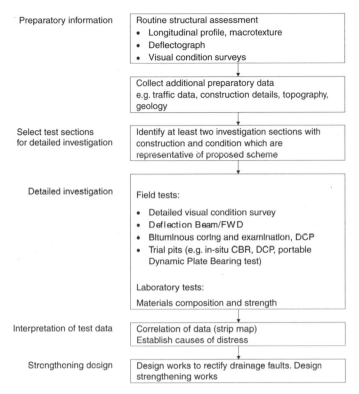

Preparatory information — Routine structural assessment
- Longitudinal profile, macrotexture
- Deflectograph
- Visual condition surveys

Collect additional preparatory data
e.g. traffic data, construction details, topography, geology

Select test sections for detailed investigation — Identify at least two investigation sections with construction and condition which are representative of proposed scheme

Detailed investigation — Field tests:
- Detailed visual condition survey
- Deflection Beam/FWD
- Bituminous coring and examination, DCP
- Trial pits (e.g. in-situ CBR, DCP, portable Dynamic Plate Bearing test)

Laboratory tests:

Materials composition and strength

Interpretation of test data — Correlation of data (strip map). Establish causes of distress

Strengthening design — Design works to rectify drainage faults. Design strengthening works

Fig. 20.2 Overview of structural assessment procedure for bituminous pavements (based on reference 6)

relevant records. In addition, it is arguable whether the relatively expensively-derived deflection data should be collected at the preparatory stage. In many such systems this would be collected solely at the detailed investigation stage.

Ideally, the condition collection portion of the *preparatory information stage* should be an annual one using relatively coarse and economic techniques. This level of survey should throw up a number of sections where the investigatory level is breached and a further detailed investigation is required.

A *detailed investigation* is generally undertaken for those roads where it has been shown that potential problems exist. The main objectives of the detailed investigation are to establish the reasons for the condition of the pavement, so that appropriate remedial and strengthening measures may be applied, and to collect sufficient data to facilitate the design of works.

At each test location a programme of field tests is undertaken which should include deflection testing by an appropriate device (see Section 20.4). Associated measurements of rut depth and surface cracking are made at each test location.

A full inspection of the surface water drainage system is also recommended, including carriageway crossfall and gradient, gullies, manholes and catchpits. Examination of outfall pipes will indicate whether they are functioning correctly. Filter edge drains should also be inspected for excessive growth or detritus over the filter media and, if necessary, a trial pit should be dug to expose the pipe for inspection.

Where it seems likely that remedial works will be required, the condition of each pavement layer and the subgrade should be assessed. The condition of the bituminous layer can be assessed by subjecting bituminous cores (150 mm diameter) to visual inspection and laboratory testing. The cores should be taken at locations exhibiting different defects (e.g. surface cracking, wheel track rutting). In addition, the condition of the granular layers and subgrade can be assessed using trial pits and associated laboratory testing. The testing programme will normally include particle size distribution, Atterberg limits, in-situ moisture content and density, dry density–moisture content relationship, in-situ CBR, and CBR at 95–98 per cent maximum dry density (soaked).

An estimate of the relative strength of each granular layer and the subgrade can be obtained using the *Dynamic Cone Penetrometer (DCP)*. This device offers a relatively rapid means of identifying any weak layers beneath the bituminous surfacing. The DCP test is usually undertaken immediately after the extraction of a bituminous core (to assess the lower pavement layers) or at the bottom of a trial pit (to assess subgrade strength under thicker pavements). The test is suitable for most granular and weakly stabilized layers, but some difficulty may be experienced in penetrating materials with large particles, strongly stabilized layers or very dense crushed stone. Details of the DCP and associated analysis software are given elsewhere[1,7].

With favourable ground conditions the DCP test can provide information on lower pavement layer thickness and in situ CBR values[1,8–10]. This can be particularly useful for detecting variations in layer strength and thickness, essential knowledge when analysing deflection test results.

Proper *interpretation of the test data* collected from the test sections should reveal the main contributions to the deterioration in overall pavement strength. The survey data can be effectively presented in the form of a strip map showing the pattern of cracking and rutting in each 100 m length of carriageway, with other relevant data shown in graphical or tabular format below the strip map. The data should include deflection results, pavement layer thicknesses, subgrade strength, projected traffic over the design life, and other relevant field and laboratory test results.

As an example, current UK recommendations[6] advise that the primary parameters which should be used to determine the causes of distress, and thereafter remedial treatments, are the observed defects and conditions of the layers. Deflections and layer thickness relative to design are of secondary importance. It is often helpful to effect a quantitative structural analysis taking account of the relative condition of each layer to overall pavement strength. Pavement model calibration may be based on measured deflection, adjusting the stiffness moduli to values appropriate to known environmental and loading conditions.

Differences in deflection measurements between test sections can indicate a change in subgrade condition if there are no significant variations in the material properties of the pavement layers. The reasons for differences in deflection measurements between test sections (or variation in deflection along a given section) should be sought first by comparison of the information presented on the strip map.

Higher deflections may be expected to coincide with areas of cracking or wheel track rutting, but this is not always the case. The overriding aim is to explain the causes of the pavement distress and design appropriate treatments. Deflections are used as an indicator of pavement performance with the knowledge that deflection–condition relationships (discussed in Section 20.5) generally exhibit considerable scatter.

When the causes of the observed defects have been identified, consideration can be given to the *strengthening design* of any necessary works. From a structural viewpoint, the desirable timing of the works will be indicated by the condition of the road (confirmed by site inspection) and any estimate of remaining life available. Additional factors such as existing and future traffic levels and the relative importance of the road under study should influence budget availability. However, for medium- and heavily-trafficked roads, whole life costing invariably demonstrates that any remedial works needed to preserve the existing structure should not be delayed.

In some instances reconstruction, or partial reconstruction, will be an option that is preferred to overlaying, e.g. if only one lane of a multi-lane carriageway requires strengthening, the additional cost of applying an overlay over extra lanes may be prohibitive. Or, where investigations indicate that there is a weakness in any of the pavement layers, it may be preferable to reconstruct to the bottom of the layer concerned rather than overlay the weakened structure.

Methods of overlay design are described in Sections 20.6 and 20.7.

20.4 Use of deflection measurements

It is generally accepted that pavement deflections can be used as an indicator of a pavement's structural performance, i.e. the magnitude of the deflection under a known load can be related to the strength of the pavement layers and subgrade[4,5,11]. By comparison with other design parameters (e.g. strain in the pavement layers and subgrade), surface deflection can be measured relatively easily. With jointed concrete pavements, load transfer across joints can be assessed by loading the slab on one side of the joint and measuring the deflection at both sides of the joint.

In France and the UK the *Deflectograph* is widely used to measure deflections. This device is essentially an automated version of the *Deflection Beam*, both of which measure the surface deflection under a rolling wheel load. Alternatively the *Falling Weight Deflectometer (FWD)* records the surface deflection profile under a stationary impact load.

The choice of deflection measurement device and operating procedure depends on the design method to be used to assess remaining life and any strengthening measures. This is because design charts developed for deflections from a specific testing device generally cannot be used with deflections measured by others. In addition, design relationships developed for one country, particularly those developed empirically, are not necessarily applicable elsewhere. When a seemingly appropriate relationship is selected, it should be calibrated to local conditions with considerable care.

20.4.1 Deflection beam and deflectograph

The *deflection beam* was originally developed by Dr A C Benkelman in the USA. The variant used in the UK was designed by the Transport Research Laboratory[12,13]; operating procedures appropriate to hotter climates are described in the literature[14].

The deflection in the pavement which is measured by the deflection beam is induced by passing a loaded dual-wheeled axle past the beam tip. The load used on the axle varies from country to country. In the USA and UK it is commonly set at 6350 kg, whereas other users set it to 8160 kg or 10 000 kg. Whilst the deflection–load relationship may be assumed linear, considerable care is required when comparing results under different axle loads.

There are two different ways of applying the axle loading. One is where the dual wheels are initially stationary at the beam tip to give the point of maximum deflection and thereafter the vehicle is moved away; this is known as the *rebound deflection*[15]. The other is where the dual wheels straddle the beam close to the pivot point at the start of the measurement cycle, and the dual wheel assembly then moves forward, passes the beam tip to give the maximum deflection, and continues moving away from the beam; this is termed the *transient deflection*[13]. The two methods can yield very different results, particularly in hot climates where the bituminous layer moves from mainly elastic to viscous behaviour.

The *Lacroix deflectograph*, which was was developed in France by the Laboratoire Central des Ponts et Chaussées (LCPC), consists of two deflection beam type mechanisms (one for each wheel track) that are mounted on a common frame located underneath a two-axle lorry. The deflectograph is driven along the road at a constant but slow speed of about 4 km/h, and measures the deflection in both wheel tracks at approximately 4 m intervals (i.e. the spacing of the measurements varies with the wheelbase of the vehicle).

A correlation can be established between deflections measured with the deflectograph and the deflection beam. The correlation shown in Fig. 20.3 is based on measurements on pavements containing crushed stone, cement-bound and bituminous-bound roadbases constructed on a range of subgrades with CBR values varying from 2.5 to 15.0 per cent. Pavements with rolled asphalt, bitumen macadam and tarmacadam surfacings were included, and the relationship was established within the temperature range 10–30°C. The relationship indicates that deflectograph deflections are generally lower than those obtained with the deflection beam; one reason for this is that neither device gives an absolute measure of deflection, and the influence of the deflection bowl on the beam supports is greater with the deflectograph.

20.4.2 Falling weight deflectometer (FWD)

The *falling weight deflectometer* was originally designed in France and introduced later in Denmark[16] and the Netherlands[17–19]. The main features of the device are given in Fig. 20.4, which shows the trailer-mounted variant; the device can also be mounted within the body of a vehicle. With the FWD an impulse load is applied to the road surface by dropping a weight onto a spring system. The weight and drop height can be adjusted to give the required impact

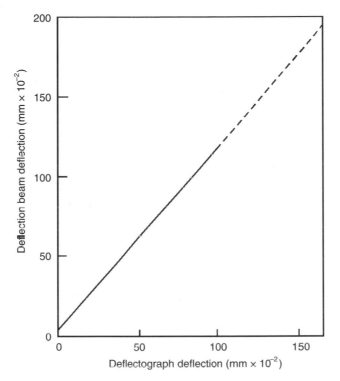

Fig. 20.3 Correlation between deflections measured with the deflection beam and deflectograph[4]

loading. The surface deflection of the road is measured by geophones located directly under the loaded area and at several offset positions. This gives the surface profile of the deflection bowl in addition to the maximum deflection under the loaded area.

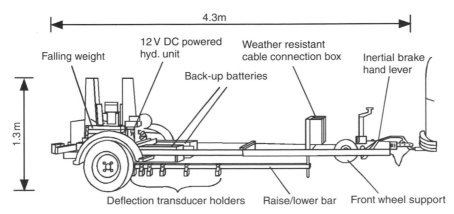

Fig. 20.4 Main features of the falling weight deflectometer[20]

The analysis of FWD deflections (to determine likely weak pavement layers, estimate remaining life and, where required, the overlay thickness) is normally based on analytical methods. The layer stiffnesses are derived by a back analysis procedure based on the FWD deflection profile, ideally assisted by a comprehensive knowledge base and auxiliary testing of the constituent layers of the pavement[21]. Strain or deflection criteria may then be used to estimate residual life and any required overlay thickness.

There is no direct correlation between FWD deflections and those measured with the deflectograph or deflection beam. Hence, FWD deflections cannot be used as input to design charts developed for other measurement devices.

20.4.3 Effect of temperature on deflection

With all deflection-based design methods, the influence of pavement temperature needs to be taken into account. The effect of higher temperature is to reduce the effective stiffness of the bituminous layer, resulting in higher deflections. (In the longer term, however, age hardening increases the effective stiffness and lessens the influence of temperature on measured deflections.) With some bituminous materials there is a tendency for the surfacing to flow plastically upwards between the dual tyres of the deflectograph (or deflection beam vehicle) as the temperature increases. It is therefore desirable to restrict deflection measurement to those months of the year and times of the day when temperatures are relatively low and pavement response can be considered elastic. Also, it is necessary to establish a relationship between measured deflection and pavement temperature (normally measured at a depth of 40 mm) so that all deflections can be 'corrected' to that which would occur at a standard reference temperature.

The need for a temperature correction can be assessed by taking repeated deflection measurements at several test positions as the pavement heats up during the morning. For each position, a deflection–temperature relationship is plotted from which a correction to a reference temperature can be deduced. A reference temperature of 20°C is normally adopted for moderate climates; 35°C has been suggested[14] as appropriate for tropical and subtropical conditions.

20.5 Use of deflection–life relationships

For most pavement types it is possible to define a relationship between deflection (under a specific loading regime), traffic damage (expressed in cumulative standard axles carried), and structural condition (usually defined in terms of surface cracking and rutting). The main features of such a relationship (Fig. 20.5) are (a) a deflection trend line which shows how deflection varies as the cumulative traffic (i.e. damage) increases, and (b) an envelope criterion curve plotted through points of similar condition at different levels of cumulative traffic. The relationship is based on the assumption that at a given level of cumulative traffic damage, higher deflections will be recorded as the structural condition worsens.

By plotting the position (point P in Fig. 20.5) corresponding to the current deflection level and traffic N_P (in cumulative standard axles) carried to-date, the remaining life N_R to the condition defined by the criterion curve can be estimated

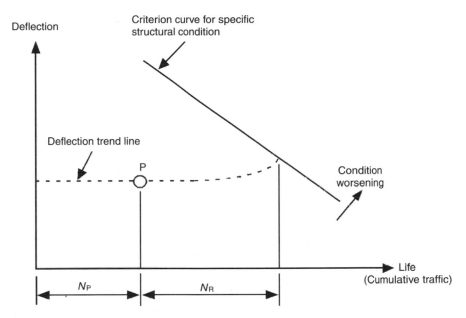

Fig. 20.5 Elements of a deflection–life relationship

by following the trend line. The criterion curve is used with certain deflection-based design procedures[4,15] to define the limiting deflection values for overlay design or new pavement design, based on traffic anticipated during the desired design life.

Various relationships of this type have been prepared for deflection data stemming from the deflectograph and FWD[4,5,11,15,22]. In the more recent variants, probability has been introduced to increase the confidence with which the road engineer may use the system. These relationships are usually obtained from field-based observations of varying robustness which will be specific to the device used for deflection measurement, its method of operation, and the materials and environment for which the relationships were originally derived. Hence, considerable care is required when assessing the suitability of a relationship in anything other than its original specification.

20.5.1 Deflection–life relationships for UK conditions

As an example of how such methods have been derived and used, the current UK process[4,5] is described.

The development of deflection histories for UK roads has involved regular monitoring of pavement deflection and associated condition. Road condition has been classified as sound, critical or failed, based on the severity and extent of wheel track rutting and wheel track cracking.

Four performance charts have been published[4] based on the deflection trends observed from construction to the onset of 'critical' conditions. Different charts have been produced for pavements with roadbases of granular, cement-bound, and

Table 20.2 Phases in the design life of a pavement[5]

Phase	Structural behaviour
1	New or strengthened pavement is stabilizing. Deflection is variable but generally decreasing.
2	Structural behaviour may be predicted with confidence. Deflection remains stable.
3	Structural behaviour becomes less predictable. Deflections may continue as Phase 2, gradually increase, or increase more rapidly. (Formerly known as the 'critical' condition[4], the term 'investigatory' is now used to emphasize the need to monitor structural behaviour.)
4	Pavement deteriorates to a failed condition from which it can be restored only by total reconstruction.

bituminous roadbases, and a further distinction has been made between granular bases which develop some form of cementing action and those which do not. The relationships are reported to be valid for flexible pavements with subgrade CBR in the range 2.5 to 15 per cent, surfaced with bituminous materials common in the UK. The charts (e.g. Fig. 20.9) indicate deflection trend lines which lead to four envelope curves defining different levels of probability that the 'critical' life of the pavement will be achieved. As in the previously described generic system (Fig. 20.5), the remaining life is determined by following the deflection trend line from a point defined by the present deflection and past traffic (in cumulative standard axles) to the envelope of the selected probability level for the particular construction type.

It has been suggested[5] that the life of a flexible pavement can be characterised as having four main phases (Table 20.2) that can be identified on a deflection-life relationship. During the early life of the pavement (Phase 1) the deflection may decrease due to compaction of the pavement layers and also moisture content changes in the granular layers and subgrade. Phase 2 is represented by a horizontal section of the deflection trend line, leading to a gradual increase of deflection at the investigatory condition (Phase 3) where it is likely that cracks are being initiated, with a more rapid deflection increase at Phase 4. The need for investigation at Phase 3 is emphasized in reference 5, so that the cause of the deterioration can be determined before the design of strengthening works.

There is evidence to suggest that for pavements with bituminous bases, oxidation of bitumen throughout the bituminous layer results in increasing stiffness and reduction in deflection[23,24]. It has been suggested[23] that, where this occurs, Phase 1 may 'spill over' into Phase 2.

More recent evidence suggests that thicker flexible pavements may not conform to the phased behaviour of Table 20.2. A study[40] of ten heavily-trafficked motorway sections in the UK has shown a trend of decreasing deflection with age and traffic. The sections under study were of flexible construction and had carried up to 56 msa. At three of the sites the trend of decreasing deflection continued well beyond the design life.

There is clearly a need for continuous monitoring of pavements to observe structural behaviour at increasing traffic levels. The use of accelerated testing methods[41] would assist in predicting such behaviour for different pavement constructions.

Table 20.3 Data required at each test site to establish a deflection–life relationship

Traffic loading	Historic traffic data
	Manual classified counts (MCC)
	Automatic traffic counts (ATC)
	Annual vehicle growth rates for each vehicle class
	Axle load data (ESA/cv)
Surface condition	Deflection under specified load
	Verified deflection–temperature relationship
	Surface cracking (type and intensity)
	Rut depth

20.5.2 Site-specific relationships

As noted previously, published deflection–life relationships generally do not have application outside the region for which they were developed. However, a relationship can be established for any given road, based on a condition monitoring programme at selected test sites.

The test sites should be selected to represent the full range of pavement condition (including no defects), traffic loading, construction thickness, subgrade strength, topography and elevation (cut/fill). Condition data collected periodically at each test site (Table 20.3) should include deflection and associated structural condition assessed in terms of surface cracking and wheel track rutting. The test sites should be monitored at least two or three times each year to assist detection of seasonal variations in the measurements taken.

An accurate assessment of traffic loading (in terms of cumulative standard axles carried) is needed at each test site. Historic loading data may be available from records or previous road study reports; failing that, it may be necessary to project back in time from current survey data. Current traffic loading should be assessed by regular manual classified counts and axle load surveys[25]. This should facilitate the assessment of the number of commercial vehicles (cv) and the commercial vehicle damage factor (number of equivalent standard axles per cv). In the longer term this should also assist in verifying the annual vehicle growth rate.

The structural condition should be classified according to the severity of the cracking and rutting in the vicinity of the deflection test position. The classification of Table 20.1 (from UK practice) may be suitable in this respect; however, it may be more appropriate to adopt a classification which measures crack intensity (i.e. length of crack per square metre), as in Table 20.4. The classification system may be based on crack intensity alone (if this is the major mode of failure[26]) or, alternatively, crack type may be an appropriate proxy for measured intensity (see Table 20.5).

For each test site, the mean deflection corresponding to each condition classification should be plotted against the number of cumulative standard axles carried. Ideally, this should result in a sufficient range of condition and associated deflection to provide a clear demarcation between deflection levels at which the road can be classified, for example, as sound, critical or failed (Fig. 20.6).

Table 20.4 A condition classification in terms of crack intensity[26]

Crack index	Crack intensity (m/m²)	Condition
0	No crack	Sound
1	0–2	Critical
2	>2	Failed

Table 20.5 A condition classification in terms of crack description and rut depth[27]

Crack index	Crack description	Rut index	Rut depth (mm)
0	No crack	0	No rut
1	Single crack	1	1–5
2	Non-interconnected	2	6–10
3	Interconnected	3	11–15
4	Block	4	16–20
5	Disintegration	5	over 20

Pavement condition index (PCI) = Crack index + (rut index – 1*)
*when rut index ≥1

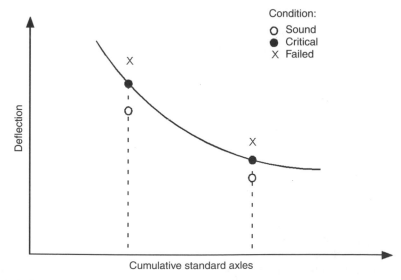

Fig. 20.6 Development of an idealized deflection–life relationship

However, in practice there may be considerable scatter of results, in which case a set of criterion curves should be established to represent different levels of probability that a given life can be achieved before the onset of the defined 'critical' conditions.

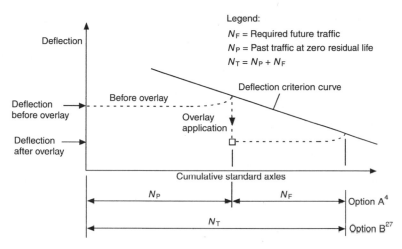

Fig. 20.7 Effect of overlay application on position within deflection–life relationship

In the longer term, the deflection data from the test sites should be used to establish deflection trend lines from the time of new construction to the onset of the predefined limiting condition. In addition, the influence of strengthening overlays should be assessed. The philosophy of reference 4 suggests that a pavement may be treated 'as new' following a structural overlay, i.e. the deflection after overlaying becomes the early life deflection, and the overlay is designed for future traffic, N_F, as shown for Option A in Fig. 20.7. A more conservative approach[27] would be to design overlays on the basis of the sum of past traffic at zero residual life, N_P, and required future traffic, N_F, as for Option B in Fig. 20.7.

20.6 Overlay design methods for flexible pavements

Overlay design methods for flexible pavements can be broadly classified[28] according to whether they are based on component analysis, on non-destructive empirical procedures, or on analytically based procedures.

Early overlay design methods relied on component analysis involving the comparison of an existing pavement structure with a new pavement design, the difference giving an estimate of required overlay thickness. Examples of this approach are the *Asphalt Institute's effective thickness* procedure[15] and the *AASHTO structural number* concept[29], both of which require an assessment of the condition of the individual layers in the existing pavement. Current non-destructive empirical overlay design procedures include deflection-based methods such as the *Transport Research Laboratory* method[4]. *Analytical overlay design* procedures may also use deflection testing, and assume that the pavement can be analysed as a multilayer elastic system; procedures based on the falling weight deflectometer are included in this category. Analytical procedures are also applicable to concrete overlays on flexible pavements.

The different overlay design methods are demonstrated by the following examples.

20.6.1 Asphalt Institute effective thickness procedure

With the Asphalt Institute method[15], each layer in the existing pavement is converted to an equivalent thickness of 'asphaltic concrete' by applying a conversion factor, based on the layer's condition. The sum of the equivalent thicknesses for each layer is subtracted from the design thickness[30] of a new full depth asphalt pavement (for the future traffic loading) to give the required overlay thickness. The procedure is illustrated by the following example.

Problem: Assume that a flexible pavement's construction details are as follows:

Layer No	Thickness (mm)	Material	Condition	Conversion Factor*
1	100	Asphaltic concrete	Block cracking	0.5
2	200	Cement stabilized roadbase	Well graded, $PI=4$	0.3
3	150	Granular subbase	Well graded, $PI=6$	0.2
4	–	Subgrade	Resilient modulus 60 MPa	–

* MS-17, Asphalt Institute[15]

Determine the required asphaltic overlay thickness for a future life of 4 msa. The mean annual air temperature (MAAT) is 24°C.

Solution: From Fig. 20.8 the design thickness of a new full-depth asphalt concrete pavement, T_n, is 325 mm, based on future traffic of 4 msa and a subgrade modulus of 60 MPa. The effective thickness of the existing pavement, T_e, is given by:

$$T_e = (100 \times 0.5) + (200 \times 0.3) + (150 \times 0.2) = 140 \text{ mm}$$

Therefore, the required overlay thickness $T_o = T_n - T_e = 325 - 140 = 185$ mm

20.6.2 Overlay design based on structural number

The structural number (SN) of a pavement is empirically related to the thickness and condition of the constituent layers[29], thus:

$$SN = a_1 D_1 + a_2 D_2 m_2 + a_3 D_3 m_3 \qquad (20.1)$$

where a_1, a_2 and a_3 are layer strength coefficients, and D_1, D_2 and D_3 are the thicknesses (inches) of the surfacing, roadbase and subbase respectively; m_2 and m_3 are drainage coefficients for the roadbase and subbase respectively. The basis for overlay design using the structural number is to compare the design SN (required for future traffic loading) with the existing SN and make up the difference with a bituminous overlay, i.e. the required overlay thickness is given by:

$$\text{Overlay} = \frac{\text{Design SN} - \text{Existing SN}}{a_1} \qquad (20.2)$$

where a_1 is the layer coefficient for the overlay material. The design SN is obtained by use of the AASHTO design charts based on future traffic loading requirements. The existing SN is estimated using Equation 20.1 with coefficients a_1, a_2, a_3, m_2 and

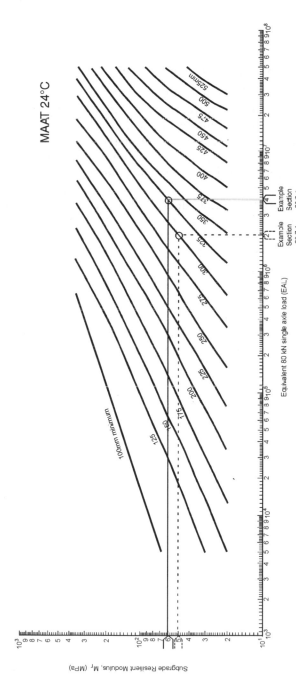

Fig. 20.8 Asphalt Institute design chart for full depth asphalt concrete pavements[30].

m_3 assigned on the basis of the condition and drainage characteristics of the existing layers. (For further information on layer coefficients, and details of overlay design methods based on Structural Number, see the comprehensive AASHTO design guide[29].)

20.6.3 Transport Research Laboratory method

The TRL method[4] is based on deflections measured with the deflection beam under a 3175 kg wheel load, and corrected to a standard temperature of 20°C. For overlay design purposes a representative deflection is assigned to sections of road (design lengths) which exhibit uniform deflection levels. The design lengths should have consistent construction and traffic levels and may be identified from a study of the deflection profile (showing individual deflections) from the road under investigation.

In UK practice the representative deflection is generally taken as the 85th percentile value, which is used in association with a selected criterion curve (Fig. 20.9) to give an estimate of remaining life in terms of cumulative standard axles. (The 85th percentile is that value of deflection within a design length at or below which 85 per cent of the deflections are found.) If the predicted remaining life is less than that required, a strengthening overlay will be considered. If this is appropriate (in terms of possible physical constraints such as kerb levels and bridge clearances), the thickness needed is determined by reference to the appropriate overlay design chart (Fig. 20.10). The charts specify the thickness of rolled asphalt needed to give the desired future life, taking into account the remaining life of the existing pavement. The procedure is illustrated by the following example.

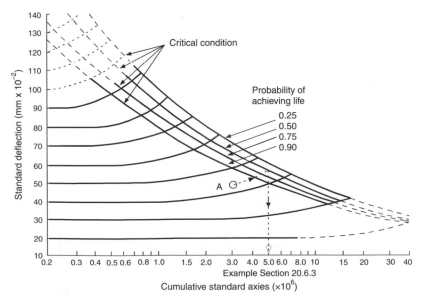

Fig. 20.9 Relation between standard deflection and life for pavements with bituminous roadbases[4]

Problem: Consider a highway pavement that has the following characteristics:

Pavement course	Thickness (mm)	Details
Surfacing	100	Rolled asphalt
Roadbase	150	Dense coated macadam
Subbase	200	Crushed rock
Subgrade	–	CBR = 6%

The highway has carried 3 msa since being newly constructed and appears to require major maintenance. Determine the overlay requirements of a design length having an 85th percentile deflection beam deflection (corrected to 20°C) of 48×10^{-2} mm, in order to give a future life of 10 msa with a 0.50 probability of achieving that life.

Solution The pavement is classified as having a bituminous-bound roadbase as more than 150 mm of bituminous materials are present.

First, determine the remaining life of the design length by plotting the position corresponding to the 85th percentile deflection and the past traffic (3 msa) on the deflection–life relationship for pavements with bituminous roadbases (A in Fig. 20.9), and following the deflection trend line to the 0.50 critical curve. A deflection of 48×10^{-2} mm gives a remaining life of 5 minus 3 = 2 msa. An overlay is therefore needed to achieve the desired life.

Then, determine the overlay thickness of rolled asphalt required for a deflection of 48×10^{-2} mm by plotting the future life (10 msa) and the 85th percentile deflection on the overlay design chart (A in Fig. 20.10) for pavements with bituminous roadbases (0.5 probability). This shows that a deflection of 48×10^{-2} mm requires an overlay of 50 mm.

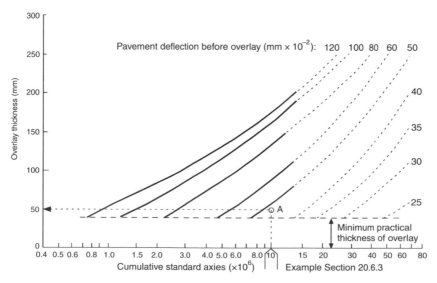

Fig. 20.10 Overlay design chart (0.50 probability) for pavements with bituminous roadbases[4]

It should be noted that while the TRL method[4] is based on deflections measured with the deflection beam, the variant currently adopted for UK trunk roads[5,6] makes use of deflections measured with the deflectograph.

20.6.4 Analytical methods

Analytical methods of overlay design generally make use of a layered structural analysis and associated design criteria (e.g. deflection–life or strain–life relationships) appropriate to observed pavement behaviour. The basic components of this approach are as follows[26]: (1) select a suitable method of elastic analysis, capable of applying a single or dual wheel load; (2) as shown in Fig. 20.11 (a), represent the existing pavement and subgrade by a layered elastic system in which each layer is characterized by an elastic modulus (E) and Poisson's ratio (ν); (3) calibrate the pavement model to existing site conditions, e.g. adjust layer properties, within

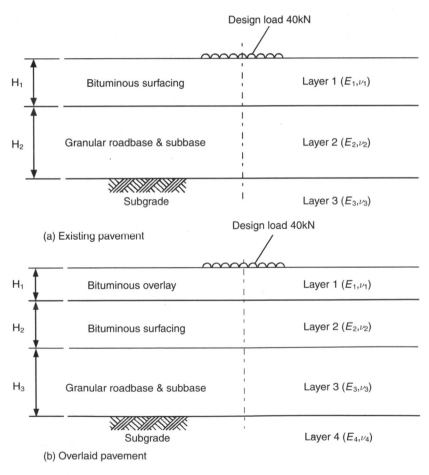

Fig. 20.11 Typical pavement models for elastic analysis: (a) existing pavement, and (b) overlaid pavement

limits, until computed deflections/strains agree with measured values; (4) calculate critical parameters (deflection, strains) with increasing overlay thickness (Fig. 20.11(b) shows a typical pavement model); (5) use an appropriate deflection–life or strain–life relationship, valid for the site conditions, to estimate the future life given by each overlay increment; and (6) present the results of analyses in the form of overlay design charts, giving the overlay thickness required to achieve different values of future life.

An example of design charts produced by analytical means is seen in the *Shell Design Manual*[18,19], produced for world-wide application. In addition the analytical approach may be used to extend (in terms of cumulative standard axles) design charts produced by empirical means over a limited range of traffic loading.

20.6.5 Falling weight deflectometer (FWD) analysis procedure

Overlay design systems based on falling weight deflectometer deflections generally make use of the full deflection profile, together with available layer thickness information, to assign layer properties (elastic modulus, E, and Poisson's ratio, ν) to an idealized elastic model of the pavement (e.g. Fig. 20.11). An analytical approach is adopted for the estimation of remaining life and design of overlay requirements. Information on existing layer thicknesses is usually obtained from asphalt coring, dynamic cone penetrometer testing, and trial pits. In addition, ground penetrating radar techniques offer a rapid means of assessing the layer thickness of bound layers.

The main elements of the FWD analysis procedure are shown in Fig. 20.12. The procedure can be used for flexible and rigid pavements, provided appropriate design criteria, calibrated to local conditions, are available. The analysis may be applied to each FWD test position or to be identified as uniform sections.

Considerable research effort has been spent in the search for a reliable method of determining the pavement layer moduli from the FWD deflection bowl. A three- or four-layer pavement model is often used. Empirical studies have shown that the outer FWD deflections (i.e. those furthest from the centre of loading) are influenced principally by the subgrade modulus. As the offset from the centre of loading decreases, the influence of the upper pavement layers increases, and the central deflection is a function of the moduli of all pavement layers and subgrade[5].

Evaluation of the pavement layer moduli generally relies on a back-analysis procedure whereby the modulus value for each layer of the model is adjusted (within specified limits) until a close correlation is achieved between the deflection bowl measured under the FWD loading and that computed for the pavement model. In practice, FWD analysts may use one or more software packages (proprietary or written to specification) to evaluate the moduli. However, sound engineering judgement is needed when using such software, as their main function is to introduce efficiency, logic and organization to the numerical analyses identified in Fig. 20.12.

An earlier study[31] which compared the results produced by different back analysis programs (all of which used curve-fitting procedures) found that the different programs gave significantly different results and that most methods were highly sensitive to the assumed layer thicknesses. Current UK advice[5] states that an overestimate of thickness by as little as 15 per cent can result in a 50 per cent

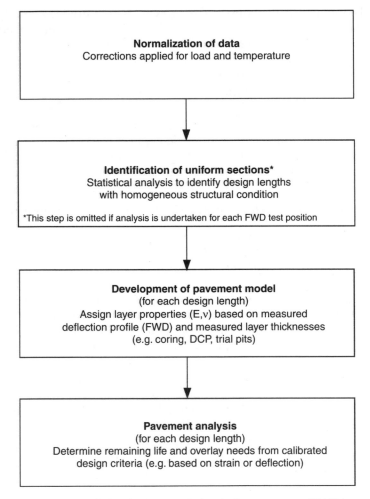

Fig. 20.12 Overview of the falling weight deflectometer (FWD) analysis procedure

underestimate in moduli values, which is enough to give the impression of poor integrity in a sound layer. As this aspect is generally well known, most FWD operators routinely measure pavement layer thicknesses at regular intervals; if ground radar techniques are adopted, regular physical measurements are still required for calibration purposes. It cannot be over-emphasized that accurate measurement and location of cores, DCP tests, trial pits, etc. is essential. Also, with any analysis software, it is useful to undertake a sensitivity study[32] to assess the influence of variations in all input data on the critical output parameters.

As noted previously, the moduli assigned to each layer in the pavement model will clearly have boundary limits imposed by the material type and its condition. In addition, the method of elastic analysis adopted may impose further conditions (e.g. by specifying a minimum layer thickness in terms of applied loading radius, or by specifying a minimum modular ratio between adjacent layers). The analysis

Table 20.6 Concrete overlay options and treatments to existing pavements used in the UK[6]

		Existing pavement			
		Flexible or flexible composite	**URC or JRC**	**CRCP**	**CRCR**
Overlay	URC	2	3	1	2
	JRC	2	3	1	2
	CRCP	1	4	1	1
	CRCR	1	2	1	1

Notes:
1. Acceptable and no surface treatment other than remedial works is normally necessary.
2. Separation membrane required.
3. No treatment other than remedial works is normally necessary, but joints should occur above one another.
4. This combination not normally appropriate.
5. URC = Unreinforced concrete;
   JRC = Jointed reinforced concrete;
   CRCP = Continuously reinforced concrete pavement; and
   CRCR = Continuously reinforced concrete roadbase.

process, therefore, comprises an analytical procedure, the results of which are modified in the light of engineering experience. The combination of analytical technique and engineering experience is seen in the development of knowledge-based systems[21] which aim to combine the concepts presently used separately in most evaluation procedures.

20.6.6 Concrete overlays to flexible pavements

Design charts for the selection of concrete overlay thicknesses to both flexible and rigid pavements are given in Figs 20.14 to 20.17. Recommended treatments to existing surfaces are given in Table 20.6. The charts are discussed in subsection 20.7.2.

20.7 Overlay design methods for concrete pavements

Overlays to concrete pavements can be applied as a bituminous overlay or as a concrete overlay (overslabbing). The overlaying of concrete pavements generally presents greater problems than is the case with bituminous pavements, due to the presence of joints or wide cracks in the underlying concrete. With bituminous overlays, this can result in *reflection cracking*, where the cracking pattern in the existing pavement propagates through the overlay material and produces a similar pattern at the surface of the overlay. With concrete overlays, the new slab is required to accommodate differential movement across any existing joints or cracks; this is

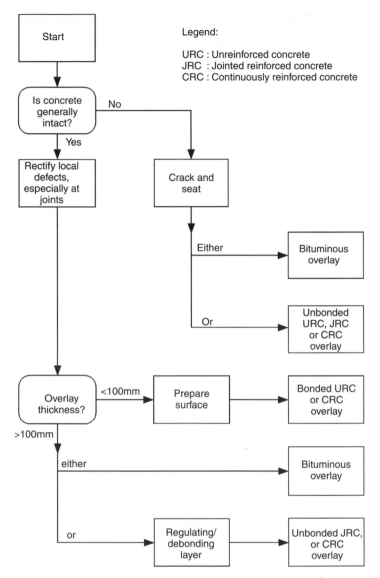

Fig. 20.13 Overlay construction options for rigid pavements[6]

achieved[33] by forming joints in the overlay at the same location as the existing joints or cracks in the underlying pavement, or by separating (debonding) the overlay from the existing pavement.

A useful summary of the overlay construction options recommended for rigid pavements is given in Fig. 20.13. With all overlay options, the existing concrete pavement should provide a stable platform of uniform strength to receive the overlay. As with bituminous pavements, drainage failures can lead to weakening of the unbound layers and subgrade, so it is important that any drainage faults are rectified before assessing overlay requirements. Any loose or rocking slabs should be stabilized or replaced, whilst any vertical movements at joints or cracks should be stabilized either by pressure grouting or vacuum grouting[34,35]; surface levels should also be regulated, where necessary, and any spalling at joints rectified.

20.7.1 Bituminous overlays on rigid pavements

There is no uniformly accepted method for assessing the thickness of a bituminous overlay required to strengthen a rigid pavement. If the existing pavement has significant deterioration it may be appropriate to crack and seat it prior to overlaying. As an example, guidance on bituminous overlay thickness may then be obtained from UK design charts for new pavements[36] by assuming that the existing pavement is either a granular subbase or a cement-bound roadbase.

In addition to structural considerations, the bituminous overlay thickness will be governed by the need to delay the development of reflection cracking initiated from joints or cracks in the underlying pavement.

As noted previously, the falling weight deflectometer (FWD) analysis procedure may be used to assess the required thickness of bituminous overlay on a rigid pavement, provided appropriate design criteria (calibrated to local conditions) are available.

The *Asphalt Institute effective thickness procedure*[15], described previously for overlays to flexible pavements, can also be applied to rigid pavements. As with bituminous pavements, each layer of the existing rigid pavement is converted to an equivalent thickness of asphaltic concrete by applying an appropriate conversion factor. A minimum thickness of 100 mm is recommended[15] for bituminous overlays placed directly on concrete pavements; if the design overlay thickness is 175–225 mm, the use of a crack-relief layer may also be considered. The procedure is illustrated by the following example.

Problem: Consider a rigid pavement that has the following construction details:

Layer	Thickness (mm)	Material	Condition	Conversion Factor*
1	200	Jointed Portland cement concrete	Cracked, seated on subbase	0.5
2	100	Sand and gravel subbase	Well graded	0.2
3	–	Subgrade	Resilient modulus 50 MPa	–

* From MS-17, Asphalt Institute[15]

Determine the required asphaltic overlay thickness for a future life of 2 msa. The mean annual air temperature is 24°C.

Solution   From Fig. 20.8 the design thickness of a new full depth asphaltic concrete pavement, T_n, is 300 mm, based on future traffic of 2 msa and subgrade modulus 50 MPa. The effective thickness of the existing pavement, T_e, derived using factors from the Asphalt Institute, is given by:

$$T_e = (200 \times 0.5) + (100 \times 0.2) = 120 \, \text{mm}$$

The required overlay thickness $T_o = T_n - T_e = 300 - 120 = 180$ mm.

The Asphalt Institute has also produced a design chart which provides an indication of the asphaltic concrete overlay needed on a Portland cement concrete pavement; this chart relates overlay thickness to slab length and temperature change. Suggested overlay thicknesses range from 100 mm to 225 mm for slab lengths that vary from 3 m to 18 m and temperature change ranges from 17°C to 44°C. (The temperature change is defined as the difference between the highest normal daily maximum temperature and the lowest normal daily minimum temperature for the hottest and coldest months, respectively, based on a 30 year average.) Further details are given in reference 15.

Reflection cracking generally refers to a cracking pattern which appears at the surface of a bituminous layer due to differential vertical movement (under wheel loading) across joints or cracks in the underlying layer. The surface cracks may take several years to develop, and eventually reflect the pattern of cracking in the underlying layer. This form of cracking commonly develops after a bituminous overlay is placed on a concrete pavement, or when a bituminous surfacing is laid on a cement-bound roadbase. If the differential vertical movement at joints/cracks is kept to a minimum there will be a lower probability of reflection cracking developing. One source[15] advises that the differential deflection should be limited to 0.05 mm.

Several methods of controlling the development of *reflection cracking in bituminous overlays* have been proposed[15,33,37,42] as follows.

1. Use of the 'crack and seat' process, whereby fine vertical transverse cracks are induced in the existing concrete slab. This reduces the load spreading capability of the slab, but assists in controlling the development of reflection cracking in the overlay by two mechanisms, i.e. by reducing the horizontal strain level between adjacent concrete elements and by minimizing the relative vertical movement between concrete elements under traffic loading.
2. Using an 'asphaltic crack relief layer' to provide a medium through which differential movements in the underlying slabs are not easily transmitted. The Asphalt Institute recommends that this layer should be a 90 mm thick, coarse, open-graded asphaltic basecourse with 25–35 per cent interconnected voids that is placed on the prepared concrete pavement. The asphaltic overlay (at least 90 mm thick) is then placed on top of the crack relief layer in two lifts. The overall minimum thickness of the overlay system is therefore 180 mm. (It is suggested that a comparable result could be obtained by replacing the bituminous relief layer with an appropriately graded aggregate layer of the same thickness.)
3. Using a 'fabric interlayer' such as a polypropylene continuous filament between the pavement and overlay. A heavy tack coat should be applied to the prepared pavement surface to prevent any horizontal movement of the cracked surface being transmitted to the overlying material. Overlay paving operators may follow immediately behind. (Various interlayers have been suggested, e.g. a stress absorbing membrane interlayer (SAMI) which is a thin rubber-asphalt layer that is applied to the surface of the pavement prior to overlaying.)
4. Using a pre-tensioned polymer reinforcement grid in the lower layer of the bituminous overlay to resist any horizontal movements in the cracked layer.

5. Using a modified binder to improve the elastic recovery and fatigue resistance of the bituminous overlay material.
6. Use of the 'saw cut and seal' method, which involves creating a 'joint' in the bituminous overlay by cutting a slot above each joint in the concrete, applying bond breaker tape, and applying a sealant. The performance of this method is being assessed[43] in the UK, with a view to reducing the 180 mm minimum overlay thickness recommended[6,33] for UK trunk roads to delay the development of reflective cracking.

20.7.2 Concrete overlays on rigid pavements

The curing time required for concrete slabs has, in the past, discouraged their use as overlays. However, concrete with standard rapid-hardening Portland cements can achieve a strength of $25\,N/mm^2$ in less than 18 hours, with a cement content of $400\,kg/m^3$ or more. Factors which influence early strength development are the cement type and content, water–cement ratio and curing time. Current UK advice on materials and methods for rapid construction is given in reference 38. It is noted that concrete overlays have been successfully used in the USA.

Three bond conditions can be identified between a concrete overlay and the underlying concrete pavement, i.e. fully bonded, unbonded or partially bonded.

Fully bonded cement overlays are only appropriate if the existing concrete pavement is in good condition and the required concrete overlay is relatively thin (i.e. 50–100 mm). With this form of construction the two concrete layers should behave as a monolithic slab; careful preparation of the existing concrete surface is therefore needed to ensure a good bond is achieved. Surface preparation may consist of cleaning by gritblasting or shotblasting. Existing joints should be cleaned and resealed prior to overlaying. With bonded overlays the new joints should be located directly above those in the underlying pavement, to prevent reflection cracking. A cement grout may be applied immediately before the concrete overlay is placed, to assist in creating the bond. Bonded concrete overlays are generally unreinforced concrete. The use of limestone aggregate in the overlay concrete mix and also a fabric reinforcing mesh have been suggested[33] as a means of controlling shrinkage induced cracking.

Unbonded (or debonded) concrete overlays are separated from the existing concrete pavement by a positive separation course which may be lean concrete, bituminous material (e.g. seal coats, slurry seals or a 40–50 mm thick bituminous regulating layer), or polythene film in a double layer. Their use is most appropriate when the existing concrete pavement is extensively cracked. The overlay may be either unreinforced, jointed reinforced, or continuously reinforced. The overlay joints need not be matched to the same locations as with the underlying existing slab. For unbonded concrete overlays, a minimum practical thickness of 150 mm is recommended[6] for both unreinforced and reinforced slabs, whilst a minimum thickness of 200 mm is suggested for continuously reinforced concrete overlays.

Partially bonded concrete overlays are placed directly on the existing pavement with no attempt made to produce a bond. The existing surface is cleared of loose debris and excess joint seal. Partially bonded overlays are not to be recommended as, in practice, they are difficult to specify or produce in a uniform manner.

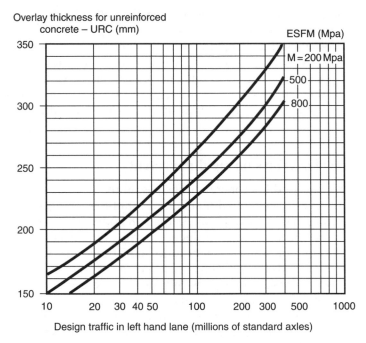

Fig 20.14 Design thickness for unreinforced concrete overlay – URC[6]

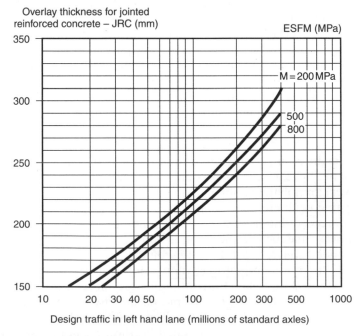

Fig. 20.15 Design thickness for jointed reinforced concrete overlay – JRC[6]

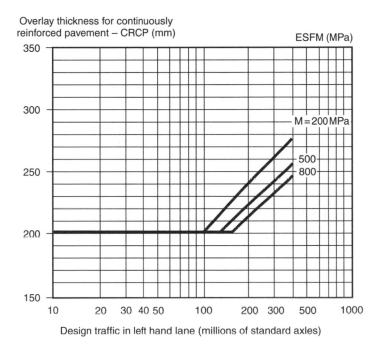

Fig. 20.16 Design thickness for continuously reinforced concrete pavement overlay – CRCP[6]

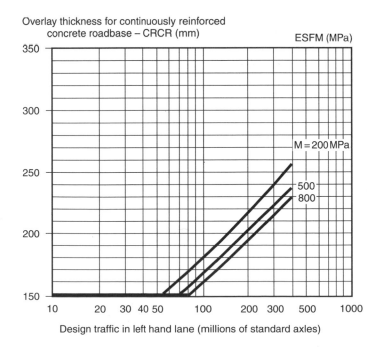

Fig. 20.17 Design thickness for continuously reinforced concrete roadbase overlay – CRCR[6]

The concrete overlay design options permitted in the UK are given in Table 20.6. (Further details of each type of concrete overlay are given in reference 36.) The overlay design charts used with these options are given in Figs 20.14 to 20.17. Note that for each type of concrete overlay a chart indicates the overlay thickness as a function of the required pavement life (expressed in millions of standard axles carried in the 'slow' lane), and the equivalent surface foundation modulus (or ESFM, denoted by 'M' in the charts). The ESFM is a measure of the strength of the existing road structure and is defined[39] as the modulus of a uniform elastic foundation that would give the same deflection under the same wheel load as that of the existing structure. The ESFM can be evaluated from the results of FWD deflection tests.

20.8 References

1. Transport and Road Research Laboratory, *A Guide to the Structural Design of Bitumen-surfaced Roads in Tropical and Sub-tropical Countries*, Overseas Road Note 31 (4th edition). Crowthorne, Berkshire: The Transport Research Laboratory, 1993.
2. Rolt, J., Smith, H.R. and Jones, C.R., The design and performance of bituminous overlays in tropical environments. *Proceedings of the International Conference on the Bearing Capacity of Roads and Airfields*, held at Plymouth, England, 1986.
3. Cedergren, H.R., *Drainage of Highway and Airfield Pavements*, Florida: R.E. Krieger Publishing Company, 1987 (Reprint).
4. Kennedy C.K. and Lister N.W., *Prediction of Pavement Performance and the Design of Overlays*, TRRL Report LR833. Crowthorne, Berkshire: The Transport and Road Research Laboratory, 1978.
5. Department of Transport, Structural assessment methods, HD 29/94, *Design Manual for Roads and Bridges, Vol 7: Pavement Design and Maintenance*. London: HMSO 1994 (revised Nov 1999).
6. Department of Transport, Structural assessment procedure, HD 30/99, *Design Manual for Roads and Bridges, Vol 7: Pavement Design and Maintenance*. London: HMSO, 1999.
7. Transport and Road Research Laboratory, *A Users Manual for a Program to Analyse Dynamic Cone Penetration Data*, Overseas Road Note 8. Crowthorne, Berkshire: The Transport and Road Research Laboratory, 1990.
8. Kleyn, E.G., *The Use of the Dynamic Cone Penetrometer (DCP)*, Report 2/74. Pretoria, The Materials Branch, Transvaal Roads Department, July 1975.
9. Kleyn, E.G. and Van Heerden, D.J., *Using DCP Soundings to Optimise Pavement Rehabilitation*, Report LS/83. Pretoria, The Materials Branch, Transvaal Roads Department, July 1983.
10. Harison, J.A., Correlation between California Bearing Ratio and dynamic cone penetrometer strength measurement of soils, *Proceedings of the Institution of Civil Engineers*, 1987, **83**, Part 2, pp. 833–844.
11. Lister, N.W., *Deflection Criteria for Flexible Pavements*, TRRL Report LR375. Crowthorne, Berkshire: The Transport and Road Research Laboratory, 1972.

12. Kennedy, C.K., Fevre, P. and Clarke, C., *Pavement Deflection: Equipment for Measurement in the United Kingdom*, TRRL Report LR834. Crowthorne, Berkshire: The Transport and Road Research Laboratory, 1978.

13. Kennedy, C.K., *Pavement Deflection: Operating Procedures for Use in the United Kingdom*, TRRL Report LR835. Crowthorne, Berkshire: The Transport and Road Research Laboratory, 1978.

14. Smith, H.R. and Jones, C.R., *Measurement of Pavement Deflections in Tropical and Sub-tropical Climates*, TRRL Report LR935. Crowthorne, Berkshire: The Transport and Road Research Laboratory, 1980.

15. Asphalt Institute, *Asphalt Overlays for Highway and Street Rehabilitation*, Manual Series No. 17 (MS-17). Maryland: The Asphalt Institute, 1983.

16. Bohn, A., Ullidtz, P., Stubstad, R. and Sorensen, A., Danish experiments with the French Falling Weight Deflectometer. *Proceedings of the 3rd International Conference on the Structural Design of Asphalt Pavements*, Vol 1. Ann Arbor, University of Michigan, 1972.

17. Claessen, A.I.M., Valkering, C.P. and Ditmarsch, R., Pavement evaluation with the Falling Weight Deflectometer, *Proceedings of the Association of Asphalt Paving Technologists*, 1976, **45**, pp. 122–157.

18. Claessen, A.I.M. and Ditmarsch, R., Pavement evaluation and overlay design – The Shell Method. *Proceedings of the 4th International Conference on the Structural Design of Asphalt Pavements*, Vol 1. Ann Arbor, University of Michigan, 1977.

19. *Shell Pavement Design Manual – Asphalt Pavements and Overlays for Road Traffic*. London: Shell International Petroleum Company Limited, 1978 (and Addendum 1985).

20. Sorensen, A. and Hayven, M., The Dynatest 8000 Falling Weight Deflectometer test system. *Proceedings of the International Symposium on Bearing Capacity of Roads and Airfields*, Trondheim: Tapir Publishers, 1982.

21. Evdorides, H.T. and Snaith, M.S., A knowledge-based analysis process for road pavement condition assessment, *Proceedings of the Institution of Civil Engineers: Transport*, 1996, **117**, pp. 202–210.

22. Norman, P.J., Snowdon, R.A. and Jacobs, J.C., *Pavement Deflection Measurements and Their Application to Structural Maintenance and Overlay Design*, TRRL Report LR571. Crowthorne, Berkshire: The Transport and Road Research Laboratory, 1973.

23. Croney, D. and Croney, P., *The Design and Performance of Road Pavements*, 2nd edn. London: McGraw Hill, 1991.

24. Kerali, H.R., Lawrance, A.J. and Awad, K.R., Data analysis procedures for the determination of long term pavement performance relationships, *Transportation Research Record 1524*, September 1996.

25. Transport and Road Research Laboratory, *A Guide to the Measurement of Axle Loads in Developing Countries using a Portable Weighbridge*, Road Note 40. London: HMSO, 1978.

26. McMullen, D., Snaith, M.S., May, P.H. and Vrahimis, S., A practical example of the application of analytical methods to pavement design and rehabilitation. *Proceedings of the 3rd International Conference on Bearing Capacity of Roads and Airfields*. Trondheim: Tapir Publishers, 1990.

27. Snaith, M.S. and Hattrell, D.V., A deflection based approach to flexible pavement design and rehabilitation in Malaysia, *Proceedings of the Institution of Civil Engineers: Transport*, 1994, **105**, pp. 219–225.

28. Finn, F.N. and Monismith, C.L., Asphalt overlay design procedures, *NCHRP Synthesis of Highway Practice 116*. Washington DC: Transportation Research Board, 1984.

29. American Association of State Highway and Transportation Officials, *Guide for Design of Pavement Structures*. Washington, DC: AASHTO, 1993.

30. Asphalt Institute, *Thickness Design: Asphalt Pavements for Highways and Streets*, Manual Series No. 1 (MS-1). Lexington, Kentucky: Asphalt Institute, February 1991.

31. Rwebangira, T., Hicks, R.G. and Truebe, M., Sensitivity analysis of selected back calculation procedures, *Transportation Research Record 1117*, 1987.

32. McMullen, D., Snaith, M.S. and Burrow, J.C., Back analysis techniques for pavement condition determination. *Proceedings of the 2nd International Conference on Bearing Capacity of Roads and Airfields*, Plymouth, England, 1986.

33. Department of Transport, Maintenance of concrete roads, HD 32/94, *Design Manual for Roads and Bridges, Vol 7: Pavement Design and Maintenance*. London: HMSO, 1994.

34. Mildenhall, H.S., and Northcott, G.D.S., *Manual for the Maintenance and Repair of Concrete Roads*. London: HMSO, 1986.

35. Transport and Road Research Laboratory, *A Guide to Concrete Road Construction*. London: HMSO, 1979.

36. Department of Transport, Pavement design, HD 26/94, *Design Manual for Roads and Bridges, Vol 7: Pavement Design and Maintenance*. London: HMSO, 1994 (revised Feb 1996).

37. Sherman, G., Minimising reflection cracking of pavement overlays, *NCHRP Synthesis of Highway Practice 92*. Washington, DC: Transportation Research Board, 1982.

38. Department of Transport, Pavement construction methods, HD 27/94, *Design Manual for Roads and Bridges, Vol 7: Pavement Design and Maintenance*. London: HMSO 1994 (revised March 1995).

39. Mayhew, H.C. and Harding, H.M., *Thickness Design of Concrete Roads*, TRRL Report RR87. Crowthorne, Berkshire: The Transport and Road Research Laboratory, 1987.

40. Nunn, M.E., Structural design of long-life flexible roads for heavy traffic, *Proceedings of the Institution of Civil Engineers: Transport*, 1998, **129**, pp. 126–133.

41. Brown, S.F., Developments in pavement structural design and maintenance, *Proceedings of the Institution of Civil Engineers: Transport*, 1998, **129**, pp. 201–206.

42. Potter, J.F., Dudgeon, R. and Langdale, P.C., Implementation of crack and seat for concrete pavement maintenance. *Proceedings of the 4th RILEM International Conference on Reflective Cracking in Pavements*, Ottawa, 2000.

43. Burtwell, M.H., Lloyd, W. and Carswell, I., Performance of the saw cut and seal method for inhibiting reflective cracking. *Proceedings of the 4th RILEM International Conference on Reflective Cracking in Pavements*, Ottawa, 2000.

Index